Intellectual Property and Emerging Technologies

QUEEN MARY STUDIES IN INTELLECTUAL PROPERTY

The Centre for Commercial Law Studies at Queen Mary, University of London, has teamed up with international social science publisher, Edward Elgar Publishing Ltd, to re-launch its prestigious book series, Queen Mary Studies in Intellectual Property. Already established as an important forum for acute and relevant debate in the field of intellectual property, the series aims to build further upon its success, publishing books by renowned guest editors who will commission original and often provocative work by both leading scholars and specialist practitioners on key subjects and issues in IP. Combining volumes that are theoretical in their analysis with others that seek to explore the issues from a more practical perspective, all will be rigorous, as well as international in origin, outlook and appeal. The scope of the series will be wide-ranging and its approach will be to synthesize insights from a range of disciplinary backgrounds. The books in this series will not only provide unique sources of reference for researchers and students in IP, but will offer a level of insight to lawyers, policymakers and legislators that is consistent with the reputation of the Centre.

Titles in the series include:

US Intellectual Property Law and Policy
Edited by Hugh Hansen

Emerging Issues in Intellectual Property
Trade, Technology and Market Freedom: Essays in Honour of Herchel Smith
Edited by Guido Westkamp

An Emerging Intellectual Property Paradigm
Perspectives from Canada
Edited by Ysolde Gendreau

Intellectual Property and Emerging Technologies
The New Biology
Edited by Matthew Rimmer and Alison McLennan

Intellectual Property and Emerging Technologies

The New Biology

Edited by

Matthew Rimmer

Australian Research Council Future Fellow, Associate Professor, The Australian National University College of Law, Associate Director, ACIPA

Alison McLennan

PhD candidate and Vice Chancellor's Scholar, The Australian National University College of Law

QUEEN MARY STUDIES IN INTELLECTUAL PROPERTY

Edward Elgar

Cheltenham, UK • Northampton, MA, USA

Published by
Edward Elgar Publishing Limited
The Lypiatts
15 Lansdown Road
Cheltenham
Glos GL50 2JA
UK

Edward Elgar Publishing, Inc.
William Pratt House
9 Dewey Court
Northampton
Massachusetts 01060
USA

A catalogue record for this book is available from the British Library

Library of Congress Control Number: 2011932890

FSC www.fsc.org
MIX
Paper from responsible sources
FSC® C018575

ISBN 978 1 84980 246 8

Typeset by Columns Design XML Ltd, Reading
Printed and bound by MPG Books Group, UK

Contents

Contributors

Amina Agovic is a Post-doctoral researcher at Malmö University, Sweden. She received her PhD in law from Helsinki University for her dissertation on 'Patents, Ethics and Stem Cell Research: The Case of hESC Innovation in Australia, Europe and the United States'. Amina is currently teaching intellectual property rights and working on developing an E-learning stem cell system for stem cell scientists and relevant parties. She is an Australian qualified solicitor with a common law background and specialisation in European and Anglo-American intellectual property laws. To date, she has published a number of publications; participated in many conferences in respect of stem cell sciences and intellectual property rights and contributed to a number of European Union Framework Program (FP) 6 and 7 Projects. In addition Amina has spent a year as a visiting scholar at the Faculty of Medicine, Lund University, Sweden.

Adam Bostanci is a Senior Research Fellow at St Vincent's Hospital in Melbourne and an Honorary Fellow in History and Philosophy of Science at the University of Melbourne. Previously, he was a lecturer in social science at the Hughes Hall Centre for Biomedical Science in Society at the University of Cambridge. He did his doctoral training at the ESRC Centre for Genomics in Society at the University of Exeter (UK). He also holds master's degrees in natural sciences and in the history and philosophy of science, both from the University of Cambridge. Recent research projects have focused on the development of techniques for the inference of likely ethnic identification of a subject from a DNA sample (sometimes discussed as 'racial profiling'), and the development of techniques for non-invasive prenatal diagnosis (NIPD).

Jane Calvert is an RCUK Academic Fellow at the ESRC Innogen Centre, University of Edinburgh. She has a BSc in human sciences (Sussex), an MSc in the history and philosophy of science (LSE), and a DPhil in science and technology policy research (Sussex). She worked at the ESRC Centre for Genomics in Society at the University of Exeter, where she became interested in the insights that the philosophy of biology can bring to the social study of genomics. Her work focused on intellectual property issues, showing that ideas about the nature of genes and genomes have implications for patentability. At Edinburgh, Jane's broad area of research is the sociology of the life sciences, with particular attention to systems and synthetic biology. She is interested in interdisciplinary interactions, intellectual property and open source, and design and aesthetics in synthetic biology. She is principal investigator on the Synthetic Aesthetics Project (www.syntheticaesthetics.org).

Graham Dutfield is Professor of International Governance at the School of Law, University of Leeds. His approach to the law and policy of intellectual property and innovation is interdisciplinary, containing elements of anthropology, economics, geography, history (of science and pharmaceutical intellectual property), development studies, and the life sciences including genetics. His most recent works include *Intellectual Property Rights and the Life Science Industries: Past, Present and Future*, and *Intellectual Property and Human Development: Current Trends and Future Scenarios* (edited with Tzen Wong). He has affiliations with the Intellectual Property and Technology Program, Osgoode Hall Law School, York University, Toronto, and with the Centre for Studies of Intellectual Property Rights, Zhongnan University of Economics and Law, Wuhan, China. In addition, he is on the Scientific Advisory Board of PhytoMetaSyn, a large synthetic biology project funded by Genome Canada. He has a DPhil in geography from the University of Oxford.

Donna M. Gitter is an Associate Professor of Law at Baruch College's Zicklin School of Business, City University of New York, in New York City. She has published numerous articles relating to judicial and legislative responses in the European Union and the United States to issues at the intersection of technology and ethics, such as biotechnology patenting and biomedical ethics. These articles have appeared in publications such as the *Berkeley Journal of International Law*, the *Columbia Science and Technology Law Review*, the *Duke Journal for Comparative and International Law*, the *New York University Law Review*, and the *Washington and Lee Law Review*. Professor Gitter has won several research grants and awards, including The Ralph C. Hoeber Memorial Award for Excellence in Business. She has been invited to present her work at international conferences in Europe and Asia, as well as in the United States. Professor Gitter graduated Phi Beta Kappa and *cum laude* from Cornell University's College of Arts and Sciences, where she earned a BA in government in 1989. In 1994, she earned her JD from the University of Pennsylvania Law School.

Richard Gold is a James McGill Professor at McGill University's Faculty of Law where he was the founding director of the Centre for Intellectual

Property Policy. He teaches in the area of intellectual property, property and innovation. His research centres on the nexus between innovation, development and commerce, particularly with respect to biotechnology. He has provided advice to Canadian federal and provincial governments, the Organisation for Economic Cooperation and Development, the World Health Organization, the World Intellectual Property Organization and UNITAID.

Francis Hemmings completed his bachelor's degree in biotechnology engineering before switching gears and starting his law degree at the University of Montreal in 2008. Compelled by an interest in ethics, biotechnology and law he works part-time, during his studies, at the Center of Genomics and Policy. At the CGP he conducted an exhaustive analysis on the limits of what modern society considers patentable material. This work ultimately led to a report for the World Intellectual Property Organization (WIPO). In a separate project, he investigated the recognition of intellectual property rights in one's own biological samples and the problem this poses for publicly funded biobanks.

Sarah Holcombe has a disciplinary background in social anthropology, undertaking her PhD field research in remote central Australia. Half of her 20 years experience working with Indigenous Australians has been as an applied anthropologist in the Northern Territory. More recently this work has involved developing resources, such as protocols and guidelines, for government agencies and research bodies to systematise an ethical and collaborative approach to working with Indigenous knowledge holders in research. She has, for the last nine years, been engaged in a diverse range of research projects at the Australian National University (ANU) through the Centre for Aboriginal Economic Policy Research (CAEPR), including as Social Science Coordinator for the Desert Knowledge Cooperative Research Centre. This research has ranged from working in the Pilbara region of Western Australia on the socio-economic sustainability of mining for Indigenous communities, to research in central Australia with community councils on Indigenous governance issues, issues of land tenure and reterritorialisation, and the governance of Indigenous knowledge in research. She is currently an Adjunct Fellow in the Research School of Humanities and the Arts, ANU and a Senior Research Officer at the recently established Aboriginal and Torres Strait Islander Healing Foundation.

Terri Janke is the solicitor director of Terri Janke and Company Pty Ltd, a Sydney based law firm that specialises in Indigenous cultural and intellectual property. A significant part of Terri's client base involves the advancement of protection for Indigenous Australian people in respect of intellectual property matters, and she has represented a diverse range of clients across many fields of the arts and biodiversity knowledge management. She has written many reports, protocol documents and articles on Indigenous cultural and intellectual property. See www.terrijanke.com.au for more information. Terri is enrolled in a PhD at the National Centre for Indigenous Studies at the Australian National University (ANU).

Pierre-Benoit Joly, economist and sociologist, is Directeur de Recherche at the National Institute of Agronomic Research (INRA) in France. He holds a degree in agronomy (1982), a PhD in economics (1987) and the habilitation à diriger les recherches (1995). He is Director of the IFRIS (French Institute for Research, Innovation and Society). Since 1996, his research activities have focused on the governance of collective risks, socio-technical controversies, the use of scientific advice in public decision making and the forms of public participation in scientific activities. He was a member of the expert group Science and Governance, at the European Commission; and he is a member of the Council of European Association for the Study of Science and Technology (EASST) and chairs the Scientific Council of the Programme on GMOs at the French Ministry for Ecology. He has published about a hundred articles (of which more than 50 in refereed journals) and three books, and he has coordinated five special issues of social science journals. He lectures at Ecole des Hautes Etudes en Sciences Sociales (EHESS) on science, expertise and public debate and at Sciences Po Paris on risk governance.

Yann Joly, PhD (DCL), is a lawyer and Assistant Professor at the Faculty of Medicine, Department of Human Genetics at McGill University, as well as an ethics and legal consultant in the private sector. His research activities lie at the interface of intellectual property, health law (biotechnology and other emerging health technologies) and bioethics. In recent years, he has worked as a lecturer teaching intellectual property and biotechnology law at the law faculties of both McGill University and the Université de Montreal. He has served as a legal advisor on several ethics committees of the MUHC and is currently a member of the Independent Ethics Advisory Committee for the biotechnology company Genizon Biosciences. Assistant Professor Joly is a co-author of the recent report from the Canadian Academy of Health Sciences on the topic of the returns on investments in health research. He has authored more than 50 articles on the legal issues raised by the development, integration and use of new health technologies in leading legal and scientific journals. Recently, he has co-authored a report for the World Intellectual Property Organization.

Alison McLennan is a PhD candidate and Vice-Chancellor's Scholar at the Australian National University College of Law. She studied science (cellular and molecular biology) and law as an undergraduate at ANU. She is interested in the regulation of emerging technology, particularly biotechnology. Alison is currently researching legal issues relating to synthetic biology. She is exploring the regulatory, intellectual property and biosecurity issues raised by recent developments in this field. Alison is also interested in health and medical law, and the history and philosophy of science.

Dianne Nicol is Professor in the Law Faculty at the University of Tasmania and is a deputy director of the Centre for Law and Genetics (CLG). The broad theme of the CLG's research is the regulation of biotechnology, human genetics and stem cell technology. Dianne is particularly interested in the legal issues associated with the commercialisation of genetic knowledge and patenting of genetic inventions. She has a PhD in cell biology as well as her qualification in law. Her primary teaching responsibilities are in the area of law and technology, including intellectual property, biotechnology, information technology and media law.

Matthew Rimmer is an Australian Research Council Future Fellow, an Associate Professor at the ANU College of Law, and an Associate Director of the Australian Centre for Intellectual Property in Agriculture. He holds a BA (Hons) and a University Medal in literature, and an LLB (Hons) from the Australian National University. Dr Rimmer received a PhD in law from the University of New South Wales for his dissertation on *The Pirate Bazaar: The Social Life of Copyright Law*. He is a member of the ANU Climate Change Institute and a director of the Australian Digital Alliance. He has published widely on copyright law and information technology, patent law and biotechnology, access to medicines, clean technologies, and traditional knowledge. Dr Rimmer is the author of *Digital Copyright and the Consumer Revolution: Hands Off My iPod* (Edward Elgar, 2007), *Intellectual Property and Biotechnology: Biological Inventions* (Edward Elgar, 2008), and *Intellectual Property and Climate Change: Inventing Clean Technologies* (Edward Elgar, 2011). He edited the collection *Patent Law and Biological Inventions* (Federation Press, 2006) and co-edited the collection *Incentives for Global Public Health: Patent Law and Access to Essential Medicines* (Cambridge University Press, 2010). Dr Rimmer was a chief investigator in an Australian Research Council Discovery Project, 'Gene Patents in Australia: Options for Reform' (2003–2005), and an Australian

Research Council Linkage Grant, 'The Protection of Botanical Inventions' (2003). He is currently a chief investigator in an Australian Research Council Discovery Project, 'Promoting Plant Innovation in Australia' (2009–2011), and an Australian Research Council Future Fellowship, 'Intellectual Property and Climate Change: Inventing Clean Technologies' (2011–2015).

Joshua D. Sarnoff is Associate Professor at DePaul University College of Law, where he teaches patent law, advanced patent law, and intellectual property and climate change in the Center for Intellectual Property Law and Information Technology. He was previously a professor at the Washington College of Law, American University, in the Glushko-Samuelson Intellectual Property Law Clinic, and at the University of Arizona College of Law. He is a registered patent attorney and a member of the bars of Washington DC and California, a former member of the Board of Governors of the Federal Circuit Bar Association, and a member of the boards of directors and advisory boards of various non-profit organisations. He has written numerous articles and book chapters on patent law and has been involved in a wide range of intellectual property legal and policy disputes. He has submitted testimony on domestic patent law reform bills, has filed numerous amicus briefs in the United States Supreme Court and in the Court of Appeals for the Federal Circuit on important patent law issues, has been a pro bono mediator for the Federal Circuit, and has been a consultant to the United Nations Conference on Trade and Development on international intellectual property, trade and environmental issues.

Eva Hemmungs Wirtén is Professor in Library and Information Science (also Associate Professor (Docent) in Comparative Literature) at Uppsala University, Sweden. Her research is focused on international copyright and the history of the public domain. She is the author of *No Trespassing: Authorship, Intellectual Property Rights, and the Boundaries of Globalization* (2004) and *Terms of Use: Negotiating the Jungle of the Intellectual Commons* (2008), both University of Toronto Press. Forthcoming in 2011 is the essay 'A Diplomatic *Salto Mortale*: Translation Trouble in Berne, 1884–1886' for *Book History*, **14**, 2011, and a small volume on copyright and translation entitled *Cosmopolitan Copyright: Law and Language in the Translation Zone*. Her new book project, on *The Intellectual Properties of Marie Curie*, is being funded by HERA (Humanities in the European Research Area) between 2010 and 2013. Website: www.abm.uu.se/evahw

Preface

This collection is part of the Queen Mary Studies in Intellectual Property series published by Edward Elgar. We were invited by Professor Graham Dutfield to edit a collection on intellectual property and emerging technologies in the life sciences. Having explored gene patents and stem cell patents in our research, we were particularly keen to explore new challenges posed by paradigm shifts in biology. In 2009, the National Research Council of the National Academies deployed the term 'New Biology' to capture the convergence and integration of the various disciplines of biology:

> The essence of the New Biology, as defined by the committee, is integration – re-integration of the many sub-disciplines of biology, and the integration into biology of physicists, chemists, computer scientists, engineers, and mathematicians to create a research community with the capacity to tackle a broad range of scientific and societal problems. Integrating knowledge from many disciplines will permit deeper understanding of biological systems, which will both lead to biology-based solutions to societal problems and also feed back to enrich the individual scientific disciplines that contribute new insights.[1]

We invited a number of the pre-eminent researchers and scholars to explore future trends and developments in a cluster of emerging scientific disciplines – including biodiscovery, genomics, medicine, bioinformatics, synthetic biology, RNA interference, stem cell research, nanotechnology, traditional knowledge, genetically modified crops, and clean technologies. Given the interdisciplinary nature of the topic, the contributors are drawn from an array of fields. Moreover, given the international, collaborative nature of large-scale biology projects, we have gathered together researchers from the four corners of the globe.

This book was written at the Australian National University (ANU) College of Law. We are thankful for the outstanding support and leadership of this institution – including the Dean, Professor Michael Coper, the Associate Dean, Professor Fiona Wheeler, and the Head of School, Professor Stephen Bottomley. We have also benefitted from the guidance of

---

1 National Research Council of the National Academies (2009), *A New Biology for the 21st Century*, Washington DC: The National Academies Press, 3.

Professor Tim Bonyhady, the director of the Australian Centre for Environmental Law and the Centre for Climate Law and Policy, and Professor Kim Rubenstein, the director of the Centre of International and Public Law. We also acknowledge the legacy of Emeritus Professor Dennis Pearce who established the intellectual property concentration at the ANU College of Law. We have also learnt much from our fellow intellectual property researchers at the ANU – including Dr Miranda Forsyth, Dr Sarah Holcombe, Professor Peter Veth, Dr Judith Bannister (now at the University of Adelaide), Hafiz Aziz ur Rehman, Gusman Siswandi, Terri Janke and Alan Hui.

Matthew Rimmer is a member of the research centre, the Australian Centre for Intellectual Property in Agriculture (ACIPA), which is based at the Australian National University and Griffith University. Versions of a couple of chapters were presented at events hosted by this centre. Matthew is grateful for the intellectual camaraderie of Professor Brad Sherman, Associate Professor Charles Lawson, Jay Sanderson, Associate Professor Leanne Wiseman, Stephen Hubicki and Dr Kathryn Adams; and for the administrative support of Carol Ballard.

Alison is a PhD candidate and would like to thank her PhD supervisors Professor Tim Bonyhady and Associate Professor Matthew Rimmer, and her advisor Associate Professor Barbara van Leeuwen. She is grateful for the support and guidance of the Director of Higher Degree Research at the ANU College of Law, Dr Mark Nolan, and for the administrative assistance of Dinah Irvine. Alison also appreciates the excellent training provided by the University's Centre for Educational Development and Academic Methods, particularly Dr Beth Beckman.

The editors are grateful for a number of contributions during the making of the book. Lisa Bradley, science librarian at the Hancock Library at the Australian National University, provided assistance in locating research on convergence in the biological sciences and literature on the regulation of nanotechnology. Arjuna Dibley provided invaluable research assistance for Matthew Rimmer's work in Chapter 12 on climate-ready crops. Katherine Phillips provided research assistance to Matthew Rimmer for his work in Chapter 13 on the Doomsday Vault.

We are grateful for the support and help of the publisher, Edward Elgar, and his team, including Tim Williams, John-Paul McDonald and Rebecca Hastie, and the series editor Graham Dutfield. Over the last decade, the publishing house has created an outstanding list of books on intellectual property. We are happy to be in such fine company.

The editors would like to thank their families and friends for their support and inspiration. Matthew Rimmer would particularly like to pay tribute to his parents, Professor Peter Rimmer and Dr Susan Rimmer;

Grandmother Ford; his siblings, Joe Rimmer and Rachel Rimmer; his wife, Dr Susan Harris Rimmer; and his children, Marina Rimmer and Joshua Rimmer. He would also like to pay respects to his uncle, Dr Ralph Rimmer (1926–2011), who was a glorious storyteller, a fine photographer and an outstanding Taid.

Alison would like to thank her parents, Margaret McCulloch and Dennis McLennan. She also appreciates the support and encouragement of her fellow research students at the ANU College of Law, particularly Elinor Jean, Brad Jessup, Mary Spiers Williams and Fanny Thornton.

This collection represents the law at July 2011.

Introduction: Inventing life: intellectual property and the New Biology

Alison McLennan and Matthew Rimmer

The Human Genome Project represented the first foray into 'Big Science' by the medical and the biological science communities.[1] The initiative was promoted as 'the moon shot of the life sciences', the 'holy grail of man', 'the code of codes', and 'the book of life'.[2] On 14 March 2000, United States President Bill Clinton and British Prime Minister Tony Blair made a joint announcement of the sequencing of the human genome.[3] In February 2001, *Nature* and *Science* published rival papers reporting the sequence of the 3.2 billion base pair human genome. The *Nature* paper was by the publicly funded International Human Genome Sequencing Consortium,[4] while the *Science* paper was by the private company Celera Genomics – led by J. Craig Venter.[5]

So what has changed in biological research since this milestone? In 2003, Francis Collins and his collaborators reflected that the success of the Human Genome Project has inspired a raft of new large-scale biology projects: 'Large-scale genomic enterprises now extend well beyond straightforward sequencing, as witnessed by the recent launch from our own funding agencies of other large-scale biology initiatives involving functional genomics, structural biology, microbial genomics and proteomics, and haplotype

1 Galison, Peter and Bruce Hevly (eds) (1992), *Big Science: The Growth of Large-Scale Research,* Stanford, CA: Stanford University Press.
2 Cook-Deegan, R. (2001), 'Hype And Hope', *American Scientist*, **89**, 62.
3 Clinton, W. and T. Blair (2000), 'Joint Statement by President William Clinton and Prime Minister Tony Blair of the United Kingdom', http://ipmall.info/hosted_resources/ippresdocs/ippd_44.htm.
4 The International Human Genome Sequencing Consortium (2001), 'Initial Sequencing and Analysis of the Human Genome', *Nature*, **409**, 860–921.
5 Venter, J.C. et al. (2001), 'The Sequence of the Human Genome', *Science*, **291**, 1301–4.

mapping focusing on human populations.'[6] They noted: 'Other candidates for the "big science" approach to biology include interagency and international approaches to biological database creation and maintenance, the construction of public small-molecule libraries for use by basic scientists in their efforts to chart biological pathways, and the large-scale application of microarray technologies with the potential for applications in a wide range of biological research settings.'[7] The Human Genome Project has been followed by such large-scale biology research projects as the Human Genome Diversity Project, the International HapMap Project, the 1000 Genomes Project and the Rice Genome Project.[8] The World Health Organization has facilitated international research on infectious diseases – such as HIV/AIDS, the SARS virus and avian influenza.[9] Public funding agencies, philanthropic foundations and private entities have supported large-scale research, such as the Sorcerer II Expedition and the Genographic Project.[10]

Reflecting on the decade of research since the announcement in 2000, Collins observes a 'breathtaking acceleration in genome science'.[11] The cost

---

6 Collins, F., N. Morgan and A. Patrinos (2003), 'The Human Genome Project: Lessons from Large-Scale Biology', *Science*, **300** (5617), 286–90, at 290.

7 Ibid.

8 Gitter, D. (2007), 'Resolving the Open Source Paradox in Biotechnology: A Proposal for a Revised Open Source Policy for Publicly Funded Genomic Databases', *Houston Law Review*, **43** (4), 1476–1521; and Rimmer, M. (2005), 'Japonica Rice: Intellectual Property, Scientific Publishing, and Data-Sharing', *Prometheus*, **23** (3), 325–47.

9 See Dukes, Graham and Frederick Abbott (2009), *Global Pharmaceutical Policy: Ensuring Medicines for Tomorrow's World*, Cheltenham and Northampton, MA: Edward Elgar; Rimmer, M. (2004), 'The Race to Patent the SARS Virus: The TRIPS Agreement and Access to Essential Medicines', *Melbourne Journal of International Law*, **5** (2), 335–74; Pogge, Thomas, Matthew Rimmer and Kim Rubenstein (eds) (2010), *Incentives for Global Public Health: Patent Law and Access to Essential Medicines*, Cambridge: Cambridge University Press.

10 On the Genographic Project, see Wells, Spencer (2006), *Deep Ancestry: Inside the Genographic Project*, New York: Penguin Books and National Geographic; Rimmer, M. (2007), 'The Genographic Project: Traditional Knowledge and Population Genetics', *Australian Indigenous Law Review*, **11** (2), 33–55; and Grant Allen, J. (2010), 'Group Consent and the Nature of Group Belonging: Genomics, Race and Indigenous Rights', *Journal of Information, Law, and Science*, **20** (2), 28–59; and on the Sorcerer II Expedition, see Venter, J. Craig (2007), *A Life Decoded*, London: Allen Lane and Penguin Books, 342–9; and Rimmer, M. (2009), 'The Sorcerer II Expedition: Intellectual Property and Biodiscovery', *Macquarie Journal of International and Comparative Environmental Law*, **6**, 147–89.

11 Collins, Francis (2010), 'Has the Revolution Arrived?' *Nature*, **464,** 674–5, at 674.

of DNA sequencing, that is, reading the sequence of chemical bases encoded by DNA, has reduced by a factor of about 14 000.[12] This has allowed researchers to begin comparing whole genome sequences in projects such as the 1000 Genomes Project, discussed in this volume.[13] Through Genome Wide Association Studies, we are now able to look across the whole genome for variations that are relevant to disease risk.[14] The decreasing cost and increasing efficiency of genome sequencing and genotyping can also be seen to have given rise to 'personal genomics', in which customers can purchase the service of having their own DNA analysed in order to illuminate their ancestry and risk of various diseases.[15]

However, the Human Genome Project has been criticised for not yielding the scientific answers and medical leaps promised. For example, Richard Lewontin has observed: 'The big irony of the sequencing of the human genome is that the result turns out not to provide the answer to the chief question that motivated the project.'[16] Collins' view is that, although the clinical outcomes have so far been 'modest', genome science has had 'an exceptionally powerful enabling role in biomedical advances' and that 'the best is yet to come'.[17]

Indeed, we have come to understand that the genome *sequence* is only part of the story. Although we have now elucidated the genome sequence of various organisms, 'our understanding of the complicated network that takes place inside… cells is far from complete'.[18] As Bostanci, Calvert and Joly note in their chapter in this volume, the recognition of the complexity of biology has led to the emergence of new fields of research such as systems biology and epigenetics. Systems biology is the study of the overall patterns of metabolic activity in cells. Epigenetics is the study of modifications to DNA that occur within the cell without changing the actual genetic sequence.[19] These processes of modification could help us to understand

---

12 Ibid.

13 Ibid.

14 Ibid.

15 See 23andMe Inc. https://www.23andme.com/ and Ng, P., S. Murray, S. Levy and J.C. Venter (2009), 'An Agenda for Personalized Medicine', *Nature*, **461**, 724–6.

16 Lewontin, R. (2001), 'They got the Wrong Key of Life', *The Sunday Times*, 8 July.

17 Collins, Francis (2010), 'Has the Revolution Arrived?' *Nature*, **464**, 674–5.

18 Costa, F. (2008), 'Noncoding RNAs, Epigenetics and Complexity', *Gene*, **410** (1), 9–17, at 9.

19 Ibid.

how cells use the information contained in genes, and how the same genetic *sequence* can produce different biological results in different circumstances.[20]

Bio-mathematics specialist Eric Schadt gave this perspective in the genome's 10th anniversary edition of *Science*:

> We have learned that the human genome is much more dynamic than previously thought. Elucidating its complexity will require a more systems-level approach, including comprehensive integration with other data dimensions, such as RNA, metabolite, protein, and clinical data. For me, although this past decade has exposed many amazing aspects of the genome, it has revealed the existence of a world about which we know very little.[21]

The new field of synthetic biology can also be seen to have developed from our increased appreciation of biological complexity. Synthetic biology aims to use engineering tools and principles to reduce the complexity of biological systems.[22] These complexities, and how much we are yet to understand about biology generally, were strong themes in evidence from synthetic biologists to the 2010 inquiry of the United States Presidential Commission for the Study of Bioethical Issues into synthetic biology.[23]

In addition to the enabling technology of DNA sequencing, new fields of biological research such as synthetic biology have also been enabled by the increasingly powerful technology of DNA synthesis. At the same time as our ability to sequence DNA has increased exponentially,[24] we have developed the capacity to build DNA from its components in the laboratory. Our ability to do this is also increasing exponentially, while the costs fall rapidly.[25] According to Robert Carlson, 'The rate at which DNA synthesis capacity is changing is thus a measure of the improvement in our ability to manipulate biological systems and biological information.'[26]

---

20 Ibid.

21 Schadt, E. (2011), 'First Steps on a Long Road', *Science*, **331** (6018), 691.

22 Endy, D. (2005), 'Foundations for Engineering Biology', *Nature*, **438** (7067), 449–53.

23 See the Presidential Commission for the Study of Bioethical Issues (2010), 'Synthetic Biology Meeting Transcripts', http://www.bioethics.gov.

24 Carlson, Robert (2010), *Biology is Technology: The Promise, Peril, and New Business of Engineering Life*, Cambridge, MA and London: Harvard University Press, 66–7.

25 Ibid.

26 Carlson, Robert (2003), 'The Pace and Proliferation of Biological Technologies', *Biosecurity and Bioterrorism: Biodefense Strategy, Practice, and Science*, **1,** 203–14, at 203.

DNA synthesis is significant because 'the ability to specify genetic instructions by writing DNA from scratch provides, in principle, the opportunity to specify the behaviour of biological systems'.[27] This opens up many new fields of biological research and potential industrial applications.

Increased capacity to sequence and compare DNA has also given rise to a great deal of experimental data. The development of powerful bioinformatics tools for data analysis has become increasingly important. We have seen the development of new professions such as curation of biological data.[28] Underlining the importance of learning to manage and analyse huge sets of data, Eric Schadt suggests: 'We will have to become masters of information if we ever hope to go from the big data sets coming to dominate biology to knowledge and to understanding.'[29]

In a 2003 report for the United States National Science Foundation, *Converging Technologies for Improving Human Performance*, Mihail Roco and William Sims Bainbridge predicted that there would be a convergence of four emerging technologies – nanoscience and nanotechnology; biotechnology, biomedicine and genetics; information technology; and cognitive sciences.[30] The pair invented the shorthand 'NBIC' to capture these converging technologies:

> In the early decades of the 21st century, concentrated efforts can unify science based on the unity of nature, thereby advancing the combination of nanotechnology, biotechnology, information technology, and new technologies based in cognitive science. With proper attention to ethical issues and societal needs, converging technologies could achieve a tremendous improvement in human abilities, societal outcomes, the nation's productivity, and the quality of life. This is a broad, cross-cutting, emerging and timely opportunity of interest to individuals, society and humanity in the long term.[31]

---

27 Carlson, Robert (2010), *Biology is Technology: The Promise, Peril, and New Business of Engineering Life*, Cambridge, MA and London: Harvard University Press, 73.

28 See e.g. Sanderson, Katharine (2011), 'Curation Generation', *Nature*, **470**, 295–6.

29 Schadt, E. (2011), 'First Steps on a Long Road', *Science*, **331** (6018), 691.

30 Roco, Mihail and William Sims Bainbridge (2003), *Converging Technologies for Improving Human Performance: Nanotechnology, Biotechnology, Information Technology and Cognitive Science*, AA Dordrecht: Kluwer Academic.

31 Ibid.

There has been much critical discussion about whether the phrase NBIC describes a historical fact or a future aspiration.[32]

In 2009, the National Research Council of the National Academies released a report on *A New Biology for the 21st Century*.[33] The council preferred the term 'New Biology' to capture the convergence and integration of the various disciplines of biology. The National Research Council stressed: 'The essence of the New Biology, as defined by the committee, is integration – re-integration of the many sub-disciplines of biology, and the integration into biology of physicists, chemists, computer scientists, engineers, and mathematicians to create a research community with the capacity to tackle a broad range of scientific and societal problems.'[34] They define the 'New Biology' as 'integrating life science research with physical science, engineering, computational science, and mathematics'.[35] The National Research Council reflected:

> Biology is at a point of inflection. Years of research have generated detailed information about the components of the complex systems that characterize life – genes, cells, organisms, ecosystems – and this knowledge has begun to fuse into greater understanding of how all those components work together as systems. Powerful tools are allowing biologists to probe complex systems in ever greater detail, from molecular events in individual cells to global biogeochemical cycles. Integration within biology and increasingly fruitful collaboration with physical, earth, and computational scientists, mathematicians, and engineers are making it possible to predict and control the activities of biological systems in ever greater detail.[36]

The National Research Council contended that the New Biology could address a number of pressing challenges. First, it stressed that the New Biology could 'generate food plants to adapt and grow sustainably in changing environments'.[37] Second, the New Biology could 'understand and sustain ecosystem function and biodiversity in the face of rapid change'.[38] Third, the New Biology could 'expand sustainable alternatives to fossil

---

32 See e.g. Schummer, Joachim (2010), 'From Nano-Convergence to NBIC-Convergence', in Kaiser, Mario, Monika Kurath, Sabine Maasen and Christoph Rehmann-Sutter (2010), *Governing Future Technologies Nanotechnology and the Rise of an Assessment Regime*, Dordrecht: Springer, 47–72 at 67.

33 National Research Council of the National Academies (2009), *A New Biology for the 21st Century*, Washington, DC: The National Academies Press.

34 Ibid., 3.

35 Ibid., 22.

36 Ibid., 12.

37 Ibid., 4.

38 Ibid.

fuels'.[39] Moreover, it was hoped that the New Biology could lead to a better understanding of individual health: 'The New Biology can accelerate fundamental understanding of the systems that underlie health and the development of the tools and technologies that will in turn lead to more efficient approaches to developing therapeutics and enabling individualized, predictive medicine.'[40]

Biological research has certainly been changing direction in response to changing societal problems. Over the last decade, increasing awareness of the impacts of climate change and dwindling supplies of fossil fuels can be seen to have generated investment in fields such as biofuels, climate-ready crops and storage of agricultural genetic resources. In considering biotechnology's role in the twenty-first century, biological future-predictor Carlson's firm Biodesic states: 'The problems the world faces today – ecosystem responses to global warming, geriatric care in the developed world or infectious diseases in the developing world, the efficient production of more goods using less energy and fewer raw materials – all depend on understanding and then applying biology as a technology.'[41]

This collection considers the roles of intellectual property law in regulating emerging technologies in the biological sciences. Stephen Hilgartner comments that patent law plays a significant part in social negotiations about the shape of emerging technological systems or artefacts:

> Emerging technology – especially in such hotbeds of change as the life sciences, information technology, biomedicine, and nanotechnology – became a site of contention where competing groups pursued incompatible normative visions. Indeed, as people recognized that questions about the shape of technological systems were nothing less than questions about the future shape of societies, science and technology achieved central significance in contemporary democracies. In this context, states face ongoing difficulties trying to mediate these tensions and establish mechanisms for addressing problems of representation and participation in the sociopolitical process that shapes emerging technology.[42]

---

39 Ibid.
40 Ibid.
41 Biodesic (2010), 'Company', http://www.biodesic.com/Company.html.
42 Hilgartner, S. (2009), 'Intellectual Property and the Politics of Emerging Technology: Inventors, Citizens, and Powers to Shape the Future', *Chicago-Kent Law Review*, **84**, 197, 205–6.

The introduction to the collection will provide a thumbnail, comparative overview of recent developments in intellectual property and biotechnology – as a foundation to the collection.[43] Section I of this introduction considers recent developments in United States patent law, policy and practice with respect to biotechnology – in particular, highlighting the *Myriad Genetics* dispute and the decision of the Supreme Court of the United States in *Bilski* v. *Kappos*. Section II considers the cross-currents in Canadian jurisprudence in intellectual property and biotechnology. Section III surveys developments in the European Union – and the interpretation of the *European Biotechnology Directive*. Section IV focuses upon Australia and New Zealand, and considers the policy responses to the controversy of Genetic Technologies Limited's patents in respect of non-coding DNA and genomic mapping. Section V outlines the parts of the collection and the contents of the chapters.

I THE UNITED STATES OF AMERICA

After experimenting with *sui generis* regimes, such as the *Plant Patent Act* 1930 (US) and the *Plant Variety Protection Act* 1970 (US), the United States courts took an expansive approach to the patenting of life forms. The decision of the majority of the Supreme Court of the United States in *Diamond* v. *Chakrabarty* to allow for the patenting of an oil-eating bacterium was a critical ruling.[44] Alain Pottage and Brad Sherman discuss how the decision of the Supreme Court of the United States has had the effect of 'mechanizing nature': 'Just as the mechanical and chemical sciences instrumentalized inanimate nature, so biotechnology instrumentalizes animate nature, and turns organisms into manufactures.'[45] In other words, analogies have been drawn between mechanical inventions and a host of forms of biotechnology – such as nanotechnology and synthetic biology – in order to expand the limits of patentable subject matter.

---

[43] For a comprehensive introduction to intellectual property and biotechnology, see Magnus, David, Arthur Caplan and Glenn McGee (eds) (2002), *Who Owns Life?*, Amherst, NY: Prometheus Books; Rimmer, Matthew (2008), *Intellectual Property and Biotechnology: Biological Inventions*, Cheltenham and Northampton, MA: Edward Elgar; Dutfield, Graham (2009), *Intellectual Property Rights and the Life Science Industries: Past, Present and Future*, Singapore: World Scientific; and Rai, Arti (ed.) (2011), *Intellectual Property Law and Biotechnology: Critical Concepts,* Cheltenham and Northampton, MA: Edward Elgar.

[44] *Diamond* v. *Chakrabarty*, 447 US 303 (1980).

[45] Pottage, Alain and Brad Sherman (2010), *Figures of Invention: A History of Modern Patent Law*, Oxford: Oxford University Press, 181–2.

In the 1990s, there were both public and private efforts to sequence the human genome. A number of biotechnology and pharmaceutical companies filed patent applications in respect of genes and gene sequences, which were of medical significance. In his candid autobiography, *A Life Decoded*, J. Craig Venter reflects that the hectic 'gold rush' by private biotechnology companies for gene patents was not necessarily profitable:

> Because a few human genes were, in fact, worth billions, it was widely assumed that there were hundreds or thousands more genes that would be equally lucrative. The logic was simple and simpleminded. The biotech companies [Human Genome Sciences] and Incyte took the lead in human gene patents, but today their stocks trade below their cash value despite their vast human gene patent portfolios. Most now understand what I have always believed: that human gene patents usually have less value than the cost to pursue them. Of the twenty-three thousand or so human genes, fewer than a dozen have generated real value for the businessman or the patent.[46]

In part, this statement seems something of a post-facto rationalisation – given that Venter pursued patents in respect of express sequence tags at the National Institutes of Health, and gene patents, while engaging in shotgun sequencing at Celera Genomics.

In the United States, the American Civil Liberties Union and the Public Patent Foundation filed a lawsuit in 2009 against the validity of patents related to BRCA1 and BRCA2 held by Myriad Genetics Inc. and the University of Utah Research Foundation.[47] The action was brought on behalf of pathologists, medical practitioners, human genetics researchers, patient organisations and, most importantly, individual women affected by breast and ovarian cancer. Raising arguments about human rights and bioethics, the plaintiffs alleged that the patents were unconstitutional because they impinged upon freedom of speech by restricting medical research. They also argued that the patents were invalid because the genes were 'products of nature'.

A district court judge, Sweet J, was sympathetic to the challenge, expressing reservations about taking a broad approach to patentable subject

---

46 Venter, J.C. (2007), *A Life Decoded*, London: Allen Lane and Penguin Books, 257. See also Venter, J.C. (2011), 'The Human Genome at 10: Successes and Challenges', *Science*, **331** (6017), 546–7.

47 American Civil Liberties Union (2009), 'BRCA: Genes and Patents', http://www.aclu.org/free-speech/brca-genes-and-patents.

matter.[48] The judge was particularly concerned about the impact of gene patents upon patient care, medical research and the administration of health care: 'The resolution of the issues presented to this Court deeply concerns breast cancer patients, medical professionals, researchers, caregivers, advocacy groups, existing gene patent holders and their investors, and those seeking to advance public health.'[49] The judge held:

> It is concluded that DNA's existence in an 'isolated' form alters neither this fundamental quality of DNA as it exists in the body nor the information it encodes. Therefore, the patents at issue directed to 'isolated DNA' containing sequences found in nature are unsustainable as a matter of law and are deemed unpatentable subject matter under 35 U.S.C. § 101.[50]

The initial ruling of the District Court has been much contested by members of the biotechnology and pharmaceutical industries.

The United States Court of Appeals for the Federal Circuit overturned the initial ruling of Sweet J in 2011.[51] In the lead judgment, Lourie J followed past broad, expansionist precedents on patentable subject matter and held that the composition claims covering isolated DNA sequences were patentable subject matter: 'We therefore reject the district court's unwarranted categorical exclusion of isolated DNA molecules.'[52] He held that 'the patent eligibility of an isolated DNA is not negated because it has similar informational properties to a different, more complex natural material that embodies it'.[53] The judge further held that: 'If the law is to be changed, and DNA inventions excluded from the broad scope of § 101 contrary to the settled expectation of the inventing community, the decision must come not from the courts, but from Congress.'[54] Moreover, the method claim for screening potential cancer therapeutics via changes in cell growth rates was patentable.[55] However, Lourie J held that the method claims for comparing claims for comparing or analysing isolated DNA sequences were not patentable: 'Myriad's claimed methods of comparing

48 *Association for Molecular Pathology* v. *United States Patent and Trademark Office, Myriad Genetics Inc., and the directors of the University of Utah Research Foundation*, 702 F. Supp. 2d 181 at 185 (2010).

49 Ibid.

50 Ibid., 185.

51 *Association for Molecular Pathology et al.* v. *United States Patent and Trademark Office, Myriad Genetics et al.* 653 F.3d 1329, 2011 WL 3211513 (C.A. Fed. (N.Y.)), 99 USPQ 2d 1398.

52 Ibid., 18.

53 Ibid., 20.

54 Ibid., 18.

55 Ibid., 23.

or analyzing nucleotide sequences fail to satisfy the machine-or-transformation test, and are instead directed to the abstract mental process of comparing two nucleotide sequences.'[56]

Moore J filed an opinion concurring in part. The judge, though, provided a somewhat circumspect line of reasoning on the issue of the patent-eligibility of DNA sequences – ultimately being swayed by the desire not to disrupt settled business expectations. The judge observed:

> If I were deciding this case on a blank canvas, I might conclude that an isolated DNA sequence that includes most or all of a gene is not patentable subject matter. Despite the literal chemical difference, the isolated full length gene does not clearly have a new utility and appears to simply serve the same ends devised by nature, namely to act as a gene encoding a protein sequence. This case, however, comes to us with a substantial historical background. Congress has, for centuries, authorized an expansive scope of patentable subject matter. Likewise, the United States Patent Office has allowed patents on isolated DNA sequences for decades, and, more generally, has allowed patents on purified natural products for centuries. There are now thousands of patents with claims to isolated DNA, and some unknown (but certainly large) number of patents to purified natural products or fragments thereof. As I explain below, I believe we must be particularly wary of expanding the judicial exception to patentable subject matter where both settled expectations and extensive property rights are involved. Combined with my belief that we should defer to Congress, these settled expectations tip the scale in favor of patentability.[57]

Bryson J dissented on this issue, holding that isolated human genes were products of nature: 'The isolation of the naturally occurring genetic material does not make the claims to the isolated BRCA genes patent-eligible.'[58]

The question of patentability of emerging technologies has also been considered in the *Bilski* litigation. Ostensibly, the litigation in *Bilski* v. *Kappos* concerned business methods. Bernard Bilski had applied for a patent in respect of a method for hedging commodities. The patent examiner rejected the application on the grounds that the invention was merely an abstract idea, and solved a purely mathematical problem. The Board of Patent Appeals and Interferences agreed with this verdict. The Court of Appeals for the Federal Circuit reviewed the patentability of business methods. In the lead judgment, Michel J stressed:

> A claimed process is surely patent-eligible under § 101 if: (1) it is tied to a particular machine or apparatus, or (2) it transforms a particular article into a different state or thing. We agree that future developments in technology and the

---

56 Ibid., 22.
57 Ibid., 31.
58 Ibid., 42.

> sciences may present difficult challenges to the machine-or-transformation test, just as the widespread use of computers and the advent of the Internet has begun to challenge it in the past decade. Thus, we recognize that the Supreme Court may ultimately decide to alter or perhaps even set aside this test to accommodate emerging technologies.[59]

However, a number of dissenting judges expressed reservations about the impact of the 'machine-or-transformation' test upon emerging technologies.

In *Bilski* v. *Kappos*, the Supreme Court of the United States held that a patent application for method of hedging risk in the field of commodities trading in the energy market was an unpatentable abstract idea.[60] Nonetheless, in the lead judgment, Kennedy J declined to restrict the breadth of patentable subject matter, reflecting:

> The machine-or-transformation test may well provide a sufficient basis for evaluating processes similar to those in the Industrial Age – for example, inventions grounded in a physical or other tangible form. But there are reasons to doubt whether the test should be the sole criterion for determining the patentability of inventions in the Information Age. As numerous amicus briefs argue, the machine-or-transformation test would create uncertainty as to the patentability of software, advanced diagnostic medicine techniques, and inventions based on linear programming, data compression, and the manipulation of digital signals. In the course of applying the machine-or-transformation test to emerging technologies, courts may pose questions of such intricacy and refinement that they risk obscuring the larger object of securing patents for valuable inventions without transgressing the public domain ... Section 101's terms suggest that new technologies may call for new inquiries.[61]

Kennedy J cites with the approval the remarks in *Gottschalk* v. *Benson*,[62] that to 'freeze process patents to old technologies, leaving no room for the revelations of the new, onrushing technology[,] ... is not our purpose'.[63]

In the remanded case of *Prometheus Laboratories Inc*. v. *Mayo Collaborative Services*, the United States Court of Appeals for the Federal Circuit applied the ruling in *Bilski* v. *Kappos* and held that Prometheus's asserted medical treatment claims were patentable subject matter.[64]

---

59 *In re Bilski*, 545 F.3d 943 (Fed. Cir., 2008).
60 *Bilski* v. *Kappos* 130 S. Ct. 3218 (2010).
61 Ibid., 3228.
62 *Gottschalk* v. *Benson*, 409 U.S. 63, 71 (1972).
63 *Bilski* v. *Kappos*, 130 S. Ct. 3218 at 3228 (2010).
64 *Prometheus Laboratories, Inc.* v. *Mayo Collaborative Services*, No. 2008–1403, slip op. (Fed. Cir. 2010).

Dan Burk and Mark Lemley have long advocated that patent offices and courts should make use of existing policy levers within patent law to address and respond to new technologies.[65] The USPTO (United States Patent and Trademark Office) and the courts have applied a number of doctrines – such as novelty, inventive step,[66] utility,[67] and written description[68] – in order to tailor patent law to the particular demands and needs of biotechnology. Dan Burk and Mark Lemley have considered future directions in intellectual property and biotechnology – such as large-scale projects in genomics, proteomics, bioinformatics and metabolomics , and the emergence of synthetic biology. The pair 'expect that as the biotechnology industry continues to evolve, new policy levers may become appropriate to allow the proper scope of innovation and reward'.[69]

There have been a number of attempts to pass specialist legislation in respect of gene patents. Congressmen Xavier Becerra (Democrat, California) and Dave Weldon (Republican) have promoted the *Genomic Research and Accessibility Act* 2007 H.R. 977. The bill provides: 'Notwithstanding any other provision of law, no patent may be obtained for a nucleotide sequence, or its functions or correlations, or the naturally occurring products it specifies.' In the wake of the Myriad Genetics decision, Becerra has promised to reintroduce legislation, placing a moratorium on gene patents: 'I will once again introduce legislation banning gene patenting to ensure patients' access to their own medical information, reduce the costs of gene tests and increase scientific research into personalized medicine.'[70] However, it is doubtful that such legislative efforts will win broader support.[71] There is, however, somewhat more hope with regard to general patent law reform. After many years of stalled attempts and efforts, the United States

65 Burk, Dan and Mark Lemley (2009), *The Patent Crisis and How the Courts Can Solve It*, Chicago and London: The University of Chicago Press.

66 *In re Kubin*, 561 F. 3d 1351 (C.A. Fed, 2009).

67 *In re Fisher*, 421 F.3d 1365 (C.A. Fed., 2005).

68 *Carnegie Mellon University* v. *Hoffman-La Roche*, 541 F 3d. 1115 (2008).

69 Burk, Dan and Mark Lemley (2009), *The Patent Crisis and How the Courts Can Solve It*, Chicago and London: The University of Chicago Press, 153.

70 Becerra, X. (2010), 'Representative Becerra Applauds Decision on Gene Patenting', http://becerra.house.gov/index.php?option=com_content&view=article&id=412%3Arep-becerra-applauds-court-decision-on-gene-patenting&catid=3%3Apress-releases&Itemid=26.

71 For a critique of the legislative proposal, see DeGiulio, J. (2010), 'The Genomic Research and Accessibility Act: More Science Fiction than Fact', *Northwestern Journal of Technology and Intellectual Property*, **8**, 292–306.

House of Representatives and the Senate passed the *America Invents Act* 2011 (US).[72] President Barack Obama signed the legislation on 16th September 2011.

II CANADA

The Supreme Court of Canada has been divided over the question of patenting life forms – split into warring factions. In the case of *Harvard College* v. *Canada (The Commissioner of Patents)*, a five–four majority of the Supreme Court of Canada held that 'higher life forms' were not patentable subject matter in Canada.[73] The leading judgment by Bastarache J noted: 'Owing to the fact that the patenting of higher life forms is a highly contentious and complex matter that raises serious practical, ethical and environmental concerns that the Act does not contemplate, I conclude that the Commissioner was correct to reject the patent application.'[74] In *Monsanto Canada* v. *Schmeiser*, a majority of the Supreme Court of Canada ruled that a patent in relation to glyphosate resistant canola was valid – because only a component of the plant was the subject of patent claims.[75] McLachlin CJ and Fish J noted: 'Inventions in the field of agriculture may give rise to concerns not raised in other fields – moral concerns about whether it is right to manipulate genes in order to obtain better weed control or higher yields.'[76] Clearly, it is difficult to reconcile the tensions in the jurisprudence of the Supreme Court of Canada in respect of such matters.

In a piece for Ysolde Gendreau's collection, *An Emerging Intellectual Property Paradigm*, Mark Perry from the University of Western Ontario surveys the competing approaches to the patenting of life in Canadian jurisprudence: 'In *Harvard*, the Supreme Court of Canada denied a patent

---

72 *Leahy-Smith America Invents Act*, Pub. L. 112–29, §§ 14(a), 125 Stat. 284, 327 (2011); see Kappos, D. (2011), 'Statement on the America Invents Act', Subcommittee on Intellectual Property, Competition and the Internet, Committee on the Judiciary, United States House of Representatives, 30 March, http://www.uspto.gov/news/speeches/2011/kappos_house_testimony.jsp; and Obama, B. (2011), 'President Obama Signs *America Invents Act*, Overhauling the Patent System to Stimulate Economic Growth, and Announces New Steps to Help Entrepreneurs Create Jobs', The White House Press Office, http://www.whitehouse.gov/the-press-office/2011/09/16/president-obama-signs-america-invents-act-overhauling-patent-system-stim..

73 *Harvard College* v. *Canada (The Commissioner of Patents)* [2002] 2 SCR 45.

74 *Harvard College* v. *Canada (The Commissioner of Patents)* [2002] 2 SCR 45.

75 *Monsanto Canada Inc.* v. *Schmeiser* (2004) SCC 34; 2004 SCC 34.

76 *Monsanto Canada Inc.* v. *Schmeiser* (2004) SCC 34; 2004 SCC 34.

on a transgenic mouse that was undoubtedly an innovation, and in *Monsanto*, the same court found infringement in the use of a transgenic cell in a plant.'[77] He reflects:

> In the last two decades the Canadian Supreme Court has been challenged with life patent scenarios, most recently with two cases based on transgenic engineering. The 5–4 split decisions in these two cases have led to what is arguably a legally coherent but illogical outcome. The Canadian approach to patent claims that are based on new forms of life has been to distinguish 'higher' lifeforms from other lifeforms, denying patents on the former.[78]

Perry notes that 'there is little objective standard to determine what is a higher lifeform, nor are there any rational means of discriminating even between the developmental stages in the same lifeform'.[79] He concludes: 'The pressure of science on the legal system, particularly in cutting edge areas such as biotechnology patents, requires legislation for patents that is clear and "technologically neutral" (as policymakers often claim of proposed legislation).'[80]

The Canadian Biotechnology Advisory Committee produced a number of reports on intellectual property and emerging technologies.[81] However, the policy body has since been disbanded. The Canadian Parliament has been reluctant to debate intellectual property and emerging biotechnologies. In the past decade, there has been a succession of minority governments, and a number of bills have lapsed, when Parliament has been prorogued for elections.

An International Expert Group on Biotechnology, Innovation and Intellectual Property led by Richard Gold of McGill University envisages that there is a new era of intellectual property upon us, marked by negotiation:

> As the current era of IP wanes, a new one is emerging. It centres on the principle of granting the right amount of IP and having the private and public sectors use

---

77 Perry, Mark (2008), 'From Pasteur to Monsanto: Approaches to Patenting Life in Canada', in Ysolde Gendreau (ed.) (2008), *An Emerging Intellectual Property Paradigm: Perspectives from Canada*, Cheltenham and Northampton, MA: Edward Elgar, 67–80.

78 Ibid., 68.

79 Ibid., 79.

80 Ibid., 80.

81 Canadian Biotechnology Advisory Committee (CBAC) (2002), *Patenting Of Higher Life Forms*, Ottawa: CBAC, June; Canadian Biotechnology Advisory Committee (2003), *Advisory Memorandum: Higher Life Forms And The Patent Act*, Ottawa: CBAC, 24 February; and Canadian Biotechnology Advisory Committee (2006), *Human Genetic Materials, Intellectual Property, and the Health Sector*, Ottawa: CBAC.

> those protections more effectively. We call this the era of New IP, an era in which IP becomes a servant to, and not master of, values such as equity and fairness.[82]

The report emphasises a number of themes – including the need to build trust and improve communication amongst stakeholders; the importance of new models of co-operation and collaboration, and scientific infrastructure; and the need for cross-cutting thinking, and data and metrics. The report suggests: 'Once we better understand what spurs innovation, we can ensure that our discussions stay focused on the potential of biotechnology to address health, agricultural and industrial needs.'[83]

In the absence of legislative reform, there has been scope for management of intellectual property rights. David Castle comments that there is a heated disagreement regarding the design and reform of intellectual property systems, and the rules, institutions and practices they support. His edited collection emphasises the need for contextual studies of intellectual property management – taking into account the variables of history, jurisdiction, actors, institutions, and technological and scientific fields.[84]

Professor Bartha Knoppers from the University of Montreal has long argued that patent law should be linked with bioethical notions of informed consent and benefit sharing in a range of contexts – agriculture, biodiversity and medicine. She has recapitulated some of her key themes in a collection, edited with Richard Gold, entitled *Biotechnology, IP and Ethics*.[85] Knoppers comments upon the debate over convergence in the biological sciences:

> The last decade has spawned what has been called the 'cogno-info-bio-nano' continuum. In the area of biotechnology, a field which spans all living organisms, this evolving continuum has not so much produced new ethical quandaries as it has refocused the locus, form, and content of the debate. Indeed, the ethical concerns that accompany novel technologies often play out through variations in the 'name-game'. From neuroethics to nutrigenomics and nanotechnologies to nuclear transfer in stem cell research, the classical nature-nurture dynamic

---

82 The International Expert Group on Biotechnology, Innovation and Intellectual Property (2008), *Toward a New Era of Intellectual Property: From Confrontation to Negotiation*, Montreal: McGill University, 14.

83 Ibid., 10.

84 Castle, David (ed.) (2009), *The Role of Intellectual Property Rights in Biotechnology Innovation*, Cheltenham and Northampton, MA: Edward Elgar.

85 Gold, Richard and Bartha Knoppers (eds) (2009) *Biotechnology, IP and Ethics*, Markham, ON: LexisNexis Canada.

resurfaces as ethical debate and frameworks attempt to address what makes us distinctively human.[86]

Knoppers notes that 'certainly, stem cell research, genetically modified organisms and nanotechnologies typify such convergence' of science, technology, politics, economics and values.[87] She contends that 'the promise of biotechnology will become empty rhetoric, with the accompanying danger of public cynicism, if there is an absence of responsibility'.[88]

The Canadian ETC Group has a long-standing interest in the legal, ethical and social implications of emerging and converging technologies:

> The ability, through nanotechnology, to manipulate matter atom by atom is enabling a new fusion of powerful technologies as nanotech, biotech, information technology and neurotechnologies (brain technologies) converge into one common technology platform. ETC refers to this convergence as BANG (Bits Atoms Neurons and Genes) since it allows flexible manipulation of the bits of information, the atoms of matter, the neurons of the brain and the genes that code for life. Examples of 'convergent' technologies include nanobiotechnology, synthetic biology, DNA computing and neuroengineering.[89]

This populist definition, again, emphasises the convergence of various emerging technologies. The ETC Group has, traditionally, been concerned about the impact of intellectual property upon agriculture and farmers' rights. Perhaps most famously, the civil society organization led a public campaign against genetic use restriction technologies – which it dubbed 'terminator technologies'. More recently, the ETC Group has expanded its horizons to focus on the granting of patents in respect of nanotechnology, synthetic biology, geo-engineering and clean technologies.

---

86 Knoppers, Bartha (2009), 'Biotechnology and Ethics', in Richard Gold and Bartha Knoppers (eds) (2009), *Biotechnology, IP and Ethics*, Markham, ON: LexisNexis Canada, 59–67, 59.

87 Ibid., 67.

88 Ibid., 66.

89 ETC Group, 'BANG: Converging Technology', http://www.etcgroup.org/en/issues/bangconvergence.

III THE EUROPEAN UNION

In the European Union, there has also been a similar reappraisal of biotechnological patents, with ongoing conflicts in the European Patent Office.[90]

Aurora Plomer and Paul Torremans have reflected that the *European Union Directive on the Legal Protection of Biotechnological Inventions* 1998 (EU) has struggled to accommodate emerging technologies – such as stem cell research: 'The emerging difficulties concerning the Directive's implementation and interpretation 10 years on are a witness to the enduring social and ethical concerns about new biotechnologies but also disclose the complex incidental legal impact of the moral exclusion clauses within and outside patent law, the full breadth of which had not been anticipated at the time.'[91] Nonetheless, the European Court of Justice has defended the validity of the *European Biotechnology Directive*, rejecting a challenge by the Netherlands Government in 2001.[92]

The role of the patent system and the proper function of patent examining bodies in dealing with issues of ethics and morality is a controversial topic. In Europe, Article 53 (a) of the *European Patent Convention* provides that patents shall not be granted for:

> (a) inventions the commercial exploitation of which would be contrary to 'ordre public' or morality; such exploitation shall not be deemed to be so contrary merely because it is prohibited by law or regulation in some or all of the Contracting States ...[93]

---

[90] Hawkins, N. (2010), 'Human Gene Patents and Genetic Testing in Europe: A Reappraisal', *SCRIPTed*, **7** (3), 453–73, http://www.law.ed.ac.uk/ahrc/script-ed/vol7–3/hawkins.asp.

[91] Plomer, Aurora and Paul Torremans (eds) (2009), *Embryonic Stem Cell Patents: European Patent Law and Ethics*, Oxford: Oxford University Press, xxxvii.

[92] *Netherlands* v. *European Parliament* (2001) 3 CMLR 49; Scott, A. (1999), 'The Dutch Challenge to the Bio-Patenting Directive', *European Intellectual Property Review*, **21** (4), 212–15; Moore, S. (2002), 'Challenge to the Biotechnology Directive', *European Intellectual Property Review*, **24** (3), 149–54; and Beyleveld, D. and R. Brownsword (2002), 'Is Patent Law Part of the EC Legal Order? A Critical Commentary on the Interpretation of Article 6(1) of Directive 98/44/EC in Case C-377/98', *Intellectual Property Quarterly*, **1**, 97–110.

[93] *Convention on the Grant of European Patents ('European Patent Convention')* of 5 October 1973, Official Journal Special Edition 1/2007 (entered into force 7 October 2000).

There has been much debate in the European Union over the meaning of *ordre public* (public morality and order).[94]

In the Harvard 'OncoMouse' litigation in the European Union, there was significant debate as to what test was appropriate in these circumstances.[95] In the 'OncoMouse' decision, the Examining Division initially declined to consider the issue, taking the view that it was inappropriate for technocrats to adjudicate upon the morality of the invention.[96] The Board of Appeal suggested that there was a need to engage in a balancing exercise between the benefits and risks associated with an invention.[97] In a further consideration of the Harvard OncoMouse in 2006, the Technical Board of Appeals reiterated that the balancing test 'clearly allows the scope or extent of, on the one hand, the animal suffering and/or environmental risk and, on the other hand, the usefulness to mankind to be considered'.[98]

In the case of *Plant Genetic Systems/Glutamine Synthetase Inhibitors*, the Technical Board of Appeals provided further consideration of the requirement that a patent application is not contrary to either *ordre public* or morality.[99] Also noteworthy is the *Relaxin* case, in which the Green Party unsuccessfully objected to the Howard Florey Institute's patent application in respect of Relaxin.[100] There has also been much discussion about the ethics of patenting human embryonic stem cells.[101] Further, in the *Euthanasia Compositions* case, the Board held that when considering whether the commercial exploitation of an invention is contrary to morality or *ordre public*, 'exploitation' should be interpreted as the 'normal avowed use indicated in the patent'.[102] The patent should only be denied under Article 53 (a) if that intended use would infringe *ordre public* or morality.[103]

---

94 See for instance Bently, Lionel and Brad Sherman (2009), *Intellectual Property Law*, 3rd edn, Oxford: Oxford University Press, 455–63; and MacQueen, Hector, Charlotte Waelde and Graeme Laurie (2008), *Contemporary Intellectual Property: Law and Policy*, Oxford: Oxford University Press, 424–7.

95 For an extended discussion of the transnational litigation involving the Harvard OncoMouse, see Rimmer, Matthew (2008), *Intellectual Property and Biotechnology: Biological Inventions*, Cheltenham and Northampton, MA: Edward Elgar, 82–107.

96 *The Harvard/Onco-mouse* [1989] OJ EPO 451.

97 *The Harvard/Onco-mouse* [1990] OJ EPO 490 (TBA).

98 *Harvard/Transgenic Animals,* T315/03 [2006] OJ EPO at 98.

99 *Plant Genetic Systems/Glutamine Synthetase Inhibitors* [1995] EPOR 357.

100 *Howard Florey/Relaxin* T74/91 [1995] EPOR 541 (Op Div).

101 Rimmer, Matthew (2008), *Intellectual Property and Biotechnology: Biological Inventions*, Cheltenham and Northampton, MA: Edward Elgar, 263–72.

102 *Euthanasia Compositions/Michigan State University* (2005), T 0866/01; 5.7.

103 *Euthanasia Compositions/Michigan State University* (2005), T 0866/01; 5.8.

In the area of environmental bioprospecting, the EPO (European Patent Office) has also struggled to deal with claims of 'biopiracy' – especially in relation to patents filed in respect of the neem tree and Nap Hal wheat.[104]

The EPO seems to regard engaging with moral and ethical issues as outside its remit. Aurora Plomer has discussed the future of general morality exclusions with respect to biotechnology patents:

> To the extent that 'human dignity' and more generally fundamental moral norms are implicated in the construction of the scope of exclusion of patents on biotechnological inventions, there is a clear alignment between the jurisprudence of the European Court of Justice and the jurisprudence of the European Court of Human Rights recognizing a wide margin of discretion to Member States in deference and respect for the plurality and diversity of moral cultures and state autonomy on morally sensitive questions.[105]

In 2010, the European Court of Justice considered the question of patent infringement and biotechnology in the case of *Monsanto Technology LLC* v. *Cefetra BV*, ruling that the importation of soybean meal did not amount to an infringement of a gene patent.[106] Some commentators fear that 'the Cefetra judgment may unintentionally inflict serious economic harm on the European biotechnology industry because countless biotechnology products may no longer be protected by what are otherwise valid and enforceable patent claims'.[107]

The European Parliament has repeatedly expressed concerns about the breadth of patents granted by the EPO in respect of emerging technologies. In 2001, the European Parliament passed a resolution protesting against the broad patents granted in respect of the BRCA1 and BRCA2 genes.[108] In 2005, the European Parliament called 'on the European Patent Office and the Member States to grant patents on human DNA only in connection with a concrete application and for the scope of the patent to be limited to

---

104 Mgbeoji, Ikechi (2006), *Global Biopiracy: Patents, Plants, and Indigenous Knowledge*, Vancouver: University of British Columbia Press.

105 Plomer, Aurora (2009), 'Human Dignity, Human Rights, and Article 6 (1) of the EU Directive on Biotechnological Inventions', in Aurora Plomer and Paul Torremans (eds) (2009), *Embryonic Stem Cell Patents: European Patent Law and Ethics*, Oxford: Oxford University Press, 203–26, at 223–4.

106 *Monsanto Technology LLC* v. *Cefetra BV and others*, No. C-428/08, ¶¶ 33–50, European Court of Justice (6 July 2010).

107 Mohan-Ram, V., R. Peet and P. Vlaemminck 2011, 'Biotech Patent Infringement in Europe: The "Functionality" Gatekeeper', *John Marshall Review of Intellectual Property Law*, **10**, 540–52.

108 *European Parliament Resolution on the Patenting of BRCA1 and BRCA2 ('Breast Cancer') Genes*, Texts Adopted by Parliament, Provisional Edition: 04/10/2001, B5–0633, 0641, 0651 and 0663/2001.

this concrete application so that other users can use and patent the same DNA sequence for other applications (purpose-bound protection)'.[109] At the same time, the European Parliament emphasised in the resolution that the patenting of procedures involving human embryonic stem cells or cells that are grown from human embryonic stem cells was a violation of the *European Biotechnology Directive*.[110] In a resolution adopted in November 2007, the European Parliament suggested that there needed to be greater leeway for the compulsory licensing of environmentally necessary technologies.[111] In 2009, the European Parliament passed a resolution on the regulation of nanotechnology, noting fears that the 'broad application of patents to nanomaterials as well as the excessive cost of patenting and the absence of patent access facilities for very small businesses and small and medium-sized enterprises ... could stifle further innovation'.[112]

In its 2007 report on *Scenarios for the Future*, the EPO predicted that patent law, policy and practice would have to grapple with new scientific revolutions.[113] The EPO explicitly cited the report of the National Science Foundation and United States Department of Commerce, which envisaged two overlapping processes of convergence – 'the computer and communications revolution in progress today and a second, just beginning, representing the NBIC revolution, i.e. the combination of nanotechnology, biotechnology, information technology and cognitive sciences'.[114] The EPO reflected on the implications of converging technologies for patent law, policy and administration:

> Increasing 'interdisciplinarity' – for example, the combination of nanotechnology, biotechnology, information technology (IT) and cognitive sciences (NBIC) – is also increasing the challenges for the patent offices when it comes to

---

109 *European Parliament Resolution on Patents for Biotechnological Inventions*, P6_TA (2005) 0407, 26 October 2005, http://eur-lex.europa.eu/LexUriServ/LexUriServ.do?uri=OJ:C:2006:272E:0440:0442:EN:PDF.

110 Ibid.

111 *European Parliament Resolution on Trade and Climate Change*, 29 November 2007, (2007/2003 (INI)), http://www.europarl.europa.eu/sides/getDoc.do?Type=TA&Reference=P6-TA-2007–0576&language=EN.

112 *European Parliament Resolution of 24 April 2009 on Regulatory Aspects of Nanomaterials*, http://www.europarl.europa.eu/sides/getDoc.do?type=TA&reference=P6-TA-2009–0328&language=EN.

113 European Patent Office (2007), *Scenarios for the Future: How Might IP Regimes Evolve by 2025? What Global Legitimacy Might Such Regimes Have?*, Munich: European Patent Office, http://documents.epo.org/projects/babylon/eponet.nsf/0/63A726D28B589B5BC12572DB00597683/$File/EPO_scenarios_bookmarked.pdf.

114 Ibid., 90.

> expertise. Multidisciplinary technologies create a new challenge for the people charged with administering patents: how can you expertly assess a patent application for a concept that requires top-level understanding of not one, but perhaps four different disciplines?[115]

The EPO noted that, in addition to questions of patent law, policy and practice, there were larger questions of regulations of emerging technologies: 'In the foreseeable future – and even now – ethical issues concerning certain technological developments could force us to question our moral values and the rules that ought to govern human conduct.'[116]

IV AUSTRALIA AND NEW ZEALAND

In Australia, there have been cycles of debate and controversy over gene patents and stem cell patents. In the 1990s, Senators John Coulter and Natasha Stott Despoja from the Democrats – a minor party in the Australian Parliament, since defunct – introduced bills seeking to prohibit gene patents. However, such measures did not win support from the major parties. Independent Senator Brian Harradine was successful in introducing an amendment to the *Patents Act* 1990 (Cth). The resulting section 18 (2) provides that 'Human beings and the biological processes for their generation are not patentable inventions.' IP Australia has sought to divine the meaning of this strange, ambiguous and indeterminate prohibition.

In the 2000s, the public debate over gene patents has re-emerged. A Melbourne company, called Genetic Technologies Limited (GTG), obtained broad patents on a range of scientific inventions arising out of the work of Malcolm Simons. Most significantly, the USPTO awarded U.S. Patent No. 5,612,179 to GTG for an invention entitled, 'Intron sequence analysis method for detection of adjacent and remote locus alleles as haplotypes'.[117] The USPTO also issued U.S. Patent No. 5,851,762 to GTG for an invention entitled 'Genomic mapping method by direct haplotyping using intron sequence analysis'.[118] The company has embarked upon a

---

115 Ibid., 91.

116 European Patent Office (2007), *Scenarios for the Future: How Might IP Regimes Evolve by 2025? What Global Legitimacy Might Such Regimes Have?*, Munich: European Patent Office, http://documents.epo.org/projects/babylon/eponet.nsf/0/63A726D28B589B5BC12572DB00597683/$File/EPO_scenarios_bookmarked.pdf.

117 Simons, M. (1992), 'Intron Sequence Analysis Method for Detection of Adjacent and Remote Locus Alleles as Haplotypes', U.S. Patent No: 5,612,179.

118 Simons, M. (1994), 'Genomic Mapping Method by Direct Haplotyping Using Intron Sequence Analysis', U.S. Patent No: 5,851,762.

vigorous licensing program – obtaining, in the process, licences in Australasia to Myriad's patents in respect of genetic testing. In 2008, GTG announced that the company would enforce its rights in respect of these patents against public, as well as private, providers of genetic testing.[119] Its press release sought to explain the motivations behind this: 'Given that Genetic Technologies now offers an excellent service and has considerable excess capacity, the Company has made a commercial decision to enforce the exclusive rights granted to it by Myriad Genetics to perform diagnostic testing of the BRCA1 and BRCA2 genes in Australia and New Zealand.'[120]

In response to widespread criticism of GTG, the company changed the composition of its Board of Directors, and withdrew its demands. Its somewhat more conciliatory press release noted: 'The Company looks forward to working positively with all its partners, including other public and private testing laboratories, to continue providing these world class testing services.'[121] Nonetheless, despite its position in Australia, GTG has continued to pursue litigation against biotechnology companies in other jurisdictions. In 2010, GTG filed a patent infringement suit in respect of its non-coding DNA technologies against nine parties in the United States District Court, Western District of Wisconsin. The parties were Beckman Coulter Inc., Orchid Cellmark Inc., Gen-Probe Inc., Interleukin Genetics Inc., Molecular Pathology Laboratory Network Inc., Monsanto Inc., PIC USA Inc., Sunrise Medical Laboratories and Pioneer Hi-Bred International Inc.[122] In 2011, GTG filed a lawsuit in the United States District Court for the Western District of Wisconsin against a number of clinical pathology laboratories associated with Sonic Healthcare Limited.[123]

---

119 Genetic Technologies Limited (2008), 'Genetic Technologies to Enforce Rights to BRCA Patents', Press Release, 11 July, http://www.gtglabs.com.au/announcements/genetic-technologies-to-enforce-rights-to-brca-patents.

120 Ibid.

121 Genetic Technologies Limited (2008), 'New Position re BRCA Testing', Press Release, 2 December, http://www.gtglabs.com.au/announcements/new-position-re-brca-testing.

122 *Genetic Technologies Limited* v. *Berkman Coulter Inc. et al.* (2010), United States District Court for the Western District of Wisconsin, Civil Action No. 10 CV 0069; and Genetic Technologies Limited (2010), 'Genetic Technologies Moves to Assert Its Intellectual Property Rights and Files Patent Infringement Suit in USA', Press Release, 16 February, http://www.gtglabs.com.au/images/stories/pdf/ASX%20Announcements/Assertion_of_property_rights_February_16_2010.pdf.

123 *Genetic Technologies Ltd.* v. *American Esoteric Laboratories et al.* (2011), United States District Court for the Western District of Texas, 1:11-cv-00057; Genetic Technologies Limited (2011), 'Genetic Technologies Files Second Patent Infringement Suit in the USA', Press Release, 20 January.

In June 2010, Cancer Voices Australia and Yvonne D'Arcy brought an action in the Federal Court of Australia against the validity of a BRCA1 patent – held by Myriad Genetics Inc., the Centre de Recherche du Chul, the Cancer Institute of Japan and Genetic Technologies Limited.[124] The applicants made the following arguments:

> Genes and the information represented by human gene sequences are products of nature universally present in each individual, and the information content of a human gene sequence is fixed. Genetic variations or mutations are products of nature. The isolation of the BRCA1 gene mutation from the human body constitutes no more than a medical or scientific discovery of a naturally occurring phenomenon and does not give rise to a patentable invention.[125]

In the alternative, the applicants argued that 'the alleged invention is not a patentable invention in that, so far as claimed in claims 1–3, it is not a manner of manufacture within the meaning of s 6 of the *Statute of Monopolies*'.[126] The applicants maintained that 'the alleged invention is a mere discovery'.[127] Furthermore, the applicants contended that 'the alleged invention of each of claims 1–3 is not a patentable invention because they are claims for biological processes for the generation of human beings'.[128]

As a result of the controversy over GTG and Myriad Genetics, the Australian Law Reform Commission was instructed to undertake a review of intellectual property rights over genes and genetic and related technologies, with a particular focus on human health issues. The Commission released an issue paper in July 2003, a discussion paper in January 2004, and a final 678-page report in June 2004, which was tabled in Parliament in August 2004.[129] The Australian Law Reform Commission reflected upon how patent law accommodated successive waves of emerging technologies:

---

124 *Cancer Voices Australia et al.* v *Myriad Genetics Inc. et al.* (2010), Federal Court of Australia, Statement of Claim, 8 June.

125 Ibid.

126 Ibid.

127 Ibid.

128 Ibid.

129 Australian Law Reform Commission (2003), *Gene Patenting and Human Health, Issues Paper 27*, Sydney: Australian Commonwealth, http://www.austlii.edu.au/au/other/alrc/publications/issues/27/; Australian Law Reform Commission (2004), *Gene Patenting and Human Health, Discussion Paper 68*, Sydney: Australian Commonwealth, http://www.austlii.edu.au/au/other/alrc/publications/dp/68/; Australian Law Reform Commission (2004), *Genes and Ingenuity: Gene Patenting and Human Health, Report 99*, Sydney: Australian Commonwealth, http://www.austlii.edu.au/au/other/alrc/publications/reports/99/.

'Once patent examiners have become familiar with genetics, they will, no doubt, be met with a new range of challenges from emerging disciplines such as bioinformatics, pharmacogenomics, proteomics and nanotechnology.'[130] Emphasizing the need for 'technological neutrality', the Commission did 'not believe that concerns about the patenting of inventions involving genetic materials and technologies should be addressed by provisions in the *Patents Act* dedicated only to these types of inventions'.[131] The final report contained another 50 recommendations. The key recommendation was that the Australian Government should recognise a statutory defence under patent law for experimental use. The Advisory Council on Intellectual Property conducted an independent inquiry into the desirability of a defence of experimental use in patent law. The reform body was of the view that there should be a research exemption.

In November 2010, the Senate Standing Committees on Community Affairs Reference Committee released its long awaited report on gene patents.[132] As the executive summary indicates, the report focused upon gene patents in relation to health and medicine.[133] The majority of the committee were reluctant to impose a prohibition on patents being granted in the field of biotechnology, observing that it 'heard conflicting evidence as to whether a prohibition on the patenting of genes and other biological materials (a) would be effective and (b) would not lead to unforeseen consequences in other fields of technology, particularly biotechnology research and development'.[134]

A number of considerations persuaded the Committee that it would not, at this point in time, recommend that the Act be amended to expressly prohibit the patenting of genes – including uncertainty about the effectiveness of any such prohibition; international developments such as the litigation over Myriad Genetics' patents in the United States; and national developments. Instead, the majority of the committee recommended a number of procedural and substantive reforms to improve the quality of patents granted – particularly in the area of biotechnology. However, a minority of the committee – Senators Helen Coonan and Bill Heffernan – were of the view that the *Patents Act* 1990 (Cth) should be amended to ban

---

130 Australian Law Reform Commission (2004), *Genes and Ingenuity. Report 99*, Sydney: Australian Government, 64.

131 Ibid.

132 Senate Standing Committees on Community Affairs Reference Committee (2010), *Gene Patents*, Canberra: Australian Parliament, http://www.aph.gov.au/senate/committee/clac_ctte/gene_patents_43/report/index.htm.

133 Ibid., xi.

134 Ibid.

the grant of patents over biological materials that are identical or substantially identical to such materials as they exist in nature.[135]

A private members' bill has been introduced into the Australian Parliament, entitled the *Patent Amendment (Human Genes and Biological Materials) Bill* 2010 (Cth). The sponsors include the Liberal Party members, Senators Heffernan and Coonan; Greens Senator Rachel Siewert; and Independent Senator Nick Xenophon. The rather crudely drafted bill proposes a broad prohibition: 'The following are not patentable inventions: (a) human beings, and the biological processes for their generation; and (b) biological materials including their components and derivatives, whether isolated or purified or not and however made, which are identical or substantially identical to such materials as they exist in nature.' Such a provision is ambiguous, uncertain and indeterminate in its breadth and scope. There is no accompanying definition of 'biological materials'. There has been substantive debate over the topic in the committee and the Parliament.[136] Given the weaknesses of the legislative proposal, it is doubtful that such a bill will win the support of the major parties in the Australian Parliament.

The policy debate in Australia has failed to canvass larger questions about how patent law deals with the full range of emerging technologies – not just gene patents; but also nanotechnology, stem cell research, synthetic biology and clean technologies.

In New Zealand, there has been a lively debate about the implications of the Human Genome Project for legal regulation.[137] There has also been controversy over GTG seeking licence fees from the Auckland District Health Board, and Crown Research Institutes. The *Patents Bill* 2008 (NZ) 235–2 put forward a number of proposals for exclusions of subject matter from patent protection. Section 14 stipulated that 'an invention is not a patentable invention if the commercial exploitation of the invention, so far as claimed in a claim, is contrary to – (a) public order (which in this section

---

135 Ibid.

136 Senate Standing Committees on Legal and Constitutional Affairs (2011), *Patent Amendment (Human Genes and Biological Materials) Bill 2010*, Canberra: Australian Parliament, http://www.aph.gov.au/senate/committee/legcon_ctte/patent_amendment/index.htm.

137 Henaghan, Mark (ed.) (2006), *Choosing Genes for Future Children: Regulating Preimplantation Genetic Diagnosis*, Dunedin: Human Genome Research Project and New Zealand Law Foundation; Henaghan, Mark (ed.) (2007), *Genes, Society, and the Future*, Dunedin: Human Genome Research Project and New Zealand Law Foundation; and Henaghan, Mark (ed.) (2009), *Findings from the Law Foundation Sponsored Human Genome Research Project*, Dunedin: Human Genome Research Project and New Zealand Law Foundation.

has the same meaning as the term *ordre public* as used in Article 27.2 of the *TRIPS agreement*); or (b) morality'. Section 15 proposed excluding a number of subject matter from patentability – including 'human beings, and biological processes for their generation'; methods of human treatment; methods of diagnosis, computer programs and plant varieties.

V OUTLINE OF CHAPTERS

Rather than revisit the well-documented debate over gene patents,[138] this collection instead explores the impact of intellectual property upon 'emerging technologies' in the life sciences.[139] In April 2011, the Nuffield Council on Bioethics, based in the United Kingdom, released a consultation paper on emerging biotechnologies.[140] The Council considered the meaning of the term 'emerging technologies', and its derivative, 'emerging biotechnologies':

> Emerging biotechnologies ... are those emerging technologies with a biological basis or use. Without settling on a definition of emerging biotechnologies for the time being we can nevertheless suggest certain features that may characterise them (although not necessarily all of them) and by which they might be recognised. The developmental pathway followed by these technologies may vary considerably. Some emerging biotechnologies that were conceived decades ago have taken a long time or have not (yet) achieved their original promise (e.g. xenotransplantation) while others have been implemented relatively quickly

---

138 See Magnus, David, Arthur Caplan and Glenn McGee (eds) (2002), *Who Owns Life?*, Amherst, NY: Prometheus Books; Rimmer, Matthew (2008), *Intellectual Property and Biotechnology: Biological Inventions*, Cheltenham and Northampton, MA: Edward Elgar; Dutfield, Graham (2009), *Intellectual Property Rights and the Life Science Industries: Past, Present and Future*, Singapore: World Scientific; and Rai, Arti (ed.) (2011), *Intellectual Property Law and Biotechnology: Critical Concepts*, Cheltenham and Northampton, MA: Edward Elgar.

139 Susan Cozzens and her colleagues identified a number of themes in a literature review of the usage of the term 'emerging technologies': 'The concepts reflected in the definitions of emerging technologies, however, can be summarised four-fold as follows: (1) fast recent growth; (2) in the process of transition and/or change; (3) market or economic potential that is not exploited fully yet; (4) increasingly science-based.' Cozzens, S. et al. (2010), 'Emerging Technologies: Quantitative Identification and Measurement', *Technology Analysis and Strategic Management*, **22** (3), 361–76.

140 Nuffield Council on Bioethics (2011), *Emerging Biotechnologies*, London: Nuffield Council on Bioethics, April, http://www.nuffieldbioethics.org/sites/default/files/files/Nuffield_Council_Bioethics_Emerging_Biotechnologies_consultation_April-June_2011.pdf.

> (e.g. preimplantation genetic diagnosis); some are in early stages of use with uncertain prospects (e.g. genomic medicine) while others are at early stages of research (e.g. synthetic biology) with an uncertain range of uses.[141]

The report notes that 'Emerging biotechnologies often have an interdisciplinary character, drawing on knowledge originating in a variety of more established fields.' Indeed, such technologies could be characterised by '[n]ovelty (of approach and perspective), convergence (between fields and practices), and divergence (from an established "parent" field)'.[142] The Consultation Paper reflects that 'emerging biotechnologies have the capacity to generate significant controversy' – particularly 'because of uncertainty or disagreement about the nature and level of associated risks'.[143]

Each contributor in this collection considers a particular 'scientific field', to use a phrase of Pierre Bourdieu.[144] Such an approach will reveal commonalities between emerging technologies; as well as divergences in law, policy and practice. In this collection, scholars and researchers consider a panoply of emerging technologies. There is an exploration of genomics, biomedicine, bioinformatics, information technology, RNA interference, synthetic biology, stem cell research, nanotechnology, clean technologies, biodiscovery and traditional knowledge. Interestingly enough, under the NBIC categories, nanotechnology, biotechnology and information technology are well represented in the collection; but cognitive sciences and neurosciences do not figure highly. The fields of scientific research have multiple applications, covering agriculture, research, medicine and the environment. The contributors are a cosmopolitan group. The scholars hail from a variety of jurisdictions – including the United States, Canada, Great Britain, France, Sweden, Finland and Australia. This lends to the collection a decidedly comparative approach to understanding intellectual property and emerging technologies. The researchers consider not only intellectual property law, but also the adjoining areas of health law and bioethics, environmental law and climate law, international law, and Indigenous legal issues. Hilgartner suggests that a purely doctrinal approach to patent law is insufficient to address the politics of emerging technologies: 'Traditional doctrine, with its emphasis on innovation and its focus on a relatively narrow group of actors, is poorly equipped for considering questions of democratic representation in decision making

141 Ibid.
142 Ibid.
143 Ibid.
144 Bourdieu, Pierre (2004), *Science of Science and Reflexivity*, Cambridge: Polity Press.

about the shape of future technological and social orders.'[145] Accordingly, the scholars here consider not only law, but also an array of intersecting disciplines. The contributors address the science of various fields, as well as the history, sociology and philosophy of science; technology and innovation policy; politics and governance; social anthropology and Indigenous studies. The researchers apply a range of methodologies to shed light upon their respective fields.

In the preliminary part of this volume, Eva Hemmungs Wirtén examines the flow and control of knowledge of plants and their potential pharmaceutical applications. She begins in the mid eighteenth century, when European countries such as Spain were colonising South America and sending botanists forth to the new colonies to explore the native flora. She traces the journey of plants and seeds, and knowledge of their useful effects, from the colonies of South America to Europe. Hemmungs Wirtén argues that, rather than being motivated only by scientific curiosity, these explorations and the knowledge they yielded were part of the colonial expansion agendas of European powers. Hemmungs Wirtén uses the cinchona tree, from which quinine is extracted to treat malaria, as her key example. This discussion raises debates about the ownership of plant resources and knowledge of their effects, as well as the preservation and stewardship of those resources. Hemmungs Wirtén argues that the struggles over the control of knowledge of useful plants, and the role of the various actors, are far from black and white. Hemmungs Wirtén shows that, in more recent times, use of plants by the pharmaceutical industry has caused nature to be seen as 'a lucrative medicine cabinet', emphasising the importance of the rainforest as a storehouse of genetic diversity and information. She concludes that the history described is an important part of understanding current approaches to biopiracy.

Questions about the scope of patentable subject matter are further developed in the second part of this volume.

In Chapter 2, Joly and Hemmings stress the importance of changes in subject matter patentability for the biotechnology research field. They survey United States case law on patentability of processes prior to the *Bilski* litigation, before considering the decisions in *In re Bilski* and *Bilski* v. *Kappos*[146] regarding the patentability of a method of hedging commodities.

---

145 Hilgartner, S. (2009), 'Intellectual Property and the Politics of Emerging Technology: Inventors, Citizens, and Powers to Shape the Future', *Chicago-Kent Law Review*, **84**, 197 at 223.

146 *In re Bilski*, 545 F.3d 943 (Fed. Cir., 2008); and *Bilski* v. *Kappos*, 130 S. Ct. 3218 (2010).

In the decision in *In re Bilski*, the Federal Court favoured a 'machine-or-transformation test', controversially holding that this was the definitive test to determine whether a process was patentable.[147] On appeal, the Supreme Court held in *Bilski* v. *Kappos* that the machine-or-transformation test only provides a clue about patentability and is not definitive.[148] The authors suggest that the majority judges were influenced by concerns about uncertainty over the patentability of emerging technologies such as medical diagnostic techniques. Examining the extent to which the *Bilski* decision changes the existing law, Joly and Hemmings conclude that the law has been clarified rather than altered, by making it clear that 'physicality' of invention is only a clue to its patentability. They analyse the case of *Amazon.com* v. *Canada (Attorney General)*[149] to demonstrate that even this small clarification can increase the scope of patentable subject matter. They consider the ramifications of *Bilski* in the specific context of genomics research, a growing field that both utilises many processes and is likely to yield a huge amount of innovation in processes for research and medical diagnosis.

In the third chapter, Joshua Sarnoff provides his perspective on the recent subject matter decisions, particularly regarding patentability of *applications* of unpatentable ideas or principles. He criticises what he sees to be an inconsistent approach in decisions of the Supreme Court of the United States – such as *Bilski* v. *Kappos* – and a failure to clarify the circumstances in which applications of excluded ideas will be patentable.[150] Sarnoff argues that in *Prometheus Laboratories Inc* v. *Mayo Collaborative Services*,[151] the United States Court of Appeals for the Federal Circuit erroneously allowed a patent over a process of administering a drug, then measuring metabolites and using this measurement to determine whether to alter the drug dosage. The Court failed to treat the correlation involved as prior art and did not look for inventiveness in the specific application of the correlation at issue. The author then turns to the recent litigation in the *Myriad* decision[152] over the patenting of genes implicated in breast cancer

147 *In re Bilski*, 545 F.3d 943 (Fed. Cir., 2008).

148 *Bilski* v. *Kappos*, 130 S. Ct. 3218 (2010).

149 *Amazon.com* v. *Canada (Attorney General)* (2010), F.C.J. 1209.

150 *Bilski* v. *Kappos*, 130 S. Ct. 3218 (2010).

151 *Prometheus Laboratories Inc.* v. *Mayo Collaborative Services*, No. 04-CV-1200, 2008 WL 878910 (S.D. Cal. Mar. 28, 2008); and *Prometheus Laboratories, Inc.* v. *Mayo Collaborative Services*, No. 2008–1403, slip op. (Fed. Cir. 2010).

152 *Association for Molecular Pathology et al.* v. *United States Patent and Trademark Office, Myriad Genetics, et al.*, 702 F. Supp. 2d 181, (S.D.N.Y. 2010), appeal docketed, No. 2010–1406 (Fed. Cir. June 22, 2010); and *Association for Molecular Pathology et al.* v. *United States Patent and Trademark Office,*

and related methods of genetic diagnosis. Overall, Sarnoff identifies the need for a greater understanding of the degree of creativity required for such medical and genetic applications to be patentable.

In Chapter 4, Graham Dutfield investigates the role of language in interpreting patentable subject matter. He explores the use of analogies and metaphors in the life sciences, demonstrating the move from a discourse of 'life as chemistry' to one of 'life as information'. Dutfield considers the use of the terms 'machine' and 'mechanism' to describe biological phenomena. He notes that reasoning by analogy is one of the approaches courts faced with unfamiliar patentable subject matter questions may take. In *Diamond* v. *Chakrabarty*,[153] Dutfield argues, the Court was able to find the modified bacterium patentable by analogising it with inanimate chemicals, whereas in the Harvard OncoMouse™ case[154] a modified higher organism did not fit this analogy. He then turns to the use of engineering concepts and terminology in the emerging field of synthetic biology, asking whether the metaphors and analogies being deployed by synthetic biologists are sufficient to legitimise the grant of product patents. Dutfield argues that when we are truly able to create a living organism in the laboratory, life forms will indeed become manufactured devices and the engineering analogy will 'finally fit perfectly'.

In the third part, we consider several institutional models for research in the life sciences.

In Chapter 5, Dianne Nicol and Richard Gold present a discussion of intellectual property and ethics in biobanks, providing case studies of the United Kingdom Biobank, Generation Scotland, and the Canadian CARTaGENE project. The authors identify five aspects of biobanking that form competing considerations: access to the biobank's resources for research; downstream development of research and private sector involvement; privacy and autonomy of participants; participant and public trust; and securing funding. Throughout the discussion, the authors emphasise the importance of maintaining public trust in biobanking through transparency and accountability. They also note the need to ensure benefit sharing and adequate information about benefits for participants. Nicol and Gold consider the possibilities for 'open source' or 'open access' arrangements in biobanking, concluding that there are practical reasons why open source is more difficult to apply in biobanking than in the software environment.

Myriad Genetics, et al. 653 F.3d 1329, 2011 WL 3211513 (C.A. Fed. (N.Y.)), 99 USPO 2d 1398.

153 *Diamond* v. *Chakrabarty*, 447 U.S. 303 (1980).

154 *Harvard College* v. *Canada (Commissioner of Patents)*, 2002 SCC 76.

In Chapter 6, Donna Gitter explains that genomic research is now generating large amounts of data, so that scientific databases and policies for data access and sharing have increased in importance. She examines the 1000 Genomes Project, in which the genomes of 1000 people from across the world will be sequenced and compared with a view to obtaining a more detailed picture of how genetic variation affects disease susceptibility, response to drugs and reactions to the environment. Gitter traces the development and effectiveness of consensus principles for data sharing from the *Bermuda Statement* 1996 to the *Toronto Statement* 2009, noting that sharing tends to play out differently in large publicly funded projects from within the 'small science' context. She canvasses some options for giving due recognition to the original contributors of shared scientific data. The discussion touches on the roles of funding bodies and scientific journals in promoting and managing data sharing. Gitter also notes the importance of infrastructure for ongoing maintenance of data and specimens. Finally, she identifies certain data that are not suitable for open access models.

In Chapter 7, Alison McLennan picks up some of the intellectual property issues around synthetic biology research raised earlier in the volume. She analyses the structures established to advance synthetic biology research by the BioBricks Foundation, a group of scientists in the United States. Sharing of information and resources in synthetic biology research is facilitated by the Registry of Standard Biological Parts, which is supported by a culture of sharing in the synthetic biology community. McLennan analyses the draft legal arrangements that have been proposed to underpin these sharing activities. She considers how models for sharing in technology development, including 'open source', 'user driven innovation' and 'open biotechnology', apply to synthetic biology. She concludes that this case study can be seen as an example of a growing trend towards openness and sharing in biological research.

In the fourth part, we consider the current state of intellectual property debates in three areas of research: RNA (ribonucleic acid) interference, embryonic stem cell research and nanotechnology.

In Chapter 8, Adam Bostanci, Jane Calvert and Pierre-Benoit Joly explore recent trends in research and patenting in the field of RNA interference. RNA is an important intermediate message-carrying molecule in the process of reading and carrying out the instructions contained in the genome (gene expression). The information encoded in DNA is carried by complementary RNA molecules to the cellular machinery that produces proteins. These proteins carry out various functions within the cell. Scientists have found that introducing certain RNAs into cells can reduce or block expression of the matching gene. This effect is known as RNA interference (RNAi). The authors identify a 'gold rush' of patent activity in this area and describe several significant patents including the Carnegie patent and the Tuschl patents. These patents are placed in the context of preceding controversies over gene patenting. Issues that arise include enablement problems and establishing therapeutic utility. Through this survey, the authors demonstrate the ongoing presence of 'gene centrism' in RNAi patent applications, as well as in the United States patenting requirements. They argue that RNAi technologies may have greater utility as research tools than as medical treatments.

In Chapter 9, Amina Agovic considers interactions between patent law and bioethics in the context of embryonic stem cell research. She compares the ethical debates taking place in the United States and the European Union regarding destruction of human embryos, creation of human embryos for use in research, and creation of mixed species embryos. Agovic analyses the European Patent Office rulings rejecting the Wisconsin Alumni Research Foundation and California Institute of Technology patent applications.[155] She identifies a lack of clarity in these decisions regarding interpretation of the morality clause, what constitutes morality and ethics in the European context, and how these issues affect the patentability of inventions involving embryonic stem cell technology. Considering recent developments in relation to the Wisconsin Alumni Research Foundation's stem cell patents, Agovic argues that both the USPTO and the United States courts are reluctant to address ethical issues, deferring them to Congress. She argues that patent law should include mechanisms for addressing ethical concerns.

In Chapter 10, Alison McLennan and Matthew Rimmer survey the field of patent law and nanotechnology – highlighting matters of nomenclature and classification; the potential for patent thickets and a 'tragedy of the anticommons'; and ethical conundrums about the potential risks posed by nanotechnology to human health, the environment and peace.

In the concluding part, we consider the ability of the intellectual property system to protect and preserve genetic resources, respond to the changing natural environment, and protect the rights of traditional holders of ecological knowledge.

In Chapter 11, by Sarah Holcombe and Terri Janke, we return to some themes raised at the beginning of the volume regarding local traditional knowledge of biological resources and appropriation of these resources by

---

155 *Stem Cells/WARF* (T 1374/04) 2007 EPO OJ 313; *California/Stem Cells* (T 522/04) 2009 EPOR 45.

others. Holcombe and Janke outline the Australian approach to implementation of the *Convention on Biological Diversity* 1992 under federal legislation and several State and Territory acts, noting some loopholes and inconsistencies between the various regimes.[156] The authors compare two case studies: Mary Kay Cosmetics' application for a patent on the use of the Kakadu plum in cosmetics; and a patent on use of the Marjarla tree for pain relief. The cosmetic company is likely to have accessed the plum without following the applicable protocols and negotiating benefit sharing with Indigenous people, demonstrating inadequacies in the legal system. By contrast, the Marjarla patent was obtained through a collaborative partnership between a university and an Indigenous community, who are to receive mutually agreed benefits. Overall, the authors argue that federal government action is needed to create a consistent Australian approach to the protection of Indigenous cultural and intellectual property. Moreover, they argue that there needs to be substantive international action – above and beyond the minimalist *Nagoya Protocol* 2010.[157]

Chapter 12 by Matthew Rimmer considers the controversies over plant breeders' rights, patent law and climate-ready crops. In the field of plant breeders' rights, plant breeders have argued that genetically modified crops will be better adapted to the environmental stresses caused by climate change. Doug Waterhouse, the registrar for Plant Breeders' Rights in Australia, observed: 'In response or in anticipation to climate change, new varieties are being developed without a chill factor requirement, they have low or no chill requirement that means that the area available for growing those plants can be moved perhaps north from where they are now.'[158] In the area of patents, technology developers have built significant patent portfolios in respect of climate-ready crops to deal with environmental stresses – such as heat, drought and floods. This trend has caused some consternation amongst farmers, environmental groups and civil society organisations, like the ETC Group. There has been a discussion about exceptions to intellectual property rights in respect of agricultural inventions – dealing with patentable subject matter, farmers' rights, patent pools, public sector research, technology transfer and compulsory licensing.

---

156 *Convention on Biological Diversity* 1992, opened for signature 5 June 1992, 1760 UNTS 143 (entered into force 29 December 1993).

157 *Nagoya Protocol on Access to Genetic Resources and the Fair and Equitable Sharing of Benefits Arising from their Utilization to the Convention on Biological Diversity*, COP 10 Decision, ('*Nagoya Protocol* 2010').

158 Dakis, S. (2008), 'PBR Helping with Climate Change', *ABC News*, 27 August, http://www.abc.net.au/rural/sa/content/2006/s2348159.htm.

Chapter 13 by Matthew Rimmer considers the threat posed to plant genetic resources by global warming and climate change. The Food and Agriculture Organization of the United Nations has been concerned with climate change, and the boom in bioenergy is affecting farming and the price of seeds. The *International Treaty on Plant Genetic Resources* 2001 aims to promote 'the conservation and sustainable use of plant genetic resources for food and agriculture' and 'sustainable agriculture and food security'.[159] The Svalbard Global Seed Vault – nicknamed the Doomsday Vault – was established and is managed by the Norwegian Government, the Global Crop Diversity Trust and the Nordic Genetic Resource Center to protect crop diversity from catastrophes, such as those caused by climate change. This 'Big Science' project is worth evaluating as a means of addressing intellectual property, climate change and food security.

---

159 *International Treaty on Plant Genetic Resources for Food and Agriculture*, adopted 3 November 2001, S. Treaty Doc. No. 110–19, ftp://ftp.fao.org/ag/cgrfa/it/ITPGRe.pdf (entered into force 29 June 2004).

PART I

A history of biodiscovery

1. Of plants, pills and patents: circulating knowledge

Eva Hemmungs Wirtén

This chapter considers the old and new 'bio-contact zones',[1] where exotic plants, pharmaceutical application and patents co-inhabit a historically contested universe. Both eighteenth-century exploratory travels into the tropics as well as instantaneous (or stalled) sharing of bio-information in the global biotech industry of today constitute, I argue, a space for circulating knowledge. The geopolitics of movements within this sphere, the travels, routes, and flows that enable (or disable) encounters and exchanges, must not, however, be interpreted only in terms of a conventional one-way only model of centre and periphery, neither when it comes to the British Empire during the height of Queen Victoria's reign, nor when situated within the intricacies of the present-day Empire of the World Trade Organization (WTO) or the *TRIPS Agreement* 1994.[2] Exchanges of biological material and the knowledge of their uses enter a new phase with Western expansionism, but have transpired between tropical countries as much as from them to temperate zones only.[3] Utter depletion at one end and total accumulation on the other is too simple a précis of the movements within this colonial exchange.[4]

1 Drawing on Mary Louise Pratt's well-known notion of 'contact zone', Londa Schiebinger coins 'biocontact zones' to describe the exchange of plants and their cultural uses. Schiebinger, Londa (2004), *Plants and Empire: Colonial Bioprospecting in the Atlantic World*, Cambridge, MA: Harvard University Press, 83.

2 *Marrakesh Agreement Establishing the World Trade Organization*, opened for signature 15 April 1994, 1867 UNTS 3 (entered into force 1 January 1995) Annex 1C ('*Agreement on Trade-related Aspects of Intellectual Property Rights*' – *TRIPS Agreement* 1994).

3 Brush, S.B. (1993), 'Indigenous Knowledge of Biological Resources and Intellectual Property Rights: The Role of Anthropology', *American Anthropologist*, **95** (3), 653–71 at 657.

4 Kloppenburg Jr., Jack (1988), *First the Seed: The Political Economy of Plant Biotechnology 1492–2000*, Cambridge: Cambridge University Press, 155.

Yet, despite the presence of competing trajectories, there have been and continue to be weaker and stronger directions in the flows of plants and in the knowledge about their uses. Plants themselves are actually less important to this narrative than the knowledge of what biological material does and can be made to do in terms of pharmaceutical application. Consequently, the overarching purpose of this chapter is to consider the circulation, and, even more importantly, the control of knowledge in relation to plants, pills and patents. It is a story that begins mid eighteenth century, when the exploration of the globe had taken on a new direction and momentum. Up until that time scientific interest as well as imperial conquest had been a maritime endeavor, but the attention of the European 'waterborne parasites'[5] increasingly turned inland.[6]

Even though Arctic destinations, such as Lapland, Siberia, or Newfoundland, were among those previously uncharted regions that now became the target of intense study, no place was more associated with endless profusion and difference than the tropics.[7] It was in warmer climates that raw material for the benefit of empires was routinely found. Bark, rubber, spices, cotton, plants: 'the Rest' provided an extensive list of accoutrements for the entertainment or survival of 'the West'. In the conflicting and co-producing rationales of scientific exploration and imperial conquest and colonization, diminutive plants occupied a far weightier role than their relative smallness would seem to suggest.[8]

5 Drayton, Richard (2000), *Nature's Government: Science, Imperial Britain, and the 'Improvement' of the World*, Cambridge, MA: Harvard University Press, 92.

6 Pratt, Mary Louise (1992), *Imperial Eyes: Travel Writing and Transculturation*, London: Routledge, 23.

7 Arnold, David (1996), *The Problem of Nature: Environment, Culture and European Expansion*, London: Blackwell, 141–68.

8 In making the connection between empire and botany, I am influenced by the works of scholars such as Casid, Jill H. (2005), *Sowing Empire: Landscape and Colonization*, Minneapolis: University of Minnesota Press; Schiebinger, Londa (2004), *Plants and Empire: Colonial Bioprospecting in the Atlantic World*, Cambridge, MA: Harvard University Press; Drayton, Richard (2000), *Nature's Government: Science, Imperial Britain, and the 'Improvement' of the World*, Cambridge, MA: Harvard University Press; and Miller, Philip David and Peter Hanns Reill (eds) (1996), *Visions of Empire: Voyages, Botany, and Representations of Empire*, Cambridge: Cambridge University Press.

I THE PLANT

One of the most coveted of tropical specimens was the South American cinchona tree.[9] Extractable from its bark was a febrifuge working with immediate and sensational effect – although nobody knew why – on those who suffered from (and invariably succumbed to) the intermittent hot and cold fever cycles known then as the 'ague', or, more commonly today, malaria. The chronicle of how the countess of Chinchón, wife to the viceroy of Peru, was cured from such a fever by drinking a concoction made from the bark of a local tree had by the end of the seventeenth century become the equivalent of a modern urban legend. Linnaeus helped circulate the erroneous tale by naming the tree after the Countess – using an incorrect spelling – in 1742. The famous botanist had also a more direct and personal experience of the disease, which had claimed one of his most promising disciples. Destined for Venezuela by way of Spain, Pehr Löfling wrote to his parents in 1754 that he had contracted an irritating fever in Barcelona before setting out for Venezuela. After he waded through a swamp in the South American jungle, a place he otherwise seems to have considered lush and beautifully green, the fever returned. In 1756, at age 27, Löfling died of malaria at a Venezuelan missionary station.[10] If combing the Americas for exotic therapeutic plants was a dangerous gamble for explorers, their presence was even more hazardous for indigenous populations who had little or no protection against the new and unknown germs carried on European bodies.[11]

What caused this dangerous affliction, however, remained a mystery. Contrary to what one might perhaps expect, it was not an ailment limited to distant colonies (as in the mal' aria, or 'bad air', of territories like Africa

---

9 There is no shortage of books on the history of cinchona and malaria. For an account of the early uses of the plant, see Jarcho, Saul (1993), *Quinine's Predecessor: Francesco Torti and the Early History of Cinchona*, Baltimore and London: The Johns Hopkins University Press. Recent general introductions include Honingsbaum, Mark (2001), *The Fever Trail: The Hunt for the Cure for Malaria*, London: Macmillan; and Rocco, Fiammetta (2003), *The Miraculous Fever-Tree: Malaria and the Quest for a Cure that Changed the World*, New York: HarperCollins. Hobhouse, Henry (2002), *Seeds of Change: Six Plants that Transformed Mankind*, London: Pan, is an excellent popular account on cinchona and five other, equally revolutionary plants.

10 Sörlin, Sverker and Otto Fagerstedt (2004), *Linné och hans apostlar*, Stockholm: Natur och Kultur, 101.

11 The body as a weapon in the service of colonial expansion is a central trope in Crosby, Alfred (1986), *Ecological Imperialism: The Biological Expansion of Europe, 900–1900*, Cambridge: Cambridge University Press.

where the climate worked against imperialist expansionist politics), but a chronic curse striking indiscriminately even at home.[12] In a missive sent to the Spanish King Charles III, José Celestino Mutis, whom the monarch had dispatched to the New World in the 1760s expressly in search of the plant, reported that while the Americas were rich in precious stones there was also an abundance of other natural riches. First among these was 'quinine, a priceless possession of which your Majesty is the only owner and which divine Providence has bestowed upon you for the good of mankind. It is indispensable to study the cinchona tree so that only the best kind will be sold to the public at the lowest price.'[13] Mutis relied on flattery and the King's vanity when he underlined that the Spanish sovereign was the sole rightful owner of this remarkable tree. What he did not argue was that such ownership was without distinct obligations; instead, he accentuated both how essential it was to consider its impact on the 'good of mankind', and how central the dissemination of the best quality at the 'lowest price' was to that first objective. Thus began the hunt for a tree that for the next century would be the Holy Grail for all European colonial powers of the time.

A The Royal Botanic Gardens, Kew

The British Empire had one unsurpassable trump card in the cinchona hunt: the Royal Botanic Gardens, Kew. Between 1841 and 1873 and under the stewardship of Sir William Hooker and later his son Joseph Hooker, Kew became the key node in an elaborate network geared towards the accumulation of botanical specimens. Civil servants, adventurers, and botanists were all quick to dispatch plants from their various assignments back home and supply Kew with exotic flora. The setting up of botanical gardens interacting with Kew in Asia, the Caribbean, the Indian subcontinent and the Pacific world indicated that in the beginning of the nineteenth century botanical gardens were integral to the British colonial project.[14]

---

12 During the seventeenth and eighteenth centuries, the entire Mediterranean area was plagued by outbursts of the fever, and parts of London were infamous for being rife with the disease. Rocco, Fiammetta (2003), *The Miraculous Fever-Tree: Malaria and the Quest for a Cure that Changed the World*, New York: HarperCollins, 96–7, 168.

13 Mutis quoted in Honingsbaum, Mark (2001), *The Fever Trail: The Hunt for the Cure for Malaria*, London: Macmillan, 58.

14 Drayton, Richard (2000), *Nature's Government: Science, Imperial Britain, and the 'Improvement' of the World*, Cambridge, MA: Harvard University Press, 121.

Kew initiated expeditions searching for plants, but also acted as a hub within an elaborate scheme of colonial relocation. The valuable cinchona seeds traveled from their South American point of origin back to London. Receiving tender care in Kew's conservatories, their final destination was Malayan or Indian soil, colonies under British rule considered excellent substitutes for the original terrain. Greenhouses were crucial to the success of such transfers because they enabled replication of environments, or the conditions of environments rather than complete ecosystems.[15] The trajectory moves along three axes of power relations: acquisition; concentration and control; and finally recirculation and regulation. From start to finish, plants did indeed travel laterally. They did so, however, by way of a detour to London. There, Kew functioned as a critical node in the network, a 'centre of calculation' where botanists were not only shielded from the dangers of the jungle but had the charts, maps, and instruments enabling them to see the plants differently. Accumulation of specimens paired with imperial sanctioned knowledge practices allowed botanical networks like Kew to act at a distance, dominating the periphery both spatially and chronologically.[16]

In their search for these valuables, however, European botanists were lost without the knowledge of local guides, guides that sometimes favored sabotage to ensure that samples would not make it back to European botanical gardens. If the expedition in question was not incapacitated en route, guerrilla tactics continued on board ship, for instance, by attempts to bore holes in the cases carrying the plants across the Atlantic and pour boiling water on them.[17] Large-scale expeditions through the Amazon and South America often resulted in staggering amounts of specimens, although most never made it back to Europe. Crossing the Atlantic posed a formidable obstacle in terms of adequate transportation. The Wardian case solved the problem and proved the importance of new innovations for the purposes of colonial transfer.[18] In this closely fitted glass case, plants were suddenly not only mobile enough to cross the ocean safely, but they also arrived at their end destination in Kew stable enough to yield maximum

---

15 Parry, Bronwyn (2004), *Trading the Genome: Investigating the Commodification of Bio-Information*, New York: Columbia University Press, 23–4.

16 For an extended discussion on the premises of centres of calculation, see Latour, Bruno (1987), *Science in Action: How to Follow Scientists and Engineers through Society*, Cambridge, MA: Harvard University Press, 215–57.

17 Honingsbaum, Mark (2001), *The Fever Trail: The Hunt for the Cure for Malaria*, London: Macmillan, 135.

18 For an introduction see Headrick, Daniel (1981), *The Tools of Empire: Technology and European Imperialism in the Nineteenth Century*, Oxford: Oxford University Press.

information.[19] The Wardian case made knowledge transfer possible, but technological innovation did not guarantee the success of such transfer, however valuable the resource and knowledge might be. In her book, *Plants and Empire: Colonial Bioprospecting in the Atlantic World*, Londa Schiebinger argues that uses of the abortifacient peacock flower (*Poinciana pulcherrima*) constituted an active form of resistance to slavery in the West Indies. Rather than see their children grow up and fall victim to the cruelty of the colonizers, women chose to terminate pregnancies by digesting this plant.[20] Compared with the swiftness by which the use of cinchona reached Spain, England, and France, and despite the fact that the peacock flower grew in many European botanical gardens, knowledge of its abortifacient qualities never reached European *materia medica*. Circulating the plant itself and circulating information about what the plant does are two very different things.

Cinchona was only one of many indigenous South American plants that followed a route of the kind I have just outlined; rubber is another famous example. Essential for insulation, packing, and hosing, India rubber enabled the late nineteenth-century boom in communication and transportation.[21] The turbulent political climate of the time no doubt worked in favor of those who came to South America in pursuit of the sought-after bark, and, perhaps more importantly, gave them the opportunity to couch as selfless philanthropy what others considered a euphemism for state-funded skullduggery. Botanist Richard Spruce noted that the hunt for the red bark caused overexploitation. Concerned that the trees were on the verge of extinction, he provided the perfect alibi for intervention and management by the Empire's emissaries.[22]

In the classic imperialist terminology of waste and improvement, the British Empire justified its actions by stressing their superior know-how as

---

19 Latour, Bruno (1987), *Science in Action: How to Follow Scientists and Engineers through Society*, Cambridge, MA: Harvard University Press, 223.

20 Schiebinger, Londa (2004), *Plants and Empire: Colonial Bioprospecting in the Atlantic World*, Cambridge, MA: Harvard University Press, 128–9.

21 Drayton, Richard (2000), *Nature's Government: Science, Imperial Britain, and the 'Improvement' of the World*, Cambridge, MA: Harvard University Press, 194. Dean, Warren (1987), *Brazil and the Struggle for Rubber: A Study in Environmental History*, Cambridge: Cambridge University Press, is the standard work on the rubber trade. For a fascinating story on Henry Ford's ambition to secure rubber for car tyres by establishing a presence in the Amazon, see Grandin, Greg (2010), *Fordlandia: The Rise and Fall of Henry Ford's Forgotten Jungle City*, London: Icon Books.

22 Spruce, quoted in Honingsbaum, Mark (2001), *The Fever Trail: The Hunt for the Cure for Malaria*, London: Macmillan, 114.

key to the extinction of malaria. In his 1880 account of cinchona and its implementation into India, *Peruvian Bark: A Popular Account of the Introduction of Chinchona Cultivation into British India*, Sir Clements R. Markham, once a cinchona hunter himself and future president of the Royal Geographical Society, commented on accusations of theft and bemoaned the way in which South Americans have destroyed cinchona trees rather than see them fall into the wrong hands. Such direct action was not the only reason why harvesting failed. When Charles Ledger, a trader/adventurer ambitious to make his fortune on cinchona, came across a group of splendid *cinchona calisaya* trees in Bolivia in 1851, together with his native companion Manuel Inca Mamani, the trees were in full bloom and no seeds could be extracted. More than a decade went by, years Ledger spent in Australia where he turned his attention to raising alpaca, before Mamani in 1865 gathered a batch of similar high-quality seeds. By then, Ledger had returned to Peru and, when Mamani showed up on his doorstep with the prized seeds, Ledger must have jumped at the chance to redeem himself for his previous setbacks. What followed then was a sequence of unfortunate events worthy of a Victorian melodrama.

Anxious to get the seeds to London Ledger sent them off to his brother George, who, in turn, immediately took the shipment to Kew. Inconceivably, Kew rejected his proposal. Fully aware that the seeds deteriorate by the hour, George Ledger desperately searched for a buyer. He wrote a letter to the Under-Secretary of State for the Colonies, offering to sell him the priceless bounty. However, the British Government, disappointed with what they had received from funding costly cinchona expeditions so far and faced with Ledger's unsolicited tender, chose silence and ignored him. In a supreme twist of irony, George Ledger managed to convince a competing imperial power – the Dutch consul-general in London – to purchase a meager one pound of his supply to send to their Javanese plantations at the bargain price of £20. Proving an ideal location for Ledger and Mamani's seeds, the Javanese plantations produced trees with the highest content of quinine ever encountered before, generating as much as 14 per cent.[23] In the end, the rest of the seeds in that ill-fated batch found their way into British

[23] My account on the trials and tribulations of Ledger and Mamani relies substantially on Honingsbaum, Mark (2001), *The Fever Trail: The Hunt for the Cure for Malaria*, London: Macmillan, 85–6, 162–5; and Rocco, Fiammetta (2003), *The Miraculous Fever-Tree: Malaria and the Quest for a Cure that Changed the World*, New York: HarperCollins, 242–9.

India but did not germinate as hoped, partly because of different techniques.[24] By 1930, that one pound purchase ensured the Dutch an almost complete monopoly, or 95 per cent of the manufacture of the world's quinine.[25]

The *Cinchona calisaya* that Mamani gave to Ledger in 1865 today goes by the name *Cinchona ledgeriana*. Recognition of Mamani's efforts are long overdue, but the seeds he found for Ledger still do their work for Pharmakina, the last remaining cinchona plantation in Bukavu, on the Rwanda–Congo border.[26]

B Rationales for Scientific Explorations

The Botanists dispatched by Kew or sent out as probes by the reluctant traveler Linnaeus knew perhaps – and sometimes experienced first-hand – that their actions were not completely above board. Still, the question of to whom nature belonged, or of what right they had to the plants in question, caused them few sleepless nights.[27]

Rationales for scientific explorations proved from the very beginning to revolve around a complex, even contradictory, set of incentives. Curiosity alone was not a good enough excuse for setting in motion sometimes fatal and always long journeys of discovery. To the legions of seventeenth century French naturalists who embarked on such voyages it was far more important to call attention to the potential results these held for the benefit of mankind and the public good.[28] The lust for exotic adventures spoke of a new identity politics, one that would have been impossible without significant advances in material technologies. Last but certainly not least, beyond personal reasons lay political ones; the surveying and documentation of

---

24 For a more detailed description of the techniques deployed in this transferral, with particular emphasis on the plantations in Sri Lanka, see Webb, James L.A. (2002), *Tropical Pioneers: Human Agency and Ecological Change in the Highlands of Sri Lanka, 1800–1900*, Athens: Ohio University Press, 116–28.

25 Rocco, Fiammetta (2003), *The Miraculous Fever-Tree: Malaria and the Quest for a Cure that Changed the World*, New York: HarperCollins, 282.

26 Ibid., 303–14. See also Pharmakina's website, www.pharmakina.com.

27 In her book *Plants and Empire*, Londa Schiebinger seems to contradict herself on this point. On page 17 she writes, 'Few agonized over who owned nature', whereas on page 45 the question of who owned nature in the eighteenth century 'was a question Europeans were concerned with'. Schiebinger, Londa (2004), *Plants and Empire: Colonial Bioprospecting in the Atlantic World*, Cambridge, MA: Harvard University Press.

28 Spary, E.C. (2003), 'Peaches which the Patriarchs Lacked: Natural History, Natural Resources, and the Natural Economy in France', *History of Political Economy*, **35**, 15–16.

new territories and new resources was an integral part of empire building, conquest, and nation-state formation. The Spanish King may have taken his role as the 'only owner' of cinchona quite literally, but with privileges came obligations. As benefactor to the nation, the king had a duty to find the best-yielding trees in order to produce a cheap and effective febrifuge. For Clements Markham, quinine ensured the British administration an upper hand in managing and improving the Indian subcontinent. Its proper distribution and use was the ultimate proof of the benevolent Empire. Charles Ledger, on the other hand, was an adventurer who failed to make a living on either alpaca or cinchona. We can never condense the interests of these men in pursuing cinchona into a single objective. Aspirations overlap and intersect: the desire for imperial expansion, the search for new scientific discoveries, and the ambition to build a personal fortune.

II THE PILL

What Ledger, Spruce, and Markham suspected, we now know. Thanks to the samples, notations, and drawings brought back to the centers of calculation by these four and many others like them, any lingering idea of the tropics and jungle as nothing more than a 'trackless waste'[29] has by 2010 been thoroughly disproved. We credit the Amazon rainforest with supplying 20 per cent of the world's oxygen. More than half of the estimated 10 million species of plants, animals, and insects on this globe live in the tropical rainforests. One fifth of the world's fresh water is in the Amazon Basin.[30] While the tropical regions of the world make up only 6 per cent of the world's surface, they contain the greatest diversity of the world's flora. Of the earth's 250 000 species of higher plants, we find 20 per cent in the tropical rainforests of the Amazon. However, less than half of 1 per cent of all flowering plant species has been studied for pharmacological

---

29 Dean, Warren (1987), *Brazil and the Struggle for Rubber: A Study in Environmental History*, Cambridge: Cambridge University Press, 11.

30 These numbers are available at: http://www.rain-tree.com/facts.htm. Clearly, a plethora of Internet sites offers more or less reliable statistics on the matter. Kemal Baslar quotes figures stating that the Amazon provides a full 50 per cent of the world's oxygen in Baslar, Kemal (1998), *The Concept of the Common Heritage of Mankind in International Law*, The Hague: Martinus Nijhoff Publishers, 140, note 38.

use.[31] Cinchona may be the paradigm case confirming just how therapeutically valuable exotic plants can be, but it is neither the first nor the last such example.

One of the most durable representations of the jungle or rainforest from Löfling's early travels and up to our own time is precisely as a lucrative medicine cabinet. The actual likelihood that plants will end up as pills on the shelves of your local pharmacy is far from certain, but the potential is powerful enough to create a scenario where the tropics – despite the fact that only 13 of the 1223 new drugs marketed between 1975 and 1997 were specifically developed to treat tropical diseases[32] – hold the key to relief from epidemic health costs. Frequently quoted figures in support of the significance of plant pharmaceuticals state that, as at 1995, 118 of the top 150 prescription drugs in the United States derive from plants or animals. Yet, the proportion of plant species investigated for their medical properties is only 1100 out of 365 000. On average, one important drug has been produced for every 125 plant species studied, whereas the equivalent rate for chemical compounds is one in 10 000.[33] With steady growth of around 7 per cent per year since 2002, the industry organization IMS Health set the global sales of pharmaceuticals at $808 billion in 2009.[34]

Providing additional evidence of how 'gene-rich' the South is and how correspondingly 'gene-poor' the North is, germ plasm from the West Central Asian and Latin American regions has historically made the largest genetic contribution to feeding the world, whereas none of the world's 20 most important food crops is indigenous to North America (or Australia). The less developed regions of the world contribute plant genetic material, providing the base for fully 95.7 per cent of global food crop production.[35] Centuries of accumulated knowledge have made the abundance of the

31 Schiebinger, Londa (2004), *Plants and Empire: Colonial Bioprospecting in the Atlantic World*, Cambridge, MA: Harvard University Press, 16.

32 Drahos, Peter and John Braithwaite (2002), *Information Feudalism: Who Owns the Knowledge Economy?* London: Earthscan, 189.

33 Veash, N. (2000), 'Biopiracy – A New Threat to Amazon Rainforest's Treasures', *The Independent*, London, 16 October. For additional statistics in this field, see Alves, R.R.N. and I.M.L. Rosa (2007), 'Biodiversity, Traditional Medicine and Public Health: Where Do They Meet?', *Journal of Ethnobiology and Ethnomedicine*, **3**, 14.

34 IMS Health (2010), 'Top-Line Market Data', http://www.imshealth.com/portal/site/imshealth/menuitem.a46c6d4df3db4b3d88f611019418c22a/?vgnextoid=e599410b6c718210VgnVCM100000ed152ca2RCRD&cpsextcurrchannel=1.

35 Kloppenburg Jr., Jack (1988), *First the Seed: The Political Economy of Plant Biotechnology 1492–2000*, Cambridge: Cambridge University Press, 181.

Amazon and other biodiverse regions of the world famous, but, since only a fraction of its flora and fauna has been identified for their pharmaceutical potential, the rainforest still remains a vast storehouse of 'information *not yet catalogued* and thus with a value that can *only* be imagined'.[36]

Well, not altogether. Today, patents are extremely important tools by which the knowledge contained in this storehouse of information and its value is controlled and fixed.

III THE PATENT

The rationale for patents rests on rewarding and providing incentives for innovation.[37] Patents can be issued for a product as well as a process, but used to be limited to end products, or 'the culmination of research, not the fountain from which it sprung'.[38] Rewarding disclosure with exclusive and limited rights, patents are, at least in theory, commensurable rather than incommensurable with ideas of open science, access to and sharing of knowledge. Indeed, it has been argued that the '[s]tudy of patented information is essential to the creation of new knowledge, thereby achieving further scientific and technological progress'.[39] Patents in the context of

---

36 Hayden, Cori (2003), *When Nature Goes Public: The Making and Unmaking of Bioprospecting in Mexico*, Princeton: Princeton University Press, 57. See also Michael Jeffrey's emphasis on the economical significance of biodiversity and its potential value to industry in Jeffrey, M. (2002), 'Bioprospecting: Access to Genetic Resources and Benefit-Sharing under the Convention on Biodiversity and the Bonn Guidelines', *Singapore Journal of International and Comparative Law*, **6**, 747–808 at 756.

37 For a comprehensive international and historical overview of the patent system, see Mgbeoji, I. (2003), 'The Juridical Origins of the International Patent System: Towards a Historiography of the Role of Patents in Industrialization', *Journal of the History of International Law*, **5**, 403–22; and Mgbeoji, I. (2006), *Global Biopiracy: Patents, Plants, and Indigenous Knowledge*, Vancouver: University of British Columbia Press. The history of patents is less discussed by scholars in the humanities than copyright, but includes studies by Coulter, Maureen (1991), *Property in Ideas: The Patent Question in Mid-Victorian Britain*, Kirksville: The Thomas Jefferson University Press; and Pettitt, Clare (2004), *Patent Inventions: Intellectual Property and the Victorian Novel*, Oxford: Oxford University Press.

38 Dreyfuss, R.C. (2004), 'Protecting the Public Domain of Science: Has the Time for an Experimental Use Defense Arrived?', *Arizona Law Review*, **46**, 457–72 at 462.

39 Justice Pauline Newman, dissenting opinion in *Integra Life Sciences I Ltd.* v. *Merck KgA* 331 F. 3d 860 at 876 (2003). Cited in Rimmer, Matthew (2008), *Intellectual Property and Biotechnology: Biological Inventions*, Cheltenham and Northampton, MA: Edward Elgar, 176.

this chapter, then, conform in more ways than one to Bruno Latour's notion of 'inscription devices' that 'transform pieces of matter into written documents'.[40]

There are, however, many reasons why a scenario of the kind I have briefly outlined appears like wishful thinking rather than reality. Legendary cases such as *Diamond* v. *Chakrabarty*[41] and the controversial U.S. Patent 4,736,866 for Harvard's OncoMouse™[42] have come to typify a gradual corrosion of principles and distinctions that patent offices used to, if not hold sacrosanct, at least abide by. One such demarcation was – and I choose the past tense very consciously – between discovery and invention.[43] Traditionally, so-called 'products of nature' were considered beyond the reach of patents. Described as not patentable were 'the laws of nature, physical phenomena, and abstract ideas [...] Thus a new mineral discovered in the earth or a new plant found in the wild is not patentable subject matter.'[44] If you happen to trek through the rainforest and accidentally come across a rare orchid nobody has laid eyes on before, this discovery alone does not make the flower patentable. Infuse labor in the laboratory and the isolation of a valuable gene sequence in that orchid then that discovery becomes human creation by innovation, and hence patentable.

Increasingly sophisticated technology put at the disposal of the scientist makes it possible, even logical, to view the flower or plant as a carrier of bio-information from which valuable data can be extracted and isolated

40 Latour, Bruno and Steven Woolgar (1979), *Laboratory Life: The Social Construction of Scientific Facts*, London: Sage, 51. See Swanson, K. (2007), 'Biotech in Court: A Legal Lesson on the Unity of Science', *Social Studies of Science*, **37** (3), 357–84, for an overview of similar ways to consider patents. See also Myers, G. (1995), 'From Discovery to Invention: The Writing and Rewriting of Two Patents', *Social Studies of Science*, **25** (1), 57–105.

41 *Diamond* v. *Chakrabarty*, 447 U.S. 303 (1980).

42 Leder, P. and T. Stewart (1988), 'Transgenic Non-Human Mammals', U.S. Patent No: 4,736,866. For a general introduction to the OncoMouse, see Haraway, Donna J. (1997), *Modest_Witness@Second_Millennium. Femaleman©_Meets_Oncomouse™*, New York: Routledge, 79–85 and Kevles, D. (2002), 'Of Mice and Money: The Story of the World's First Animal Patent', *Daedalus*, **131** (2), 78–88.

43 For a good overview, see Rimmer, Matthew (2008), *Intellectual Property and Biotechnology: Biological Inventions*, Cheltenham and Northampton, MA: Edward Elgar, 110–37.

44 *Diamond* v. *Chakrabarty*, 447 U.S. 303 (1980), at 309. Note, however, Ikechi Mgbeoji's discussion on fifteenth-century Tyrolean mining patents being issued more in terms of 'discovery' than 'invention'. Mgbeoji, I. (2003), 'The Juridical Origins of the International Patent System: Towards a Historiography of the Role of Patents in Industrialization', *Journal of the History of International Law*, **5**, 403–22 at 410.

rather than a 'mere' product of nature. In order to arrive at a position where isolation replaces innovation as a primary criterion for patentability the judiciary had to overcome three obstacles: (1) genetic material is made from products of nature; (2) one can make a case that they are hardly 'novel' in the traditional eyes of the intellectual property system; and (3) they can be labeled as 'discoveries' rather than 'innovations'.[45] In other social and cultural contexts where other forms of knowledge accrue, the same hindrances cannot be surmounted and therefore fail to register on the intellectual property radar.[46] The long-lived Western ideology of improvement has been counterproductive in recognizing traditions where the regulation of biodiversity depends upon a continued use of a biological heritage, a heritage constituted by a bundle of relationships and therefore irreducible to a bundle of economic rights.[47] Jack Kloppenburg Jr. crucially notes that the common representation of agriculture in the South as 'just' nature – a pristine, child-like condition – effectively hides the reality of thousands of years of day-to-day agricultural modification.[48]

A Biopiracy

When today we speak of 'biopiracy'[49] as an appropriate label for the history I have outlined above, we think not only of a geopolitical South–

---

[45] Parry, Bronwyn (2004), *Trading the Genome: Investigating the Commodification of Bio-Information*, New York: Columbia University Press, 85.

[46] A particularly contentious aspect is the so-called prior art principle, or the possibility that the invention in question is already patented or used somewhere else. As patent examiners turn to relevant databases in order to verify the novelty of a patent claim they are unlikely to come across information that pertains to plant uses in folk medicine, simply because these have never been documented. Brown, Michael F. (2003), *Who Owns Native Culture?*, Cambridge, MA: Harvard University Press, 133. See also Halbert, Debora J. (2005), *Resisting Intellectual Property*, London and New York: Routledge, 148; Shiva, Vandana (2001), *Protect or Plunder? Understanding Intellectual Property Rights*, London: Zed Books, 17–18 and Fecteau, L. (2001), 'The Ayahuasca Patent Revocation: Raising Questions About Current U.S. Patent Policy', *Boston College Third World Law Journal*, **21** (1), 69–104.

[47] Shiva, Vandana (2001), *Protect or Plunder? Understanding Intellectual Property Rights*, London: Zed Books, 46.

[48] Kloppenburg Jr., Jack (1988), *First the Seed: The Political Economy of Plant Biotechnology 1492–2000*, Cambridge: Cambridge University Press, 185–6.

[49] Biopiracy is the 'privatization and unauthorized use of biological resources by entities (including corporations, universities and governments) outside of a country which has pre-existing knowledge'. Definition from Wikipedia, http://en.wikipedia.org/wiki/Biopiracy (accessed 7 April 2006). When accessing on 16

North movement of plants, but of a geopolitical South–North movement of knowledge. Already in 1787, when Lapérouse came to Sakhalin and knew nothing of the area, he had to make permanent the writings made for him by the natives in the Sakhalin sand, or see what he had learned lost forever. It was a process whereby, as Bruno Latour notes, 'the *implicit* geography of the natives is made *explicit* by geographers; the *local* knowledge of the savages becomes the *universal* knowledge of the cartographers; the fuzzy, approximate and ungrounded *beliefs* of the locals are turned into a precise, certain and justified *knowledge*'.[50] Lapérouse perished somewhere on the return voyage to France, but the documents, maps, and drawings had fortunately been transferred to another, English ship, and reached Versailles intact. Because they did, new ships could be sent out to collect even more data and specimens that in extension widened the great divide between the knowledge at Sakhalin and the one in Paris.[51]

We can consider this historical anecdote in light of Vandana Shiva's argument that if Third World communities charged 2 per cent royalty on the development of their biological diversity these countries would stand to gain $300 million on farmers' seeds and $5 billion in unpaid royalties for medicinal plants.[52] As Shiva and others have noted, when patents are issued on biological material that has been circulating freely for centuries there may be severe costs for those whose long-standing uses make that very patenting possible to begin with.[53] The point is that it is the local knowledge

---

September 2010 there is a redirect from this page to another entry, this time entitled 'Commercialization of traditional medicines'.

50 Latour, Bruno (1987), *Science in Action: How to Follow Scientists and Engineers through Society*, Cambridge, MA: Harvard University Press, 216.

51 Ibid., 221.

52 Shiva, Vandana (2001), *Protect or Plunder? Understanding Intellectual Property Rights*, London: Zed Books, 29.

53 The case of the rosy periwinkle (*catharanthus roseus*), usually described as a flower native to Madagascar, which contains the two powerful anti-cancer compounds Vinblastine and Vincristine, is the classic example of biomedical exploitation. See Kadidal, S. (1993), 'Plants, Poverty, and Pharmaceutical Patents', *The Yale Law Journal*, **103** (1), 223–4. Brown, Michael F. (2003), *Who Owns Native Culture?* Cambridge, MA: Harvard University Press, 136–8, offers a somewhat different interpretation of the Madagascar claim for compensation. Other cases involving patents on plants with long-standing local uses include the Amazon ayahuasca vine and the Indian neem tree. See Fecteau, L. (2001), 'The Ayahuasca Patent Revocation: Raising Questions about Current U.S. Patent Policy', *Boston College Third World Law Journal*, **21** (1), 69–104 at 86. On the neem tree, see for instance Marden, E. (1999), 'The Neem Tree Patent: International Conflict over the Commodification of Life', *Boston College International and Comparative Law Review*, **22** (2), 279–95 at 283; and

of the medicinal properties of plants that has directly led to the patenting of said plants in Western universities or corporate laboratories.[54] Processes of consultation or 'cultural prescreening' of substances may increase the success ratio of clinical trials substantially.[55] Only at the second stage, where laser-sharp improvement awaits under the microscope, subjected to the knowledge of the scientist, isolated and patented, and then published in the most prestigious of academic journals, does the true and measurable value begin.

Consequently, when natural resources convert into more intangible assets, an almost completely inverted geopolitical tendency becomes visible. By far one of the most biologically diverse regions in the world, when the percentage share of world techno-scientific literature was considered in 1993, Central and South America represented only 1.5 per cent of all biomedical articles, whereas the United States accounted for 38.9 per cent.[56] In 2005, the combined share of biotechnology patents of Europe, the United States, and Japan was 82.7 per cent.[57]

Shiva, Vandana (2001), *Protect or Plunder? Understanding Intellectual Property Rights*, London: Zed Books, 57.

54 For an overview of the substantial body of scholarly works on traditional knowledge, see Oguamanam, Chidi (2006), *International Law and Indigenous Knowledge: Intellectual Property, Plant Biodiversity, and Traditional Knowledge*, Toronto: University of Toronto Press.

55 Parry, Bronwyn (2004), *Trading the Genome: Investigating the Commodification of Bio-Information*, New York: Columbia University Press, 133; and Safrin, S. (2004), 'Hyperownership in a Time of Biotechnological Promise: The International Conflict to Control the Building Blocks of Life', *The American Journal of International Law*, **98** (4), 641–85, note 169.

56 Haraway, Donna J. (1997), *Modest_Witness@Second_Millennium. Femaleman©_Meets_Oncomouse™*, New York: Routledge, 58. Recent statistics from the US National Science Institute (NSI) indicate that the number of science and engineering articles credited to Latin American authors published in academic journals tripled between 1988 and 2001, showing an increase of almost 200 per cent between the same years. See Hill, D. (2004), 'Latin America Shows Rapid Rise in S&E Articles', *InfoBrief*, National Science Foundation, August, 1.

57 Organisation for Economic Co-operation and Development (OECD) (2008), 'Compendium of Patent Statistics 2008', 19, http://www.oecd.org/dataoecd/5/19/37569377.pdf. For a comparison between the patent systems of these three, see Chambers, J. (2002–3), 'Patent Eligibility of Biotechnological Inventions in the United States, Europe, and Japan: How Much Patent Policy Is Public Policy?' *George Washington International Law Review*, **34**, 223–46.

B Universities

Nineteenth-century botanists did not necessarily crave exclusivity in their findings. Indeed, for a long time 'the advantage of exchanging plant materials freely was immensely greater than maintaining exclusivity'.[58] One significant change in this circulation of knowledge considered so far has occurred within what we sometimes refer to as the 'science commons', and within one institution in particular: universities. The infamous United States *Bayh-Dole Act* from 1980,[59] which paved the way for federally funded research institutions, for instance universities, to patent their research, is often described as a watershed moment in the balance between public and private interests in academia, a shift that is far from limited to the United States.[60] The biomedical sciences are especially conspicuous in this regard, as they represent a domain dependent on basic research but with significant commercial potential through pharmaceutical application.[61] Judging by statistics, the *Bayh-Dole Act* clearly achieved its intended purpose of increasing the willingness of universities to patent; in 1979 U.S. universities received 264 patents, in 1997 2436, and in 2001 more than 3200.[62] Critics of the *Bayh-Dole Act* note that many of these patents were

---

58 Dean, Warren (1987), *Brazil and the Struggle for Rubber: A Study in Environmental History*, Cambridge: Cambridge University Press, 31.

59 *University and Small Business Patent Procedures Act* 1980 (United States) (the '*Bayh-Dole Act*') 35 U.S.C. § 200–212 (1980).

60 However, the discussions in United States academia on patenting have a longer history. For an excellent overview, see Metlay, G. (2006) 'Reconsidering Renormalization: Stability and Change in 20th-Century Views on University Patents', *Social Studies of Science*, **36** (4), 565–97. For a good overview of the background to the *Bayh-Dole Act*, see Washburn, Jennifer (2005), *University Inc.: The Corporate Corruption of Higher Education*, New York: Basic Books; and McSherry, Corynne (2001), *Who Owns Academic Work? Battling for Control of Intellectual Property*, Cambridge, MA: Harvard University Press, 146–9. See also Jasanoff, Sheila (2007), *Designs on Nature: Science and Democracy in Europe and the United States*, Princeton: Princeton University Press, for a comparative analysis on the different approaches between the US and Europe vis-à-vis these tendencies.

61 Eisenberg, R. and R. Nelson (2002), 'Public vs. Proprietary Science: A Fruitful Tension', *Daedalus*, **131**, 89–101 at 91.

62 Rai, A. and R. Eisenberg (2003), 'Bayh-Dole Reform and the Progress of Biomedicine', *Law and Contemporary Problems*, **66** (1&2), 289–314 at 292. The data on 2001 is from Washburn, Jennifer (2005), *University Inc.: The Corporate Corruption of Higher Education*, New York: Basic Books, 70. A Swedish dissertation from 2006 demonstrates inconclusive findings on patents as incentives in higher education in Sweden and Germany, and ends on a note of caution when considering patents as supplemental income for universities; see

actually issued for basic research discoveries that previously used to be shared freely or at symbolic cost in the academic community, but now have become subject to licensing agreements that may be exorbitantly pricey. In addition, they may take forever to negotiate and hence delay research. The fact that protection now includes genetic sequences 'valued for what they say and not, as with the typical consumer product, primarily for what they do'[63] means that intellectual property rights creep further and further into the building blocks of science, where they curtail use by creating a 'patent thicket' or 'anti-commons' tragedy in biomedical research.[64] Usually defined as 'a property regime in which multiple owners hold effective rights of exclusion in a scarce resource',[65] such a tragedy occurs when one person or group holds the right to exclude others to the possible detriment of everybody's greater good. The consequence of increased segmentation of control is under- rather than overuse.

In the case of the pro-vitamin A-rich, so-called 'Golden Rice', 70 intellectual property or tangible property rights belonging to 32 companies and universities were identified in developing the rice line.[66] So many stakeholders are involved in fragmented patents that transaction costs in time and money actually deter continued research. The problem is particularly acute when it comes to so-called 'upstream patenting', which may cause severe speed bumps in terms of hindering innovation.[67] Even the downstream and upstream terminology, then, frames the movement of

---

Sellenthin, Mark O. (2006), *Beyond the Ivory Tower: A Comparison of Patent Rights Regimes in Sweden and Germany* (Linköping: Linköping Studies in Arts and Science, No 355).

63 Dreyfuss, R.C. (2004), 'Protecting the Public Domain of Science: Has the Time for an Experimental Use Defense Arrived?', *Arizona Law Review*, **46**, 457–72 at 466.

64 Rai, A. and R. Eisenberg (2003), 'Bayh-Dole Reform and the Progress of Biomedicine,' *Law and Contemporary Problems*, **66** (1&2), 289–314 at 297.

65 Heller, M. (1998) 'The Tragedy of the Anticommons: Property in the Transition from Marx to Markets', *Harvard Law Review*, **111** (3), 621–88 at 668. See also Heller, Michael (2008), *The Gridlock Economy: How Too Much Ownership Wrecks Markets, Stops Innovation and Costs Lives*, New York: Basic Books; Heller, Michael (ed.) (2010), *Commons and Anticommons*, vols I and II, Cheltenham and Northampton, MA: Edward Elgar.

66 Gepts, P. (2004), 'Who Owns Biodiversity, and How Should the Owners Be Compensated?' *Plant Physiology*, **134**, 1295–1307 at 1302.

67 Dinwoodie, Graeme and Rochelle Cooper Dreyfuss (2006), 'Patenting Science: Protecting the Domain of Accessible Knowledge', in Bernt Hugenholtz and Lucie Guibault (eds), *The Future of the Public Domain*, Amsterdam: Kluwer Law International, 17. See also the Royal Society's similar arguments on 'research tools' in the Royal Society of the United Kingdom (2003), *Keeping Science Open: The Effects of Intellectual Property Policy on the Conduct of*

knowledge in familiar waterway terms, recalling the old principal mode of transfer discussed previously, albeit now in a more abstract sense. Also part of this history is malaria, an excellent example of the tragedies that can occur in the science commons because of such over-patenting. Malaria was a mystery when Pehr Löfling succumbed to the fever in 1756. Today, we know that the *Anopheles* mosquito family carries the disease and we have synthetic drugs to protect us from catching it. Not by a long shot, however, have we eradicated malaria. The World Health Organization estimates that some 500 million people are infected and that 3 million die of it every year, mostly small children. This translates into 1 person falling victim to malaria every 15 seconds.[68] One region is particularly hard hit: more than 80 per cent of malaria deaths occur in Africa south of the Sahara.[69] Research on malaria could result in a vaccine socially beneficial to millions of people. But when the primary market lies in cash-poor developing nations the 34 different groups of upstream patent rights that apply in this case will do little to increase the motivation to find a permanent solution to the problem.[70]

However, for creative works in any domain – be it science or arts – to come about, it must be possible to exchange and draw on data and information freely. Rochelle Dreyfuss lists 'spillover effects' as one of the distinctive features of scholarly output, and continues to stress how a commitment 'to a system of open science, where results are shared, criticized, and ultimately, utilized', relied for a very long period on a sense of responsibility, or propriety, between universities and commercial actors.[71] During the last three decades, however, radical changes in the make-up of patents have reallocated weight from the first imperative to the second.

Science, London: The Royal Society, http://royalsociety.org/Keeping-science-open-the-effects-of-intellectual-property-policy-on-the-conduct-of-science.

68 Rocco, Fiammetta (2003), *The Miraculous Fever-Tree: Malaria and the Quest for a Cure that Changed the World*, New York: HarperCollins, 23.

69 World Health Organization (2005), *World Malaria Report 2005*, Geneva: World Health Organization, http://rbm.who.int/wmr2005/html/exsummary_en.htm.

70 Rai, Arti (2005), 'Proprietary Rights and Collective Action: The Case of Biotechnology Research with Low Commercial Value', in Keith Maskus and Jerome H. Reichman (eds), *International Public Goods and Transfer of Technology under a Globalized Intellectual Property Regime*, Cambridge: Cambridge University Press, 295.

71 Dreyfuss, R.C. (2004), 'Protecting the Public Domain of Science: Has the Time for an Experimental Use Defense Arrived?', *Arizona Law Review*, **46**, 457–72 at 463–4.

CONCLUSION: FROM HOT TO COLD

Cataloguing, surveying, classifying, and storing remain essential tasks in the new informational scenario, much as they did for those who ventured into the Amazon in the eighteenth and nineteenth centuries. The botanical inventories described at the beginning of this chapter have been compared to censuses helping the state control civil society. Such censuses were projects in the service of the colonial powers, as most of the early classification and documentation took place through the intervention and benefit of Western explorers.

From the first part of this chapter to the last, however, attention has slowly but surely veered from 'the expansive to the interior aspects of the natural world and from an extensive to an intensive project of exploration'.[72] Currently, the organism itself is all but eradicated at the expense of the bio-information hidden within it; data has been there all the time but only in the last 30 years has it morphed into another phase and form, where it is not only easily extracted but also reproduced, sampled, and commodified with increasing complexity and speed. Life sciences are increasingly becoming *information* sciences.[73]

Figuratively speaking, we can describe anti-commons tragedies, wherever they occur, as a freezing of information. The 'chilling effects' of anti-commons tragedies cause us not to use or act upon information simply because it is too cumbersome and/or costly to try. Freezing is also a more literal description of what happens to biological material today. Perhaps the most important innovation in respect to biological resources is not more sophisticated instruments in the laboratory, but the development of hyper-technological storage facilities housing cell lines, extracted DNA, or blood lines from humans, animals, and plants. These giant cryogenic freezers, or, in somewhat different terminology, cold archives of bio-information, can hold such biological material in perfect condition for at least several hundred years.[74] Under the *Convention on Biological Diversity* 1992 bioprospecting must involve prior informed consent and include

---

72 Parry, Bronwyn (2004), *Trading the Genome: Investigating the Commodification of Bio-Information*, New York: Columbia University Press, 49.

73 Rajan, Kaushik Sunder (2006), *Biocapital: The Constitution of Postgenomic Life*, Durham: Duke University Press, 3.

74 See also Catherine Waldby's work on the National Library of Medicine's Visible Human Project as a way to 'render human bodies as compendia of data, *information archives* [my italics], which can be stored, retrieved, and rewritten in digital and/or genetic modes'. Waldby, C. (1999) 'Iatrogenesis: The Visible Human Project and the Reproduction of Life', *Australian Feminist Studies*, **14**, 77–90.

protocols of benefit sharing.[75] These provisions are there to ensure the equitable distribution of likely profits and to forestall continued immoral and unethical uses. While calls for compensation were raised as early as the eighteenth century, contemporary bioprospecting ventures represent a contested activity that involves intricate networks of people and knowledge, and rights and obligations.[76] Some of these agents are still botanical gardens; others include museums and universities, acting on their own or sometimes on behalf of a 'master-collector', for instance the United States National Institutes of Health (NIH).[77]

Stipulations on benefit sharing and compensation schemes are invalid in any time period, though, if there is not a nature out there to mine. One of the effects of these extremely sophisticated freezers on contemporary collecting practices is that they can make the continued search for biological material obsolete. A potential large nail in the coffin of bioprospecting ventures and benefit sharing, cryogenic storage facilities do away with both spatial and temporal constraints. The enormous material already collected and stored may act as a disincentive to continue programs of that kind in the future.[78] From biological reproduction to laboratory-based facsimiles, from being dependent on the conditions that govern biological processes to being independent of such constrictions, good old-fashioned plants might prove to be less and less necessary to the discovery of new pharmaceuticals.

I opened this chapter by attempting to visualize how European botanists began their journeys into the Americas, and by examining the conflicting incentives and rationales that prompted them in their search for exotic specimens intended for pharmaceutical purposes. Such collection, concentration, and later redistribution of biological material operated within the grid of the Victorian centers of calculation. An extended historical discussion of this kind is important in understanding some of the processes whereby the contemporary notion of biopiracy has formed. As we drew closer to our own time, a decisive expansion of patentability to include

---

75 *Convention on Biological Diversity* 1992, opened for signature 5 June 1992, 1760 UNTS 143 (entered into force 29 December 1993).

76 See Londa Schiebinger's discussion on Thiery de Menonville in Schiebinger, Londa (2004), *Plants and Empire: Colonial Bioprospecting in the Atlantic World*, Cambridge, MA: Harvard University Press, 45. On the complexities of current benefit-sharing programs, see Hayden, C. (2007), 'Taking as Giving: Bioscience, Exchange, and the Politics of Benefit-Sharing', *Social Studies of Science*, **37**, 729–58 at 734.

77 For an overview of these contemporary collecting endeavours, see Parry, Bronwyn (2004), *Trading the Genome: Investigating the Commodification of Bio-Information*, New York: Columbia University Press, 107–13.

78 Ibid., 171.

living organisms and biological material cemented the patent system's already skewed privileging of specific forms of knowledge and use. Yet, for all the injustices that have been committed throughout the centuries, what this chapter seems to indicate is that the ownership paradigms at play during the entire period have been far from black and white. Public institutions like universities have been in the forefront when it comes to patenting their research, but they are also active defenders of a science commons, the vitality of which remains the backbone of the same institutions.[79] Private enterprises may see the dangers of over-patenting better than academia and thus ultimately serve public purposes;[80] sovereign control over biological material – as well-meaning and logical as, for instance, the *Convention on Biological Diversity* 1992 can be under the historical circumstances – can result in anti-commons tragedies not unlike those that occur by over-patenting. Centuries of struggles concerning the discovery, control, and use of plants involve an intricate web of individual actors, private and public institutions, and ownership paradigms that are, in their very complexity, emblematic of the hybridity of ownership as well as knowledge.

---

79 Compare for instance the tendencies in Washburn, Jennifer (2005), *University Inc.: The Corporate Corruption of Higher Education*, New York: Basic Books, with those described by Willinsky, John (2006), *The Access Principle: The Case for Open Access to Research and Scholarship*, Cambridge, MA: The MIT Press.

80 Washburn suggests that Merck & Co.'s opposition to the NIH patent on anonymous gene fragments is one such example. Washburn, Jennifer (2005), *University Inc.: The Corporate Corruption of Higher Education*, New York: Basic Books, 149.

PART II

Medicine, biotechnology and genomics

2. *Bilski* v. *Kappos* and biotechnology patents: back to the future?

Yann Joly and Francis Hemmings

New processes will undoubtedly play a crucial role in the continuing development of biotechnology. In this chapter, developments in process patentability that have been generated by the 2010 case of *Bilski* v. *Kappos*[1] will be examined with particular emphasis laid on the concept of balancing of rights: employing a patent regime as a means to stimulate innovation requires a delicate balance to be struck between the rights granted to inventors on the one hand and the rights of access of the public on the other. In most patent regimes, that balance is achieved in part by allowing inventors to patent inventions, but not 'laws of nature, physical phenomena, and abstract ideas' [hereinafter referred to as *fundamental principles*].[2] While this balancing act appears to be simple, in practice the application of the distinction between the two presents a considerable challenge and the topic remains the subject of ongoing debate.

Slight changes to the scope of patentable subject matter can have substantial consequences for processes in the biotechnology field, as exemplified by the case of *Association for Molecular Pathology* v. *United States Patent and Trademark Office*, where a claim for patent protection of a diagnostic method for breast cancer was refused at first instance.[3] The Court said that it sought to protect a fundamental principle.[4] On appeal, the Court of Appeals for the Federal Circuit partially overturned the first instance decision, with the majority of the Court upholding the validity of

1 *Bilski* v. *Kappos* 130 S. Ct. 3218 (2010).
2 *Diamond* v. *Chakrabarty*, 447 U.S. 303 (1980).
3 *Association for Molecular Pathology et al.* v. *United States Patent and Trademark Office, Myriad Genetics, et al.* 702 F. Supp. 2d 181 at 234, (S.D.N.Y. 2010), *appeal docketed,* No. 2010–1406 (Fed. Cir. June 22, 2010).
4 Van Overwalle, G. (2010), 'Turning Patent Swords into Shares', *Science*, **330**, 1630.

Myriad Genetics' claims to isolated genes.[5] The considerable attention this case received in the biotechnology community[6] is indicative of the importance of questions of subject-matter patentability to the biotechnology field. Moreover, consequences of such changes are not strictly limited to a specific jurisdiction; the trend towards setting common minimal intellectual property standards at the international level renders any change to subject-matter patentability in one jurisdiction relevant to other jurisdictions. Hence, differences in subject-matter patentability amongst jurisdictions might create hurdles for commerce and research collaborations; developments in process patentability in the United States, which have been generated in *Bilski* v. *Kappos*, could have an important impact on biotechnology from an international perspective.

Bilski v. *Kappos*[7] sheds new light on subject-matter patentability. The central issue in the case is how the distinction between patentable and non-patentable processes is to be made. This chapter will examine the issues surrounding process patentability as they are addressed in this case, as well as the impact of the judgment on the law of process patentability in the field of genomics.

Section I will begin by defining the state of the law on process patentability prior to *Bilski* v. *Kappos.* Next, the controversial judgment rendered by the Court of Appeals for the Federal Circuit in this case will be examined. Lastly the state of the law post *Bilski* v. *Kappos* will be examined and a similar Canadian case will be used to illustrate the consequences when slight changes in process patentability occur.

Section II of the chapter looks at the possible ramifications of the *Bilski* v. *Kappos* decision, particularly in the context of genomics. The first part of the section illustrates the importance of processes for the advancement of genomic research (and clinical integration). Part II analyzes case law pertaining to similar methods, identifies commonalities between cases and looks at the evolution of the law. Part III discusses the potential consequences of *Bilski* v. *Kappos* for genomic innovation.

---

5 *Association for Molecular Pathology et al.* v. *United States Patent and Trademark Office, Myriad Genetics, et al.* 653 F.3d 1329, 2011 WL 3211513 (C.A. Fed. (N.Y.)), 99 USPQ 2d 1398.

6 Ibid.

7 *Bilski* v. *Kappos* 130 S. Ct. 3218 (2010).

I PROCESS PATENTABILITY

A State of the Law Prior to *In re Bilski*

A balancing of rights has been established in American intellectual property law in order to promote innovation. Accordingly, not all inventions will be patentable: first, a general limitation to patentability is contained within the *United States Constitution* itself: art. I, § 8, cl. 8 states that Congress has power 'to promote the progress of science and useful arts [...]'. According to this constitutional provision, Congress may not grant rights that will, in the end, impede, rather than promote, innovation.[8] A balance ought to be struck because granting unjustified monopolies may impede innovation.[9]

Further limitations to patentability are set out in statutory provisions. According to 35 U.S.C. §101, only a 'process, machine, manufacture, or composition of matter, or any [...] improvement thereof [...]' may be patented. Thus, in order to be patentable, an invention must fall within one of these four categories.

As regards what constitutes a 'process', the Supreme Court of the United States has interpreted the term broadly,[10] as '... process, art, or method', including 'a new use of a known process, machine, manufacture, composition of matter, or material'.[11] However, the observed breadth of the definition of 'process' is more than simply a matter of fact; it is also a function of the intent of Congress: 'The Committee Reports accompanying the 1952 Act inform us that Congress intended statutory patentable subject matter to include anything under the sun that is made by man.'[12] Hence, the term 'process' has been defined broadly in order to promote a liberal interpretation of the statute.

To summarize, the *United States Constitution* sets limits on patentability so as to establish a balance regarding the rights of inventors and the rights of the public so as to stimulate innovation. Patent law has contributed

---

8 *Graham* v. *John Deere*, 383 U.S. 1 (1966), cited in Walterscheid, E. (2002), 'To Promote the Progress of Science and Useful Arts: the Anatomy of a Congressional Power', *IDEA*, **43**, 1–82 at 30.

9 *Bonito Boats* v. *Thunder Craft Boats*, 489 U.S. 141 at 146 (1989), cited in Walterscheid, E. (2002), 'To Promote the Progress of Science and Useful Arts: The Anatomy of a Congressional Power', *IDEA*, **43**, 1–82 at 31.

10 *Parker* v. *Flook*, 437 U.S. 584 at 589 (1978).

11 35 U.S.C. §100.

12 *Diamond* v. *Chakrabarty*, 447 U.S. 303 at 308–9 (1980).

further to the striking of this balance by requiring that patentable inventions fall within one of four categories. One of these categories is 'process'. While this term must be interpreted liberally, there remain some significant limits to what may be patentable as a process: given these limitations, the term 'process' does not seem, as the Supreme Court of the United States originally stated in *Diamond* v. *Chakrabarty*, to comprise 'laws of nature, physical phenomena, and abstract ideas'.[13]

Despite attempts to define what constitutes a patentable process, a good deal of uncertainty remains. One of the main sources of uncertainty is the fact that the frontier between fundamental research and applied research has become increasingly blurred during the past century.[14] Including in the definition of 'process' a border delimited by fundamental principles has not clarified this imprecision.

The Supreme Court of the United States has, however, given some additional indications as to what defines a patentable process. In *Gottschalk* v. *Benson*, the Court provided a clue that would help determine the patentability of process patents; transforming an object '"to a different state or thing" is the clue to the patentability of a process claim that does not include particular machines [...]'.[15] Hence, a claim that incorporates a fundamental principle may be patented if it transforms an object or if it is related to a machine. This indication is useful because most inventions involve applications of fundamental principles. By requiring a *specific* application of these principles to make them patentable, the Supreme Court prevents only the patenting of fundamental principles per se.

B Decision of the Court of Appeals for the Federal Circuit in *In re Bilski*, 2008

The state of the law regarding process patentability prior to *In re Bilski* was that a process could employ fundamental principles but only in the context of *a specific application*. By limiting their use to a specific application, fundamental principles can be used in other patent applications. The law was then temporarily revisited by the Court of Appeals for the Federal Circuit in *In re Bilski*.[16]

---

13 Ibid.

14 Organisation for Economic Co-operation and Development (OECD) (2010), *The Patent System and Research Freedom: a Comparative Study*, Working Paper, Doc. No. SCP/15/3, Annex VI.

15 *Diamond* v. *Diehr*, 450 U.S. 175 (1981).

16 *In re Bilski*, 545 F.3d 943 (Fed. Cir., 2008).

Bernard L. Bilski and Rand A. Warsaw applied for a patent wherein they claimed a method to hedge risk in commodity markets.[17] Their claims were rejected by the patent examiner for three reasons: (1) the methods described were considered to be abstract ideas (e.g. general principles of hedging); (2) these ideas were not tied to a specific apparatus; and (3) the claimed methods were not technological arts.[18] The Board of Appeal also rejected the patent application because it sought to protect, in essence, general principles of hedging and because it did not produce a 'useful, concrete and tangible result'; the Board did, however, reject the technological art and apparatus criterion used by the patent examiner.[19] The case then went to the Court of Appeals for the Federal Circuit.

Given the difficulties in defining the term 'process', the majority of Court of Appeals (9 out of 12 judges, led by Chief Judge Michel) deemed it necessary to review what was comprised within that term.[20] In the eyes of the court, the key issue that needed to be addressed in this case was whether or not the applicants sought to claim fundamental principles with a process.[21]

In making its determination the Court referred to the decision in *Diamond* v. *Diehr*,[22] which drew a distinction between those claims that 'seek to pre-empt the use of' a fundamental principle, on the one hand, and those claims that seek only to foreclose others from using a *particular application of that fundamental principle*, on the other.[23] Hence, the boundary between what is and what is not a patentable process involving the application of fundamental principles[24] appears to be whether granting a patent on the process would result in the foreclosure of these fundamental elements for a specific application only.

Up to this point, the Court's analysis is largely in accordance with the state of the law prior to Bilski. The Court went on, however, to distance itself from the prior state of the law by addressing the ambiguities that become apparent when courts attempt to determine whether or not a fundamental principle is effectively limited to a specific application.[25] The Court of Appeals inferred the new law from past jurisprudence of the

---

17 Ibid. at 949.
18 Ibid. at 950.
19 Ibid.
20 Ibid. at 951.
21 Ibid. at 952.
22 *Diamond* v. *Diehr*, 450 U.S. 175 (1981).
23 *In re Bilski*, 545 F.3d 943 at 953 (Fed. Cir., 2008), citing *Diamond* v. *Diehr*, 450 U.S. 175 (1981).
24 Ibid. at 952.
25 Ibid. at 954.

Supreme Court of the United States.[26] Citations from the Supreme Court pointing in a different direction were considered as caveats.[27]

In an attempt to resolve these ambiguities, the Court of Appeals conducted an in-depth analysis of the Supreme Court jurisprudence from which the Court of Appeals concluded that what separates unpatentable ideas and principles from patentable processes is the machine-or-transformation test.[28] According to this test, a process is patentable if (1) 'it is tied to a particular machine or apparatus, or (2) it transforms a particular article into a different state or thing'.[29] The Court of Appeals deemed this test faithful to the Supreme Court's rulings and held it to be definitive.[30]

In addition, the Court of Appeals highlighted two other criteria from existing case law regarding the test for patentability: first, an 'insignificant postsolution activity will not transform an unpatentable principle into a patentable process'.[31] Otherwise, a skilled patent agent could use an insignificant application to circumvent the prohibition to patent a fundamental principle per se.[32] Second, a claim must be analyzed as a whole; what is not patentable when isolated may become patentable when, in the same claim, it is tied to a machine or transforming an object.[33]

Lastly, the majority concluded by declaring both the Freeman-Walter-Abele test and the 'useful, concrete, and tangible result' test, two tests previously developed for the same purpose by the Court of Appeals for the Federal Circuit, to be inadequate.[34]

To recapitulate, the majority of the Court of Appeals in *In re Bilski* largely follows the prior state of the law, which, as already stated, is that (a) even though the term 'process' must be interpreted liberally, a process patent application should not seek to patent fundamental principles except in the context of a specific application thus not precluding their use in other applications; and that (b) claims as a whole must not apply fundamental principles in insignificant post-solution activities in order to circumvent the prohibition.

---

26 Ibid. at 956.

27 Ibid.

28 *In re Bilski*, 545 F.3d 943 at 955 (Fed. Cir., 2008).

29 Ibid. at 954.

30 Ibid.

31 *Diamond* v. *Diehr*, 450 U.S. 175 (1981), cited at *In re Bilski*, 545 F.3d 943 at 957 (Fed. Cir., 2008).

32 *Parker* v. *Flook*, 437 U.S. 584 (1978) cited at *In re Bilski*, 545 F.3d 943 at 957 (Fed. Cir., 2008).

33 *In re Bilski*, 545 F.3d 943 at 958 (Fed. Cir., 2008).

34 Ibid. at 959.

The Court of Appeals modified the prior state of the law, however, by stating that the *definitive* test to determine whether or not a post-solution activity is insignificant is the machine-or-transformation test. Moreover, prior tests used to determine whether or not a post-solution activity is insignificant (e.g. the Freeman-Walter-Abele test and the 'useful, concrete and tangible result') were discarded.

The adoption of the machine-or-transformation test resulted in the Court of Appeals' refusing to grant patent protection to Bilski and Warsaw.

Many criticisms of this decision have been voiced, arguing that the Supreme Court precedents have not, in fact, implemented the machine-or-transformation test as definitive, but rather considered this element as only one amongst many to be considered in determining process patentability.[35] Indeed, the dissenting minority had raised this very issue when they referred to *Gottschalk* v. *Benson*: it is possible that inventions that fail the machine-or-transformation test could, perhaps, be patented nonetheless.[36] Justice Newman (dissenting) also raised the fact that this same point had been reaffirmed in other cases. Moreover, it was held that §101 must be interpreted liberally in order to avoid endangering certain types of patent.

> The Court thus excludes many of the kinds of inventions that apply today's electronic and photonic technologies, as well as other processes that handle data and information in novel ways. [...] The innovations of the 'knowledge economy' – of 'digital prosperity' – have been dominant contributors to today's economic growth and societal change. Revision of the commercial structure affecting major aspects of today's industry should be approached with care, for there has been significant reliance on the law as it has existed, as many amici curiae pointed out. Indeed, the full reach of today's change of law is not clear, and the majority opinion states that many existing situations may require reassessment under the new criteria.[37]

C Decision of the Supreme Court of the United States in *Bilski* v. *Kappos*

The applicants appealed the decision to the Supreme Court that affirmed the ruling of the Court of Appeals. The Supreme Court, however, disagreed with the Court of Appeals' interpretation of what constitutes a patentable

---

35 Patrick, A. (2009), 'Patent Eligibility and Computer-Related Processes: A Critique of *In re Bilski* and the Machine-or-Transformation Test', *Virginia Journal of Law and Technology*, **14**, 181—-211; and Carney, J.J. (2009), 'Retreat from the Brink of Clarity: Why the Federal Circuit Got *In re Bilski* Wrong, and What Can Be Done about It', *University of Illinois Journal of Law Technology and Policy*, 473–91.

36 *Gottschalk* v. *Benson*, 409 U.S. 63 (1972).

37 *In re Bilski*, 545 F.3d 943 at 976−7 (Fed. Cir., 2008).

process. This section of the chapter will analyze the case, focusing on similarities and divergences between the reasoning of the Court of Appeals for the Federal Circuit and that of the Supreme Court.

Justice Kennedy delivering the opinion for the majority (five out of nine judges), began by pointing out that patentable subject matter is defined by § 101 of the *Patent Act*:[38] it prescribes four categories of patentable subject matter, all of which should be considered as having a broad scope in order to stimulate innovation.[39] Hence, as did the Court of Appeals, the Supreme Court started its analysis by recalling the wide scope of subject matter eligible for patent protection.

The Supreme Court confirmed that, even though the scope of patentable subject matter is wide, it is not without limits: laws of nature, physical phenomena and abstract ideas are not patentable.[40] Also, according to the Court, applications that are attempts to patent fundamental principles should not be granted.[41] More precisely, claims employing fundamental principles are to be considered valid unless they pre-empt the use of a fundamental principle in other applications. In addition, the application used to delineate a patent's scope should not be so insignificant that the claim can be determined to be an attempt to patent a fundamental principle.[42] Finally, the Court considered that each claim should be considered as a whole.[43]

Thus far, the arguments of the majority of the Supreme Court were not any different from those of the Court of Appeals: both were consistent with the prior state of the law. Where the two courts diverged, however, was in their treatment of the machine-or-transformation test. While the Court of Appeals held this test to be definitive, the Supreme Court rejected its definitive character.[44] In reaching this conclusion, the Supreme Court started by recalling that courts should not prescribe limitations not expressed by Congress.[45] While the Supreme Court had previously set a limit to patentable subject matter (e.g. regarding fundamental principles), it

---

38 *Bilski* v. *Kappos* 130 S. Ct. 3218 (2010).

39 Ibid. at 13.

40 *Diamond* v. *Chakrabarty*, 447 U.S. 303 (1980), cited in *Bilski* v. *Kappos* 130 S. Ct. 3218 at 3236–7 (2010).

41 *Bilski* v. *Kappos* 130 S. Ct. 3218 at 3258 (2010).

42 Ibid. at 28–9.

43 Ibid.

44 Ibid.; Lemley, M., M. Risch, T. Sichelman and R.P. Wagner (2011), 'Life After *Bilski*', *Stanford Law Review*, **63**, 1315–47.

45 *Diamond* v. *Diehr*, 450 U.S. 175 (1981), cited in *Bilski* v. *Kappos* 130 S. Ct. 3218 at 3226 (2010).

had not given other courts the authorization to set additional limits to patentable subject matter;[46] this would violate principles of statutory interpretation.[47] These principles led the majority of the Supreme Court to the conclusion that the machine-or-transformation test is only a clue to patentability, and is not a definitive test. This situation is similar to that expressed in the jurisprudence prior to this case.[48]

According to Justice Kennedy, emerging technologies have been challenging the application of patent law and they will probably continue to do so. A definitive machine-or-transformation test may 'create uncertainty as to the patentability of software, advanced diagnostic medicine techniques, and inventions based on linear programming, data compression, and the manipulation of digital signals'.[49] Hence, it appears that Justice Kennedy and his concurring colleagues (Justice Scalia did not concur with this *obiter*) were also motivated by the dangers of precluding the patentability of valuable emerging technologies.

The minority, led by Justice Stevens and joined by Justices Ginsburg, Breyer and Sotomayor, concurred with the majority decision. The minority decision, however, was based on an analysis of the constitutional and legislative history. After discussing the history of business method patents[50] and recalling the 'constitutionally mandated purpose' of patent law,[51] the minority concluded that general business methods do not fall within the scope of patentable processes.

To conclude section I of this chapter, it is possible to make the following comment about the post *Bilski* state of the law regarding process patentability in the United States: while the term 'process' must be interpreted liberally, a process patent employing laws of nature, natural phenomena or abstract ideas must not preclude their use in other circumstances. When considered as a whole, claims must not use insignificant post-solution activity in order to circumvent this prohibition. Finally, the *physicality* of an invention may be considered as a clue for sufficient reduction to practice but not as a *definitive* criterion.

The question that must now be answered is whether the state of the law on the patentability of processes post *Bilski* v. *Kappos* is any different from that prior to the case. It could be argued that the law post *Bilski* v. *Kappos* is what it has been since 1981 following the judgment in *Diamond* v. *Diehr*. On

---

46 *Bilski* v. *Kappos* 130 S. Ct. 3218 at 3226 (2010).
47 Ibid. at 3226.
48 Ibid. at 3227.
49 Ibid.
50 Ibid. at 3250–52.
51 Ibid. at 3252.

the other hand, it might also be argued that since the case law during the last 30 years has often, if not always, relied on *physicality* as a reference to determine whether or not an invention is sufficiently reduced to practice, the Supreme Court judgment in *Bilski* v. *Kappos* has in fact created new law. The reality, however, is likely somewhere between these two poles; while the law has not been changed, a crucial aspect of process patentability has been clarified in the eyes of the intellectual property community.

Had the machine-or-transformation test been characterized as the definitive test in *Bilski* v. *Kappos*, the scope of patentable subject matter would likely have been significantly reduced. The importance of the *Bilski* v. *Kappos* case is illustrated in *Amazon.com* v. *Canada (Attorney General)*.[52] This Canadian case exemplifies not only the influence of the *Bilski* v. *Kappos* case outside of the United States; it also demonstrates that slight changes in the role of physicality for subject matter enquiries can result in an increased scope of patentable subject matter. In *Amazon.com* v. *Canada (Attorney General)*, a business patent, initially deemed invalid, was subsequently deemed valid by the Federal Court, after specifying the role of physicality in subject matter enquiries.

In this Canadian case, Amazon.com had applied for a 'business patent' described in the following way:

> The customer visits a website, enters address and payment information and is given an identifier stored in a 'cookie' in their computer. A 'server' [...] is able to recognize the 'client' [...] and recall the purchasing information which is now stored in the vendor's computer system. The customer can thus purchase an item with a 'single click' – the order is made without the need to 'check out' or enter any more information.[53]

The Board of Appeal of the Patent Commissioner rejected the application, arguing the claims were unpatentable subject matter.[54] More precisely, the Board focused on the necessity of a 'change of character or condition in a *physical object*' to deem the process unpatentable.[55] The Canadian Federal Court analyzed the same question as seen in *Bilski* v. *Kappos*.

The case recalled a previous definition of 'art'. The Court states that, while the subject matter and utility requirements do prevent the patenting of mere ideas, these requirements do not imply the need for a physical

---

52 *Amazon.com* v. *Canada (Attorney General)* (2010), F.C.J. 1209.
53 Ibid.
54 Ibid., para. 28.
55 Ibid., para. 49.

character per se.[56] Rather, emphasis must be laid on practical applications of fundamental principles.[57] With respect to the debates they settle and the solutions they propose, *Amazon.com* v. *Canada (Attorney General)* and *Bilski* v. *Kappos* are thus very similar in outcome. This Canadian case demonstrates that even slight changes in the role of physicality for subject-matter enquiries can result in an increased scope of patentable subject matter.

II APPLICATION TO GENOMICS

Gene patents have raised controversy within the academic and scientific community for more than two decades. While investors and private bio-pharmaceutical companies believe they are necessary to promote research and commercialization in a cost intensive development field,[58] academic researchers and patient representatives have raised numerous concerns about the ethical, commercial and scientific impact of such patents.[59]

This section will look at the impact of the *Bilski* v. *Kappos* decision on the patentability of processes in the field of genomics. From this overview it will be easier to determine the impact this Supreme Court of the United States judgment is likely to have beyond specific research areas and the United States geographical boundaries.

The first part of this section describes the importance of processes in genomic research and development; the second part presents an overview of the way in which medical processes have been dealt with in the case law to date and examines the impact *Bilski* v. *Kappos* might have on the patentability of genomic processes.

A Processes in Genomic Research

The genome 'carries the information for all the proteins and RNA molecules that [an] organism will ever synthesize'.[60] Because proteins play such

---

[56] Ibid. at paras 50–53.

[57] Ibid. at paras 53, 60.

[58] Stott, M. and J. Valentine (2004), 'Gene Patenting and Medical Research: A View from a Pharmaceutical Company', *Nature Reviews Drug Discovery*, **3**, 364–8.

[59] Joly, Yann (2009) 'Patents on Genetic Research Tools: Recent Trends and Future Outlook', in Bartha Knoppers and Richard Gold (eds), *Biotechnology, IP and Ethics*, Markham, ON: LexisNexis Canada, 39–57.

[60] Alberts, Bruce (2008), *Molecular Biology of the Cell*, 5th edn, San Francisco: Garland Science, 200.

an important role from a structural and transformational (e.g. biochemical dynamics) point of view in all living organisms,[61] the study of genomes enables the scientific community to gain a better understanding of hereditary determinants of life. This understanding derives not only from the study of individual gene sequences, but also from the study of gene interactions, epigenetics (inheritable determinants distinguishable from DNA sequences, but influencing DNA expression), and cancer genomics (cancer cells are genetically altered and contain mutations that drive tumorgenesis),[62] and so on. Hence, the study of genomics will greatly enhance our knowledge of biology in general and it is hoped that progress in genomics will generate significant improvements in population health.

Many processes will play important roles in genomic research. In particular, research technologies and computational biology will evolve as the market for these fundamental research tools expands. For example, there is a need for more accessible genotyping methods, for 'modulation' methods of gene expression, for new imaging methods, and so on.[63] In addition to these potential new technologies, further innovation will be necessary in computational biology: methods to analyze large amounts of information generated by new sequencing tools, methods to analyze information and 'methods to elucidate the effects of environmental factors and of gene-environment interaction' will be necessary.[64] In short, a great deal of innovation pertaining to genetic methods will be needed.

Genomic research is not, however, the only endeavor necessitating the development of new processes. To benefit also from fundamental genomic discoveries in the clinical context, new diagnostic methods and new treatment methods (including genome-based methods) will need to be developed.[65] In addition, new uses of existing methods or molecules might be developed.[66] Therefore, while many new processes will be developed in the field of genomics, some of these processes may well neither be carried out by a machine nor involve the transformation of a physical element as required by the machine-or-transformation test from *In re Bilski*.

---

61 Ibid. at 196.

62 Stratton, M., P. Campbell and A. Futreal (2009), 'The Cancer Genome', *Nature*, **458** (8), 719-–24.

63 Collins, F., E. Green, A. Guttmacher and M. Guyer (2003), 'A Vision for the Future of Genomics Research', *Nature*, **422**, 835–47.

64 Ibid.

65 Ibid.

66 Minssen, T. and R. Schwartz (2011), 'US Patent Eligibility in the Wake of *Bilski* v. *Kappos:* "Business as Usual" in an Age of New Technologies?' *Biotechnology Law Report*, **30** (1), 3–56.

B Process Patentability in Biotechnology: Case Law in the United States

The first part of this chapter demonstrated the important role that processes will play in the future of genomics. Since patents give inventors the right to exclude others from making, selling, using or importing protected inventions,[67] a potential increase in subject-matter patentability would have a significant impact on genomic research. Will the scope of patentable biotechnology processes change, in light of *Bilski* v. *Kappos*?

By looking at *ratios* used in past cases pertaining to biotechnology processes, it is possible to determine whether *Bilski* v. *Kappos* will affect fields of innovation other than business methods. In this part of the chapter, patent applications that were held invalid by the courts are reviewed first, while those that were deemed valid will be reviewed after. To complete this part, commonalities between the two types of decisions will also be discussed.

In the first decision we review, *In re Grams*,[68] M. Grams tried to patent a method the first step of which explains, in general terms, how to conduct a series of clinical laboratory tests, with the remaining steps prescribing how to associate values obtained from the tests in order to make a diagnosis.[69] The first step was characterized as a data-gathering step and the remaining steps as a mathematical algorithm.[70]

Referring to *Gottschalk* v. *Benson*, the Court stated that while a mathematical formula per se is not patentable, not all processes involving mathematical formulas will be held to be unpatentable;[71] for example, if a claim applies a mathematical formula to a physical element, such an application of a mathematical formula will not render the claim invalid. In other words, a process that incorporates an application of a mathematical formula as one, but not its sole, element may be deemed patentable.[72] While a physical element may render a formula patentable, the test of *physicality* is not necessarily definitive.[73]

The Court determined that the only physical element of Claim 1 in *In re Grams* was the data-gathering step.[74] Nevertheless, the step was considered

---

67 *United States Code* (U.S.C.) Title 35 – Patents. 35 U.S.C. 271.
68 *In re Grams*, 888 F.2d 835 (1989).
69 Ibid.
70 Ibid.
71 Ibid. at 837.
72 Ibid.
73 Ibid. at 839.
74 Ibid.

insignificant, because no specific test was disclosed.[75] The claim was deemed to cover, in essence, a mathematical formula only and was therefore deemed invalid. The conclusion in *In re Grams* is similar to that in *In re Meyer* (cited in *In re Grams*):[76] 'the Board [. . .] found that the claim included an algorithm and a data-gathering step, the Board rejected it as "an attempt to patent a mathematical algorithm rather than a process for producing a product," [. . .]'.[77] However, in *In re Grams*, the Court did not hold that all claims composed of a data-gathering step followed by a mathematical formula are necessarily unpatentable.[78]

A more recent noteworthy decision is *Labcorp* v. *Metabolite*.[79] In this 2006 case, a correlation 'between elevated levels of total homocysteine and a deficiency in either cobalamin or folate'[80] had been discovered and a patent was issued for a related method: claim 13 of the patent application sought protection for a method by which a physician would first make an assay of total homocysteine and then correlate any elevated levels to a vitamin deficiency (colabalmin or folate).[81] The patent was first licensed by Competitive Technologies Inc. to Metabolite, who then sublicensed to Labcorp.[82] From 1992 to 1998, Labcorp used the total homocysteine test under license from Metabolite. Then, in 1998, Labcorp switched to an 'assay developed by Abbott Laboratories' and ceased paying royalties to Metabolite.[83]

The District Court of Colorado held the patent to be valid and in consequence that Labcorp was guilty of infringing the patent and breaching the terms of its contract with Metabolite.[84] This decision was upheld by the Court of Appeals for the Federal Circuit. In neither of these decisions was the eligible subject-matter criterion analyzed. That discussion took place two years later at the Supreme Court level.

---

75 Ibid. at 840.

76 *In re Meyer*, 688 F.2d 789, (CCPA 1982), cited in *In re Grams*, 888 F.2d 835 (1989).

77 *In re Grams*, 888 F.2d 835 (1989), citing *In re Meyer*, 688 F.2d 789, (CCPA 1982).

78 *In re Grams*, 888 F.2d 835 at 840 (1989).

79 *Laboratory Corporation of America Holdings* v. *Metabolite*, 548 U.S.C. 124 (2006).

80 *Metabolite Laboratories, Inc.* v. *Laboratory Corporation of America Holdings*, 370 F.3d 1354 (2004).

81 Ibid.

82 Ibid.

83 Ibid.

84 *Laboratory Corporation of America Holdings* v. *Metabolite*, 548 U.S.C. 124 at 125 (2006).

In 2006 the Supreme Court granted certiorari to Labcorp to determine 'whether a method patent [...] directing a party simply to "correlate" test results can validly claim a monopoly over a basic scientific relationship [...]'. The writ was dismissed on technical grounds. It was, however, the dissenting opinion of Justice Breyer (joined by Justice Stevens and Souter) that addressed the issue of what subject matter is patentable.

Justice Breyer recalled that laws of nature, abstract ideas and natural phenomena cannot be patented. While the discovery of these fundamental principles may be difficult, time-consuming and expensive they are, nonetheless, not patentable because patent law seeks to avoid overprotecting the primary resources of innovation.[85] This statement is consistent with the constitutional patent clauses previously described.

Justice Breyer conceded that, while the boundary distinguishing unpatentable abstract ideas, mental processes, natural phenomenon and patentable subject matter is often difficult to establish, here the subject matter in question does not fall even close to the boundary line. He therefore had no difficulty whatsoever concluding that 'there can be little doubt that the correlation between homocysteine and vitamin deficiency set forth in claim 13 is a "natural phenomenon"'. He added that, although 'a process is not unpatentable simply because it contains a law of nature',[86] respondents had simply described the natural law at issue in the abstract patent language of a 'process'.

In addition, even though Justice Breyer referred to the non-physical nature of the invention,[87] the dissenting judgment was based on the fact that the claim is an attempt to patent a 'natural phenomenon'.[88] Justice Breyer's opinion cannot, therefore, be said to be necessarily based on the *physical or non-physical* nature of the invention.

One of the most recent patent-invalidating cases is *Classen* v. *Biogen*.[89] Classen Immunotherapies 'developed and patented methods for evaluating and improving the safety of immunization schedules'.[90] In this case, Classen sued Merck for patent infringement. In response, Merck alleged that the three patents were invalid because, amongst other things, they attempted to protect unpatentable subject matter.

---

[85] Ibid. at 126.
[86] Ibid., citing *Parker* v. *Flook*, 437 U.S. 584 at 589 (1978).
[87] Ibid. at 136.
[88] Ibid. at 138.
[89] *Classen* v. *Biogen*, 304 Fed. Appx. 866 (2008).
[90] *Classen* v. *Biogen*, WDQ-04–2607 (2006).

The District Court of Maryland determined that patent 283 claimed a method involving (1) immunization of a treatment group; and (2) a comparison of the incidence of chronic immune mediated disorders in the treatment group relative to a control group.[91] As for patents 139 and 739, they described an immunizing method comprising two steps: (1) comparing the incidence of immune mediated disorders in treatment groups with different vaccination schedules; and (2) immunizing patients on a schedule identified as low risk.[92]

All three claims were deemed invalid, not because their subject matter was not *physical* (this factor was not addressed by the District Court), but because they were considered to relate to a natural phenomenon.[93] It is only when the Court of Appeals for the Federal Circuit affirmed the decision that reference to the *physicality* of the invention was made.[94] This can be explained by the fact that the Court of Appeals had rendered the decision *In re Bilski*, wherein the machine-or-transformation test had been articulated.[95] As explained in the first section of this chapter, the machine-or-transformation test was, at that time, a definitive test for patentability.

Apart from the Federal Circuit decision that followed the *In re Bilski* judgment, all three cases analyzed up to this point did not use *physicality* as the cornerstone of their *ratios*. The last important case (*Prometheus* v. *Mayo*)[96] did refer to physicality in order to validate a patent.

In *Prometheus* v. *Mayo*, two patents claimed methods for calibrating the proper dosage of thiopurine drugs, which are used for treating both gastrointestinal and non-gastrointestinal autoimmune diseases.[97] These were commercialized by Prometheus and Mayo had purchased the test. When Mayo developed its own test and started selling it to other hospitals, Prometheus filed a suit against Mayo for patent infringement. In response, Mayo argued that the patents were invalid because they sought to protect a natural phenomenon. The district court determined that 'the claims have three steps: (1) administer the drug to a subject; (2) determine metabolite levels; and (3) [receive warning] that an adjustment in dosage may be required'.[98] The same district court defined these as natural phenomena and deemed the patents invalid.

---

[91] Ibid.
[92] Ibid.
[93] *Classen* v. *Biogen*, 304 Fed. Appx. 866 (2008) at 15.
[94] Ibid.
[95] Ibid.
[96] *Prometheus Laboratories Inc.* v. *Mayo Collaborative Services*, 581 F.3d 1336 (2009).
[97] Ibid.
[98] Ibid.

This judgment was reversed by the Court of Appeals for the Federal Circuit, which deemed the patent to satisfy its unique machine-or-transformation test set forth in *In re Bilski*. The Court based its decision on the fact that the patent does transform a physical element (e.g. the human body).[99]

As these four cases demonstrate, courts have relied heavily on *physicality* as a reference. Nevertheless, *physicality* is not used in *ratios* to determine whether or not a process patent falls within the ambit of patentable subject matter. Only decisions from the Court of Appeals for the Federal Circuit rendered after *In re Bilski* and before *Bilski* v. *Kappos* have relied on the *physicality* criterion (*Prometheus* v. *Mayo* and *Classen* v. *Biogen*).

The state of the law prior to *In re Bilski* has been followed by these decisions, with the exception of decisions rendered by the Court of Appeals for the Federal Circuit after *In re Bilski* and before *Bilski* v. *Kappos* (*Prometheus* v. *Mayo* and *Classen* v. *Biogen*). Therefore, it is reasonable to expect that the scope of patentable subject matter will remain the same as it was prior to *In re Bilski*. In this respect, a case from the Court of Appeals has been reviewed (patentability upheld)[100] and two other soon will be, in light of *Bilski* v. *Kappos*.[101] This includes the case of *Association for Molecular Pathology* v. *United States Patent and Trademark Office* where the District Court used the machine-or-transformation test to invalidate a process claim; and the Court of Appeals for the Federal Circuit also applied this approach.[102]

It is not clear to what extent a definitive machine-or-transformation test (as in *In re Bilski*) would have reduced the scope of patentable subject matter. Indeed, the Court of Appeals for the Federal Circuit had not defined the precise contours of the test, leaving this task to future cases. In

---

99 Ibid. at 1345. The matter is currently under appeal before the Supreme Court of the United States in *Mayo Collaborative Services* v. *Prometheus Laboratories, Inc.*, *cert. granted*, __ U.S. __, 2011 WL 973139 (U.S. June 20, 2011) (No. 10–1150).

100 *Prometheus* v. *Mayo*, 2010 U.S. App. LEXIS 25956 (2010).

101 *Classen* v. *Biogen*, 2010 U.S. LEXIS 5533 (2010).

102 *Association for Molecular Pathology et al.* v. *United States Patent and Trademark Office, Myriad Genetics, et al.* 702 F. Supp. 2d 181, (S.D.N.Y. 2010), *appeal docketed,* No. 2010–1406 (Fed. Cir. June 22, 2010); *Association for Molecular Pathology et al.* v. *United States Patent and Trademark Office, Myriad Genetics, et al.* 653 F.3d 1329, 2011 WL3211513 (C.A. Fed. (N.Y.)), 99 USPQ 2d 1398; and Van Overwalle, G. (2010), 'Turning Patent Swords into Shares', *Science*, **330**, 1630.

addition, its interpretation of the test seemed to be liberal, since it considered an 'electronic transformation of [data] into a visual depiction'[103] as patentable. The Court of Appeals even reversed a decision of a lower court by applying the machine-or-transformation test in *Prometheus* v. *Mayo.*[104] In any case, the Supreme Court of the United States has settled the issue by characterizing the test as not definitive. Therefore, the scope of patentable processes will remain similar to what it was prior to *Bilski*.

C Differences amongst Jurisdictions in Process Patentability and Consequences for Genomics from an International Perspective

Part I demonstrated how important processes are in the developing field of genomics. Such processes include research tools (including genotyping methods), methods in computational biology, methods to analyze large quantities of information, diagnostic methods, modulation methods of gene expression, new uses of existing pharmaceutical molecules and other treatment methods. In the second part of section II, past cases pertaining to methods similar to those listed above were analyzed. We have seen that courts have relied on *physicality* as one clue to patentability but not as a decisive criterion. Because the current state of the law is similar to that prevailing prior to *In re Bilski*, the scope of patentable subject matter will likely remain the same. This state of the law may well differ significantly from that which prevails in a number of other countries.

All countries that are signatories to the *Trade Related Aspects of Intellectual Property Rights Agreement* (the *TRIPS Agreement* 1994) are required to patent processes.[105] However, section 27 of the *TRIPS Agreement* 1994 allows member countries to exclude some technologies from the patent realm, including those pertaining to 'diagnostic, therapeutic and surgical methods for the treatment of humans or animals'.[106] Many countries have indeed elected to exclude some of these methods. For instance, Canada[107]

---

[103] *In re Bilski*, 545 F.3d 943 at 963 (Fed. Cir., 2008).

[104] *Prometheus* v. *Mayo*, 581 F.3d 1336 (2009).

[105] *Marrakesh Agreement Establishing the World Trade Organization*, opened for signature 15 April 1994, 1867 UNTS 3 (entered into force 1 January 1995) annex 1C ('*Agreement on Trade-related Aspects of Intellectual Property Rights*' – *TRIPS Agreement* 1994).

[106] Ibid., article 27 (3) (a).

[107] Canadian Intellectual Property Office, *Manual of Patent Office Practice*, chapter 17.02.03, citing *Industries Ltd.* v. *Commissioner of Patents*, [1986] 3 F.C. 40 and *Tennessee Eastman Co.* v. *Commissioner of Patents* [1974] S.C.R. 111.

and New Zealand[108] currently prohibit patenting of therapeutic methods (surgical methods, dental methods, etc.).

In the *European Patent Convention*, Europe has decided to exclude from patentability 'methods for treatment of the human or animal body by surgery or therapy and diagnostic methods practiced on the human or animal body'.[109] The term 'surgery' refers to: (1) a physical intervention (2) on a human or an animal body (3) to maintain 'the life and health of the subject'.[110] The term 'therapy' is defined as 'the curing of a disease or malfunction' (including prophylactic treatments).[111] As for the expression 'diagnostic methods', it refers to methods comprising a data-gathering step, followed by comparison of values, identification of the deviation and final diagnosis. Any diagnostic method not comprising one of these four steps is patentable.[112] In addition, those methods that are not applied on or in a human body, that is *in vitro* methods,[113] are also patentable.

Similarly, the European exclusion of 'programs for computers' from patentable subject matter has also been weakened by jurisprudence[114] and many have been granted software patents.[115]

To summarize, the scope of patentable subject matter in other jurisdictions does not seem as wide as it is in the United States. Some provisions in these jurisdictions forbid the patenting of medical inventions. However, the effective scope of these provisions is often weakened by case law.

These jurisdictional differences in the scope of patentability between countries might affect genomics in two ways. First, differences in process patentability might impact international research collaboration, since there could be important processes patentable in the United States but not in other jurisdictions. In addition, as patent regimes tend towards harmonization, the prospect of increasing the breadth of patentability outside the United States raises the issue of fostering a scientific anti-commons. The term 'anti-commons' generally refers to the patenting of a set of 'upstream'

---

108 Keim, B. (2007), 'Patent Eligible Subject Matter in the Biotechnological Arts', Berkeley Electronic Press, http://works.bepress.com/benjamin_keim/1.
109 *European Patent Convention*, 29 November 2000, art. 53 (c).
110 Organisation for Economic Co-operation and Development (OECD) (2010), *The Patent System and Research Freedom: A Comparative Study*, Working Paper, Doc. No. SCP/15/3, Annex VI, 10.
111 *European Patent Convention*, 29 November 2000, art. 53 (c).
112 Ibid.
113 Ibid.
114 *European Patent Convention*, 29 November 2000, art. 52 (2).
115 Ballardini, R.M. (2008), 'Software Patents in Europe: the Technical Requirement Dilemma', *Journal of Intellectual Property Law and Practice*, **3** (9), 563–75.

crucial inventions within a research field (i.e. research tools).[116] If accessing these inventions becomes too difficult, 'downstream' innovation might be hindered.[117] Any increase in patentable subject matter might result in an exacerbation of this problem. On the other hand, the applicability of the anti-commons theory to the field of biotechnology remains to be confirmed by evidence,[118] particularly since the opposite argument can also be made, that is, that widening the scope of subject matter patentability would actually provide a greater incentive to invent.

CONCLUSION

In the first section of this chapter, the state of the law regarding process patentability was described. Under the *United States Constitution*, Congress may grant patent rights, *in order to stimulate* innovation. The constitutionally-embedded and required balance was further defined by Congress and the courts: to be patentable, an invention must fall within one of four categories, one of which is process. The courts interpreted this last term as excluding 'laws of nature, physical phenomena, and abstract ideas'.[119] More precisely, if patent claims apply laws of nature, physical phenomena or abstract ideas they must do so in such a way as to not pre-empt their use in other applications.

From prior Supreme Court cases, the Court of Appeal for the Federal Circuit inferred in *In re Bilski* the existence of a definitive test (machine-or-transformation test)[120] to distinguish what is patentable from what is not. As explained by the Court, 'a claimed process is surely patent-eligible under § 101 if: (1) it is tied to a particular machine or apparatus, or (2) it transforms a particular article into a different state or thing'.[121] This test relies on the *physicality* of inventions to separate what is patentable from what is not.

---

116 Gold, R., Y. Joly and T. Caulfield (2005), 'Genetic Research Tools, the Research Exception and Open Science', *GenEdit,* **3** (2), 1–8.

117 Eisenberg, R. (2008), 'Noncompliance, Nonenforcement, Nonproblem? Rethinking the Anticommons in Biomedical Research', *Houston Law Review*, **45**, 1059–99.

118 Ibid.; Caulfield, T., R. Cook-Deegan, F.S. Kieff and J. Walsh (2006), 'Evidence and Anecdotes: An Analysis of Human Gene Patenting Controversies', *Nature Biotechnology*, **24** (9), 1091–4.

119 *Diamond* v. *Chakrabarty*, 447 U.S. 303 at 309 (1980).

120 *In re Bilski*, 545 F.3d 943 (Fed. Cir., 2008).

121 Ibid. at 954.

The Supreme Court of the United States in *Bilski* v. *Kappos* recognized the value of the machine-or-transformation test as an important clue to patentability, but not as a definitive test. Hence, some inventions that are not tied to a machine or do not transform a physical element could be patented. *Bilski* v. *Kappos* thus clarified a crucial aspect of process patentability.

The second section of the chapter focused on the potential impact of this decision on international biotechnology research. Many processes will have to be developed to carry out this type of research; hence, process patentability is important for biotechnology. Changing the scope of patentable processes could impact the rate of innovation.

Prior case law pertaining to biomedical patents was analyzed in order to determine whether or not *Bilski* v. *Kappos* had changed the scope of patentable subject matter for medical and biotechnology patents. The four selected court cases demonstrate that, prior to *In re Bilski,* courts relied on *physicality* as a factor in *obiter* but not in *ratio*. Hence, the state of law seems similar now to what it was back before *In re Bilski* in the United States.

The second section ended with an international perspective. In the field of medical patents, the United States seems to have a wider scope of patentable subject matter than Europe, Canada and New Zealand. Moreover, the lack of statutory exclusion concerning biotechnology patents in the United States could be alleviated by case law. This raises issues pertaining to research collaborations and potential science anti-commons.

Lack of data concerning what constitutes an ideal balance between inventors' rights and users' rights renders the pursuit of the constitutionally embedded purpose very challenging. In *Bilski* v. *Kappos*, the Supreme Court took into account this lack of knowledge by excluding only the patenting of fundamental principles per se. A majority of scholars agree that promoting innovation requires a similar exclusion in patent law.

It also appears that, on the one hand, striving for more clarity in patent law through more specific tests could be a desirable objective. On the other hand, flexible patent statutes can respond more easily to emerging technologies. Reconciling these two opposing requirements with a view to maximizing innovation has proved very difficult. While some tests and clues could be used to achieve a better balance, other mechanisms, such as protections for practitioners or researchers, or exceptions for specific fields, could also be used to promote research and development in the biotechnology field.

3. The current state of patent eligibility of medical and biotechnological inventions in the United States

Joshua D. Sarnoff

This chapter briefly explains the current (and conflicting) doctrinal standards in the United States for exclusions from patent eligibility that apply to the medical and biotechnology fields in light of the decision of the Supreme Court of the United States in *Bilski* v. *Kappos.*[1] The United States Court of Appeals for the Federal Circuit (Federal Circuit), albeit reluctantly, and the United States Patent and Trademark Office (USPTO), perhaps more enthusiastically, have begun to employ these standards to claims of invention that apply new discoveries of medicine and biotechnology. Unless and until a better theory of patent eligibility is supplied, the results will likely remain both uncertain and unsatisfying to all concerned.[2]

In *Bilski,* the Supreme Court held that business methods may be eligible subject matter for patents, so long as the claimed, practical applications are not merely 'abstract ideas' and so long as other claim limitations are not merely 'field of use' limitations or 'token post-solution components' added to the abstract idea.[3] In so holding, the Court reiterated as a matter of long-standing precedent and *stare decisis* that the patent system categorically excludes 'laws of nature, physical phenomena, and abstract ideas'[4] (science, nature, and ideas), despite the broad categorical language recited

1 *Bilski* v. *Kappos*, 130 S. Ct. 3218 (2010).

2 I provide such a theory in Sarnoff, J. (2011), 'Patent Eligible Inventions after *Bilski:* History and Theory', *Hastings Law Journal*, **63** (forthcoming). Much of this chapter was published originally in Sarnoff, J. (2011), 'Patent Eligible Inventions after *Bilski, Prometheus* and *Myriad*', *Texas Intellectual Property Law Journal*, **19** (3), 393–418. The author thanks the many people who have contributed to these works.

3 *Bilski* v. *Kappos,* 130 S. Ct. at 3228–29, 3231 (2010).

4 Ibid. at 3225. See *Diamond* v. *Chakrabarty*, 447 U.S. 303, 309 (1980) (citing, inter alia, *O'Reilly* v. *Morse*, 56 U.S. (15 How.) 62, 112–21 (1854); *Le Roy* v. *Tatham*, 55 U.S. (14 How.) 156, 175 (1853)).

in Section 101: 'Whoever invents or discovers any new and useful process, machine, manufacture, or composition of matter [...] may obtain a patent therefor.'[5] The Court thus required the courts and the USPTO to engage in complex and under-theorized efforts to draw lines between ineligible natural discoveries and eligible inventions applying those discoveries. In regard to the specifically excluded category of abstract ideas, the Court also invited the Federal Circuit to specify narrower categories or classes that would provide the public with greater certainty of what can qualify as eligible and what cannot.[6]

A burgeoning set of eligibility decisions is now issuing from the Federal Circuit in regard to a wide range of practical and useful medical and biotechnology applications, such as the *Prometheus Laboratories Inc.* v. *Mayo Collaborative Services* medical treatment method case[7] and the *Association for Molecular Pathology et al.* v. *United States Patent and Trademark Office, Myriad Genetics, et al.* (*Myriad*) isolated genetic sequence and diagnostic method case.[8] Notwithstanding the Supreme Court's invitation, the Federal Circuit under Chief Judge Rader has signaled its desire to avoid relying on patent eligibility doctrine to exclude claimed inventions from the patent system, by requiring 'recognition that [a] disqualifying characteristic should exhibit itself so manifestly as to override the broad statutory categories of eligible subject matter and the statutory context that directs primary attention on the patentability criteria of the rest of the Patent Act'.[9] The patentability criteria referred to are novelty, non-obviousness (inventive step) and adequacy of the disclosure in describing the invention and enabling others to use it.[10] Implicit in the Federal Circuit's effort to avoid eligibility exclusions is the view that Section 101 largely duplicates patentability criteria in preventing 'the issuance of bad patents' – bad in the sense of not being 'really innovative'[11] – and that

5 35 U.S.C. § 101.

6 See *Bilski* v. *Kappos*, 130 S. Ct. 3218, 3229 (2010).

7 *Prometheus Laboratories Inc.* v. *Mayo Collaborative Services* No. 2008–1403, 2010 WL 5175124 (Fed. Cir. Dec. 17, 2010).

8 *Association for Molecular Pathology et al.* v. *United States Patent and Trademark Office, Myriad Genetics, et al.* 702 F. Supp. 2d 181, (S.D.N.Y. 2010), *appeal docketed,* No. 2010–1406 (Fed. Cir. June 22, 2010).

9 *Research Corp. Techs., Inc.* v. *Microsoft Corp.*, 627 F.3d 859, 868 (Fed. Cir. 2010).

10 See 35 U.S.C. §§ 102 (novelty), 103(a) (non-obviousness), 112, para. 1 (written description and enablement).

11 Fusco, S. (2010), '*In re Bilski*: A Conversation with Judge Randall Rader and a First Look at the BPAI's Patent Cases', *Albany Law Journal of Science and Technology*, **20**, 123, 144. See Risch, M. (2009–10), 'Forward to the Past', *Cato Supreme Court Review*, 333, 348.

there is no field of scientific, technological, or other functional endeavor for which the patent system would categorically impede rather than promote innovation.[12] But even under the Federal Circuit's restrictive approach, lines will still need to be drawn between ineligible discoveries and eligible inventions, and the Federal Circuit's approach lacks any theory to inform such line-drawing exercises.

The rest of the chapter first discusses the *Bilski* holding and its theory of exclusions from patent eligibility, providing useful comparisons to different approaches in other jurisdictions. It then briefly analyzes the Federal Circuit's decision in the *Prometheus* case, and relates it to the earlier Supreme Court *Laboratory Corporation of American Holdings* v. *Metabolite Laboratories, Inc.* (*LabCorp*) case,[13] which addressed a medical diagnostic patent but did not result in an issued decision. The chapter then describes the *Myriad* case – pending in the Federal Circuit at the time of writing – and the issues that it raises.

Throughout the chapter, the focus is on how the current doctrine applies to eligible and ineligible *applications* of categorically excluded science, nature, and ideas, which provides insights into the difficulties in drawing the required lines regarding eligibility of claimed inventions, systemic benefits of employing eligibility exclusions, and the utilitarian and deontological moral concerns (including social and innovation harms) that have been raised in regard to such applications. The chapter concludes with a brief projection of the continuing, contested future of medical and biotechnology eligibility determinations, and the recognition that patentable subject matter eligibility will remain a controversial area – in the United States and around the globe.

I *BILSKI*

The Supreme Court in *Bilski* not only reaffirmed the existence of the categorical exclusions from eligibility for science, nature, and ideas, but also reiterated the long-standing requirement to treat them as if they were

---

12 See Fusco, S. (2010), '*In re Bilski*: A Conversation with Judge Randall Rader and a First Look at the BPAI's Patent Cases', *Albany Law Journal of Science and Technology*, **20**, 123 at 144.

13 *Laboratory Corporation of American Holdings* v. *Metabolite Laboratories, Inc.* (*LabCorp*) 548 U.S. 124 at 127–28 (2006) (Breyer, J., dissenting from dismissal of certiorari).

already 'a familiar part of the prior art',[14] even when they are newly discovered by the patent claimant. This legal fiction exists because such discoveries are the 'basic tools of scientific and technological work',[15] and as 'part of the [public domain] storehouse of knowledge of all men [.... they must remain] free to all men and reserved exclusively to none'.[16] The patent system is not supposed to reward such basic scientific or conceptual discoveries,[17] no matter how much money, effort, creativity, and disclosure go into developing and disseminating that knowledge. Nor does patent law exist to reward such discoveries and recoup the investments of money, effort, and creativity in making them *through* eligible inventions that apply the discoveries.[18] Rather, patent claimants must invent and disclose some 'other inventive concept' than a merely novel, physically limited application of the new discovery.[19]

As a result of the prior art status of categorically excluded science, nature, and ideas, the human creativity involved in discovering them should not contribute to assessing the nature, eligibility, *or* patentability of the claimed 'invention'[20] in their application.[21] As the Court repeated in *Bilski,* once an excluded algorithm is 'assumed to be within the prior art, the application, considered as a whole, [may] *contain no patentable invention*'.[22] Stated differently, for an eligible and patentable invention to exist, there

---

14 *Parker* v. *Flook*, 437 U.S. 584 at 592 (1978). See ibid. at 591 (citing *Mackay Radio & Tel. Co.* v. *Radio Corp. of America*, 306 U.S. 86 at 94 (1939); *Funk Brothers. Seed Co.* v. *Kalo Inoculant Co.*, 333 U.S. 127 at 130 (1948)); and *Bilski* v. *Kappos*, 130 S.Ct. 3218, 3230 (2010).

15 Ibid. at 591 (quoting *Gottschalk* v. *Benson*, 409 U.S. 63, 67 (1972)).

16 *Funk Brothers Seed Co.* v. *Kalo Inoculant Co.*, 333 U.S. 127 at 130 (1948).

17 See e.g. Holmes, H.M. (1932), 'Patent Rights for Scientific Discoveries by C.J. Hamson', *Harvard Law Review*, **45**, 1431–3 at 1432. *In re Alappat*, 33 F.3d 1526 at 1553–54 (Fed. Cir. 1994) (en banc) (Archer, C.J. dissenting). See generally Hamson, C.J. (1930), *Patent Rights for Scientific Discoveries*, Indianapolis: Bobbs-Merill Co., 20–29.

18 But cf. *Ariad Pharmaceuticals., Inc.* v. *Eli Lilly & Co.*, 598 F.3d 1336 at 1359 (Fed. Cir. 2010) (en banc) (Newman, J., additional views).

19 *Parker* v. *Flook*, 437 U.S. 584 at 594 (1978).

20 Cf. Mohammed, E.A.C. (2008), 'What Is an Invention? A Review of the Literature on Patentable Subject Matter', *Richmond Journal of Law and Technology*, **15**, 1; and Vaver, D. (2003), 'Invention in Patent Law: A Review and Proposal', *International Journal of Law and Information Technology*, **11**, 286 at 289–90.

21 See e.g. *O'Reilly* v. *Morse*, 56 U.S. (15 How.) 62 at 115–16 (1854); and *Morton* v. *New York Eye Infirmary,* 17 F. Cas. 879 at 881–82 (C.C.S.D.N.Y. 1862) (No. 9,865).

22 *Bilski* v. *Kappos*, 130 S. Ct. 3218 at 3230 (2010) (quoting *Parker* v. *Flook*, 437 U.S. 584, at 594 (1978)) (emphasis added).

must be invention (human creativity) *in* the application of excluded discoveries, and not merely creativity in identifying the discovery that makes the application possible. For this reason, the Court in *Bilski* repeated language from its most recent (and claimant-friendly) eligibility case, *Diamond* v. *Diehr*,[23] stating that 'the prohibition against patenting abstract ideas "cannot be circumvented by limiting the[ir] use [...] to a particular technological environment" or [by] adding "insignificant postsolution activity"'.[24]

This approach in the United States differs substantially from other approaches, such as in Australia and under the *European Patent Convention* ('EPC').[25] Australian decisional law explicitly refuses to treat new discoveries as publicly known prior art when considering the creativity of claimed inventive applications of them, although the discoveries themselves ('as such' under the EPC,[26] which excludes from being 'regarded as inventions' 'discoveries, scientific theories and mathematical methods'[27]) remain categorically ineligible.[28] Thus, an applicant's:

> claim for a patent is not validly answered by telling him that although there was ingenuity in his discovery [...] no ingenuity was involved in showing how the discovery, once it has been made, might be applied. The fallacy lies in dividing up the process that he puts forward as his invention.[29]

Under the EPC, in contrast, a 'contribution' approach to eligibility similar to that of the United States' approach was initially adopted, under which the creativity of categorically excluded subject matter could not contribute novelty to claimed applications. Although the EPO later abandoned that approach to eligibility, the contributed knowledge of the categorically ineligible discovery remains excluded from the consideration of an inventive step (although it is not necessarily treated as prior art) when evaluating the 'technical contribution' of the applicant. The EPC thus currently permits to be considered eligible inventions claims that employ a 'technical means' or that are a 'technical product', even if all of the creative novelty lies in the excluded discovery. But it requires that any 'technical effect' for

---

23 *Diamond* v. *Diehr*, 450 U.S. 175 (1981).

24 *Bilski* v. *Kappos*, 130 S. Ct. 3218 at 3230 (2010) quoting *Diamond* v. *Diehr*, 450 U.S. 175 at 191–2 (1981).

25 *Convention on the Grant of European Patents* 1973, art. 52(3), Oct. 5, 1973, 1065 U.N.T.S. 255, 13 I.L.M. 270, http://www.european-patent-office.org/legal/epc/e/ma1.html (EPC).

26 Ibid., art. 52(3).

27 Ibid., art. 52(2)(a).

28 See *National Research Development Corporation* v. *Commissioner of Patents* (1959) 102 CLR 279.

29 Ibid. at 252.

inventive step be reflected in a 'technical character' found in 'all the features together' (and thus in the application of the discovery). Where the only creative and novel feature is non-technical (i.e., in the categorically excluded subject matter), the claim will not be patentable.[30] As the EPO's own Board recognized, many have criticized as 'distasteful' the choice to permit discoveries to contribute to eligibility given that they do not contribute to patentability.[31] The United States 'prior art' approach avoids having the creativity of new discoveries contribute to either.

The Supreme Court of the United States, however, has not been consistent in approaching eligibility and has not provided clear guidance regarding what qualifies as categorically excluded subject matter (particular 'abstract ideas')[32] and what applications are eligible in light of them, or in relating its practical decisions to the theoretical grounds for making them. Thus, the Court in *Bilski* held to be ineligible abstract ideas various independent claims for a method of hedging risks from unexpected events that create fluctuating volumes for fixed-price-contract purchased commodities. These claims were both practical and somewhat more specific than the underlying fundamental idea that the claims applied (hedging risks), but did not require the use of any specifically identified machines or artifacts.[33] In reaching its decision, the Court continued to express a concern articulated in a 1972 decision that patents may not issue that 'would wholly pre-empt the [ineligible discovery] and in practical effect would be a patent on the [discovery] itself'.[34] The Court also failed to explain adequately why the more specific dependent claims at issue – which limited the methods to commodities and energy markets and which required the use of well-known techniques as inputs – added only 'field of

---

30 *Opinion of the Enlarged Board of Appeal of 12 May 2010 in relation to a point of law referred by the President of the European Patent Office pursuant to Article 112(1)(b) of the EPC*, Case G 003/08 at ¶¶ 10.3, 10.4, 10.5, 10.7.1, 10.13.1, 12.2.1, 12.2.2, and 13.5.1.

31 Ibid. at ¶ 10.13.

32 See *Bilski* v. *Kappos*, 130 S. Ct. 3218 at 3236 (2010) (Stevens, J., concurring in the judgment). Cf. Eleven Law Professors and the American Association for Retired Persons (2009), 'Brief of Eleven Law Professors and AARP as *Amici Curiae* in Support of Respondent in *Bilski* v. *Kappos*', 2009 WL 3167954, 2 October, at 23–4, ('AARP *Bilski* Brief') (citing *Rubber-Tip Pencil Co.* v. *Howard,* 87 U.S. (20 Wall.) 498 at 507 (1874)); and Samuelson, P. and J. Schultz (2011), '"Clues" for Determining Whether Business and Service Innovations Are Unpatentable Abstract Ideas', *Lewis and Clark Law Review*, **15**, 110–31.

33 See *Bilski* v. *Kappos*, 130 S. Ct. 3218 at 3223–4 (2010).

34 Ibid. at 3230 (quoting *Gottschalk* v. *Benson*, 409 U.S. 63, 72 (1972)). See ibid. at 3231.

use' limits or 'token post-solution components' that 'did not make the concept patentable'.[35]

The Supreme Court also rejected the Federal Circuit's effort to create clearer rules of eligibility based on Supreme Court precedents and dicta that applications involving particular machines or accomplishing specific physical transformations are eligible.[36] The Supreme Court overruled the Federal Circuit's holding that this 'machine-or-transformation test [... is] the sole test for what constitutes a[n eligible] "process" (as opposed to just an important and useful clue)',[37] preserving the potential for expansion of the patent system to intangible and information technologies. But the Supreme Court in *Bilski* did not apply its own clue, and focused solely on the 'abstract idea' exclusion. Nevertheless, the Court's discussion of the field of use limits and token post-solution activity in regard to the dependent claims suggests that it did not view any physical implementations of the abstract idea implied by the claim language as requiring the use of particular machines or as accomplishing sufficient physical transformations.[38]

In contrast, earlier Supreme Court cases have recognized (based on the prior art approach) the need for an eligible invention to possess a sufficient kind and degree of creativity (or 'sophistication'[39]) *in the application* of categorically excluded science, nature, or ideas to accomplish a practical result.[40] This requirement generates and explains various linguistic formulas (developed in other Supreme Court cases) to assess the eligibility of claimed products and processes. These tests would find particular, physical, and scope-limited novel applications of discoveries to be ineligible unless the claimed products (derived from ineligible 'products of nature', i.e. physical phenomena) have 'markedly different characteristics',[41] or unless the claimed processes (employing ineligible laws of nature or abstract ideas)

---

35 Ibid. at 3231 (citations omitted).

36 See e.g. *Gottschalk* v. *Benson*, 409 U.S. 63 at 70 (1972); and *Cochrane* v. *Deener*, 94 U.S. 780, 788 (1887).

37 *Bilski* v. *Kappos*, 130 S. Ct. 3218 at 3236 (2010). See ibid. at 3227; *In re Bilski*, 545 F.3d 943 at 954–55 (Fed. Cir. 2008) (en banc).

38 *Bilski* v. *Kappos*, 130 S. Ct. 3218, 3231 (2010).

39 Winston, E. (2011), 'The Technological Edge', *Akron Intellectual Property Law Journal*, 1–25 at 6–7, http://papers.ssrn.com/sol3/papers.cfm?abstract_id=1692836.

40 See e.g. *Evans* v. *Eaton*, 8 F. Cas. 846 at 851 (C.C.D. Pa. 1816) (No. 4,559); see also *Boulton* v. *Bull*, (1795) 2 H. Bl. 463 at 495 (Opinion of Lord Eyre, C.J.); ibid. at 485 (Opinion of Buller J.); at 478 (Rooke, J.); ibid. at 482 (Heath, J.).

41 *Diamond* v. *Chakrabarty*, 447 U.S. 303 at 310 (1980). See *Hartranft* v. *Wiegmann*, 121 U.S. 609, 615 (1887); and *Morton* v. *New York Eye Infirmary* 17 F. Cas. 879 at 881–82 (C.C.S.D.N.Y. 1862) (No. 9,865).

reflect non-'analogous' uses.[42] Merely novel but insufficiently creative applications of ineligible discoveries are not eligible inventions. But once a sufficiently creative application has been invented, *that* invention (not the discovery it employs) may be patented and may thereby pre-empt *its* full scope of application; such pre-emption may include all means of accomplishing a particular end, even if the *inventive* application is the only practical means of using the discovery to accomplish the desired result.[43] Thus, the horse of determining the existence of an inventive application must precede the cart of assessing the over-breadth of claim scope compared to that application.

These are the standards that establish the current, messy state of patent eligibility law in the United States. The decisions that have been and will be issued by the Federal Circuit in applying these standards to medical and biotechnology inventions have been and likely will be similarly unharmonious. Yet further conflicts may develop if the United States Congress becomes involved in creating exclusions from patent eligibility, either by restricting entire areas of endeavor from the patent system or by adjusting the level of creativity found by the courts to be sufficient for eligibility of applications of ineligible discoveries (similar to how the United States will now have to evaluate claims applying methods for reducing, avoiding, or deferring tax liability by treating the methods themselves as prior art for novelty and obviousness (inventive step)).[44] Similar conflicts would attend future legislation to extend eligibility to areas that the courts may hold are excluded, or to levels of creativity the courts may find are insufficient. In the latter case, constitutional concerns may also arise regarding whether any limits exist on the Congress's power to grant patents to such 'inventions', including retrospectively to those that have fallen into the public domain.[45]

---

42 *Ansonia Brass & Copper Co.* v. *Electrical Supply Co.,* 144 U.S. 11 at 18 (1892).

43 *Dolbear* v. *American Bell Telephone Co*., 126 U.S. 1, 534–35 (1888).

44 *Leahy-Smith America Invents Act*, Pub. L. 112−29, §§ 14(a), 125 Stat. 284, 327 (2011).

45 See e.g. Ten Law Professors (2008), 'Brief of *Amici Curiae* Ten Law Professors In Support of Appellee Director of the United States Patent and Trademark Office in the matter of *In re Bilski*', 2008 WL 1842283 (Fed. Cir. 2008) (en banc), 6–12. Compare *Graham* v. *John Deere Co*., 383 U.S. 1 at 6 (1966) with *Eldred* v. *Ashcroft*, 537 U.S. 186, 203 (2002), and *Golan v. Holder,* 609 F.3d 1076 (10th Cir. 2010), *cert. granted* (7 Mar. 2010, No. 10–545). See generally Walterscheid, Edward (2002), *The Nature of the Intellectual Property Clause: A Study in Historical Perspective*, Buffalo: William S. Hein & Co.

II *PROMETHEUS*

On remand from the Supreme Court following *Bilski,* the Federal Circuit in the *Prometheus* case distinguished the 'essence' of the medical treatment claims at issue as physically 'transformative' (of humans) from the '"mere" data-gathering steps or "insignificant extra-solution activity"' of clinical diagnostic claims that the Federal Circuit had earlier found to be ineligible in the *Grams* case.[46] As one commentator put it immediately after the decision, the panel's attempt to distinguish *Grams* was 'less than convincing'.[47] (That commentator, on behalf of a group of law professors, had filed a brief arguing that synthetic drug breakdown products and the human-made correlations they generate are *not* categorically ineligible natural phenomena.)[48]

Specifically, the treatment claim addressed a multi-step method of 'optimizing therapeutic efficiency for treatment of an immune-mediated gastro-intestinal disorder' requiring following the steps of: (1) administering a drug containing a particular synthetic chemical (6-thioguanine, or 6TG) to a person; and (2) determining the level of 6TG in the person, where a level of 6TG at or below a specific concentration indicates a need to increase the amount of drug administered (to assure efficacy) and a different level at or above a specific concentration indicates a need to decrease the amount (to avoid toxicity).[49] (Another claim of that patent dispenses with the requirement to administer a thiopurine drug, relying only on the determining step, and a claim of a different patent is substantially the same as the first claim,

---

46 *Prometheus Laboratories., Inc.* v. *Mayo Collaborative Services.*, No. 2008–1403, 2010 WL 5175124 (Fed Cir. Dec. 17, 2010), at *9 (distinguishing *In re Grams*, 888 F.2d 835 (Fed. Cir. 1989)).

47 Holman, C. (2010), 'On Remand, Federal Circuit (Once Again) Decides *Prometheus* v. *Mayo* in Favor of Patent Eligibility for Methods of Treatment and Diagnostic Tests', Holman's Biotech IP Blog, 17 December, http://holmansbiotechipblog.blogspot.com/2010/12/on-remand-federal-circuit-once-again.html.

48 See Interested Patent Law Professors (2003), 'Brief of *Amici Curiae* Interested Patent Law Professors in Support of Neither Party in *Prometheus Laboratories, Inc.* v. *Mayo Collaborative Services.*', No. 2008–1403, 2009 WL 462602, 21 January at 2–13.

49 *Prometheus Laboratories, Inc.* v. *Mayo Collaborative Services.*, No. 2008–1403, 2010 WL 5175124 (Fed. Cir. Dec. 17, 2010), at *2 (quoting Claim 1 of Seidman, E. and Y. Theoret (1999), 'Method of treating IBD/Crohn's disease and related conditions wherein drug metabolite levels in host blood cells determine subsequent dosage', United States Patent No. 6,355,623).

adding only a requirement to also determine another metabolite's level.)[50] The first step necessarily requires physical activity to 'administer' the specific drug (a composition of matter) and the second step may be constructed to implicitly require some physical method (but not any specific method) of gathering data and of performing an analysis (even in the mind) to determine the level. But the claim as a whole does not require taking any action in response to the mental step of determining a person's level of 6TG. Mayo originally used the diagnostic test technology sold by Prometheus, but later abandoned it for its own test that employed different indicator levels of 6TG for evaluating thiopurine administration.[51]

The human metabolic pathway of converting synthetic thiopurine drugs into mercaptopurines and thiopurine nucleotides was well known, as was the use of such drugs to treat auto-immune and inflammatory bowel diseases. These metabolic products were also known to cause serious adverse side effects (including death), and thus medical practitioners were already engaged in calculating effective doses that would minimize the risks of side effects.[52] Prometheus exclusively licensed the patent from its owners, who had statistically observed the blood levels of these conversion products across a range of patients and derived an association of the blood levels with regard both to effectiveness and to avoidance of toxicity.[53] The claims reflected the particular levels of the statistical associations that were observed.

The District Court found that the claims were not eligible subject matter, as: (1) they were merely the combination of a data-gathering step and a

---

50 Ibid. (describing Seidman, E. and Y. Theoret (1999), 'Method of treating IBD/Crohn's disease and related conditions wherein drug metabolite levels in host blood cells determine subsequent dosage', United States Patent No. 6,355,623 Claim 46), Seidman, E. and Y. Theoret (2001), 'Methods of optimizing drug therapeutic efficacy for treatment of immune-mediated gastrointestinal disorders', United States Patent No. 6,680,302, Claim 1).

51 Ibid.

52 Appellate Brief (2009), 'The Brief of the Appellant in *Prometheus Laboratories Inc.* v. *Mayo Collaborative Services.*, No. 2008–1403, 2009 WL 327984, 9 January, 6–7, ('Prometheus Appellants Brief'); and American College of Medical Genetics et al. (2009), 'Corrected Brief for the American College of Medical Genetics in *Prometheus Laboratories., Inc.* v. *Mayo Collaborative Services.*, No. 2008–1403, 2009 WL 1307167, 6 April, 13.

53 See Appellate Brief (2009), 'The Brief of the Appellant in *Prometheus Laboratories Inc.* v. *Mayo Collaborative Services.*, No. 2008–1403, 2009 WL 327984, 9 January at 6–11, ('Prometheus Appellants Brief'); and American College of Medical Genetics et al. (2009), 'Corrected Brief for the American College of Medical Genetics in *Prometheus Laboratories., Inc.* v. *Mayo Collaborative Services.*, No. 2008–1403, 2009 WL 1307167, 6 April at 8–12.

mental step, without requiring any actual physical treatment (implying that they were not transformative under the machine-or-transformation framework); and (2) that the claims essentially recited correlations that were categorically ineligible natural phenomena (products of nature) that the applicants did not invent, and that the claims wholly pre-empted all uses of those correlations.[54] The Federal Circuit originally reversed, on the basis both of its en banc decision in *Bilski*,[55] which had imposed the machine-or-transformation framework as the conclusive test, and of its different understanding of the claims and the invention from that of the District Court. The claims were held to be eligible, because '[1] the "administering" and "determining" steps were [physically] transformative and not merely data-gathering steps […] and [2] as such the claims [which must be considered as a whole, even if they included a separately ineligible mental step, and as they were limited to the particular physical steps involved] did not wholly preempt the use of the recited correlations [the specific indicators] between metabolite levels and drug efficacy or toxicity'.[56]

On remand in *Prometheus* after the Supreme Court's *Bilski* decision, the Federal Circuit first held that: the Court had not disavowed the machine-or-transformation framework but had only avoided making it an exclusive test; and the machine-or-transformation framework continued to establish the eligibility of the claims at issue. The claims were for methods of treatment, 'which are always transformative when one of a defined group of drugs is administered to the body to ameliorate the effects of an undesired condition'.[57] In other words, the human body itself is transformed by treatment, even if the transformations result from natural bodily processes. In contrast to the District Court, which apparently focused on the fact that the claims did not require any *subsequent* action following recognition of indication levels, and thus the claimed method could not by itself actually optimize therapeutic efficiency without taking the next, unclaimed step of adjusting the amounts administered, the Federal Circuit apparently focused on the fact that the claim as a whole (based on the prior art administrating step) was still a method of treatment. (Alternatively, the

---

54 See *Prometheus Laboratories Inc.* v. *Mayo Collaborative Services* No. 2008–1403, 2010 WL 5175124 (Fed Cir. Dec. 17, 2010), at *3 (discussing *Prometheus Laboratories Inc.* v. *Mayo Collaborative Services.*, No. 04-CV-1200, 2008 WL 878910 (S.D. Cal. Mar. 28, 2008)).

55 Ibid. at *4 (discussing *Prometheus Laboratories., Inc.* v. *Mayo Collaborative Services.*, 581 F.3d 1336, 1345–49 (2009)).

56 Ibid., *Prometheus Laboratories, Inc.* v. *Mayo Collaborative Services*, 581 F.3d 1336 at 1349 (2009).

57 *Prometheus Laboratories Inc.* v. *Mayo Collaborative Services* No. 2008–1403, 2010 WL 5175124 (Fed. Cir. Dec. 17, 2010), at *6.

Federal Circuit may have found the step of physically adjusting dosages to be implicit in the claimed recognition step.) Of greater significance, the Federal Circuit failed to recognize that the only *novelty* of the claimed invention (as a whole) relative to the prior art was in the step of recognizing (through unspecified but physical data gathering and, presumably, non-physical mental activity) the newly identified 'natural' correlation between metabolite levels of synthetic drugs and medical needs, which is arguably a categorically ineligible scientific and medical discovery. If so, just like in *Bilski,* the dependent claims 'as a whole, contained no patentable *invention*',[58] even if physical data-gathering steps were employed and the information gathered was useful (and likely triggered subsequent action).

On remand after *Bilski,* the Federal Circuit also reiterated that the claims were drawn to a *particular* application of an ineligible natural phenomenon, not to the phenomenon (a 'law of nature') itself,[59] and the transformations achieved by these specific steps were 'central to the purpose of the claimed process'.[60] Specifically, the Court noted that 'the steps involve a particular application of the natural correlations: the treatment of a specific disease by administering specific drugs and measuring specific metabolites. As such, the claims did not preempt all uses of the natural correlations; they utilized them in a series of specific steps.'[61] The Court therefore rejected the arguments that the claims pre-empted all uses and that any machine implementation or physical transformation involved in the administering and data-gathering steps were merely 'insignificant post-solution activity'; although the court agreed that the mental recognition step would not be eligible on its own, that did not preclude the method (viewed as a whole) from eligibility.[62] The treatment claims did 'not preempt all uses of the natural correlations' as other 'drugs might be administered to optimize the therapeutic efficacy of the claimed treatment'.[63]

The Federal Circuit's remand decision failed adequately to explain why the actual invention – the *essence* of the treatment claim (or its 'gist' or

---

58 *Bilski* v. *Kappos*, 130 S. Ct. 3218, 3230 (2010) (quoting *Parker* v. *Flook*, 437 U.S. 584 at 594 (1978)) (emphasis added).

59 *Prometheus Laboratories Inc.* v. *Mayo Collaborative Services* No. 2008–1403, 2010 WL 5175124 (Fed. Cir. Dec. 17, 2010) at *7.

60 Ibid. (quoting *In re Bilski*, 545 F.3d 943 at 962 (Fed. Cir. 2008) (en banc)).

61 Ibid.

62 Ibid. at *6. See Ibid. at *8–*10.

63 Ibid. at *7.

'heart' or 'point of novelty'[64]) – lay in the physically transformative administering step (or the claim as a whole) and not in the correlation employed by it. This was because the Federal Circuit did not recognize the need for any creative invention *in the application* of the new medical discovery (treating that discovery as if it were prior art). Once that discovery was treated as known, using the *specific correlation* in an existing process of administering drugs and testing blood levels of metabolites may have been new but certainly was not *inventive.* Lacking any invention (but claiming a novel combination method), the claim also could not survive patentability evaluations for obviousness so long as that correlation were treated as prior art; the claim would necessarily lack any non-obvious invention without performing some non-analogous function when combining the medical fact that had been discovered with the prior art method of treatment.[65] The Federal Circuit thus improperly allowed the newly discovered but ineligible correlation to contribute to the 'invention' assessed for eligibility, for claims that also should be inherently obvious if the correlations are treated as prior art (even under the European approach). And it did so by focusing on the claim as a whole (rather than the inventive contribution that the claim as a whole reflected), on the additional (but uncreative) claim limits of the otherwise-known method, and on the physical nature of the administration (initial treatment) steps in the claim.

In particular, the Federal Circuit's decision failed to provide any convincing account of why the physical drug-administering or metabolite level-determining steps were not merely insignificant extra-solution activity for the claimed uses of the newly discovered correlation. The Supreme Court in *Bilski* had found the specific antecedent data-gathering and information-input steps to the hedging method claims at issue to constitute only token post-solution activity to the abstract ideas claimed, and thus held such claims as a whole to be ineligible.[66] Such insignificant steps will also avoid pre-emption, by preventing the more specific uses of newly discovered

---

64 See e.g. Lemley, M. (2011), 'Point of Novelty', Stanford Public Law Working Paper, 2, http://papers.ssrn.com/sol3/papers.cfm?abstract_id=1735045 (quoting *Loctite Corp.* v. *Ultraseal Ltd.* 781 F.2d 861, 875 (Fed. Cir. 1985) and citing Liivak, O. (2010), 'Rescuing the Invention from the Cult of the Claim', working paper, http://www.law.berkeley.edu/files/bclt_IPSC2010_Liivak%282%29.pdf; and Collins, K. (2010), 'Semiotics 101: Taking the Printed Matter Doctrine Seriously', *Indiana Law Journal*, **85**, 1379–443 at 1381.

65 See e.g. *KSR International. Co.* v. *Teleflex, Inc.*, 505 U.S. 398 at 416–17 (2007) (discussing *Sakraida* v. *Ag-Pro, Inc.* 425 U.S. 273 (1976) and *Anderson's-Black Rock, Inc.* v. *Pavement Salvage Co.*, 396 U.S. 57 (1969)).

66 See *Bilski* v. *Kappos*, 130 S. Ct. 3218 at 3231 (2010).

natural phenomena from excluding other unclaimed uses, and (for eligible inventions) claim scope concerns can be addressed through enablement doctrine.[67]

The Federal Circuit's belief that the specific steps of administering the drugs avoided pre-emption of uses of the correlations was also confused. The fact that other applications might be found for correlations that exist between *other non*-thiopurine drugs and the recited metabolite levels did not change the fact that the claims recited (and pre-empted all uses of) the *specific* correlations between the thiopurine drugs and metabolite levels actually discovered and claimed. Given so-called and controversial 'absolute protection' (excluding all making and uses, including those that are not contemplated by inventors),[68] every patented claim (in the United States) inherently pre-empts all applications of whatever the claimed inventive principle applies to. This was long ago recognized to be permissible, even for very broad claims, in the seminal *Alexander Graham Bell Telephone* case, so long as some *inventive* principle exists *in* applying a scientific discovery or natural phenomenon (and not in the discovery being applied).[69]

Further, the additional, initial physical-treatment and data-gathering steps may not necessarily change the essence of claims that apply a new discovery of a natural medical phenomenon. This was recognized by Justice Breyer in a non-precedential statement (with two of his colleagues who have now departed the Supreme Court) in the *Laboratory Corporation* diagnostic method case that preceded *Bilski* and that was argued to the Court but dismissed without opinion:[70] 'aside from the unpatented [prior art data-acquisition] test, [the steps] embody only the correlation between homocysteine and vitamin deficiency that the researchers uncovered. In my view, that correlation is an unpatentable "natural phenomenon," and I can

---

67 Cf. Holman, C. (2010), 'On Remand, Federal Circuit (Once Again) Decides *Prometheus* v. *Mayo* in Favor of Patent Eligibility for Methods of Treatment and Diagnostic Tests', Holman's Biotech IP Blog, 17 December, http://holmansbiotechipblog.blogspot.com/2010/12/on-remand-federal-circuit-once-again.html.

68 See 35 U.S.C. § 271(a); Case C-428/08, *Monsanto Technology LLC* v. *Cefetra BV and others*, No. C-428/08 at ¶¶ 33–50, European Court of Justice (6 July 2010). See generally Dorries, H. Ulrich (2001), *Patentanspruche auf DNA-Sequenzen: ein Hindernis fur die Forschung? Anmerkungen zum Regierungsentwurf fur ein Gesetzzur Umsetzung der Richtlinie* 98/44/EG, MITT. 15 at 16; and White, A. (2000–1) 'Gene and Compound Per Se Claims: An Appropriate Reward?' *Bio-Science Law Review*, **6**, 239 at 240.

69 See *Dolbear* v. *American Bell Telephone Co.*, 126 U.S. 1 at 534–35 (1888).

70 *Laboratory Corporation of America Holdings* v. *Metabolite Laboratories, Inc.*, 548 U.S. 124 (2006) (No. 04-607); (Breyer, J., dissenting from dismissal of certiorari).

find nothing in claim 13 that adds anything more of significance.'[71] Justice Breyer specifically rejected the arguments that 'the correlation is nonetheless patentable because claim 13 packages it in the form of a "process" for detecting vitamin deficiency, with discrete testing and correlating steps', and 'that claim 13 is a patentable "application of a law of nature" because, considered as a whole, it (1) "entails a physical transformation of matter," namely, the alteration of a blood sample during whatever test is used [...] and because it (2) "produces a 'useful, concrete, and tangible result,'" namely, detection of a vitamin deficiency [....]'[72] These arguments (rejected by Justice Breyer) are remarkably similar to the basis for the Federal Circuit's post-*Bilski* holding in *Prometheus*. Even if the Supreme Court does not reach out in *Prometheus* to resolve whether Justice Breyer's or the Federal Circuit's approach is to be the master, the issues and differences of approach will continue to present disputes and petitions for certiorari in other cases, given the large numbers of treatment and diagnostic claims that have issued and that reflect such applications of newly discovered natural medical phenomena (including those induced synthetically).

The *Laboratory Corporation* case is also significant for the public policy concerns that it raised and that Justice Breyer went out of his way to discuss as a reason for seeking to have the case decided rather than dismissed, even if the Court were to resolve the case against his views, as a clear decision would then allow Congress to weigh in if it felt the need to change the law.[73]

> To fail to do so threatens to leave the medical profession subject to the restrictions imposed by this individual patent and others of its kind. Those restrictions may inhibit doctors from using their best medical judgment; they may force doctors to spend unnecessary time and energy to enter into license agreements; they may divert resources from the medical task of health care to the legal task of searching patent files for similar simple correlations; they may raise the cost of healthcare while inhibiting its effective delivery.[74]

Although the United States *Patent Act* contains an exception from remedies for medical practitioners and their institutions performing patented medical methods,[75] the potential indirect liability of the clinical laboratory

71 Ibid. at 137–8.
72 Ibid. at 135–6 (citations omitted).
73 Ibid. at 138.
74 Ibid.
75 See 35 U.S.C. § 287(c).

in assisting the doctors posed serious First Amendment free speech and medical communication concerns.[76]

Many of these concerns were restated by various medical organizations in the *Prometheus* case, who argued that non-inventive applications of basic medical discoveries 'interfere with the practice of medicine, constraining the ability of physicians to make informed treatment decisions based on the latest scientific knowledge, are likely to stifle innovation, and will serve only to increase the cost and decrease the effectiveness of treatment for serious diseases'.[77] And these organizations further argued that precluding such claims will not interfere either with the development of personalized medicine nor with 'incentives necessary for medical innovation'.[78] Of course, the existence of these harms and the sufficiency of non-patent incentives for making such medical discoveries and patentable applications of them are highly contested.[79] Thus, Prometheus argued that invalidating claims like the one at issue 'would destroy the entire field of medical treatment and diagnostic patents. Thousands, if not tens of thousands, of such patents have been granted, and they have become the essential underpinning of a vibrant and innovative industry of inestimable value to mankind.'[80]

Perhaps more interestingly, the medical organizations raised deontological moral arguments for invalidating patents for such non-creative applications – that they reflect doctors' violations of their ethical duties to share information freely.[81] As the medical organizations noted:

---

76 See e.g. *Metabolite Laboratories, Inc.* v. *Laboratory Corporation of America Holdings*, 370 F.3d 1354 at 1365 (Fed. Cir. 2004); American Association for Retired Persons (2005), 'Brief Amicus Curiae of AARP in Support of Petitioner in *Laboratory Corp. of America Holdings v. Metabolite Laboratories, Inc.*', 2005 WL 3597809, 23 December, 7–8, *Laboratory Corporation of America Holdings* v. *Metabolite Laboratories, Inc.*, 548 U.S. 124 (2006) (No. 04-607); and Burk, D. (2000), 'Patenting Speech', *Texas Law Review*, **79**, 99–162 at 115–17, 136–54.

77 American College of Medical Genetics et al. (2009), 'Corrected Brief for the American College of Medical Genetics in *Prometheus Laboratories., Inc.* v. *Mayo Collaborative Services*', No. 2008–1403, 2009 WL 1307167, 6 April, 13.

78 Ibid. at 27–32.

79 See e.g. Appellate Brief (2009), 'The Brief of the Appellant in *Prometheus Laboratories Inc.* v. *Mayo Collaborative Services*', No. 2008–1403, 2009 WL 327984, 9 January at 46–50.

80 Ibid. at 46–47.

81 See American College of Medical Genetics et al. (2009), 'Corrected Brief for the American College of Medical Genetics in *Prometheus Laboratories., Inc.* v. *Mayo Collaborative Services*', No. 2008–1403, 2009 WL 1307167, 6 April at 27.

> Physicians have an obligation to share their knowledge and skills and to report the results of clinical and laboratory research [...]. The intentional withholding of new medical knowledge, skills, and techniques from colleagues for reasons of personal gain is detrimental to the medical profession and to society and is to be condemned.[82]

Such moral arguments are gaining increasing importance in political debates over the scope of patent rights, but raise concerns that are incommensurable with the innovation policy concerns and may be even less susceptible to theoretical and empirical resolution. Further, normative beliefs regarding scientific obligations to share knowledge have been changing[83] – in large part as a result of permitting scientists and their institutions to retain patents with government funds.[84] Unlike in the United States, moreover, most countries entirely prohibit patents on medical methods of diagnosis and treatment.[85] The politics of these moral disputes over the proper scope of the patent system and the obligations of discoverers to freely share their useful knowledge will remain contested,[86] just as it is for

---

82 Ibid. at 26 (quoting Opinion 9.8 of the American Medical Association's Code of Medical Ethics, http://www.ama-assn.org/ama1/pub/upload/mm/369/ceja_3i07.pdf).

83 See e.g. Rai, A. (1999), 'Regulating Scientific Research: Intellectual Property Rights and the Norms of Science', *Northwestern University Law Review*, **94**, 77–152; Eisenberg, R. (1996), 'Public Research and Private Development: Patents and Technology Transfer in Government-Sponsored Research', *Virginia Law Review*, **82**, 1663–727. But see e.g. Kieff, F.S. (2001), 'Facilitating Scientific Research: Intellectual Property Rights and the Norms of Science – A Response to Rai and Eisenberg', *Northwestern University Law Review*, **95**, 691–705.

84 See e.g. McManis, C. and S. Noh (2006), 'The Impact of the Bayh-Dole Act on Genetic Research and Development: Evaluating the Arguments and Empirical Evidence to Date', The University of California, Berkeley, 13 August, 1, www.law.berkeley.edu/files/mcmanis(1).doc; and Rai, A. (2001), 'Evolving Scientific Norms and Intellectual Property Rights: A Reply to Kieff', *Northwestern University Law Review*, **95**, 707–14. See generally Rai, A. and R. Eisenberg (2003), 'Bayh-Dole Reform and the Progress of Biomedicine,' *Law and Contemporary Problems*, **66** (1&2), 289–314.

85 See, e.g., article 53 (c) of the *Convention on the Grant of European Patents* 1973, art. 52(3), Oct. 5, 1973, 1065 U.N.T.S. 255, 13 I.L.M. 270, http://www.european-patent-office.org/legal/epc/e/ma1.html (EPC).

86 See generally Parthasarathy, S. (2010), 'Breaking the Expertise Barrier: Understanding Activist Strategies in Science and Technology Policy Domains', *Science and Public Policy*, **37**, 355–67.

software,[87] sports moves,[88] tax planning methods,[89] human organisms, cloned organisms,[90] and the many important products of biotechnological research and development, to which we now turn.

III *MYRIAD*

In 2009, the American Civil Liberties Union and the Public Patent Foundation (collectively the ACLU) brought suit on behalf of numerous medical organizations, doctors, scientists and patients to challenge various specific claims that had been obtained on two genes associated with breast cancer (BRCA1 and BRCA2) and methods of diagnosing genetic mutations thereof in a person's gene sequences. As the ACLU noted, gene patents raise civil liberties concerns by 'unreasonably restraining free speech and scientific research', and by violating rights to freedom of research, thought and expression possessed by scientific researchers, clinical geneticists and genetic counselors, and the public.[91]

The claims followed on the efforts of an international consortium that was in the process of sequencing the breast cancer genome, had identified the chromosome on which it was located, and had intended to place the sequence in the public domain. But one of the researchers, Mark Skolnick at the University of Utah, departed from the consortium and founded Myriad Genetics to commercialize the gene once it was sequenced. Skolnick was the first to specifically locate both the BRCA1 and BRCA2 genes, based on his access to Mormon genealogical records, which he compared with Utah state public health records using computational analysis techniques (and federal funds and the assistance of a federal researcher at the National Institutes of Health). Skolnick then sequenced the genes using

---

87 See, e.g., Chatterjee, A. (2004), 'Europe Struggles Over Software Patents', *IEEE Spectrum* (September), http://spectrum.ieee.org/at-work/innovation/europe-struggles-over-software-patents.

88 See, e.g., Magliocca, G. (2009), 'Patenting the Curve Ball: Business Methods and Industry Norms', *Brigham Young University Law Review* 875 at 877 (2009).

89 See, e.g., *Leahy-Smith America Invents Act*, Pub. L. No. 112–29, §§ 14(a), 33(a), 125 Stat. 284, 327, 340 (2011).

90 See, e.g., National Institutes of Health (2011), 'Legislative Updates: Cloning', Office of Legislative and Policy Analysis Legislative Updates, 3 February, http://olpa.od.nih.gov/legislation/107/pendinglegislation/7cloning.asp; 148 *Congressional Record* S5579 (daily ed. June 14, 2002).

91 See American Civil Liberties Union (2009), 'BRCA: Genes and Patents', 27 May, http://www.aclu.org/free-speech/brca-genes-and-patents#12.

well-known biotechnological techniques, and the University of Utah obtained patents on the isolated sequences and their diagnostic method uses.[92]

After the Myriad patents were granted, they were subjected to substantial public criticism in the United States (principally on utilitarian grounds) for interfering with research, for raising the costs of breast cancer diagnostic treatment, and for preventing the use of a better (more comprehensive) diagnostic test that had been developed by others and was being used in Europe.[93] These concerns led to a widely publicized editorial by the fiction writer Michael Crichton in the *New York Times*, which criticized gene patents generally and encouraged legislation to ban gene patents in the United States.[94] The ACLU's suit followed.

The ACLU challenged – as being beyond the USPTO's statutory authority to grant and as being unconstitutional – specific claims to genetic sequences and to methods employing them in the various patents. The first set of claims was for 'isolated DNA' or 'isolated DNA molecules' coding for BRCA1 or BRCA2 proteins, or short sequences or mutations thereof. Although this language could in theory encompass pure information, the claims were construed to apply to physical DNA.[95] The second set of claims was for methods of 'analyzing a sequence', 'detecting' a mutation, or 'comparing' a sequence to the normal (disclosed) sequence. These claims did not recite any specific method for acquiring sequence information, and

---

92 See, e.g., Paradise, J. (2004), 'European Opposition to Exclusive Control Over Predictive Breast Cancer Testing and the Inherent Implications for U.S. Patent Law and Public Policy: A Case Study of the Myriad Genetics' BRCA Patent Controversy', *Food and Drug Law Journal*, **59**, 133–54 at 143–4; Williams-Jones, B. (2002), 'History of a Gene Patent: Tracing the Development and Application of Commercial BRCA Testing', *Health Law Journal* **10**, 123–46 at 131; Brown, P. and K. Kleiner (1994), 'Patent Row Splits Breast Cancer Researchers', *New Scientist* 44, September 24; and Appellate Brief (2010), 'Brief for the Appellants in the *Association for Molecular Pathology et al.* v. *United States Patent and Trademark Office et al.*, No. 2010–1406 (Fed. Cir), 2010 WL 4600106, 22 October, ('*Myriad* Appellants Brief').

93 See, e.g., Andrews, L. and J. Paradise (2005), 'Gene Patents: The Need for Bioethics Scrutiny and Legal Change', *Yale Journal of Health Policy Law and Ethics*, **5**, 403–12 at 407–9; and Lewin, T. (1996), 'Move to Patent Cancer Gene is Called Obstacle to Research', *The New York Times*, 21 May, 14.

94 See Crichton, M. (2007), 'Patenting Life', *The New York Times*, 13 February; and H.R. 977, 110th Cong. (2007).

95 See *Association for Molecular Pathology et al.* v. *United States Patent and Trademark Office et al.*, 702 F. Supp. 2d 181 at 212–13 (2010), *appeal docketed,* No. 2010–1406 (Fed. Cir. June 22, 2010) ('*Myriad*').

(like the recognition step of the Prometheus claim) arguably could be performed solely in the mind.[96]

The District Court avoided reaching the constitutional issues by finding the claims to be unauthorized by the statute.[97] The Court issued a broad doctrinal holding that merely isolating genetic sequences (even if the resulting sequences are also minimally chemically modified) does not alter their status under the *Patent Act* as unpatentable 'products of nature'.[98] Long-standing precedents had distinguished ineligible products of nature from creative 'human-made inventions'[99] based on whether the object created from the natural product was 'a new and different article, having a distinctive name, character, or use'.[100] These precedents held that merely isolating an existing substance to increase its purity while using it for its natural functions, or merely creating a synthetic analogue to the natural product, was not sufficiently creative to be considered a patent eligible invention.[101]

As the Supreme Court noted in *American Fruit Growers, Inc.* v. *Brogdex Co.*, in order to be eligible the new creation had to 'possess a new or distinctive form, quality, or property [....] There [must be a] change in the name, appearance, or general character of the [thing from which it was created].'[102] The Court in *American Fruit Growers* addressed a novel, non-natural, human-made combination – borax-treated fruit – having a property (mold resistance) that was not possessed by the natural article alone; the Court held the claimed product to be an ineligible product of nature.[103] Following *American Fruit Growers*, the subsequent Commissioner of the Patent Office (who also became one of the principal drafters

---

96 See ibid. at 213–14; American Medical Association et al. (2010), 'Brief for Amici Curiae American Medical Association. et al. In Support of Plaintiffs' Opposition to Defendants' Motion to Dismiss and in Support of Plaintiffs' Motion for Summary Judgment, in *Association for Molecular Pathology et al.* v. *United States Patent and Trademark Office et al.* (S.D.N.Y. Apr. 5), 8–9.

97 See *Association for Molecular Pathology et al.* v. *United States Patent and Trademark Office, Myriad Genetics, et al.* 702 F. Supp. 2d 181 at 237–8, (S.D.N.Y. 2010), *appeal docketed,* No. 2010–1406 (Fed. Cir. June 22, 2010).

98 See ibid. at 222–32.

99 *Diamond* v. *Chakrabarty*, 447 U.S. 303, 306, 311, 313 (1980) (citations omitted).

100 *Hartranft* v. *Wiegmann,* 121 U.S. 609, 615 (1887).

101 See, e.g., *American Wood-Paper Co.* v. *Fibre Disintegrating Co.,* 90 U.S. (23 Wall.) 566 at 593–94 (1874); and *Cochrane* v. *Badische Anilin & Soda Fabrik*, 111 U.S. 293 at 311 (1884).

102 *American Fruit Growers, Inc.* v. *Brogdex Co.*, 283 U.S. 1 at 11–12 (1931). See also ibid. at 12–13.

103 See ibid. at 14.

of the current *Patent Act*) acknowledged that the Patent Office had improperly granted patents for biological organisms and chemicals that were merely isolated or purified from naturally occurring materials (such as Pasteur's isolated yeast patent and a purified adrenaline (takemine) patent, which was the subject of a famous opinion by Judge Learned Hand): 'it may now be doubted that the subject-matter is capable of being patented'.[104]

Similarly, the Supreme Court in the famous *Diamond* v. *Chakrabarty* case[105] had upheld the eligibility of a living, synthetic organism by distinguishing its earlier holding in *Funk Brothers Seed Co.* v. *Kalo Inoculant Co.*,[106] which had found ineligible a synthetic, man-made combination of bacteria that merely 'serve[d] the ends nature originally provided'.[107] The former, but not the latter, had '*markedly different characteristics* from any found in nature and [...] the potential for significant utility'.[108] Since both claimed products had significant utility, the distinction between eligible and ineligible inventions must have related to those characteristics. Further, the markedly different characteristic standard for products corresponded to the Court's standard for eligibility of processes, which required new uses of existing things or processes to be non-analogous (i.e., not merely different – and thus novel but similar).[109] Although the United States Congress in the 1952 *Patent Act* codified a definition of process that included new uses of known things or processes (and then included such processes in the categories of eligible subject matter),[110] the legislative history makes clear that Congress intended only to restore the non-analogous use standard in light of a conflicting lower court opinion.[111]

Returning to *Myriad,* the District Court implicitly overruled the long-standing policy (since at least 2001) of the USPTO that isolated or purified

---

104 See, e.g., Federico, P. (1937), 'Louis Pasteur's Patents', *Science*, **86**, 327 (October 8, 1937) (citing *American Fruit Growers, Inc.* v. *Brogdex Co.*, 283 U.S. 1 (1931), and *Parke-Davis & Co.* v. *H. K. Mulford Co.,* 189 F. 95 (S.D.N.Y. 1911), affirmed 196 F. 496 (2d Cir. 1912)).

105 *Diamond* v. *Chakrabarty* 447 U.S. 303, 310 (1980).

106 *Funk Brothers Seed Co.* v. *Kalo Inoculant Co* 333 U.S. 127 (1948).

107 See ibid. at 131.

108 See *Diamond* v. *Chakrabarty,* 447 U.S. at 310 (emphasis added).

109 See *Ansonia Brass & Copper Co.* v. *Electrical Supply Co.,* 144 U.S. 11 at 18 (1892).

110 See 35 U.S.C. §§ 100(b), 101.

111 See, e.g., H.R. Rep. No. 82–1923, at 6 (discussing *In re Thuau*, 135 F.2d 344 (C.C.P.A. 1943)); Federico, P. (1952 republished 1993), 'Commentary on the New Patent Act', *Journal of the Patent and Trademark Office Society*, **75**, 161–231 at 176–8; and Reisenfeld, S. (1954), 'The New United States Patent Act in the Light of Comparative Law', *University of Pennsylvania Law Review*, **102**, 291–322 at 299–300.

gene sequences are patent eligible inventions, based on an earlier grant of a patent to Louis Pasteur for isolated yeast and on a lower court opinion that isolated natural products were patent eligible.[112] As the USPTO then stated, such sequences do 'not occur in that isolated form in nature, or [are] synthetic DNA preparations [... and] their purified state is different from the naturally occurring compound'.[113]

The District Court unfortunately may have gone too far in seeking to justify its decisions on the isolated sequence claims on an exceptionalist view of genetic materials based on their information content.

> Myriad's focus on the chemical nature of DNA, however, fails to acknowledge the unique characteristics of DNA that differentiate it from other chemical compounds [....] This informational quality is unique among the chemical compounds found in our bodies, and it would be erroneous to view DNA as 'no different' than other chemicals previously the subject of patents.[114]

Given the Supreme Court precedents discussed above, the Court could readily have reached the same result of unpatentability based on the similarity of the isolated materials and the new functions that they perform (as not 'markedly different characteristics' and only 'analogous uses'). Although the new functions (such as use as probes and for diagnostic analysis) may have been novel and not performed by the naturally occurring materials (as with *American Fruit Growers*), they would remain insufficient given their reliance on the natural materials and their inherent properties. The Court's exceptionalist approach is unlikely to survive the currently pending appeal, as the Federal Circuit may not wish to acknowledge that the myriad (pun intended) claims that have issued for isolated and/or purified natural chemicals and biological materials over the last 100 years have been invalid.

Remarkably, the United States Government filed a brief in the appeal admitting that the USPTO has, for the last 20 years, lacked authority for and improperly issued thousands of claims for isolated and purified genetic sequences.[115] Although the United States Government recognized that the

---

112 *Utility Examination Guidelines*, 66 Fed. Reg. 1092 at 1093 (Jan. 5, 2001) (citing *Parke-Davis & Co.* v. *H. K. Mulford Co.*, 189 F. 95 (S.D.N.Y. 1911), *aff'd,* 196 F. 496 (2d Cir. 1912).

113 Ibid.

114 *Association for Molecular Pathology et al.* v. *United States Patent and Trademark Office, Myriad Genetics, et al.* 702 F. Supp. 2d 181 at 228, (S.D.N.Y. 2010), *appeal docketed,* No. 2010–1406 (Fed. Cir. June 22, 2010) (citations omitted).

115 United States (2010), 'Brief for the United States as Amicus Curiae in Support of Neither Party at 18, *Association for Molecular Pathology et al.* v. *United States Patent and Trademark Office et al.*', No. 2010–1406 2010 WL 4853320 (Fed. Cir.), 29 October at 18, ('USG *Myriad* Brief'). See ibid. at 17–36.

claims at issue were invalid as they applied to isolated DNA, the United States Government also argued that if the claims had been limited to cDNA (single-stranded, extra-chromosomal gene sequences with introns – non-coding regions – removed) they would be valid, because cDNA is chemically different from isolated DNA.[116] The United States Government's distinction may not make sense, given that cDNA occurs naturally within cells[117] and thus cDNA is either merely an isolated natural DNA sequence or a synthetic reproduction of such a naturally occurring sequence. Using the United States Government's own 'magic microscope' analogy, i.e., whether you could see the thing in the same form in nature but for extraction, isolated cDNA would remain a product of nature.[118] Further, the Supreme Court precedents noted above make clear that merely creating synthetic analogues of natural products does not generate eligible inventions. But notwithstanding the official position of the United States Government that the USPTO lacks legislative authority to grant patents for isolated and purified genetic sequences, the USPTO also has refused to stop issuing such claims while the case is pending.[119]

Various other approaches and distinctions that might justify eligibility have been raised in the case, but are unlikely to become the basis for an appellate court decision on the merits. For example, during oral argument some of the Federal Circuit Judges suggested a potential distinction between treating isolated natural minerals or other materials as products of nature and not treating isolated genes as such, based on the distinction of physical from chemical separation (e.g., solvent extraction from breaking of covalent chemical bonds). The Acting Solicitor General, however, argued that such a distinction could not be justified, as it would allow patents on natural

116 Ibid. at 14–17.

117 See International Human Genome Sequencing Consortium (2001), 'Initial Sequencing and Analysis of the Human Genome', *Nature*, **409**, 860–921 at 880; Gilbert, N. et al. (2005), 'Multiple Fates of L1 Retrotransposition Intermediates in Cultured Human Cells', *Molecular and Cellular Biology*, **25**, 7780–95; and *Stedman's Medical Dictionary* (2005), 513 (28th edn).

118 See, e.g., Dutra, T. (2011), 'Myriad, ACLU Battle Renewed in Fed. Cir. Oral Arguments with DOJ Joining the Fray,' *Patent, Trademark & Copyright Journal*, **81**, 1–2.

119 See, e.g., Genome Web (2010), 'US DoJ Argues Against Patenting Isolated Genes, But USPTO Will Maintain Status Quo,' *Pharmacogenetics Reporter*, 3 November, http://www.genomeweb.com/print/953728?page=show.

minerals such as lithium, which also need to be chemically separated from their natural condition in order to be useful.[120]

Similarly, some of the amici supporting Myriad focused on the human-created (synthetic) nature of the claimed isolated, extra-chromosomal sequences and the new functions such sequences can perform (e.g., gene therapy and manufacture of therapeutic proteins).[121] Others focused on the difficulty of and creative steps involved in generating the synthetic sequences.[122] But the difference of the new functions capable of being performed by isolated DNA from those performed in nature, and the difficulty of actually obtaining the sequences given contemporaneous public knowledge, was barely or was not discussed (respectively) during the oral argument. The only significant discussion of these issues occurred in regard to whether the plaintiffs could avoid other claims of Myriad's patents (directed to use of isolated sequences as probes or primers, which were not challenged by the plaintiffs), if the Court were to invalidate the isolated sequence claims and thus whether a decision would redress any asserted injury that the plaintiffs have suffered.[123] Myriad also argued in its briefs that the plaintiffs lack a sufficient threat of liability to have standing to challenge the patents, and much of the oral argument focused on the standing issue.[124]

The District Court in *Myriad* also held invalid the patents for the methods, because the claims did not require any physical acts and thus

---

120 See, e.g., Dutra, T. (2011), 'Myriad, ACLU Battle Renewed in Fed. Cir. Oral Arguments with DOJ Joining the Fray,' *Patent, Trademark & Copyright Journal*, **81**, 1–2 (discussing the magic microscope analogy of the Acting Solicitor General during oral argument).

121 See, e.g., Biotechnology Industry Organization and the Association of University Technology Managers (2010), 'Brief for the Biotechnology Industry Organization and the Association of University Technology Managers as Amici Curiae Supporting Reversal', *Association for Molecular Pathology et al.* v. *United States Patent and Trademark Office et al.*', No. 2010–1406, 2010 WL 4853322 (Fed. Cir.), 29 October, 5–12, 21–2.

122 See, e.g., Gilead Sciences, Inc., Biogenerator, and Elan Pharmaceuticals, Inc. (2010), 'Brief of Amici Curiae Gilead Sciences, Inc., Biogenerator, and Elan Pharmaceuticals, Inc. in Support of Defendants-Appellants and Urging Reversal', *Association for Molecular Pathology et al.* v. *United States Patent and Trademark Office et al.*', No. 2010–1406, 2010 WL 5306804 (Fed. Cir.), 29 October, 3–8, 18.

123 See, e.g., Dutra, T. (2011), 'Myriad, ACLU Battle Renewed in Fed. Cir. Oral Arguments with DOJ Joining the Fray,' *Patent, Trademark & Copyright Journal*, **81**, 1–2.

124 Ibid.

patented only mental steps (which are excluded as abstract ideas, although the plaintiffs had also argued that such claims are patents on thought itself and violate First Amendment freedom of expression). The method claims were held invalid based on the machine-or-transformation approach[125] applied by the en banc Federal Circuit in *Bilski,* but should theoretically fare no better under the Supreme Court's *Bilski* decision. Specifically, the District Court rejected Myriad's argument analogizing the 'analyzing' steps to the 'determining' steps of the Prometheus claim that was upheld by the Federal Circuit (before the *Bilski* remand), distinguishing the Prometheus claims as having been construed 'to include the extraction and measurement of metabolite concentrations', whereas the claims at issue 'are directed only to the abstract mental processes of "comparing" or "analyzing" gene sequences', particularly as unchallenged dependent claims recited more transformative steps.[126] (The District Court thus did not reach either the insignificant post-solution activity inquiry or pre-emption analyses, although it treated restrictions of the claims to human isolated DNA as merely further specifying the subject to be analyzed, similarly to field of use restrictions on method claims that do not supply eligibility.[127]) During oral argument in the appeal, the method claims were addressed almost as an afterthought, and were alternatively analogized to or distinguished from the claims upheld in Prometheus.[128] They are unlikely to be upheld on appeal, particularly given their broad construction by the District Court.

Returning to the theory of eligibility and contributions, the locations of the *BRCA* genes are clearly natural phenomena (medical facts). If those locations and sequences were treated as prior art (as is required by the existing eligibility doctrine), it would be apparent that little or no creativity went into isolating the genetic DNA or identifying their sequences, particularly given the advanced state of genetic technologies at the time, or into using them for comparison once the sequences were known and the molecules were isolated. Accordingly, the claims for both the isolated sequences and the methods of comparing them (which do not recite any

---

125 See *Association for Molecular Pathology et al.* v. *United States Patent and Trademark Office, Myriad Genetics, et al.* 702 F. Supp. 2d 181 at 233–7, (S.D.N.Y. 2010), appeal docketed, No. 2010–1406 (Fed. Cir. June 22, 2010).

126 *Association for Molecular Pathology et al.* v. *United States Patent and Trademark Office, Myriad Genetics, et al.* 702 F. Supp. 2d 181 at 234, (S.D.N.Y. 2010), appeal docketed, No. 2010–1406 (Fed. Cir. June 22, 2010). (citing *Prometheus,* 581 F.3d at 1347). See ibid. at 235.

127 See ibid. at 235; *Bilski* v. *Kappos*, 130 S. Ct. 3218 at 3231 (2010).

128 See, e.g., Dutra, T. (2011), 'Myriad, ACLU Battle Renewed in Fed. Cir. Oral Arguments with DOJ Joining the Fray,' *Patent, Trademark & Copyright Journal*, **81**, 1–2.

specific steps beyond performing the analysis, detection, or comparison and do not specify any means for doing so) should not have been considered eligible inventions. And it should also be apparent that such claims are necessarily obvious, just as pharmaceutical compound claims are held to be obvious if a 'lead compound' has been identified in the art and only routine methods are needed to identify its function.[129]

Unfortunately, the decision of the District Court did not and the forthcoming decision of the Federal Circuit is unlikely to provide meaningful analysis of the degree of creativity involved in any claimed application of the sequences, or of how the functions of the isolated DNA molecules differ from or improve upon the functions performed by naturally occurring DNA. Nor is the case likely to provide a more refined exposition of the machine-or-transformation approach to method claims, or of the centrality of various physical implementation steps to the claims. However it comes out, the case is therefore unlikely to provide adequate guidance for assessing the eligibility of the many future applications of new discoveries of nature that will be claimed.

Finally, if the claims are invalidated, it is possible that legislative action will result seeking to overturn the outcome to restore eligibility. Lacking adequate judicial exposition of the theory of eligibility and responding to political pressures, Congress would also be unlikely to articulate a clear approach to eligibility. Rather, Congress is more likely to codify the eligibility of specific excluded subjects – such as isolated genetic sequences – and to leave the theory of eligibility unresolved, even if many inventions restored to eligibility should also necessarily be found obvious under a prior art contribution approach.

Unlike in the United States, the attack on the Myriad gene patents in Europe proceeded even more clearly on deontological moral grounds. Various French public health organizations and various national ministries and genetics societies and centers initiated opposition proceedings in the European Patent Office (EPO) against three of the Myriad patents,[130] leading to the revocation of one and the significant limitation of two others.

---

129 See, e.g., *Daiichi Sankyo Co., Ltd.* v. *Matrix Labs., Ltd.*, 619 F.3d 1346 at 1352–54 (Fed. Cir. 2010); *Altana Pharma AG* v. *Teva Pharmaceuticals USA, Inc.*, 566 F.3d 999 at 1006–09 (2009); and *Eisai Co. Ltd.* v. *Dr. Reddy's Labs., Ltd.*, 533 F.3d 1353 at 1359 (Fed. Cir. 2008).

130 See, e.g., Paradise, J. (2004), 'European Opposition to Exclusive Control Over Predictive Breast Cancer Testing and the Inherent Implications for U.S. Patent Law and Public Policy: A Case Study of the Myriad Genetics' BRCA Patent Controversy', *Food and Drug Law Journal*, **59**, 133, at 135–9; and Farnley, S., P. Morey-Nase and D. Sternfeld (2004), 'Biotechnology – A Challenge to the Patent System', *Current Opinion in Biotechnology*, **15**, 254, 255–6.

(The original application disclosed an incorrect sequence, and later corrections to the disclosure post-dated the publication of the correct sequence in accessible scientific databases so that the claims to the entire gene sequence were invalid for lack of novelty.[131]) Nevertheless, the EPO upheld the patent on partial BRCA1 gene sequences used as probes or vectors, rejecting arguments that the sequences were immorally obtained from cells without the consent of the donors, violated 'ordre publique' given their importance to public health, and lacked 'industrial application' given that the probes and vectors were primarily used for cloning or identification of mutated genes.[132]

The concept of industrial application has received an expansive interpretation, including industrial methods of production for uses that could be considered non-industrial and non-technological (even if patents for such uses are otherwise excluded from the patent system). Thus, the EPC of 1973 specifically excluded (as a 'legal fiction,' although it 'seemed actually to be based on socio-ethical and public health considerations') methods for diagnosis or treatment by therapy or surgery of humans or animals from being considered inventions having industrial application.[133] In contrast, the EPC of 2000 simply prohibited such patents for treatment and diagnostic methods, recognizing that such methods may be 'within the commercial sphere' even if human bodies are not.[134]

---

131 See, e.g., *Decision Revoking the European Patent* (art. 102(1), (3) EPC) (Eur. Pat. Office May 17, 2004) (revoking European Patent No. 0699754); *Decision of the Technical Board of Appeal* T1213/05 -3.3.04 (arts. 52(2), 53(a), 54, 56, 57, 83, 84, 87 to 89, 111(1),112(1)(a), 123(2)(3) EPC) (European Patent Office Sept. 27, 2007) (reviewing the Interlocutory decision of the Opposition Division of the European Patent Office of 19 September 2005 concerning European patent No. 0705902, maintained in amended form). See generally Matthijs, G. (2006), 'The European opposition against the BRCA gene patents', *Familial Cancer*, **5**, 95.

132 *Decision of the Technical Board of Appeal* T1213/05 -3.3.04 (art. 52(2), 53(a), 54, 56, 57, 83, 84, 87 to 89, 111(1),112(1)(a), 123(2)(3) EPC) (European Patent Office Sept. 27, 2007) at ¶¶ 43–57 (addressing ethical concerns, and interpreting EPC art. 52(2)(a) and EPC Rule 23(e)(2)); Ibid. at ¶¶ 60–70 (addressing industrial applicability and interpreting EPC Rule 23(e)(3)). See generally Llewelyn, Margaret (2005), 'Schrodinger's Cat: An Observation on Modern Patent Law', in Drahos, Peter (ed.), *Death of Patents*, Lawtext Pub. Ltd, 39–45 (discussing application of the Rule 53(a) morality exception, including its limited application to health and environmental risks).

133 *Nicotinic acid compositions for treating hyperlipidemia*, Case G 0002/08, EPO Enlarged Board of Appeal at ¶ 5.3 (2010) (citing EPC art. 52(4) (1973). See ibid. at ¶ 5.5.

134 See EPC art. 53(c) (2000).

Similarly, the 1998 *European Biotechnology Directive* prohibits patents that would violate 'ordre publique' or morality, and provides a non-exclusive list of things that cannot be patented, which includes processes for cloning humans and commercial uses of human embryos.[135] In particular, the Directive excludes patents for genetic sequence discoveries alone, while authorizing patents for isolated genetic sequences 'even if the structure of that element is identical to that of a natural element'.[136] This approach clearly differs from that in the United States, given that the Supreme Court precedents would require a markedly different function for chemical structures that are either identical or similar to the natural sequences from which they are derived. The Directive also finds that such sequences are industrially applicable once a concrete application for them is identified, although limits protection to the disclosed use.[137] This approach is similar to that in the United States, based on the 'new and useful' language of Section 101 of the United States *Patent Act* that precludes patents on genetic sequences until a significant utility has been identified for them.[138]

CONCLUSION

Given the substantial incentives to seek patents on applications of newly discovered but categorically ineligible science, nature, and ideas at the forefront of medical and biotechnological research and development, we can continue to expect disputes over patent eligibility to arise. Given the

---

135 See *Council Directive* 98/44/EC, art. 6, 1998 O.J. (L 213) (EC) (*European Biotechnology Directive*).

136 *Council Directive* 98/44/EC, art. 6, 1998 O.J. (L 213) (EC) (*European Biotechnology Directive*), art. 5(2). See also ibid. arts. 3(1) & 3(2).

137 See article 5 of the *Council Directive* 98/44/EC, art. 6, 1998 O.J. (L 213) (EC) (*European Biotechnology Directive*); *Eli Lilly and Co.* v. *Human Genome Sciences, Inc.*, [2008] EWHC 1903 (High Court of Justice – Patents Court) at ¶¶ 177–237 (interpreting article 5 of the *Biotechnology Directive*, U.K. law, and EPO decisions); *Monsanto Technology LLC* v. *Cefetra BV and others*, No. C-428/08, ¶¶ 33–50, European Court of Justice (6 July 2010). See generally Van Overwalle, Geertrui (2005) 'Legal and Ethical Aspects of Bio-Patenting: the EU Biotechnology Directive', in Drahos, Peter (ed.), *Death of Patents*, Lawtext Pub. Ltd, 214–25; generally Llewelyn, Margaret (2005), 'Schrodinger's Cat: An Observation on Modern Patent Law', in Drahos, Peter (ed.), *Death of Patents*, Lawtext Pub. Ltd, 28–32, (discussing industrial applicability and use in the context of gene patents).

138 35 U.S.C. § 101. See *In re Fisher*, 421 F.3d 1365 at 1369–78 (Fed. Cir. 2005). See also *Eli Lilly and Co.*, v. *Human Genome Sciences, Inc.*, [2008] EWHC 1903 at ¶¶ 218–24 (citing *In re Fisher* and *Brenner* v. *Manson*, 383 U.S. 519 at 534–5 (1966)).

long-standing historical normative commitments to protecting this public domain from piecemeal encroachment through wholesale or more limited retail patent claims to those discoveries, we can also expect those disputes to remain hot. And given the accelerating pace of scientific and technological discoveries, we can expect to confront these complex issues of eligibility in regard to a wide range of new and important products and processes, such as personalized medicine, computational genomics, synthetic agriculture, and nanotechnology.

We therefore need to develop a greater degree of understanding of (and greater consensus regarding) the degree of creativity in the applications of such newly identified, fundamental knowledge that *should* support the grant of patent rights for synthetic biological, chemical, mechanical, and digital products and processes. We know that the courts have struggled with these issues, seeking to distinguish *Chakrabarty*'s eligible synthetic bacteria from *Funk Brothers*' ineligible synthetic combinations, and *Bilski*'s ineligible hedging methods from *Diehr*'s eligible chemical treatment methods. Determining the required creativity in turn will help to determine whether we view the synthetic creations and new uses for new scientific and medical discoveries as markedly different or as non-analogous to the things and functions from which they derive. We need to chart a new and clearer relationship between newly discovered knowledge of nature and medicinal facts, and between new and synthetic applications of such knowledge.

Acknowledging the confusion in the existing doctrine and recognizing the prior art status (at least in American law) of categorically excluded subject matter and the importance of preserving the public domain is the first step towards reasoned development of approaches and better resolution of the conflicting issues. In our increasingly integrated world, we will also need to expand the dialog and to address the lack of harmonization in 'contribution' approaches. In doing so, we will have to focus on both the utilitarian and the deontological moral concerns that are involved.

CODA

As this book goes to press, the Supreme Court of the United States is receiving briefs in the *Prometheus* case; the Federal Circuit, over a vigorous dissent and with an important concurring opinion, has reversed the District Court's holding that the product claims were ineligible while affirming that the method claims are ineligible in the *Myriad* case, and a petition for panel rehearing has been denied.[139] Constitutional arguments about the limits of the patent power are being raised in *Prometheus*;[140] and if certiorari is denied or the decision of the Federal Circuit is affirmed by the Supreme Court of the United States, the constitutional claims in *Myriad* may be resurrected. Another pending case in the Supreme Court of the United States (addressing copyrights) will determine whether the Congress has the power to withdraw knowledge from the public domain,[141] which might then further support changing the requirement to treat new scientific discoveries as prior art notwithstanding its historic significance and potential constitutional status.[142] The Congress has recently enacted revisions to the patent law that, as noted above, will treat tax liability methods as prior art for obviousness and a prohibition on issuing patents for human organisms. Controversy around patent eligibility and its relationship to patentability thus continues to swirl.

The Petitioners in *Prometheus* have focused their attack by emphasizing the broad, preemptive effects of the claimed methods on the natural correlation that was discovered and the token and conventional nature of the claims' data-gathering administration and determination steps.[143] Even if these challenges are successful, it is unlikely that the Court will provide useful guidance regarding the permissible scope of claims applying ineligible discoveries or regarding what makes such data-gathering steps insignificant or peripheral to the claimed invention. The Court is unlikely to recognize and affirm the importance of prior art treatment of the new discovery for eligibility, or to establish either limiting rules or standards having clear application[144] that specify the required degree of creativity, and is likely to prefer to leave such guidance for case-by-case development over time. Further, the Court in *Bilski* treated ineligible discoveries as

---

139 See *Mayo Collaborative Services* v. *Prometheus Laboratories., Inc., cert. granted,* __ U.S. __, 2011 WL 973139 (U.S. June 20, 2011) (No. 10–1150); *Association for Molecular Pathology et al.* v. *United States Patent and Trademark Office, Myriad Genetics, et al.,* __ F.3d __, 2011 WL 3211513, (C.A. Fed. (N.Y.)), 99 USPQ 2d 1398, at *15–*23 (Fed. Cir. July 29, 2011).

140 See Appellate Brief (2011), 'Brief of Nine Law Professors as *Amici Curiae* in Support of Petitioners,' *Prometheus Laboratories, Inc.* v. *Mayo Collaborative Services.,* No. 10–1150, September 9.

141 See *Golan* v. *Holder*, 609 F.3d 1076 (10th Cir. 2010), 501 F.3d 1179 (10th Cir. 2009), *cert granted,* 131 S.Ct. 1600 (2011).

142 See generally Sarnoff, J. (unpublished draft 2011), 'Constitutional Hard Limits on the Patent Power'.

143 See Appellate Brief, 'Brief for Petitioners', at 18–19, *Mayo Collaborative Services.* v. *Prometheus Laboratories, Inc.,* No. 10–1150 (Sept. 2, 2011).

144 See Duffy, J. (2009), 'Rules and Standards on the Forefront of Patentability,' *William. & Mary Law Review*, **51**, 609–53.

'exceptions', and in light of the perceived legislative desire to extend the patent system broadly preserved these exceptions only because of their long pedigree (as a matter of *stare decisis*).[145] In contrast, as noted in dissent by the Chief Judge of the predecessor to the Federal Circuit, it is wrong to treat an exclusion from eligibility under the *Patent Act* as an 'exception', as '[d]efining patentable subject matter is the *raison d'être* of' the statutory provision.[146]

The Federal Circuit's principle opinion in *Myriad* (written by Judge Lourie) focused on the 'change in the composition's identity compared to what exists in nature', and found that the claimed isolated DNA 'exist in a distinctive chemical form ... from DNAs in the human body', and distinguished isolated DNA from purified DNA based on chemical manipulation ('chemically cleaved from their chemical combination with other genetic materials' by breaking 'covalent bond[s]').[147] Accordingly, Judge Lourie held that isolated DNA has a 'markedly different chemical structure', and emphasized the 'longstanding practice' of the USPTO to find isolated DNA an eligible invention.[148] As had been argued by some medical association amici[149] and as reemphasized in the Plaintiff's Petition for Rehearing,[150] however, isolated DNA with non-coding sequences having been excised and covalent bonds having been cleaved (including single stranded DNA identical to synthetically produced cDNA) exist naturally in cells outside of the chromosomally bound DNA. '[A]ll of the fragments that make up the BRCA1/2 genes can be found in natural maternal blood with the covalent bonds having been broken and the same chemical terminations as in Myriad's claimed isolated BRCA1/2 DNA.'[151]

In contrast to the opinion for the Court, the concurring opinion by Judge Moore recognized that '[e]ven though an invention did not previously exist

---

145 See *Bilski* v. *Kappos*, 130 S.Ct. 3218 at 3225 (2010).

146 *In re Alappat*, 33 F.3d 1526, 1553 n. 13 (Fed. Cir. 1994) (en banc) (Archer, C.J., dissenting).

147 *Association for Molecular Pathology et al.* v. *United States Patent and Trademark Office, Myriad Genetics, et al.*, __ F.3d __, 2011 WL 3211513, (C.A. Fed. (N.Y.)), 99 USPQ 2d 1398, at *17–*18.

148 Ibid. at *18–*19.

149 See Appellate Brief (2010), 'Brief for American Medical Association et al. as *Amici Curiae* in Support of Plaintiffs-Appellees and Arguing for Affirmance', at 17, *Association for Molecular Pathology et al.* v. *United States Patent and Trademark Office et al.* (Fed. Cir.) (No. 2010–1406), December 6.

150 See Appellate Brief (2011), 'Plaintiffs-Appellees' Petition for Panel Rehearing', at 3–7, *Association for Molecular Pathology et al.* v. *United States Patent and Trademark Office et al.* (Fed. Cir.) (No. 2010–1406), August 25.

151 Ibid.

in nature in exactly the claimed state, however, does not automatically mean it is patentable subject matter'.[152] Judge Moore thus based her decision to uphold the claims purportedly directed only to cDNA (having introns removed from the genetic sequence) were eligible, as cDNA sequences 'are the creation of man' and 'have a distinctive name, character, and use, with markedly different chemical characteristics' from the naturally occurring RNA sequences that they mirror.[153] In contrast, notwithstanding their '"enlargement of the range of utility" as compared to nature', Judge Moore upheld the claims for isolated DNA having the same sequence as chromosomal DNA only because (at least for DNA sequences that include 'most or all of a gene') the USPTO 'has allowed patents on purified natural products for centuries. There are now thousands of patents with claims to isolated DNA... [and] we must be particularly wary of expanding the judicial exception to patentable subject matter where both settled expectations and extensive property rights are involved.'[154] Although Judge Moore's analysis more carefully evaluates the functional considerations, it may expand the controversy by suggesting that such claims would be ineligible but for prudential concerns about investment. It also fails to supply any guidance regarding how much and what kind of additional function is sufficient for eligibility, providing a conclusion without stating a useable standard.

As Judge Moore noted, Judge Bryson's dissenting opinion actually supports the concurrence with regard to DNA sequences of small gene fragments, as Judge Bryson's analogy to the eligibility of a baseball bat based on selection of the parts of the wood would similarly apply to these sequences.[155] Judge Bryson also supported the eligibility of claims to cDNA. However, Judge Bryson rejected that chemically cleaving covalent bonds 'turns the isolated genes into "different materials"' any more than collecting portions of a wild plant 'by chemical means or by scissors' makes the portion a 'human-made invention'.[156] The dissenting opinion notes that changes to the 'nature, form, and use' of natural products may result in 'manmade manufacture[s]' that are eligible, but would have held that 'the structural differences [of isolated DNA]... are irrelevant to the claim limitations, to the functioning of the genes, and to their utility in their isolated form.... Indeed, that identity of function in the isolated gene is the

---

152 *Association for Molecular Pathology et al.* v. *United States Patent and Trademark Office, Myriad Genetics, et al.*, 653 F.3d 1329, 2011 WL 3211513, (C.A. Fed. (N.Y.)), 99 USPQ 2d 1398, at *24.

153 Ibid. at *29.

154 Ibid. at *30–*31 (citation omitted).

155 Ibid. at *30.

156 Ibid. at *39, *41.

key to its value.'[157] The dissent also appears to have failed to appreciate that no structural difference may actually exist, but in any event its approach also fails to supply a theory of what makes for a change in function that is sufficient for patent eligibility.

It is unclear whether the Supreme Court will ultimately hear the case. But two things are sure. The controversy will remain, even after the legal issue of the eligibility of isolated DNA sequences is settled. And we need better guidance about what role eligibility doctrine should play, and thus what rules or standards should be imposed to distinguish eligible inventions from ineligible applications of new scientific and natural discoveries.

157 Ibid. at *42.

4. Patent law, the emerging biotechnologies and the role of language in subject-matter expansionism

Graham Dutfield

This short chapter is about science, patent law and the use of language that supports the extension of patent claims ever deeper into the realms of nature. By language I refer in particular to the use of figures of speech, terminologies and epistemologies that both express and support powerful explanatory and justificatory conceptual systems. Undoubtedly, chemical, informational and mechanistic ways of understanding life have all been enormously helpful to scientists, as are the metaphors and analogies that frame their verbal and written forms of expression. The point of the chapter is not to undermine them but to examine critically what implications they have for patent law and policy, in particular their consequences for the positioning of boundaries between the patentable and the unpatentable.

From the mid nineteenth century, patents were regularly being granted on chemical substances in those countries, like the United States, the United Kingdom and France, that had no statutory chemical exclusions.[1] In the early twentieth century, patents claiming isolated and purified natural compounds were allowed or else found by courts to constitute acceptable subject matter. Since the 1970s an increasing number of jurisdictions have granted patents on micro-organisms, cell cultures, seeds, plants, animals and genes. Patenting in the life sciences is often criticised for inappropriate expansionism.

I would argue that some but not all such criticisms are justified, not least because the application of what philosopher Michael Ruse calls 'root

[1] Dutfield, Graham (2009), *Intellectual Property Rights and the Life Science Industries: Past, Present and Future*, Singapore: World Scientific.

metaphors'[2] has served to erase boundaries (to use a territorial metaphor) that should exist, at least in the context of patent claims. This is especially so when metaphor is put forward not as metaphor but as literal truth, so that a *like* gets treated as an *is*. (Perhaps we could call this 'the metaphoristic fallacy'.) In other words, a claim that A is like B in certain helpful respects becomes A is B, or is a subset of B. Modern English is replete with examples of this happening, and life sciences discourse is particularly susceptible to this happening unnoticed. Metaphor and analogy are central to communication and understanding and we are totally reliant upon them. However, they can also mislead, or persist beyond their useful lifespan. The question is whether the metaphors and analogies now being deployed by synthetic biologists and patent applicants are sufficient to legitimise the grant of *product* patents in this new field. Or are they still misleading and inadequate? The purpose of this chapter is to address this question.

I WHAT IS SYNTHETIC BIOLOGY?

Synthetic biology is an emerging area of both tremendous opportunity and consternation. There is a wide range of potential applications. Apart from health, a whole host of environment-related possibilities has arisen, for example in relation to sewage treatment, bioremediation, biogas production and the production of fuels and industrial chemicals like ethanol and bio-diesel from biomass. In recent years, interest has grown in the idea of using metabolically engineered micro-organisms for a number of roles, including nitrogen fixation, carbon storage, the production of industrial polymers otherwise made from oil, and the generation of hydrogen and other fuels out of sugars and cellulose.

To date, micro-organisms are the raw materials for much of this work, and for very good reason. The air we breathe is, to a large extent, a microbial product. Plants start off with carbon dioxide, water and sunlight, and after some creative combining and splitting end up with sugar for energy while releasing oxygen into the air. Animals then take that oxygen and use it for their own respiration and metabolism. Microbes are far more versatile than this. Eons ago cyanobacteria made the Earth safe for animal life by releasing so much oxygen that life able to use that oxygen to generate energy could come into being and thrive. Life at the microbial level thrives just as happily in the oxygen deficient atmosphere of billions of years ago as in the aerobic gassy soup of the present age. If the oxygen supply cuts out

2 Ruse, Michael (2010), *Science and Spirituality: Making Room for Faith in the Age of Science*, New York: Cambridge University Press.

suddenly some yeasts can instantly shift their energy-generating strategy from respiration to fermentation, an effective albeit less efficient one. This is of course a trait that all beer and wine drinkers should be grateful for. Microbes can also cope in hostile environments such as mid-oceanic hydrothermal vents and hot springs.

Like all life, microbes thrive in conditions they find hospitable and will die off if these become more hostile. Here evolution steps in. Eventually some microbes will find ways to make the most of the situation. For higher life forms, evolution needs time, and lots of it. At the microbial level this may not be the case at all. Antibiotic resistance, for example, can set in rather quickly. In this case, it is not to our advantage. But evolution can be both accelerated and directed for industrial purposes. In essence this is what humans, through animal and plant domestication, have been doing for over 10 000 years, albeit not consciously for most of that time. One of synthetic biology's great promises is that it can do it much more quickly and in a far more predictable and controlled fashion than ever before. In this regard, synthetic biology claims to represent a great leap forward, at least for investors. However, for regulatory and intellectual property purposes it is in the interests of practitioners to argue that synthetic biology is no more than a cumulative advance on what is already being done. Consequently, only existing rules and doctrines need apply, albeit in creative ways if the need arises.

II METAPHOR AND ANALOGY IN THE LIFE SCIENCES

Metaphors and analogies are integral to language and therefore to verbal and written communication.[3] Indeed, they can hardly be avoided: our ability to reason from incomplete evidence or information would be severely impoverished without them. Metaphors are words used in place of another that are literally unrelated but share some correspondence to the one being substituted for. The replacement words are selected either for practical explanatory purposes or for aesthetic ones (as in poetry). Analogies may include metaphors, but they normally entail much more than the replacement of one word with another. Frequently we explain a phenomenon, such as the way that something works, by reference to something else that is unrelated. The more complicated the phenomenon, the more likely

---

[3] Lakoff, George and Mark Johnson (1980), *Metaphors We Live By*, Chicago: Chicago University Press.

we are to have to resort to analogy for us to make sense of it. For example, in patent cases life forms may be analogised to chemical compositions or human-made machines despite the fact that they have autonomy, order and the power to metabolise and reproduce without human intervention (see below).

Metaphor usage can become highly controversial. In the present century, Steven Rose comments on 'the power of technological metaphor in biology', one consequence of which is that 'living systems become analogized to machines'.[4] About this: more later. Suffice it to say here that while not self-evidently pernicious (indeed it has in many ways proved helpful), for patent attorneys and biotechnologists this metaphor is rather convenient since new, unobvious and useful living things can thus more easily be cast as manufactures. Clearly, analogies of the kind indicated by Rose are quite long-standing and affect our language and thinking. The point is that they should not be taken for granted and that just as they can helpfully explain and illuminate the complex and obscure, they can also mislead and deceive.

A Explanation in Biology: Chemistry, Mechanism, Information and Systems Engineering

Scientifically, commercially and also in terms of patenting, much hangs on the impression that synthetic biology is an exact science. This was true also for the first generation of genetic modification that started with recombinant DNA techniques. The latter is still commonly assumed to be precise and predictable in ways that conventional techniques like classical plant breeding are not. Choosing the right metaphors, analogies and discourse is very important in creating the right impression. This leads us to a discussion on chemical, mechanistic and informational ways of explaining life, which thus far have only been alluded to.

B Life as Chemistry

A number of discoveries in the converging sciences of biology and chemistry, starting from the late eighteenth century, finally refuted some very long-held misconceptions that were holding back these sciences from emulating the amazingly steep learning curves achieved in sixteenth- and seventeenth-century astronomy, mathematics and physics. Thus, vitalism[5]

---

4 Rose, Steven (2005), *Lifelines: Life Beyond the Gene*, London: Vintage, 19.

5 Vitalism is the doctrine that living things possess some special principle that makes them more than the sum of the chemicals they are composed of and not fully explicable by the laws of chemistry or physics.

in its most extreme form, according to which only living things could produce organic chemicals,[6] was undermined by Antoine Lavoisier, who began to show that the new discipline of chemistry, which he contributed to founding, offered adequate means to describe the key processes taking place in living organisms, such as respiration. Its credibility was further weakened when scientists began to make these substances in the lab. In 1828, Friedrich Wöhler accidentally synthesised, albeit partially, a naturally-occurring biochemical, urea.[7] In time, the chemical constitution of all living things and biological processes became ever more apparent and vitalism withered away, at least among the vast majority of scientists. It turned out that life was made of the same stuff that non-life is made of, nothing more, nothing less. In other words, living organisms were made of, to put it colloquially, bits of non-life. Eventually, this allowed for the possibility of making some of these bits in a laboratory, and of using living things in industrial processes for commercial ends. It also provided a powerful means to understand, describe and communicate biological processes.

C Life as Information

The idea that biological molecules and molecular processes can be defined in informational terms is quite old. But life-as-information *technology* goes back to the post World War II era, a time when biology became especially susceptible to information and computer science analogies, metaphors and definitions.[8] In the mid 1950s, physicist George Gamow conceptualised the cell as 'a storehouse of information' and also as 'a self-activating transmitter which passes on very precise messages that direct the construction of identical new cells'.[9] This suggested to Gamow that 'the continuity of all life on our planet depends on this information system contained in the tiny cell nucleus'. A historical account of molecular biology by Michel Morange suggests that information terminology was hard-wired into the discipline's language from its inception: 'molecular biology is a result of the encounter between genetics and biochemistry, two branches of biology that

---

6 Hence the artificial distinction we still draw between inorganic and organic chemistry.

7 Hunter, Graeme (2000), *Vital Forces: The Discovery of the Molecular Basis of Life*, London and San Diego: Academic Press, 56–9.

8 Kay, Lily E. (2000), *Who Wrote the Book of Life? A History of the Genetic Code*, Stanford: Stanford University Press.

9 Gamow, G. (1955) 'Information Transfer in the Living Cell', *Scientific American*, **193**, 70–78 at 70. Quoted in Kay, Lily E. (2000) *Who Wrote the Book of Life? A History of the Genetic Code*, Stanford: Stanford University Press, 154.

developed at the beginning of the twentieth century [. . .] Strictly speaking, *molecular biology is not a new discipline, but rather a new way of looking at organisms as reservoirs and transmitters of information.*'[10]

The life-as-information idea has obvious contemporary appeal. This is understandable given the digital nature of so many modern technologies, including those used by molecular biologists (as in the field of bioinformatics), and of course the apparent similarity between raw DNA sequence data and computer code. Scientists are thus encouraged to assume that information technology provides the best means for us to understand how life 'works'. Indeed, it hardly seems a stretch to regard molecular biology as a branch of information science with the enormous amount of genetic sequence data now available to be processed and analyzed. From this, many are inclined to take the next step, which is to understand cells and organisms as highly sophisticated information processing systems that are capable of being harnessed, reconstructed and improved for our benefit. According to one account, 'an organism's physiology and behaviour are dictated largely by its genes. And those genes are merely repositories of information written in a surprisingly similar manner to the one that computer scientists have devised for the storage and transmission of other information.'[11]

Terms like 'genetic code', 'translation', 'transcription' and 'messenger RNA' are well established.[12] Such language is popular in journalism too, hence the title of a 2007 *Newsweek* feature on synthetic biology: 'Life 2.0'.[13] Presumably, Life 1.0 is the now rather obsolete version comprising DNA that humans did not write! However, this language is based on what the synthetic biologists are themselves saying.

One of the key problems some critics have identified with all this information talk, especially in the context of genetics, is the woolliness and inconsistency in how some scientists use words like 'information' and (genetic) 'code', and a tendency to over-rely on their explicatory power.[14]

---

10 Morange, Michel (1998) *A History of Molecular Biology*, Cambridge: Harvard University Press, 1–2 (emphasis added).

11 *The Economist* (1999), 'Drowning in Data', *The Economist*, June 26, 97–8.

12 Van den Belt, Henk (2009), 'Philosophy of Biotechnology', in Anthonie Meijers (ed.), *Philosophy of Technology and Engineering Sciences*, Amsterdam and London: Elsevier, 1301–40.

13 Silver, L. (2007), 'Life 2.0', *Newsweek*, 4 June, 41–5.

14 Sarkar, Sahotra (1996), 'Biological Information: A Sceptical Look at Some Central Dogmas of Molecular Biology', in Sahotra Sarkar (ed.), *The Philosophy and History of Molecular Biology: New Perspectives*. Dordrecht: Kluwer Academic Publishers, 187–232, 187, quoted in Griffiths, P.E. (2001), 'Genetic

According to one especially critical commentator, 'there is no clear, technical notion of "information" in molecular biology. It is little more than a metaphor that masquerades as a theoretical concept and [...] leads to a misleading picture of possible explanations in molecular biology.' Similarly, Griffiths comments that 'the claim that biology "is, itself, an information technology" [...] is on a par with the claim that the planets compute their orbits around the sun or that the economy computes an efficient distribution of goods and resources'.[15]

D From Machine to Mechanism

Is a living thing a biochemical machine? The idea of organisms being nothing more than soulless self-copying automatons goes at least as far back as the materialist philosopher Thomas Hobbes, who held that even the human mind is little more than a calculating machine: 'reason is nothing but reckoning'.[16] This might now seem a bit outdated, but organism-as-machine language and thinking are still with us, especially in the study of brains and consciousness.[17] Admittedly, in the context of the cellular level, a word more commonly encountered in the scientific literature is not 'machine' but 'mechanism'.

One might suppose the notion of cellular mechanisms to derive directly from Cartesian or Hobbesian organisms-as-machines thinking, or else from Galileo or Newton, whose world was like a clockwork machine obeying universal physical laws. In truth, 'mechanics' had drifted away from its origins in the study of machines as early as the seventeenth century.[18] As to the present context, it is in fact misleading to derive 'mechanism' from 'machine', notwithstanding their common etymological ancestry. As William Bechtel explains in his history of modern cell biology:

> The key to the mechanistic approach was not the analogy of physiological systems to human made machines, but the quest to explain the functioning of whole systems in terms of the operations performed by their component parts [...]. Increasingly, biology became a science in which phenomena were explained

---

Information: A Metaphor in Search of a Theory', *Philosophy of Science*, **68**, 394–412 at 395.

15 Griffiths, P.E. (2001), 'Genetic Information: A Metaphor in Search of a Theory', *Philosophy of Science*, **68**, 394–412 at 395.

16 From Hobbes's *Leviathan*, quoted in Ball, Philip (2004), *Critical Mass: How One Thing Leads to Another*, London: William Heinemann, 17.

17 Ruse, Michael (2010), *Science and Spirituality: Making Room for Faith in the Age of Science*, New York: Cambridge University Press, 142.

18 Ibid., 53.

> by discovering the organized parts and operations by which a mechanism performed its function.[19]

An inference that one may draw from this differentiation is this: old organism-as-machine analogies fail in that, while human made machines are designed and produced for a purpose, the higher purpose of life forms is clearly a matter for philosophers and theologians. For biological scientists, postulating a purpose for any type of life form is decidedly tricky. Biological mechanisms are not designed; they just appear to have been designed. Beneficial mechanisms tend to persist, and in some cases have done so for eons. In actual fact, mechanism talk in biology is rather recent, only becoming common in the twentieth century.[20] In part this is because hitherto cell biology was not sufficiently advanced for such language to be of much service. The quality of microscopes has improved rather a lot in the last four centuries!

Of course, to the extent that the meaning of the word mechanism does not require its linkage to a human artefact or connote design, we may have ceased to deal in analogy but in a correct usage of the term.[21] *The New Shorter Oxford English Dictionary*'s 1993 edition is not entirely conclusive in this regard. Of the five definitions given, here are two which taken together suggest that the word's meaning has been stretched to some extent to justify biological usage:

1. The structure or way of working of the parts in a machine or natural system; the mode of operation of a process.
2. A system of mutually adapted parts working together (as) in a machine; a piece of machinery; a means by which a particular effect is produced.

In addition, 'mechanism' is also defined in the dictionary as 'the doctrine that all natural (esp. biological or mental) phenomena are produced by mechanical forces'. Overall, then, it seems as though calling a functional element of a cell a type of 'mechanism' is no radical semantic stretch, and is indeed a perfectly helpful application of the word and the assumptions behind its use.

---

19 Bechtel, William (2006), *Discovering Cell Mechanisms: The Creation of Modern Cell Biology*, New York: Cambridge University Press, 4.

20 Ibid., 3.

21 For discussion, see Ruse, Michael (2010), *Science and Spirituality: Making Room for Faith in the Age of Science*, New York: Cambridge University Press, 57–8.

E Engineering

The legacy of mechanical philosophy combined with the ambitions of synthetic biology encourages the adoption of engineering language.[22] In fact, discourse borrowed from engineering, particularly electrical engineering,[23] is now very common. It affects increasingly how the science is described and reported on, how it is carried out, *and of course how patent specifications and patent claims are going to be drawn up*. Since engineering in its various technological guises and electronics are very well established patenting fields, this is just as well if you are in the business of making money out of the science.

Some leading synthetic biologists, having to mitigate the inherent unpredictability of barely understood complex biological systems, are inspired by semiconductor engineering and applying very similar terminology. Accordingly, 'if individual transistors are the basic components of electronic circuits, then their biological equivalents are genes'.[24] Hitherto, as they claim, applying engineering approaches to molecular biology has not been possible owing to lack of standardisation in the genetic 'parts' available and a biology mind-set that tends to stress complexity as opposed to the simplification of biological systems desired by the synthetic biologists. Much of this is inevitable. As historian of science Evelyn Fox Keller explains, in the present context 'the practice of analysis and synthesis in conceptual terms is generally referred to as science; but when it is performed in terms of concrete physical components and involves literal reconstruction of the system in question, it is generally called engineering'.[25] Literal reconstruction of DNA means straight copying of an extant sequence. But with metabolic engineering, the challenge is far greater, involving the reconstruction of whole metabolic pathways from, say, a plant, to a completely different organism like a bacterium or yeast in order to scale up production of the metabolite. It is not just about reconstructing naturally occurring processes but also about rebuilding them in different

---

22 Brooke, John Hedley (1991), *Science and Religion: Some Historical Perspectives*, Cambridge: Cambridge University Press, 117.

23 For example, see Kitano, H. (2008), 'Synthetic Biology: A Brief Overview', *Science*, **295**, 1662–4; and Lok, L. (2002), 'Software for Signaling Networks, Electronic and Cellular', *Science's STKE*, **122**, PE11.

24 Baker, D., G. Church, J. Collins, D. Endy, J. Jacobson, J. Keasling, P. Modrich, C. Smolke and R. Weiss (The Bio Fab Group) (2006), 'Engineering Life: Building a FAB for Biology', *Scientific American*, June, 34–9.

25 Keller, Evelyn Fox (2009), 'What Does Synthetic Biology Have To Do With Biology?' *BioSocieties*, **4** (2–3), 291–302 at 292.

biological 'platforms' so as to make them more efficient; in a sense, to *improve* on nature.

This might appear to be an arrogant notion, but it is hardly without precedent. Europe has had a 'culture of improvement' for about 1000 years, the idea of improvement being linked to technology and technological change, and also to staking ownership claims to land by cutting down trees, building fences and planting seeds or grazing animals.[26] In the Enlightenment, improving nature became associated with scientific progress and subsequently with incipient conceptions of *intellectual* property rights.[27]

Does the analogy work?

Analogies work only to the extent that the similarities are significant and the differences are not. In fact, we cannot be sure that the differences between true engineering systems and modified microbes do not matter. An article in *Scientific American* suggests that the engineering analogy has its limits:

> Electrical and mechanical machines are generally self-contained [...] But synthetic biologists are mainly interested in building genetic devices within living cells, so that the systems can move, reproduce and interact with the real world. From a cell's point of view, the synthetic device inside it is a parasite. The cell provides it with energy, raw materials and the biochemical infrastructure that decodes DNA to messenger RNA and then to protein. The host cell, however, also adds a great deal of complexity.[28]

The key issue, of course, is whether the first generation of synthetic biologists are overconfident in their ability to eliminate complexity or at least complexity's effects. Much of their biology depends on the validity of the concept of modularity, a notion imported from engineering according to which parts are interchangeable, possessing intrinsic properties and performing specific functions. Is their conviction well founded or is it a matter of faith? Calvert suggests the latter:

> For some this reduction of complexity is not merely an instrumental aim, but is based on a faith that synthetic biology will ultimately lead to 'the elucidation of the underlying simplicity' of nature (Palsson, 2000: 1149). The pervasive idea of the simplicity of nature, and its connection to truth, is found throughout the

26 Friedel, Robert (2007), *A Culture of Improvement: Technology and the Western Millennium*, Cambridge and London: MIT Press.

27 Ritvo, Harriet (1996), 'Possessing Mother Nature: Genetic Capital in Eighteenth Century Britain', in John Brewer and Susan Staves (eds), *Early Modern Conceptions of Property*, London: Routledge, 413–26.

28 Gibbs, W.W. (2004), 'Synthetic Life', *Scientific American*, May, 75–81, 78.

history of biology (often in tension with ideas about its complexity), and perhaps the most striking example of this is the iconic image of the double helix, famously described by Crick and Watson in 1953.[29]

To sum up this part of the chapter, biologically trained mechanical engineers, chemical engineers, geneticists and information scientists are keen for us to envision life according to their chosen languages and epistemologies. Experience does show that metaphors and analogies can provide considerable guidance for researchers making very important discoveries they might otherwise never have made. The real issue is not whether they are wrong or misleading *in themselves*, but whether they are still *sufficient* at least (in the present context) for the purpose of claiming patent rights. Life remains mysteriously complex and increasingly seems to require ecological and interdisciplinary approaches for us to understand what it is and what it does.

III INTELLECTUAL PROPERTY AND SYNTHETIC BIOLOGY

We now turn to the patent system. Does synthetic biology represent a true technological discontinuity that renders obsolete reasonable objections that have been made to the inclusion of life forms as patentable inventions? And if not, are the metaphors and analogies used misleading us into thinking there are genuine inventions when there are not?

Before going further, the purpose of this discussion is not to attack the epistemologies and supporting metaphors and analogies deployed by life scientists; nor is it to demand unrealistically high standards of scientific error-avoidance. Hindsight is the best determinant of scientific accuracy. Patent examination and validity assessment are not tests of proper scientific theory or description, nor should they be. Rather, our aim is to question whether patents will be allowed in this field, not because what is claimed is a true invention, but because the metaphors and analogies employed make it seem as if what is described is an invention when it is not. In other words, how far have courts been misled by language?

Experience tells us that courts facing difficult patentable subject-matter decisions adopt one of at least three approaches. They may decide on the

---

29 Calvert, J. (2010), 'Synthetic Biology: Constructing Nature?' *The Sociological Review*, **58** (1), 95–112, 99; and Palsson, B. (2000), 'The Challenges of in Silico Biology', *Nature Biotechnology*, **18**, 1147–50 at 1149. (Other reference removed from quote.)

basis of certain tests, the fulfilment of which satisfies the invention criteria. For example, in the United States, the machine-or-transformation test may be used in deciding certain process claims. In Europe, the technical character test is applied, for example, to computer-implemented inventions. Otherwise, they may resort to an analysis of what certain key words and concepts mean: is this subject matter a patentable invention according to the meaning of this word or concept? They may also, or in addition, reason by analogy, especially if the subject matter at issue has previously not commonly been protected.

By illustration, let us look briefly at a couple of important North American cases. These are interesting in that both the United States and Canada require *physical* inventions to be machines, manufactures or compositions of matter, and yet both have accepted that life forms are patentable. What is interesting is that they have done so in quite different ways despite the very similar statutory language.[30]

The United States Supreme Court's famous *Diamond* v. *Chakrabarty* decision in 1980 concerned a bacterium modified to consume oil that was held to be patentable subject matter.[31] Micro-organisms, just like any inanimate units of matter, are of course made of chemicals. But one does not have to be a vitalist to question the validity of the assumption that a modified bacterium may be just as much an invented composition of matter as, say, a new small molecule pharmaceutical substance simply because both have a molecular structure that diverges to some degree from any known one. The autonomy, complexity and self-reproducibility of the microbe are in this way being treated as irrelevant: it has become just a chemical.

The Court also affirmed that such an 'invention' could be classed as a 'manufacture'. Over the years, manufacture has escaped from its original meaning as something made by hand to include things produced by machinery. Meanings of words inevitably evolve, as they should when need arises or common sense demands. But to call a living organism an article of manufacture neatly sidesteps the immense differences between inanimate chemical substances, whether or not they are fully characterised, and even the most 'simple' unicellular life forms.

In 2002, the Supreme Court of Canada decided, in the Harvard OncoMouse case, that a line should be drawn.[32] But this was not between the inanimate and the animate, but between the lower and the higher life form.

---

30 35 USC 101; *Patent Act*, R.S.C. 1985, c. P-4, s. 2.

31 *Diamond* v. *Chakrabarty*, 447 US 303 (1980).

32 *Harvard College* v. *Canada (Commissioner of Patents)* 2002 SCC 76.

In the latter case, analogising from the statutory language was deemed to be a stretch too far for the time being. In the words of Judge Bastarache, who delivered the Court's decision, 'it is far easier to analogize a micro-organism to a chemical compound or other inanimate object than it is to analogize a plant or an animal to an inanimate object'. Moreover, 'while a mouse may be analogized to a "manufacture" when it is produced in an industrial setting, the word in its vernacular sense does not include a higher life form'. Such line-drawing based on the perceived limits of conventional analogies, I would argue, makes far less scientific or legal sense than distinguishing between the animate and the inanimate.

It would appear that, in the United States and Canada, claiming a metabolically enhanced micro-organism in a patent will not present any difficulties, as in any other jurisdiction in which micro-organisms are already patentable, including Europe and East Asia. The chemistry analogy certainly helps in this respect, *at least as long as chemicals are deemed as inherently patentable subject matter*.

With respect to method and product claims to transplanted and improved metabolic pathways, these will most likely be deemed acceptable in light of long-established patent law statutory interpretation and jurisprudence. As for genetic sequences, both the United States and Europe have been allowing isolated DNA with disclosed utility or industrial application to be patented for well over two decades. In Europe, for example, the European Patent Office's Technical Board of Appeal affirmed in the mid 1990s that 'DNA is not "life", but a chemical substance which carries genetic information and can be used as an intermediate in the production of proteins which may be medically useful'.[33]

Notwithstanding recent rule changes in both jurisdictions and a 2010 court decision in the United States making the situation less certain (see below), this remains the practice. However, there is little doubt that objections to the patenting of isolated DNA sequences, according to which they are mere discoveries or products of nature, will not hold if claimed sequences include bases other than A, C, G and T, or else if the sequences are made of other nucleic acids than DNA. Indeed, allowing such sequences while denying the patentability of isolated ACGT sequences may be good policy if the objective is to encourage synthetic biology. According to Torrance, rejecting the patentability of isolated DNA sequence, as a

---

[33] *Howard Florey Institute/ Relaxin*. Decision of Technical Board of Appeal 3.3.1 dated 30 August 1994 (1995) *Office Journal of the EPO* 6: 373–455, 400.

court in United States District Court recently did,[34] but accepting genuinely new sequences devised by scientists as patentable inventions would likely encourage research to shift towards the development and application of truly original synthetic genes.[35]

We now move to whole organisms. The reasons for denying the patentability of life forms remain relevant albeit with some modification. However, product by process claims will be easier to justify. Moreover, product claims over modified cellular mechanisms would appear to be reasonable. But as long as debate continues as to the feasibility of the simplified organism approach, one should continue to be cautious in accepting the patentability of these metabolically enhanced organisms. Indeed, for synthetic biology, the real proof of concept requires the complete construction of a life form in the laboratory:

> SynBio scientists haven't quite proven that a cell is a kind of biochemical machine, and religious biologists like [Leon] Kass and [Francis] Collins hang on tightly to this uncertainty. Proof will come when the first discrete, self-maintaining, stable organic creature – Life 2.0 – is created from scratch in the lab.[36]

Despite the hype coming from certain quarters, we are simply not there yet. When we are, the engineering analogy presumably will finally fit perfectly, thereby ceasing to be an analogy. Synthetic biology *will be* a form of engineering. Life forms will truly become manufactured mechanical devices *at least in the artificial environments in which they will be placed.*

But does it follow that opponents of patents on life forms will have no more technical or conceptual grounds to support their position? Perhaps not. Even to a materialist like the present author, until such organisms cease to be animate – and how one will determine *that* is going to fascinate future scientists, philosophers and lawyers – the present objections will still carry weight. Life *is* different from non-life, and that is relevant for patent law. As for moral and religious objections to such patenting, that is another issue – and another chapter.

---

34 *Association for Molecular Pathology* v *U.S. Patent and Trademark Office*, No. 09 Civ. 4515, 2010 WL 1233416, (S.D.N.Y. Mar. 29, 2010); but this approach was overturned, in large part, in *Association for Molecular Pathology et al.* v. *United States Patent and Trademark Office, Myriad Genetics, et al.* 653 F.3d 1329, 2011 WL 3211513 (C.A. Fed. (N.Y.)), 99 USPQ 2d 1398.

35 Torrance, A. (2010), 'Synthesizing Law for Synthetic Biology', *Minnesota Journal of Law, Science & Technology*, **11** (2), 629–65 at 640.

36 Silver, L. (2007), 'Life 2.0', *Newsweek*, 4 June, 41–5.

PART III

Biobanks, bioinformatics and BioBricks

5. Standards for biobank access and intellectual property

Dianne Nicol and Richard Gold

The Human Genome Project (HGP) and follow-on collaborative studies have provided scientists with a vast amount of information about the human genome. Despite this, and despite an increasing focus on genome-wide analysis, there are still significant gaps in our understanding of how the human organism functions as a whole and how life choices and environmental factors interact with our pre-programmed genetic constituents. These are important questions, for which answers are vital if we want to create a healthier world for future generations. Biobanks are widely recognized as vital research tools in this quest.[1] They contribute by providing collections of human tissue linked with genetic information and other health information that permit researchers to pinpoint disease targets and perfect drug development.[2] There is a wide diversity in types and forms of biobank, ranging from small-scale disease-specific collections through to more recently established large-scale population resources.[3]

---

1 Organisation for Economic Co-operation and Development (OECD) (2006), *Creation and Governance of Human Genetic Research Databases*, Paris: OECD Publishing, 35, http://www.oecd.org/document/50/0,3343,en_2649_34537_37646258_1_1_1_1,00.html ('OECD Creation and Governance Report'). See also National Health and Medical Research Council (2010), *Biobanks Information Paper 2010*, Canberra: Commonwealth of Australia, http://www.nhmrc.gov.au/your_health/egenetics/practitioners/practitioners.htm ('NHMRC Biobanks Information Paper'); and Organisation for Economic Co-operation and Development (OECD) (2009), *Guidelines on Human Biobanks and Genetic Research Databases*, Paris: OECD Publishing, http://www.oecd.org/sti/biotechnology/hbgrd ('OECD Biobank Guidelines'), particularly Principle 1.A.

2 Organisation for Economic Co-operation and Development (OECD) (2006), *Creation and Governance of Human Genetic Research Databases*, Paris: OECD Publishing, http://www.oecd.org/document/50/0,3343,en_2649_34537_37646258_1_1_1_1,00.html ('OECD Creation and Governance Report'), 35.

3 The OECD Creation and Governance Report provides summary information on a representative group of biobanks (or human genetic research databases, as

All biobanks, irrespective of size and purpose, have the common feature of constituting tools that enable future research.[4] While this research utility is essentially a biobank's *raison d'être,* it is also the feature that raises the most serious ethical concerns, particularly those relating to consent to future use of personal information and tissue and privacy.[5] The consent process is outside the normal single project paradigm with which ethics committees are used to dealing. This is because biobank resources are intended to be used for research that may not even have been conceptualized at the time that consent was sought. Ethical standards must, therefore, reflect the need for unspecified consent in some circumstances, subject to stringent limitations.[6] Further, the linkage of personal information records and the capacity to re-identify tissue and personal information also necessitates close ethical scrutiny and the use of specific technological measures

---

it refers to them) in Chapter 2: Organisation for Economic Co-operation and Development (OECD) (2006), *Creation and Governance of Human Genetic Research Databases*, Paris: OECD Publishing, 35–43, http://www.oecd.org/document/50/0,3343,en_2649_34537_37646258_1_1_1_1,00.html.

4 National Health and Medical Research Council (2010), *Biobanks Information Paper 2010*, Canberra: Commonwealth of Australia, 7, http://www.nhmrc.gov.au/your_health/egenetics/practitioners/practitioners.htm ('NHMRC Biobanks Information Paper'). Principle 1.C of the OECD Biobank Guidelines: Organisation for Economic Co-operation and Development (OECD) (2009), *Guidelines on Human Biobanks and Genetic Research Databases*, Paris: OECD Publishing, http://www.oecd.org/sti/biotechnology/hbgrd ('OECD Biobank Guidelines').

5 See Kaye, J. (2005), 'Do We Need a Uniform Regulatory System for Biobanks across Europe?' *European Journal of Human Genetics*, **14**, 245–8; Principle 1.D. of the OECD Biobank Guidelines: Organisation for Economic Co-operation and Development (OECD) (2009), *Guidelines on Human Biobanks and Genetic Research Databases*, Paris: OECD Publishing, http://www.oecd.org/sti/biotechnology/hbgrd ('OECD Biobank Guidelines'). The complexities of balancing the benefits of data sharing with participant concerns are canvassed in the draft *Oxford Statement on International Data Sharing* 2010: http://helexoxford.com/content/oxford-statement-data-sharing. The need for a statement along these lines was established at an International Data Sharing Conference held in Oxford on 20–22 September 2010.

6 National Health and Medical Research Council (2010), *Biobanks Information Paper 2010*, Canberra: Commonwealth of Australia, 25–26, http://www.nhmrc.gov.au/your_health/egenetics/practitioners/practitioners.htm ('NHMRC Biobanks Information Paper'). Otlowski, Margaret (2009), 'Developing an Appropriate Consent Model for Biobanks: In Defence of "Broad" Consent', in Jane Kaye and Mark Stranger (eds), *Principles and Practice of Biobank Governance*, Cheltenham: Ashgate, 79–92.

to protect privacy.[7] While such measures are widely seen as essential, they inevitably increase the cost of biobanking. This, in turn, raises questions concerning the adequacy of funding in the establishment and ongoing operations of biobanks.[8] Although governments and benefactors provide significant levels of support to some of the large-scale biobanks,[9] a smaller number are funded from the private sector, which raises particular legal and ethics concerns.[10] The smaller-scale biobanks generally are likely to have fairly specific research goals, mostly undertaken in-house. However, the larger entities tend to have much grander aspirations. For example, UK Biobank exhorts people to participate with the following statement:

> UK Biobank is building a national treasure trove of health information, of an impressive scale, to be used by scientists in the future. The resource will be among the largest of its kind anywhere in the world and is likely to be the richest in terms of the available information it will provide for research.[11]

In addition to the broad research goals of the large-scale biobanks, one of the most significant emerging trends in the field is the formation of biobank networks and clusters, the key purpose of which is to facilitate collaborative research using aggregated biobank tissue collections and information.[12] One example is the European Biobanking and Biomolecular Resources Research Infrastructure (BBMRI), the purpose of which is to address issues of biobank fragmentation, insecurity of funding and incompleteness across Europe.[13] Another example is the P3G, a more broadly based international consortium, the stated aim of which is to:

---

7 National Health and Medical Research Council (2010), *Biobanks Information Paper 2010*, Canberra: Commonwealth of Australia, 45–47, http://www.nhmrc.gov.au/your_health/egenetics/practitioners/practitioners.htm ('NHMRC Biobanks Information Paper').

8 Ibid., 18.

9 For example, the United Kingdom Biobank is funded by the Department of Health, the Medical Research Council, the Scottish and Welsh Governments and other regional governmental bodies, together with two major charities, the British Heart Foundation and the Wellcome Trust. See http://www.ukbiobank.ac.uk/about/backing/funders.php.

10 Anderlik, M.R. (2003), 'Commercial Biobanks and Genetic Research: Ethical and Legal Issues', *American Journal of Pharmacogenomics*, **3**, 203–15.

11 United Kingdom Biobank, http://www.ukbiobank.ac.uk/assessment/takepart.php.

12 See Kaye, Jane (2009), 'Biobank Networks – What Are the Governance Challenges?' in Jane Kaye and Mark Stranger (eds), *Principles and Practice of Biobank Governance*, Cheltenham: Ashgate, 201–13 at 201.

13 See Yuille, M. et al. (2007), 'Biobanking for Europe', *Briefings in Bioinformatics*, **9**, 14–24 at 14. For further information on BBMRI see http://www.bbmri.eu/.

> encourage collaboration between researchers and biobankers; promote harmonization of information; optimize the design, set-up and research activities of population-based biobanks; and facilitate the transfer of knowledge and provide training to those working in the field.[14]

Principle 5.E of the *OECD Guidelines on Human Biobanks and Genetic Research Databases* ('*OECD Biobank Guidelines*') encourages biobanks to ensure interoperability to facilitate such exchanges. These Guidelines were approved by OECD member countries in 2009 after a period of extensive consultation. The purpose of the Guidelines is to provide guidance to policy makers and practitioners on the establishment, governance, management, operation, access, use and discontinuation of human biobanks and genetic research databases.[15] The Guidelines provide a comprehensive set of Principles, Best Practices and Annotations, the most relevant of which for the purpose of this chapter are found in the section relating to access and the section dealing with custodianship, benefit sharing and intellectual property.

To succeed in the goal of providing a rich resource for future biomedical research, biobanks need to build and maintain an ongoing trust relationship with participants and the wider community.[16] One consequence of the broad research aspirations of biobanks and the trend towards collaboration and sharing of biobank resources is an increased cross-border transfer of information and tissue. Involvement of industry partners is also likely to be required to facilitate the translation of research into clinical practice and the development of new drugs and therapies. This creates a significant problem for biobanks, because there is increasing recognition that this crucial trust relationship may be compromised if tissue and personal information is sent offshore and if there is commercial involvement and

---

[14] P3G, http://www.p3g.org/secretariat/index.shtml.

[15] Organisation for Economic Co-operation and Development (OECD) (2009), *Guidelines on Human Biobanks and Genetic Research Databases*, Paris: OECD Publishing, 1, http://www.oecd.org/sti/biotechnology/hbgrd ('OECD Biobank Guidelines').

[16] On the importance of trust see Levitt, M. and S. Weldon (2005), 'A Well Placed Trust? Public Perceptions of the Governance of DNA Databases', *Critical Public Health*, **15**, 311–21; Lemke, A., W. Wolf, J. Herbert-Beirne and M.E. Smith (2010), 'Public and Biobank Participant Attitudes towards Genetic Research Participation and Data Sharing', *Public Health Genomics* **13**, 368–77. On the importance of an ongoing trust relationship see Hunter, Kathryn and Graeme Laurie (2009), 'Involving Publics in Biobank Governance: Moving Beyond Existing Approaches', in Heather Widdows and Caroline Mullen (eds), *The Governance of Genetic Information*, Cambridge: Cambridge University Press.

patenting of research findings.[17] The ethical, legal and social implications of internationalization and commercialization of biobanking and use of biobank resources are recurring themes in biobank governance debates.[18]

Possibly the best way to protect privacy, simplify consent and maintain trust would be to keep all research use of biobank resources in-house but this largely defeats the purpose of their establishment. As such, ways need to be found to ensure best practice in respect of access to biobank resources and management of intellectual property, which requires a careful balancing of five aspects of biobanking that are liable to come into conflict. These are as follows:

- providing access for optimal research use of biobank resources;
- facilitating downstream translation of research into practice, which may, of necessity, require the involvement of the private sector and patenting;
- protecting the privacy and autonomy of participants;
- fostering participant and broader public trust in the internationalized and commercialized biobanking environment; and
- securing sufficient levels of funding to ensure ongoing biobank operations.[19]

The aim of this chapter is to analyse the mechanisms that need to be put in place by biobanks to balance these potentially conflicting aspects of biobanking, within the context of emerging best practice relating to other aspects of biomedical research. A set of six standards for biobank access and intellectual property policies is drawn from this analysis later in this

---

17 See, for example, Webster, Andrew, Nik Brown, Conor Douglas, Graham Lewis, Jane Kaye, Richard Tutton and Nick Williams (2008), *Public Attitudes to Third Party Access and Benefit Sharing: Their Application to UK Biobank*, Final report to the UK Biobank Ethics and Governance Council, University of York: Science and Technology Studies, University of York, http://www.egcukbiobank.org.uk/assets/wtx052208.pdf.

18 Commercialization, in particular, is often the focus of specific attention: see Organisation for Economic Co-operation and Development (OECD) (2006), *Creation and Governance of Human Genetic Research Databases*, Paris: OECD Publishing, 123–30, http://www.oecd.org/document/50/0,3343,en_2649_34537_37646258_1_1_1_1,00.html ('OECD Creation and Governance Report').

19 Others specifically refer to two of these challenges: protecting the biobank and its volunteer donors and making it available to researchers. See Fortin, S., S. Pathmasiri, R. Grintuch and M. Deschenes (2010), '"Access Arrangements" for Biobanks: A Fine Line between Facilitating and Hindering Collaboration', *Public Health Genomics*, 1–11, DOI. 10.1159/000309852.

chapter. Biobank guidelines and policies are then tested to see if they meet these standards. Particular attention is focused on the access and intellectual property arrangements of three of the best-known, large-scale biobanks, UK Biobank,[20] Generation Scotland[21] and CARTaGENE[22] in Quebec, Canada.

I TENSIONS BETWEEN DATA SHARING AND INTELLECTUAL PROPERTY

A Optimal Use of Research Data: The Mantra of Data Sharing

The biomedical research community is currently engaging in an active program of data sharing, particularly for large-scale datasets of aggregated, anonymised genetic data. The classic example of this data sharing paradigm is the 'Bermuda Rules' for access to HGP sequencing information, agreed to by all parties involved in the international public sequencing effort in 1996. There, it was agreed inter alia that all human genomic sequence information generated by centres funded for large-scale human sequencing should be freely available and in the public domain.[23] Subsequently, the same commitment to data sharing was made for other large-scale data-generating projects, including the HapMap Project,[24] the 1000 Genomes Project[25] and, more recently, the International Cancer Genome Consortium.[26]

Broad sharing of large reference databanks across the fields of biology and medicine was endorsed at a workshop in Toronto in 2009.[27] Funding agencies also support data sharing. For example, the 2008 US NIH *Policy for Sharing of Data Obtained in NIH Supported or Conducted Genome-Wide Association Studies* (GWAS) requires that datasets remain available

---

20 United Kingdom Biobank, http://www.ukbiobank.ac.uk/.

21 Generation Scotland, http://www.generationscotland.co.uk/.

22 CARTaGENE, http://www.cartagene.qc.ca/.

23 HUGO (1996), Summary of Principles Agreed at the First International Strategy Meeting on Human Genome Sequencing, http://www.ornl.gov/sci/techresources/Human_Genome/research/bermuda.shtml#.

24 International HapMap Consortium (2004), 'Integrating ethics and science in the International HapMap Project', *Nature Reviews Genetics*, **5**, 467–75.

25 The 1000 Genomes Project, http://www.1000genomes.org/.

26 The International Cancer Genome Consortium, http://www.icgc.org/.

27 Toronto International Data Release Workshop Authors (2009), 'Prepublication Data Sharing', *Nature*, **461**, 168–70.

to all investigators, unencumbered by intellectual property claims.[28] Translating these policy statements calling for broad sharing of data and tissue into practice is not, however, straightforward in the biobanking context.

B Facilitating Downstream Translation: Enter Intellectual Property

Participants in the medical biotechnology industry consistently state that patents are essential for the purpose of disseminating their technological innovations and for generating revenue.[29] Unless funding is available from governments or private benefactors, room must be left for the private sector to engage with the research sector to ensure that translation from biomedical research to product development occurs, and this will usually require patenting to provide some assurance of return on investment.

On this basis, it may well be in the public interest for biobanks to facilitate downstream commercialization and patenting. However, there are two important qualifiers to this conclusion. First, given the broad research goals of biobanks, the process of commercialization should be carried out in a way that does not interfere with other research uses of biobank resources. If not, the ways in which firms obtain and license patents may get in the way of promising areas of research. There are various ways in which this goal might be achieved. Perhaps the bluntest mechanism for attempting to achieve this end is to exclude certain types of subject matter from patenting. Patent law draws a distinction between discoveries and inventions. Currently, though, the generally recognized subject-matter threshold for patenting subject matter like DNA sequences is set quite low, requiring only that they are isolated and synthetically produced.[30] However, debate about acceptable limits on patentable subject matter is intensifying. For example, a bill that attempts to craft an express exclusion for DNA sequences and other biological materials is currently before the Australian

---

28 Principle V of the National Institutes of Health (NIH) (2003), *Statement on Sharing Research Data*, 12–13, http://grants.nih.gov/grants/policy/data_sharing/. See also Gitter, D. (2010), 'The Challenges of Achieving Open Source Sharing of Biobank Data', *Biotechnology Law Report*, **29** (6), 623–35.

29 For example, from the Australian biotechnology industry perspective, see Nicol, Dianne and Jane Nielsen (2003), *Patents and Medical Biotechnology: An Empirical Analysis of Issues Facing the Australian Industry*, Hobart: Centre for Law and Genetics Occasional Paper No. 6, 84–92.

30 See, for example, Secretary's Advisory Committee on Genetics, Health and Society (2010), *Gene Patents and Licensing Practices and their Impact on Patient Access to Genetic Tests*, Bethesda, MD: Secretary's Advisory Committee on Genetics, Health, and Society, 13, http://oba.od.nih.gov/oba/sacghs/reports/SACGHS_patents_report_2010.pdf.

Parliament.[31] There is also litigation on foot in both the US and Australia that, if successful, could set precedents for interpreting existing patent legislation in such a way as to exclude some upstream subject-matter.[32]

Uncertainty as to the practical effect, scope, legitimacy and implications of excluding DNA sequences and related subject matter has deterred law reform agencies from recommending this as a primary strategy for shielding researchers from liability for patent infringement. Rather, other options have been put forward on both the legal and the policy front. Legal strategies include clarification of the law relating to exemptions from infringement for research use and facilitating compulsory licensing.[33] From the policy perspective, it is frequently recommended that access agreements should require users of biobank resources not to interfere with other research users by making one or more undertakings. These might include: not applying for upstream patents; not enforcing patent rights against other research users; or enforcing patent rights by way of broad, inexpensive, non-exclusive licences.[34] There may also be scope for engaging in collaborative licensing strategies like patent pooling or patent clearinghouses.[35]

---

31 The Patent Amendment (Human Genes and Biological Materials) Bill 2010 (Cth) was tabled in the Australian Parliament in November 2010.

32 In Australia, an action was initiated in the Federal Court in July 2010 by Cancer Voices Australia and others against the Commissioner for Patents and others challenging some of the most notorious gene patents linked with hereditary breast cancer testing. See *Cancer Voices Australia et al.* v. *Myriad Genetics Inc. et al.* (2010), Federal Court of Australia, Statement of Claim, 8 June, and see also http://www.cancervoicesaustralia.org.au/news.htm. In the United States, the first instance decision in *Association for Molecular Pathology et al.* v. *United States Patent and Trademark Office, Myriad Genetics, et al.* 702 F. Supp. 2d 181, (S.D.N.Y. 2010), *appeal docketed,* No. 2010–1406 (Fed. Cir. June 22, 2010) invalidated US patents related to the same subject matter. The ruling was largely overturned on appeal – see *Association for Molecular Pathology et al.* v. *United States Patent and Trademark Office, Myriad Genetics, et al.* 653 F.3d 1329, 2011 WL 3211513 (C.A. Fed. (N.Y.)), 99 USPQ 2d 1398.

33 See Recommendations 13–1, 26–1 to 26–3 and 27–1 of the Australian Law Reform Commission (2004), *Genes and Ingenuity. Report 99*, Sydney: Australian Government; Recommendations 11 to 13 of the Senate Standing Committees on Community Affairs Reference Committee (2010), *Gene Patents*, Canberra: Australian Parliament, http://www.aph.gov.au/senate/committee/clac_ctte/gene_patents_43/report/report.pdf; and Recommendation 1.B (research exemption) of the Secretary's Advisory Committee on Genetics, Health and Society (2010), *Gene Patents and Licensing Practices and their Impact on Patient Access to Genetic Tests*, 13, http://oba.od.nih.gov/SACGHS/sacghs_documents.html#GHSDOC_011.

34 Organisation for Economic Co-operation and Development (OECD) (2006), *Guidelines for the Licensing of Genetic Inventions*, Paris: OECD; National Institutes of Health (1999), 'Principles and Guidelines for Recipients of NIH Research Grants and Contracts on Obtaining and Disseminating Biomedical

The second qualifier is that there should be some process of accountability. The bald statement that commercialization of the results of research using biobank resources will lead to new health care developments for the benefit of future generations needs to be capable of verification. There is some support for the notion that scientific publications reporting the results of research using biobank resources should be recorded and reported back to the biobank.[36] However, there has been less discussion of the desirability of reporting commercial developments arising from that research. There is currently no legal obligation for patent applicants of inventions developed using human tissue to inform tissue donors of this action or even to verify that the research leading to the invention was compliant with ethical standards.[37] There have also been calls for transparency registers disclosing patents and patent dealings in other areas of biomedicine,[38] which could be adapted to biobanks.

It is necessary to consider how these goals of non-interference and accountability might be achieved practically.

Research Resources: Final Notice', *Federal Register*, **64**, 72090; National Research Council (1997), *Intellectual Property Rights and the Dissemination of Research Tools in Molecular Biology*, Washington, DC: National Academy Press; and the Association of University Technology Managers (2007), *In the Public Interest: Nine Points to Consider in Licensing University Technology*, http://www.autm.net/source/NinePoints/ninepoints_endorsement.cfm.

35 For example, Van Overwalle, G., E. Van Zimmern, B. Verbeure and G. Matthijs (2006), 'Models for Facilitating Access to Patents on Genetic Inventions', *Nature Review Genetics*, **7**, 143.

36 See, for example, Cambon-Thomsen, A. (2003), 'Assessing the Impact of Biobanks', *Nature Genetics*, 34, 25–6, which proposes the creation of a Bio-Resource Impact Factor as a quantitative parameter for measuring use of biobank resources. Essentially the idea behind the BioResource Impact Factor is to create a standardised method for tracking outcomes from the use of biobank resources using universal identifiers. Examples of reportable outcomes include primary and secondary research publications and clinical trials.

37 Nicol, D. (2004), 'Property in Human Tissue and the Right of Commercialisation: the Interface between Tangible and Intellectual Property', *Monash University Law Review*, **30**, 139–64 at 160–62.

38 Recommendation 3 of Secretary's Advisory Committee on Genetics, Health and Society (2010), *Gene Patents and Licensing Practices and their Impact on Patient Access to Genetic Tests*, Bethesda, MD: Secretary's Advisory Committee on Genetics, Health, and Society, http://oba.od.nih.gov/SACGHS/sacghs_documents.html#GHSDOC_011; and Carbone, J., E.R. Gold, B. Sampat, S. Chandrasekharan, L. Knowles, M. Angrist and R. Cook-Deegan (2010), 'DNA Patents and Diagnostics: Not a Pretty Picture', *Nature Biotechnology*, **28**, 784–91.

II OPEN SOURCE AND OPEN ACCESS BIOBANKS

The first instinct of those attempting to balance the goal of data sharing and the need to retain some control over downstream commercial users has been to follow the method used by the information technology community to make their ideas and knowledge freely available: open source licensing. While this method works well for software, it faces significant challenges for the large-scale data sets and collections of tissue incorporated into biobanks.

In information technology, open source software licensing relies on property rights attached to copyright to impose enforceable conditions that include viral clauses that require the propagation of openness in successive generations. A number of commentators have posited that open source-style licensing could be translated into biotechnology, using a combination of intellectual property rights and standard contracts – usually licences but, in a modulated form, non-assert clauses – to construct a web of researchers connected through legal agreements that require sharing of tissue samples and knowledge.[39] In the biobanking context, an open source biobank would need to be able to secure intellectual property over its data (database protection where this exists), database structure (copyright) and materials (trade secrets and patents). The biobank would then issue intellectual property licences and material transfer agreements to those who wish to make use of the data and tissue. One condition of these licences would be that each user agrees to license back to the biobank (and its users) his or her own modifications and additions on exactly the same terms. This 'viral' licence ensures that each person who makes use of the knowledge or technology keeps all incremental developments available to all subsequent users.

Open source works well when dealing with small data sets or software since the cost of acquiring and licensing them is virtually nothing. This is not true when dealing with patents. As the costs of patenting can quickly become substantial,[40] an open source framework would need to provide a

---

39 See, for example, Hope, Janet (2008), *Biobazaar: The Open Source Revolution and Biotechnology*, Cambridge, MA: Harvard University Press; Feldman, R. (2004), 'The Open Source Biotechnology Movement: Is It Patent Misuse?' *Minnesota Journal of Law, Science and Technology*, **6**, 117–67; and Joly, Y. (2010), 'Open Biotechnology: Licenses Needed', *Nature Biotechnology*, **28**, 417–19.

40 Berrier Jr, E.F. (1995), 'Global Patent Must Be Reduced', *IDEA*, **36**, 473–99 found that, in 1995 US dollars: '[T]he total cradle-to-grave costs in the seventeen EPO countries is $134, 401, the total cost in Japan is $30, 498 and $22, 522

mechanism to fund patent acquisition, a not insignificant problem. The costs involved include not only the direct costs of patenting, but also, in maintaining the secrecy of knowledge until a patent is filed (in order to meet patent law requirements), establishing and updating open source licences, and managing knowledge contributed under different open source licences that can quickly result in an intellectual property thicket of its own. Further, tissue is more difficult to fit within an open source paradigm than is data and software for two principal reasons. First, tissue cannot be shared as easily as is software and data: someone must store, reproduce and handle the tissue, which are not trivial tasks because they involve not only a cost but also potential liability if the tissue contains pathogens or is improperly stored or shipped. Second, use of tissue may be limited by undertakings made to donors during the initial consent process and other privacy obligations. While neither of these reasons makes it impossible to construct an open source biobank, together they considerably complicate management, increasing costs and negotiation time. But perhaps the most significant obstacle to open source biobanks is more deeply philosophical: reliance on intellectual property rights would seem to be antithetical to a movement whose primary goal is to free basic research from proprietary encumbrances.

In contrast to open source, the goal of the open access or 'free revealing' movement[41] is to make knowledge and technology easily and freely available solely through the development and enforcement of norms within the research community. While nothing legally guarantees that knowledge and technology users will contribute their improvements and additions back to the community on open terms, open access seeks to use informal enforcement such as membership within research consortia, access to philanthropic and government funding, and invitations to conferences and other knowledge-sharing venues to achieve similar goals.

Open access aims at addressing two interrelated concerns. The first of these is that intellectual property rights increase costs and lead to inefficiencies:

> The fundamental problem is that industry collectively focuses too many resources on proof-of-concept studies for too few targets, and the studies are done in a proprietary way, with little collective learning. Further, because one

---

for a patent with ten claims and two claims respectively. The total cost in the U.S. is $14,370.' These figures would be substantially higher today.

41 Baldwin, C.Y. and E. von Hippel (2009), 'Modeling a Paradigm Shift: From Producer Innovation to User and Open Collaborative Innovation', *Harvard Business School Working Paper 10–048*, http://ssrn.com/abstract=1502864.

> 'secret' failure in proof of concept is never enough to dissuade others, these studies encumber the limited resources of industry for years, thereby limiting the ability of industry to pursue new and potentially relevant drug targets.[42]

The second is that the costs of obtaining and licensing intellectual property can be so high as to divert resources away from research and toward applying for and managing intellectual property.

To achieve these advantages of open access, much work will be needed on norm development within the affected communities. Elinor Ostrom has demonstrated that such norms can be effective tools in resource management;[43] the key, however, is to have sufficient community agreement and extra-legal enforcement to create the trust that makes those norms effective.[44] Specialized biobanks are more likely to exist in an environment in which the relevant communities can agree upon and enforce these norms. This would most likely occur where the community benefits from a homogeneity of interests and uses of the data and technology. The more general or more heterogeneous the interests involved, the more difficult it will be to reach consensus around norms and their enforcement. In such circumstances, open access may not achieve its goal.

By purposely not seeking intellectual property rights, open access biobanks offer significant advantages over open source. In particular, they are far less expensive owing to the lack of patent applications and patent maintenance fees and are flexible and easy to adapt as technology and uses change. But one dilemma is that, because open access is norm-driven, it is difficult to impose enforceable obligations on users of biobank resources if they are made freely available in accordance with true open access principles.

---

42 Edwards, A, C. Bountra, D. Kerr and T. Willson (2009), 'Open Access Chemical and Clinical Probes to Support Drug Discovery', *Nature Chemical Biology*, **5**, 436–40 at 438.

43 See, for example, Ostrom's foundational work in Ostrom, Elinor (1990), *Governing the Commons: The Evolution of Institutions for Collective Action*, Cambridge: Cambridge University Press.

44 Akcomak, I.S. and B. ter Weel (2008), 'Social Capital, Innovation and Growth: Evidence from Europe', *European Economic Review*, **53**, 544–67.

III RESPECTING PARTICIPANT RIGHTS: THE COMPLICATION OF TISSUE AND PERSONAL INFORMATION

One of the biggest obstacles to both open source and open access biobanks is that neither explicitly addresses privacy and other participant concerns. The question of whether or not individuals have property rights in their own tissue remains legally uncertain and ethically contentious.[45] But this does not mean tissue donors have no say whatsoever as to whether their tissue can be used for research purposes and who can use it. Practically, donors are unlikely to be willing to voluntarily participate in biomedical research unless they receive some assurance that the tissue they donate will be used, stored and disposed of in a way that protects their dignity, privacy and autonomy. Unencumbered open access to tissue collections and personal information is unlikely to satisfy donors that due respect will be given to these matters.

Heightened concerns about privacy were raised in a Fact Sheet distributed by the US NIH in 2008 as a result of technological advances in bioinformatics. The Fact Sheet states that: 'bioinformatics techniques have progressed to the point that with enough genomic data on an individual from another source, it is now possible to determine whether that individual participated in a study by analyzing only the pooled summary data'.[46] As a result, the GWAS policy of open access was modified. Individual-level data (including both genotypes and phenotypes) are now subject to a controlled access regime, whereas summary-level information and aggregate genotype data remain openly accessible.[47]

---

45 For a summary of the issues from the Australian perspective see Australian Law Reform Commission and Australian Health Ethics Committee (2003), *Essentially Yours, the Protection of Human Genetic Information in Australia*, Report No. 96, Sydney: Australian Government, 527–9, ('ALRC/AHEC, Essentially Yours').

46 National Institutes of Health (2008), 'Background Fact Sheet on GWAS Policy Update', August 28, http://grants.nih.gov/grants/gwas/.

47 National Institutes of Health (2008), *Modifications to Genome-Wide Association Studies (GWAS) Data Access*, 28 August, http://grants.nih.gov/grants/gwas/ together with the implementation process on the dbGaP website at http://www.ncbi.nlm.nih.gov/projects/gap/cgi-bin/about.html. For another approach see Wolfson, M. et al. (2010), 'DataSHIELD: Resolving a Conflict in Contemporary Bioscience – Performing a Pooled Analysis of Individual-Level Data Without Sharing the Data', *International Journal of Epidemiology*, **39**, 1372–82.

There are a number of legal and ethical frameworks that provide various forms of protection to participants in biomedical research. Most countries have mechanisms in place imposing ethical obligations on researchers.[48] Consent is generally considered to be an essential prerequisite for research use of human tissue and personal information. Waiver of consent may be allowed but only in limited circumstances, and full ethical review and approval will still be required.[49] As such, requests for access to biobank resources for research purposes will generally be required to provide verification that the proposal for research use has ethical approval from a properly constituted ethics committee, and that the proposed research is within the scope of the initial consent, or that new consent will be sought, or that the ethics committee has accepted an application for waiver of consent.

In addition to these consent requirements, the privacy of personal information is protected by legislation in many jurisdictions.[50] Ethical guidelines also recognize the importance of maintaining the privacy of personal information in the research context, requiring that ethics committees must be satisfied that appropriate provisions are in place to protect privacy. Aside from these legal and ethical obligations, a wide range of technological measures can also be applied to protect privacy.[51]

While these privacy obligations clearly are intended to apply to health information and other personal information, including personal details attached to tissue samples, it is more doubtful that these obligations extend to the tissue itself.[52] The problem this creates is that it is not until information is extracted from tissue that the privacy regime becomes operational.

---

48 For a useful overview of consent requirements for genomic and genetic research see McGuire, A. and L. Beskow (2010) 'Informed Consent in Genomics and Genetic Research', *Annual Review of Genomics and Human Genetics*, **11**, 361–81.

49 See National Health and Medical Research Council (2010), *Biobanks Information Paper 2010*, Canberra: Commonwealth of Australia, 37–9, http://www.nhmrc.gov.au/your_health/egenetics/practitioners/practitioners.htm ('NHMRC Biobanks Information Paper').

50 In Australia, for example, the relevant legislation is the federal *Privacy Act* 1988 (Cth) ('*Privacy Act*').

51 See National Health and Medical Research Council (2010), *Biobanks Information Paper 2010*, Canberra: Commonwealth of Australia, 45, http://www.nhmrc.gov.au/your_health/egenetics/practitioners/practitioners.htm ('NHMRC Biobanks Information Paper').

52 See the discussion on this point in Australian Law Reform Commission and Australian Health Ethics Committee (2003), *Essentially Yours, the Protection of Human Genetic Information in Australia*, Report No. 96, Sydney: Australian Government, 262–4.

This may occur some time after removal of the tissue, and may be done by a person other than the person who collected the tissue. There have been some notorious examples of misuse of human tissue in the past.[53] There is now much more awareness of the sensitivities associated with retention and use of human tissue, and some jurisdictions have enacted specific legislation in response. For example, the *Human Tissue Act* 2004 (UK) provides an additional layer of protection by establishing a licensing scheme for use of human tissue. Whilst licensing schemes have been rejected in other jurisdictions, it is generally accepted that some form of oversight is required for all research use of human tissue and personal information and that it may be necessary to go beyond monitoring by human research ethics committees, particularly for biobanks and other large-scale collections.[54] Custodians, independent intermediaries, gene trustees, charitable trusts, partnerships and other forms of institutional arrangements have all been mooted as possible options.[55]

Oversight bodies of this nature ensure that researchers accessing biobank and other resources are compliant with ethical obligations and that they provide the appropriate undertakings to respect participant privacy and confidentiality.[56] However, simple assurances that privacy and confidentiality will be respected may not give sufficient comfort to participants because of the lack of any direct relationship between the participant and the researcher. Participants may expect to see more concrete legally enforceable obligations, for example in the form of binding access and material transfer agreements between the oversight body and the researcher

---

53 In particular, there were two notorious organ retention scandals in the United Kingdom at the turn of the century. See Skene, L. (2002), 'Ownership of Human Tissue and the Law', *Nature Reviews Genetics*, 3, 145–8; and Liddell, K. and A. Hall (2005), 'Beyond *Bristol* and *Alder Hey*: The Future Regulation of Human Tissue', *Medical Law Review*, 13, 170–223.

54 See National Health and Medical Research Council (2010), *Biobanks Information Paper 2010*, Canberra: Commonwealth of Australia, 54–61, http://www.nhmrc.gov.au/your_health/egenetics/practitioners/practitioners.htm ('NHMRC Biobanks Information Paper').

55 Ibid. See also the series of articles by David Winickoff developing the notion of partnership governance, most recently articulated and referenced in Winickoff, David (2009), 'From Benefit Sharing to Power Sharing: Partnership Governance in Population Genomics Research', in Jane Kaye and Mark Stranger (eds), *Principles and Practice of Biobank Governance*, Cheltenham: Ashgate, 53–66.

56 Organisation for Economic Co-operation and Development (OECD) (2006), *Creation and Governance of Human Genetic Research Databases*, Paris: OECD Publishing, 107–9, http://www.oecd.org/document/50/0,3343,en_2649_34537_37646258_1_1_1_1,00.html ('OECD Creation and Governance Report').

accessing the resource.[57] This is particularly likely to be the case when researchers accessing their resources are in the private sector, or have some other form of relationship with commercial partners or if they are located in a different jurisdiction from the biobank.

What this all means is that, practically, it is impossible to be true to the open access philosophy for resources that include personal information and tissue. Biobanks cannot simply make their resources openly available to all comers but must be convinced that appropriate ethical safeguards are in place prior to the release of any tissue or personal information.

IV MAINTAINING PUBLIC TRUST: TRANSPARENCY AND BENEFIT SHARING

So far, this chapter has explored the difficulties associated with balancing the goals of broad access to research resources, commercial product development and protection of participant rights. But, as highlighted at the start of this chapter, there are other layers of complexity that need to be considered. In particular, participation is crucial for biobanking. Simply put, without willing participants there will be no biobank. A logical extension of this conclusion is that without trust in biobanking and biobank operators there will also be no biobank.

It has long been recognized that social responsibility is a strong motivator for participation in biomedical research.[58] This means that most people would not normally expect to be paid for participation or even necessarily to receive other forms of direct benefit. But they do need to know that there is, in fact, some tangible public benefit arising from the research.[59] Alternatively, some people might be motivated to participate on the basis that their participation will lead to immediate benefits to themselves, their families or their local communities. If this is the case then specific assurances that this will occur may be required.

In a recent Australia-wide telephone survey of attitudes towards biobanking, respondents were asked to consider the types of obligation

---

57 National Health and Medical Research Council (2010), *Biobanks Information Paper 2010*, Canberra: Commonwealth of Australia, 46, http://www.nhmrc.gov.au/your_health/egenetics/practitioners/practitioners.htm ('NHMRC Biobanks Information Paper').

58 Campbell, A. (2007), 'The Ethical Challenges of Genetic Databases: Safeguarding Altruism and Trust', *King's Law Journal*, **18**, 227–46.

59 For a review of the literature on participant motivations see Ormond, K.E., M.E. Smith and W.A. Wolf (2010), 'The Views of Participants in DNA Biobanks', *Stanford Journal of Law, Science and Policy*, **1**, 80–87 at 82.

that they would like to see imposed on users of biobank resources.[60] They were asked to consider both obligations that would lead to broad sharing of benefits with society (for example, dissemination of research results and equitable access to new treatments) and obligations that would lead to more direct sharing of benefits with participants (for example, payment for participation or preferential access to new treatments). The survey also examined whether attitudes to benefit sharing were affected by the type of biobank and/or user of biobank resources (commercial and non-commercial, Australian and international). Statistical analysis of the results of the survey showed that respondents were highly in favour of being informed about what benefit sharing obligations are in place before deciding on whether or not to participate in biobank research and also that they cared about the nature of these obligations.[61]

Further analyses of the results of this survey suggest that norms of social responsibility and reciprocity both influence attitudes towards benefit sharing.[62] Three distinct classes of respondent were identified using advanced statistical methodology. Members of the largest class tended to favour payment to biobank participants (relative to the other two classes), especially for biobanks accessed by commercial users. They also agreed strongly that any treatments developed from biobank resources should be made available to biobank participants (though not more so than other Australians and those in developing countries), and that biobanks should receive financial compensation in exchange for access, particularly if commercial and offshore interests are involved. These results suggest that this class of respondents was influenced more by the norm of reciprocity than the norm of social responsibility. The second largest class appeared to be more influenced by the norm of social responsibility, with strong negative attitudes towards payment for participation and stronger support for

---

60 Reported in Nicol, Dianne and Christine Critchley (2009), 'What Benefit Sharing Arrangements Do People Want from Biobanks? A Survey of Public Opinion in Australia', in Jane Kaye and Mark Stranger (eds), *Principles and Practice of Biobank Governance*, Cheltenham: Ashgate, 18–31 ('Nicol and Critchley Benefit Sharing Arrangements') and Nicol, D. and C. Critchley (2011), 'Benefit Sharing and Biobanking in Australia', *Public Understanding of Science*, 21 April, 1–22 ('Nicol and Critchley Benefit Sharing and Biobanking').

61 Nicol, Dianne and Christine Critchley (2009), 'What Benefit Sharing Arrangements Do People Want from Biobanks? A Survey of Public Opinion in Australia', in Jane Kaye and Mark Stranger (eds) , *Principles and Practice of Biobank Governance*, Cheltenham: Ashgate, 18–31 at 27.

62 These results are reported in detail in Nicol, D. and C. Critchley (2011), 'Benefit Sharing and Biobanking in Australia', *Public Understanding of Science*, 21 April, 1–22 ('Nicol and Critchley Benefit Sharing and Biobanking').

distribution of benefits to the broader community than to participants themselves. The third class displayed no clear preferences. All classes favoured broad dissemination of research results and return of results to the biobank, in accordance with the data-sharing principles discussed earlier in this chapter. These results illustrate that clear benefit-sharing obligations are important to potential participants in deciding whether or not to participate in biobanking, and that specific sharing of benefits with participants is as important to many of them as broader population-wide benefit sharing.

Policy makers and commentators have reached a similar conclusion regarding the importance of tangible and explicit benefit-sharing obligations. For example, Best Practice 9.1 of the *OECD Biobank Guidelines* states that:

> The operators of the [biobank] should have a clearly articulated policy regarding benefit sharing. This policy should address, inter alia, whether tests or products arising from research using its resources might be shared with the community and/or the general population, and how such sharing will be effected.

Public trust and willingness to engage in biobanking are likely to be enhanced if the benefits arising from sharing data and tissue and from engaging with the private sector are clearly explained to potential participants and the broader community. It is surprising, then, that ethical guidelines generally tend to give limited guidance on the extent to which commercial involvement in research using human tissue and personal information should be disclosed or on the need to explicitly articulate how associated benefits will be distributed.[63] As has been rightly pointed out by Bartha Knoppers and Lori Sheremeta, it is time to move beyond the rhetoric of benefit sharing and move towards implementation of practical measures.[64]

V STANDARDS FOR BIOBANK ACCESS AND IP POLICIES

Based on the analysis presented so far in this chapter, a biobank governance regime might include the following six standards.

---

63 Nicol, D., M. Otlowski and D. Chalmers (2001), 'Consent Commercialisation and Benefit Sharing', *Journal of Law and Medicine*, **9**, 80–94.

64 Knoppers, B.M. and L. Sheremeta (2003), 'Beyond the Rhetoric: Population Genetics and Benefit-sharing', *Health Law Journal*, **11**, 89–117.

- Standard 1: a stated philosophy of broad access to biobank datasets and tissue collections for research purposes – including by researchers in other jurisdictions and researchers in the private sector – but with recognition that truly open access is not achievable due to the nature of the tissue and information within the biobank.
- Standard 2: establishment of a body independent of the biobank to ensure that ethical requirements have been satisfied and that best practice technological measures are in use to protect the integrity of the biobank.
- Standard 3: a robust access framework that imposes legally enforceable obligations on users of the resource to comply with conditions on access imposed by the biobank through formal access and material transfer agreements. Ideally this should be coordinated internationally to facilitate interoperability between data and tissue extracted from different biobanks.
- Standard 4: measures that facilitate sharing of the benefits of research using the resource and protection of the ongoing viability of the resource – which could be achieved by way of fees for use or by profit sharing arrangements.
- Standard 5: measures that allow engagement with commercial entities and generation of intellectual property to facilitate downstream product development, but with proper accountability and with due attention to the risk of inhibiting other research using the resource.
- Standard 6: transparency and consultation in the development and articulation of these standards, with focus on the most contentious areas, particularly international data sharing, commercial involvement and benefit sharing.

I practice, there is likely to be a range of mechanisms for incorporation of these or similar standards into biobank governance frameworks.[65] The large-scale population biobanks have the most sophisticated governance frameworks. These all include extensive opportunities for public consultation, not only during the establishment process, but throughout the life of the biobank.[66] These public consultation exercises have shown that access

---

65 On this point see Milanovic F., D. Pontille and A. Cambon-Thomsen (2007), 'Biobanking and Data Sharing: A Plurality of Exchange Regimes', *Genomics, Society and Policy*, **3**, 17.

66 See, for example, Laurie, Graeme, Ann Bruce and Catherine Lyall (2009), 'The Roles of Values and Interests in the Governance of the Life Sciences: Learning Lessons from the "Ethics+" Approach of UK Biobank', in Catherine Lyall, Theo Papaioannou and James Smith (eds), *The Limits to Governance: The*

and intellectual property are troublesome areas, requiring particular care in terms of policy development and communication. It is salutary to note that UK Biobank released a comprehensive draft Policy on Intellectual Property and Access during its establishment phase but this document was removed from its website prior to recruitment. It was not until 31 May 2011 UK Biobank released a draft document for public consultation, *Access Procedures: Application and Review Procedures for Access to the UK Biobank Resource*.[67] The public consultation closed on 6 July 2011. At the time of publication, a final version of the document had not yet been released.

Clause II.B.1 of the UK Biobank Ethics and Governance Framework incorporates Standards 1 and 2, albeit only in general terms, providing:

> UK Biobank will retain full control of all access to, and uses of, the resource. UK Biobank will not proscribe any medical or other health related research uses at the outset. However, all proposals will be reviewed by UK Biobank to ensure they are consistent with the participants' consent and this Framework, and that they have relevant ethics approval. All users, whether employed by universities, government, charities or commercial companies, will be held to the same scientific and ethical standards.

Later clauses provide that access to data and samples will be by licence and fees will be charged. There are also obligations to share data and findings – with some allowance for a confidentiality period for the purposes of publication, and pursuit of patent applications and other commercial opportunities. As such, Standards 3 and 4 are also addressed to some extent.

These provisions make good practical sense, and although they take UK Biobank outside the realm of full open access, they do so in a way that balances the value of facilitating unencumbered research with the need to protect participants and to maintain the integrity and viability of the resource as a whole. However, the only guidance provided in the Framework as to how participants will be informed of access arrangements is found in Clause 11.B.2:

> UK Biobank will explain, to participants and the public, the policies and procedures for research access. An overall policy and detailed terms of access

---

Challenge of Policy-Making for the New Life Sciences, Aldershot: Ashgate, 51–77; and Haddow, G., S. Cunningham-Burley, A. Bruce and S. Parry (2008), 'Generation Scotland: Consulting Publics and Specialists at an Early Stage in a Genetic Database's Development', *Critical Public Health*, **18**, 139–49.

67 United Kingdom Biobank (2011), *Access Procedures: Application and Review Procedures for Access to the UK Biobank Resource*, 31 May, http://www.ukbiobank.ac.uk/procedures/.

has been, and will continue to be, developed (i.e. the IP and Access Policy) which addresses fairness and transparency of decisionmaking, the handling of conflicts of interest and the prioritisation of use of samples.

By 1 November 2010, 503 316 people had participated in UK Biobank,[68] yet the Intellectual Property and Access Policy is still in development. As such, it is difficult to see how participants could have been properly informed of terms of access at the time when they consented to participate.

Benefit sharing and intellectual property provisions in clause III.C of the Framework also provide general statements on how the biobank will lead to broad social benefits and how biobank resources should not be exploited improperly or used in any way that inappropriately constrains use by others, thereby addressing Standards 4 and 5. But again, clause III.C.2 states only that explicit intellectual property and access policies are 'being developed'.

Generation Scotland fares much better in this analysis because it already has an explicit Management, Access and Publications Policy (GSMAPP) available on its website.[69] A Commercial Access and Intellectual Property Policy is annexed. Clause 3.1 of the GSMAPP makes it clear that the resource is not entirely open access but is a 'supported access Resource' – addressing Standards 1 and 2. Potential users are required to submit a collaboration proposal and, if this is approved by the Generation Scotland Resource Management and Development Committee, they have to enter into a Data and Materials Transfer Agreement – thus complying with Standard 3. Various obligations are created with regard to sharing of research results and protection of participants – addressing Standards 2 to 4. Commercial users are allowed to access the resource, but the Commercial Access and Intellectual Property Policy prescribes that they must provide regular updates to the Resource Management and Development Committee and have various other responsibilities – which addresses Standard 5 in full. That Policy also includes provisions relating to income sharing, pricing and other matters – as in Standard 4.

Together, the two Generation Scotland policies appear to achieve an appropriate balance between data sharing, commercial use and protection of participants and the ongoing viability of the resource. Three Generation Scotland projects provide access to external researchers and each has a

---

68 As reported on the front page of the UK Biobank website at: http://www.ukbiobank.ac.uk/.

69 Generation Scotland, http://www.generationscotland.co.uk/access_policy.htm.

Participation Information Leaflet providing information to potential participants.[70] Access and commercial use are canvassed to some extent in each of these leaflets. For example, the Donor DNA Databank Participation Information Leaflet explains that blood samples and genealogical information will be taken and that the samples and information will be kept and used for other research studies. The rest of the information provided to potential participants is then set out as a series of questions and answers. For example, in response to the question 'Who will do the research and where will it be done?' the leaflet provides:

> Research using the anonymous samples will be done by staff in Hospitals and Universities. Samples will be initially processed in Edinburgh in the Wellcome Trust Clinical Research Facility at the Western General Hospital. Blood and DNA samples may then occasionally be distributed to laboratories in Hospitals and Universities around the UK for further research.

On the basis of this answer alone, it might be expected that research will only be conducted in the public sector and only in the UK, which would appear to create an expectation contrary to the intentions expressed in the GSMAPP. However, in response to a later question 'What about commercialisation? Will industry be involved in any way?' the leaflet provides:

> This project is designed first and foremost to further public good. It is funded by public money but the longer term success of the Generation Scotland project depends on attracting more funding. The main aim is to create a resource which will allow further medical research and this can be best achieved by encouraging as wide a range of research as possible. So to encourage both funding and research, Generation Scotland may allow access to the resource to private companies on a commercial basis. All access will be subject to the strictest ethical standards and keeping confidentiality. It will not be possible to give access that would lead researchers to identify any participant in the project.

As a consequence, it seems that commercial use should be in the contemplation of participants when they consent to participate. However, no mention is made of research use outside the UK.

Information provided by Generation Scotland for the purpose of this chapter gives some further guidance on how these policies are working in practice.[71] A total of 58 requests for access have been received to date, all of

---

[70] These are the Donor DNA Databank; the Scottish Family Health Study; and Genetic Health in the 21st Century, http://www.generationscotland.com/access.htm.

[71] Personal communication from Robin Morton, Generation Scotland Knowledge Exchange Officer and Shona Kerr, Chair, Generation Scotland Resource Management and Development Committee by email, dated 13 December 2010.

which are from public sector applicants. Three have a commercial component, but none have been directly received from commercial organizations so far. Almost all projects are local to Scotland, with some approaches made by other UK researchers. Although there have been no requests by international applicants, some of the Scottish applicants have one or more international co-applicants as part of a collaborating group, and one of these is awaiting a decision on whether samples can be sent out of the UK. This illustrates that internationalization and commercialization are clearly within the contemplation of the Resource Management and Development Committee. It would be useful to relay this more clearly to present and future participants to maintain the level of trust that appears to exist at present. Pertinently, a personal communication provided by Generation Scotland makes the following point: 'It is perhaps too early to say whether the policy is achieving its purpose – anecdotally, most applicants seem to value the process and appreciate the effort that GS is making to try to achieve good governance.'[72]

Like Generation Scotland's GSMAPP, CARTaGENE has a Samples and Data Access Policy available on its website.[73] It also has a Samples and Data Access Committee to evaluate requests for access to samples and data. Section 5 of the Policy states that CARTaGENE may be used by public and private sector for scientific research – thereby addressing Standards 1 to 3. Although the Policy does not explicitly endorse commercialization of research results – specified in Standard 5 – it is clearly in contemplation that patents will be taken out. For example, section 7 of the Policy requires that research results are returned to CARTaGENE – addressing Standard 4 – but recognizes that the need to protect intellectual property (e.g. patents) or pre-publication results may result in corresponding constraints on public disclosure of research results/data. Similarly, while section 10 states that CARTaGENE will not participate in licensing activities of intellectual property, it encourages those that do to follow the OECD Biobank Guidelines.[74] This Policy again reflects a genuine attempt to balance the core features underlying biobank establishment. But it is difficult to see how this balancing exercise is communicated to potential

---

72 Ibid.

73 CARTaGENE, http://www.cartagene.qc.ca/index.php?option=com_content&task=view&id=94&Itemid=83.

74 On the reasons for the unwillingness of CARTaGENE to engage in ownership of intellectual property see Fortin, S., S. Pathmasiri, R. Grintuch and M. Deschenes (2010), '"Access Arrangements" for Biobanks: A Fine Line between Facilitating and Hindering Collaboration', *Public Health Genomics*, 1–11 at 8–9, DOI. 10.1159/000309852.

participants as proposed in Standard 6. Although there is an extensive set of Frequently Asked Questions on the CARTaGENE website, none explicitly address commercial access and patenting. Moreover there is again no mention of international data sharing.

The smaller-scale biobanks have, in the past, tended not to fare as well as the larger-scale entities in dealing with access and intellectual property. A summary of the literature undertaken in 2005–2006 indicates that, at that time, many lacked access and intellectual property policies completely, and others tended to skirt around the difficult issue of how to reconcile some of the potential conflicts between the six standards articulated above.[75] Few proactively attempted to resolve these difficulties. As a general rule, the attitude towards downstream intellectual property was simply to relinquish control, unless biobank researchers were actively involved in the intellectual property-generating research, in which case co-ownership was frequently required.[76] This means that Standard 5 was only addressed in part. Admittedly, though, a more cautious approach was generally taken with regard to access to tissue and personal information, usually requiring the execution of a material transfer agreement in accordance with Standard 3.[77] Other earlier research reveals that benefit sharing and Standards 4 and 6 were rarely addressed at that time.[78] There have been some promising recent developments. For example, the Western Australian Department of Health has adopted the *OECD Biobank Guidelines* in their entirety, with the intention that they will be used by both government and non-governmental organizations in that state.[79]

CONCLUSION

The biobanking era of biomedical research has been heralded with much fanfare. There is palpable excitement about the prospect that biobank

---

75 Nicol, D. (2006), 'Public Trust, Intellectual Property and Human Genetic Databanks: The Need to Take Benefit Sharing Seriously', *Journal of International Biotechnology Law*, **3**, 89–103 at 99–103.

76 Ibid.

77 Ibid.

78 Knoppers, B.M. and L. Sheremeta (2003), 'Beyond the Rhetoric: Population Genetics and Benefit-sharing', *Health Law Journal*, **11**, 89–117.

79 Government of Western Australia Department of Health (2010), *Guidelines for Human Biobanks, Genetic Research Databases and Associated Data*, Perth: Government of Western Australia, http://www.genomics.health.wa.gov.au/publications/index.cfm.

resources can be used by researchers to make genuine progress in understanding human development and finding ways to diagnose, treat and prevent human disease. There is recognition that these important research tools become even more powerful if they can be shared and compared. Data sharing, open access and open source are the new buzz words. But there are complications, many of which are already recognized and are being responded to. Governance frameworks are being established by the large-scale biobanks to ensure that binding obligations are in place to respect participant rights, and that these obligations extend to users of biobank resources as well as biobanks themselves. Intellectual property policies attempt to balance the need to protect downstream commercial investment with the need to keep upstream research open. It is acknowledged that the types of complex governance frameworks being established by the large-scale biobanks are likely to be too costly and cumbersome for smaller biobanks. Nevertheless, it is important that attention is given to some if not all of the standards proposed in Part V of this chapter.

Even some of the larger-scale biobanks, which have put considerable effort into establishment of best practice governance frameworks, could better ensure that users of those resources are transparent and accountable in reporting the outcomes of their research, particularly to research participants. Transparency and accountability are crucial if public trust in the entire biobank endeavour is to be sustained. It is not enough to have good governance of the biobank: there must also be good governance of access, use and downstream development. This may require a greater level of accountability by biobank users than they perhaps are used to, but so be it. At the very least, biobanks should put in place measures to ensure that there is compliance with the following *OECD Biobank Guidelines*:

> 9.3 Researchers should submit to the [biobank] an annual progress report and a report at the termination of a research project. Such reports should list publications, published patent applications and patents issued arising from research accessing the [biobank's] resources.
>
> 9.4 Summary results arising from research conducted using the [biobank's] resources should be made available in easily accessible forms, such as through a newsletter or website.

6. The 1000 Genomes Project

Donna M. Gitter

Scientific discourse and progress depend upon the open discussion of ideas and full disclosure of supporting facts.[1] This discussion has traditionally occurred through the process of publication, which is the primary means by which scientists achieve recognition for their work.[2] Traditionally, researchers published papers that combined in one entity both their ideas and the underlying data. With the advent of large-scale and high-throughput data analyses, however, the creation of scientific databases replaced the traditional model. Typically, for such data-intensive projects, funding agencies require that all relevant data must be made available on a publicly accessible website at the time of the paper's publication.[3]

The Human Genome Project (HGP), completed in 2003, demonstrated to the scientific community that making data broadly available before publication results in valuable benefits to the public. This is particularly true where there is a community of scientists who can use the data more quickly than the data producers themselves, and in ways not originally anticipated at the outset of the project.[4] One successor to the HGP is the 1000 Genomes Project, which provides that project data will be released quickly, prior to publication, into the public domain.[5]

While the open access approach of the 1000 Genomes Project is the norm for large-scale, publicly funded genomic databases, many smaller projects that likewise produce vast amounts of scientific data nonetheless do not embrace data sharing. Quite often, researchers involved in 'small science', meaning research 'performed by individual investigators or small and

---

1 Birney, E. and T. Hudson et al. (2009), 'Prepublication Data Sharing' (2009), *Nature*, **461** (7261), 168–70.

2 Schofield, P. et al. (2009), 'Post-Publication Sharing of Data and Tools', *Nature*, **461** (7261), 171–3.

3 Birney, E. and T. Hudson et al. (2009), 'Prepublication Data Sharing', *Nature*, **461** (7261), 168–70.

4 Ibid.

5 The 1000 Genomes Project, 'About the 1000 Genomes Project: Project Overview', http://www.1000genomes.org/about.

autonomous research groups operating outside large, organized research programs, often with non-federal sources of funding', are not as influenced by the norms of open access and sharing and instead rely on informal exchanges of data and samples.[6] Scientific researchers, ethicists, lawyers, representatives of funding agencies and journal editors are all actively engaged in discussions aimed at establishing new scientific norms in order to overcome reluctance to data sharing in the research community. Close investigation of the 1000 Genomes Project provides guidance in terms of applying the data-sharing lessons learned from large-scale, publicly funded databases to smaller projects.

I THE 1000 GENOMES PROJECT: GOALS AND CHALLENGES

Launched in January 2008, the 1000 Genomes Project aims to develop a new map of the human genome that will provide an extraordinarily detailed view of biomedically relevant DNA variations among individuals. A clearer understanding of these small genetic variations will assist in explaining individual differences in susceptibility to disease, responses to drugs and reactions to environmental factors.[7] The 1000 Genomes Project builds upon the HapMap Project, a large-scale, publicly funded genomic project that established in 2005 a public database of common genetic variants in human beings.[8] Using HapMap data and related resources, researchers have succeeded in identifying more than 100 regions of the genome containing genetic variants that are associated with risks of common human diseases such as diabetes, coronary artery disease and prostate

---

6 Reichman, J.H. and P. Uhlir (2003), 'The Public Domain: A Contractually Reconstructed Research Commons for Scientific Data in a Highly Protectionist Intellectual Property Environment', *Law and Contemporary Problems*, **66** (1 & 2), 315–462.

7 National Institutes of Health (2008), 'International Consortium Announces the 1000 Genomes Project', Press Release, 22 January, http://www.nih.gov/news/health/jan2008/nhgri-22.htm.

8 The International HapMap Consortium (2005), 'A Haplotype Map of the Human Genome', *Nature*, **437** (7063), 1299–320. For a commentary on the HapMap Project, see Gitter, D. (2007), 'Resolving the Open Source Paradox in Biotechnology: A Proposal for a Revised Open Source Policy for Publicly Funded Genomic Databases', *Houston Law Review*, **43** (4), 1476–521.

and breast cancer.[9] The goal of the 1000 Genomes Project is to study these genetic differences in greater depth, thereby increasing the ability of researchers to discern the causal relationships between genetic variations and human disease. Ultimately, the 1000 Genomes Project will expedite efforts to diagnose, treat and prevent human diseases.[10]

Like the HapMap Project, the 1000 Genomes Project relies upon a combination of public and private support, along with the expertise of an international consortium of research teams.[11] The research teams for the 1000 Genomes Project will sequence the genomes of at least 1000 people worldwide in order to create a highly detailed database cataloguing human genetic variation. As with other large-scale, publicly funded genomic projects, data from the 1000 Genomes Project will be made available as rapidly as possible to the international scientific community via release into freely accessible public databases.[12] The quantity of data will be immense, posing great challenges for the experts in the fields of bioinformatics and statistical genetics.[13] According to Dr Gil McVean of the University of Oxford in England, one of the co-chairs of the consortium's analysis group: 'At six trillion bases, the 1000 Genomes Project will generate 60-fold more sequence data over its three-year course than have been deposited into public databases over the past twenty-five years.'[14] Certainly, the 1000 Genomes Project Consortium faces a significant challenge in devising the most effective way to analyze this overwhelming quantity of data. What is more, the consortium must ensure that its data access policy will make the data available to the maximum number of users, while simultaneously permitting researchers to seek intellectual property protection that will foster innovation.

II THE SCIENCE OF THE 1000 GENOMES PROJECT

A gene is the basic physical and functional unit of heredity for humans and almost all other organisms. Genes are made up of DNA, which is composed in part of chemical compounds called bases. Genes cause the body to make molecules called proteins that guide bodily functioning. In humans,

---

9 National Institutes of Health (2008), 'International Consortium Announces the 1000 Genomes Project', Press Release, 22 January, http://www.nih.gov/news/health/jan2008/nhgri-22.htm.

10 Ibid.

11 Ibid.

12 Ibid.

13 Ibid.

14 Ibid.

genes vary in size from a few hundred DNA bases to more than 2 million bases. The Human Genome Project has estimated that humans have between 30 000 and 40 000 genes.[15]

The DNA sequence of any two people is 99.9 percent identical, with very small genetic variations proving quite significant in terms of an individual's susceptibility to disease or response to pharmaceuticals. Sites in the DNA sequence where individuals differ at a single DNA base are called single nucleotide polymorphisms (SNPs). Sets of nearby SNPs on the same chromosome are inherited in blocks called haplotypes. While a haplotype block may contain a large number of SNPs, researchers are able to uniquely identify a haplotype by using a few SNPs, called tag SNPs.[16]

The HapMap is a map of the haplotype blocks. Its value lies in the fact that the HapMap reduced the number of SNPs required to examine the entire genome for association with a phenotype from the 10 million SNPs that exist to roughly 500 000 tag SNPs. Thus, researchers can now locate much more efficiently and comprehensively the regions of the genome containing genes that affect disease, since it is not necessary to study more SNPs than necessary.[17]

The HapMap catalogue, however, identifies only those genetic variations that occur in at least 5 percent of the population. The 1000 Genomes Project will use new sequencing technologies to 'produce a catalogue of variants present at 1 percent or greater frequency in the human population across most of the genome, and down to 0.5 percent or lower within genes'.[18] This will permit researchers to gain a better understanding of some diseases that arise less often.[19]

The 1000 Genomes Project will also surpass the HapMap Project in that it will map not only SNPs, the single-letter differences in people's DNA, but also map larger differences in genome structure called structural variants.

---

15 Gitter, D. (2001), 'International Conflicts Over Patenting Human DNA Sequences in the United States and the European Union: An Argument for Compulsory Licensing and a Fair-Use Exemption', *New York University Law Review*, **76** (6), 1623–91.

16 National Human Genome Research Institute and National Institutes of Health, 'International HapMap Project Overview', http://www.genome.gov/10001688.

17 Ibid.

18 National Institutes of Health (2008), 'International Consortium Announces the 1000 Genomes Project', Press Release, 22 January, http://www.nih.gov/news/health/jan2008/nhgri-22.htm.

19 National Institutes of Health (2008), 'Three Sequencing Companies Join 1000 Genomes Project', Press Release, 11 June, http://www.genome.gov/27526680.

Structural variants are rearrangements, deletions, or duplications of segments of the human genome that are believed to be associated with predisposition to conditions such as intellectual disability and autism.[20]

In order to achieve its scientific goals, the 1000 Genomes Project will sequence the genomes of at least 1000 individuals who gave informed consent for their DNA to be analyzed and placed in public databases. In fact, the first thousand samples will come from those used for the HapMap and from additional samples in the extended HapMap set. These participants are drawn from a geographically wide-ranging group of people of African, East Asian and European descent.[21] In addition, an effort began in 2010 to expand data collection to include individuals of Hispanic and African-American descent.[22] All of the research participants will remain anonymous and will not have any individual medical information collected from them, because the project is intended to catalogue genetic variation rather than to study the disease profile of the individual research participants. In the future, researchers will be able to use the resource that is developed from the 1000 Genomes Project in order to conduct studies of people affected by various diseases.[23]

The institutions supporting and conducting the genetic sequencing work of the 1000 Genomes Project are the Wellcome Trust Sanger Institute in England, the Beijing Genomics Institute in China, and the National Human Genome Research Institute (NHGRI), part of the National Institutes of Health (NIH) in the United States.[24] Most of the funding for the 1000 Genomes Project, which will cost between US $30 and $60 million,

---

20 National Institutes of Health (2008), 'International Consortium Announces the 1000 Genomes Project', Press Release, 22 January, http://www.nih.gov/news/health/jan2008/nhgri-22.htm.

21 Ibid.

22 Karow, J. (2009), '1000 Genomes Project to Sequence Nearly 1,000 More Samples by Early 2010; New Samples Collected', Genomeweb, http://webcache.googleusercontent.com/search?q=cache:pU05R7uEqyMJ:www. genomeweb.com/sequencing/1000-genomes-project-sequence-nearly-1000-more-samples-early-2010-new-samples-co+1000+Genomes+Project+african-american+hispanic&cd=1&hl=en&ct=clnk&gl=us.

23 National Institutes of Health (2008), 'International Consortium Announces the 1000 Genomes Project', Press Release, 22 January, http://www.nih.gov/news/health/jan2008/nhgri-22.htm.

24 Ibid.

will come from the existing sequencing budgets of the participating institutes.[25] In addition, three private US firms that have pioneered development of new sequencing technologies are contributing a significant portion of the sequence data, indicating their belief in the value of the large data sets to be produced and made available through publicly accessible databases. Indeed, these firms, 454 Life Sciences, a Roche Company; Applied Biosystems, an Applera Corp. business; and Illumina Inc., like all other participants in the 1000 Genomes Project, have consented to the open access policies established by the 1000 Genomes Project Steering Committee. These policies include 'rapid public release of the data; a prohibition on project participants gaining early access to the data; an intellectual property policy that precludes any participants from controlling the information provided by the project or filing for intellectual property rights in the sequences they produce; regular progress reporting; and coordination of scientific publications with the rest of the consortium'.[26]

III THE DATA ACCESS POLICY OF THE 1000 GENOMES PROJECT

In order to advance research and encourage international scientific collaboration, participants in large-scale, publicly funded sequencing projects have adopted broad data-sharing policies. In February 1996, at the First International Strategy Meeting on Human Genome Sequencing in Bermuda, representatives of laboratories involved in human genome sequencing and of funding agencies unanimously agreed to implement the *Bermuda Statement*, an international agreement favoring release into the public domain of genetic databases achieved through public funding.[27] The *Bermuda Statement* provides that 'all human genomic sequence information, generated by centres funded for large scale human sequencing, should be freely available and in the public domain in order to encourage research and development

---

25 Kaiser, J. (2008), 'A Plan to Capture Human Diversity in 1000 Genomes', *Science*, **319** (5862), 395; and National Institutes of Health (2008), 'Three Sequencing Companies Join 1000 Genomes Project', Press Release, 11 June, http://www.genome.gov/27526680.

26 National Institutes of Health (2008), 'Three Sequencing Companies Join 1000 Genomes Project', Press Release, 11 June, http://www.genome.gov/27526680.

27 Human Genome Project and United States Department of Energy Office of Science (1996), *Summary of Principles Agreed at the First International Strategy Meeting on Human Genome Sequencing*, http://www.ornl.gov/sci/techresources/Human_Genome/research/bermuda.shtml#1.

and to maximize its benefit to society'.[28] One of the fundamental requirements of the *Bermuda Statement* is the stipulation that sequences longer than 1000 base pairs must be made freely available to the public, preferably within 24 hours of generation.[29] For the HGP, the highest profile data set rapidly released before publication, the international sequencing centers involved in the HGP retained the right to be the first to describe and analyze their complete data sets in peer-reviewed publications, in return for the early release of their data.[30]

According to its proponents, the *Bermuda Statement* data release policy fosters independent checking of the sequence data by other researchers,[31] leading to huge gains in life sciences research. With respect to the HGP, this policy resulted in the publication of new information about 30 genes associated with disease even prior to the publication of the draft sequences.[32] In addition, a broad data access policy prevents publicly funded large-scale sequencing centers from 'establishing a privileged position in the exploitation and control of human sequence information'.[33] What is more, the policy also delivers a symbolic message that 'the genome belongs to everybody'.[34] When sequencers met again in 2003 in Florida, they reaffirmed their commitment to the *Bermuda Statement* in a statement referred to as the Fort Lauderdale Principles.[35] Consequently, organizations such as the National Institutes of Health's National Human Genome Research Institute, the Department of Energy and the Wellcome Trust require compliance with this policy as a condition of receiving funding.[36]

---

28 Ibid.

29 Nelson, B. (2009), 'Data Sharing: Empty Archives', *Nature*, **461** (7261), 160–63.

30 Birney, E. and T. Hudson et al. (2009), 'Prepublication Data Sharing', *Nature*, **461** (7261), 168–70.

31 Bentley, D (1996), 'Genomic Sequence Information Should Be Released Immediately and Freely in the Public Domain', *Science*, **274** (5287), 533–4.

32 Birney, E. and T. Hudson et al. (2009), 'Prepublication Data Sharing', *Nature*, **461** (7261), 168–70.

33 Human Genome Project and United States Department of Energy Office of Science (1996), *Summary of Principles Agreed at the First International Strategy Meeting on Human Genome Sequencing*, http://www.ornl.gov/sci/techresources/Human_Genome/research/bermuda.shtml#1.

34 Marshall, E. (2001), 'Bermuda Rules: Community Spirit, with Teeth', *Science*, **291** (5507), 1192 (quoting Ari Patrinos, who was at that time Director of the United States Department of Energy's office that funds genome research).

35 Wellcome Trust (2003), 'Sharing Data from Large-Scale Biological Research Projects: A System of Tripartite Responsibility', Fort Lauderdale, January, http://www.genome.gov/Pages/Research/WellcomeReport0303.pdf.

36 Rowen, L., G. Wong, R. Lane and L. Hood (2000), 'Publication Rights in the Era of Open Data Release Policies', *Science*, **289** (5486), 1881.

The 1000 Genomes Project has announced its adherence to the Fort Lauderdale Principles on its website.[37] The site emphasizes that data producers will 'release the Project data quickly, prior to publication, in the expectation that they will be valuable for many researchers' and that 'data users may use the data for many studies, but are expected to allow the data producers to make the first presentations and to publish the first paper with global analyses of the data'.[38]

In October 2010, the 1000 Genomes Project published in the journal *Nature* an analysis of its completed pilot phase.[39] After the completion of the pilot phase, the 1000 Genomes Project began full-scale studies that will require an additional two years.[40] Data from the pilot studies and the full-scale project are freely available on the project website, www.1000genomes.org.[41]

The data release policy of the 1000 Genomes Project aims to make the genotype data freely available, while reserving to data producers the right to make the first presentations and to publish the first paper with global analyses of the data.[42] Examination of the 1000 Genomes Project data release policy is particularly important in light of the recommendation arising from the Fort Lauderdale meeting that rapid pre-publication data release be applied to other data sets created primarily as a resource for the

---

37 Wellcome Trust (2003), 'Sharing Data from Large-Scale Biological Research Projects: A System of Tripartite Responsibility', Fort Lauderdale, January, http://www.genome.gov/Pages/Research/WellcomeReport0303.pdf.

38 The 1000 Genomes Project, 'Use of the Project Data, Presentations and Publications, and Authorship', http://www.1000genomes.org/data#DataReleasePolicy.

39 United States Department of Health and Human Services (2010), '1000 Genomes Project Publishes Analysis of Completed Pilot Phase', National Institutes of Health News, Press Release, 27 October, http://www.genome.gov/27541917; and The 1000 Genomes Project Consortium (2010), 'A Map of Human Genome Variation From Population-Scale Sequencing', *Nature*, 467 (7319), 1061–73.

40 United States Department of Health and Human Services (2010), '1000 Genomes Project Publishes Analysis of Completed Pilot Phase', National Institutes of Health News, Press Release, 27 October, http://www.genome.gov/27541917.

41 The 1000 Genomes Project, 'How to Access 1000 Genomes Data', http://www.1000genomes.org/data#DataAccess.

42 See The 1000 Genomes Project, 'Use of the Project Data, Presentations and Publications, and Authorship', http://www.1000genomes.org/data#DataReleasePolicy.

scientific community.[43] Experts have observed that data producers are no longer only large genomics centers, and that, increasingly, smaller groups are able to produce a large amount of data. Although funding agencies anticipated that the policies from the HGP and 1000 Genomes Project would be automatically adopted by these smaller projects, this did not occur. For this reason, between 80 and 100 scientists, ethicists, lawyers, representatives of funding agencies and journal editors convened at a data release workshop in Toronto in May 2009 hosted by Genome Canada in order to discuss the extension of such data release policies to fields such as proteomics and biobanking.[44] The product of that meeting was a set of proposals published in *Nature* in 2009, referred to as 'the *Toronto Statement*', which aims to establish a set of 'best practices' for scientists, both data producers and data users, as well as funding agencies and journal editors.[45] The *Toronto Statement* is meant to combat the many factors that inhibit researchers from sharing their data. Particularly in the realm of small science, these include data producers' rational desire to be recognized for producing and analyzing data, thereby preserving their ability to win future research grants; the challenge of crafting appropriate publication and intellectual property policies; and the complexity of establishing a suitable infrastructure to house the data.[46]

IV OBSTACLES TO DATA SHARING AND POTENTIAL SOLUTIONS OFFERED BY THE TORONTO STATEMENT

A Lack of Recognition of Data Producers

In pursuing career advancement, data producers are understandably reluctant to grant data users access to research results achieved after devoting

---

43 Wellcome Trust (2003), 'Sharing Data from Large-Scale Biological Research Projects: A System of Tripartite Responsibility', Fort Lauderdale, January, http://www.genome.gov/Pages/Research/WellcomeReport0303.pdf.

44 Curtin, C. (2010), 'When to Share', Genomeweb, May, http://webcache.googleusercontent.com/search?q=cache:y3_E1riwjCEJ:www.genomeweb.com/when-share+what+is+a+marker+paper+toronto+statement&cd=8&hl=en&ct=clnk&gl=us.

45 Birney, E. and T. Hudson et al. (2009), 'Prepublication Data Sharing', *Nature*, **461** (7261), 168–70 (discussing post-publication data release, but equally applicable pre-publication).

46 Gitter, D. (2010), 'The Challenges of Achieving Open Source Sharing of Biobank Data', *Biotechnology Law Report*, **29** (6), 623–35.

years of hard work to 'planning a project, securing funding and ethics approval, recruiting the participants, collecting the data and materials, performing analyses, managing the collection and infrastructure, controlling technical quality, and generally nourishing the project through waves of funding and maturation'.[47] In addition, data producers fear being 'scooped' by data users who may mine data and discover relationships in it that the producer did not discern.[48] As noted in the *Toronto Statement*, an emphasis on early data release gives rise to tension between data producers and users. The former typically wish to publish a first description of a data set, while the latter desire to publish their own analyses of the data.[49] One way to reduce this tension is to ensure proper attribution and recognition for data producers.

One means of acknowledging the contributions of data producers is to track the usage and citation of data sets through the use of electronic systems similar to those used for traditional publications.[50] In order to quantify the usefulness of particular data sets, James Boyle has offered the example of a music site associated with Creative Commons,[51] known as ccMixter.[52] This site allows users to download music from the site and remix the samples into new tracks, while simultaneously maintaining a record of the credits due to each musician.[53] A comparable system would permit attribution to the appropriate data producer and also allow universities and funding agencies to track the number of uses, and therefore the value, of a researcher's data.[54] This satisfies researchers' desire for attribution, and assists them in demonstrating their productivity to funding agencies.

Another expert, Myles Axton of *Nature*, has proposed recognizing contributors via 'microattribution', meaning the tracking of researchers'

---

47 Lowrance, W. (2006), 'Access to Collections of Data and Materials for Health Research', A Report to the Medical Research Council and Wellcome Trust, 1–36, http://www.wellcome.ac.uk/stellent/groups/corporatesite/@msh_grants/documents/web_document/wtx030842.pdf.

48 Nelson, B. (2009), 'Data Sharing: Empty Archives', *Nature*, **461** (7261), 160–63.

49 Birney, E. and T. Hudson et al. (2009), 'Prepublication Data Sharing', *Nature*, **461** (7261), 168–70.

50 Nelson, B. (2009), 'Data Sharing: Empty Archives', *Nature*, **461** (7261), 160–63.

51 Creative Commons is a non-profit corporation that aims to facilitate content sharing in accordance with the law of copyright. Creative Commons, http://creativecommons.org/about/.

52 Nelson, B. (2009), 'Data Sharing: Empty Archives', *Nature*, **461** (7261), 160–63 (citing Professor James Boyle of Duke Law School).

53 ccMixter, http://ccmixter.org/about.

54 Nelson, B. (2009), 'Data Sharing: Empty Archives', *Nature*, **461** (7261), 160–63.

contributions 'down to the very smallest meaningful unit (database record, gene), as opposed to whole papers as is tradition'.[55] Microattribution could minimize the shortcomings of the Fort Lauderdale Agreement, which Acton has critiqued for failing to reward data producers with credit for uses of their data. Nor does the Agreement make any provision for crediting data generators, such as genomics 'factory' laboratories. In addition, notes Axton, the Fort Lauderdale Agreement is not truly enforceable.[56]

The *Toronto Statement* also suggests that funding agencies have a role in fostering data sharing. First, the Statement advises that funding agencies should mandate rapid pre-publication data release for projects that 'have broad utility, are large in scale, are "reference" in character and typically have community "buy-in"'.[57]

Second, according to the *Toronto Statement*, funding agencies should make explicit the requirement for pre-publication data release and also become actively involved throughout the data release process, as demonstrated by the international HapMap Project and the 1000 Genomes Project, among others.[58] Among the essential roles for funding agencies is participation in the development of effective consent, security, access and governance mechanisms that protect research participants while encouraging pre-publication data release.[59]

Third, for projects generating large data sets, the *Toronto Statement* advises that funding agencies require that grant applications include data-sharing plans and undergo peer review, and notes that this is not commonly the case at the present time.[60] Experts have recommended, in the context of post-publication data release (and it is equally applicable pre-publication) that, where they do not yet exist, clear criteria should be developed for reviewers of grants to help them evaluate applicants' plans for sharing data and material.[61] Among examples of good practice cited in this regard are the NIH,[62] the UK Wellcome Trust,[63] and the UK Medical Research

---

55 Thorisson, G. and A.J. Brookes (2009), *IRBW2009 Workshop*, 13–14 May, Meeting Minutes, http://www.gen2phen.org/system/files/private/IRBW2009%20meeting%20minutes%20v2.pdf.

56 Ibid.

57 Birney, E. and T. Hudson et al. (2009), 'Prepublication Data Sharing', *Nature*, **461** (7261), 168–70.

58 Ibid.

59 Ibid.

60 Ibid.

61 Schofield, P. et al. (2009), 'Post-Publication Sharing of Data and Tools', *Nature*, **461** (7261), 171–3.

62 National Institutes of Health (2003), *Final NIH Statement on Sharing Research Data*, http://grants.nih.gov/grants/guide/notice-files/NOT-OD-03–032.html.

Council (MRC),[64] a government-funded research agency akin to the NIH.[65] Commentators note that, for these agencies, 'data-sharing plans are required in proposals, efforts are made to facilitate sharing, such as putting investigators in touch with repositories and, for some organizations, compliance is an important consideration in funding renewal'.[66]

Fourth, in terms of compliance, funding agencies should clearly express the rewards for honoring, as well as the sanctions for ignoring, their policies relating to pre-publication data release, and also consistently enforce such policies.[67] One possibility is to condition grant renewals and promotions to some degree on the extent to which the investigator shares data.[68]

B The Challenge of Crafting Appropriate Publication Policies

In light of researchers' desires to protect the data they have gathered and advance their own careers, efforts to create a culture of data sharing are inextricably linked to the question of the optimal publication policies to support data-sharing efforts. According to the *Toronto Statement*, although many scholarly works have been published by third parties reporting research findings based upon data sets released before publication, few data producers were affected by these publications that pre-dated their own.[69] Nevertheless, the *Toronto Statement* proposes that 'this ongoing concern is best addressed by fostering a scientific culture that encourages transparent and explicit cooperation among data producers, data analysts, reviewers and journal editors'.[70]

One way of achieving such cooperation is to permit investigators who contribute data to enjoy some period of exclusivity during which they will have the sole right to publish analyses of the data. The *Toronto Statement* declares that data producers should, as early as possible, and ideally before

---

63 Wellcome Trust (2010), 'Policy on Data Management and Sharing', August, http://www.wellcome.ac.uk/About-us/Policy/Policy-and-position-statements/wtx035043.htm.

64 Medical Research Council, 'Medical Research Council Policy on Data Sharing and Preservation', http://www.mrc.ac.uk/Ourresearch/Ethicsresearchguidance/Datasharinginitiative/Policy/index.htm.

65 Medical Research Council, http://www.mrc.ac.uk/About/Structure/index.htm.

66 Schofield, P. et al. (2009), 'Post-Publication Sharing of Data and Tools', *Nature*, **461** (7261), 171–3.

67 Ibid.

68 Birney, E. and T. Hudson et al. (2009), 'Prepublication Data Sharing', *Nature*, **461** (7261), 168–70.

69 Ibid.

70 Ibid.

large-scale data generation begins, produce a citable statement or 'marker paper' in which they describe the data set and their intentions in respect of analysis and publication.[71] This statement should include 'clear details about the data set to be produced, the associated metadata, the experimental design, pilot data, data standards, security, quality-control procedures, expected timelines, data release mechanisms and contact details for lead investigators'.[72] In the event that data producers request a protected time period to allow them to be the first to publish the data set, data producers should expect this period to be used only for the publication of a global analysis of the data and to expire within one year.[73]

The *Toronto Statement* further suggests that if the citable statement is a marker paper, it should undergo peer review and publication in a scientific journal. The *Toronto Statement* also recognizes other citable sources, such as digital object identifiers to specific pages on well-maintained funding agency or institutional websites. Data producers benefit from the creation of a citable reference, as it can later be used to quantify the impact of the data sets.[74]

The NIH and the UK Medical Research Council have both adopted a period of data exclusivity for data producers. The NIH Policy for Sharing of Data Obtained in NIH Supported or Conducted Genome-Wide Association Studies (GWAS) declares that investigators who contribute data to a NIH GWAS data repository will retain the exclusive right to publish analyses of the data set for a maximum of 12 months following its release via the NIH GWAS data repository. During this period of exclusivity, the NIH grants data access through data access committees to other investigators, who may analyze the data, but not submit for publication their analyses or conclusions until the expiration of the exclusivity period. The NIH also expects all investigators who access GWAS data sets to acknowledge in all publications the data producers who conducted the original study, along with the funding organization that supported the work and the NIH GWAS data repository.[75] Similarly, the UK Medical Research Council provides for a period of interim exclusivity for data producers.[76] The

---

71 Ibid.

72 Ibid.

73 Ibid.

74 Ibid.

75 National Institutes of Health (2007), 'Policy for Sharing of Data Obtained in NIH Supported or Conducted Genome-wide Association Studies (GWAS)', 28 August, http://grants.nih.gov/grants/guide/notice-files/NOT-OD-07–088.html.

76 Lowrance, W. (2006), 'Access to Collections of Data and Materials for Health Research', A Report to the Medical Research Council and Wellcome Trust,

MRC data-sharing policy provides that '[a] limited, defined period of exclusive use of data for primary research is reasonable [...]'.[77] The UK Biobank follows suit, asking researchers to recruit 500 000 participants and collect their data and samples, in return for which the researchers will gain exclusive access for a period of time or for research into certain specialty areas.[78] In general, the purpose of the period of exclusivity is to provide the data producer sufficient time to produce, organize, document, verify and analyze the data in preparation for publication in a scientific journal. After publication, the data is made available more broadly, either via a publicly available database or via an application system.[79]

The *Toronto Statement* emphasizes that data users should honor the 'scientific etiquette' that allows data producers to publish the first global analyses of their data set. Participants at the Toronto meeting urged, in cases where the publication plans of data users overlap the data producers proposed analyses, that the data users approach the producers with the goal of 'creating a mutually agreeable publication schedule, such as co-publication or inclusion within a set of companion papers'.[80] In addition, data users should acknowledge any data or materials used and the originating sources. Some of the potential methods to achieve this include the addition of metadata tags linking to data and bioresources, or a digital object identifier for resources in public repositories, which would facilitate the searching of the literature for specific bioresources. These approaches also offer appropriate recognition of data producers who comply with data release and deposition policies, making it possible to offer incentives such as funding contingent upon such sharing.[81]

1–36, http://www.wellcome.ac.uk/stellent/groups/corporatesite/@msh_grants/documents/web_document/wtx030842.pdf.

77 Medical Research Council, 'Medical Research Council Policy on Data Sharing and Preservation Policy', http://www.mrc.ac.uk/Ourresearch/Ethicsresearchguidance/Datasharinginitiative/Policy/index.htm#P16_1349.

78 Lowrance, W. (2006), 'Access to Collections of Data and Materials for Health Research', A Report to the Medical Research Council and Wellcome Trust, 1–36, http://www.wellcome.ac.uk/stellent/groups/corporatesite/@msh_grants/documents/web_document/wtx030842.pdf.

79 Reichman, J.H. and P. Uhlir (2003), 'The Public Domain: A Contractually Reconstructed Research Commons for Scientific Data in a Highly Protectionist Intellectual Property Environment', *Law & Contemporary Problems*, **66** (1 & 2), 315–462.

80 Birney, E. and T. Hudson et al. (2009), 'Prepublication Data Sharing', *Nature*, **461** (7261), 168–70.

81 Schofield, P. et al. (2009), 'Post-publication Sharing of Data and Tools', *Nature*, **461** (7261), 171–3.

Like data users, journal editors and reviewers also have an important role in the data-sharing process. Participants at the Toronto meeting recommended that journals engage actively in the dialogue about rapid prepublication data release, in both their formal guide to authors and their informal instructions to reviewers.[82] Journal editors should work in tandem with reviewers to consider carefully the specific policies regarding citation and use of the data that may be associated with certain large-scale data sets. This will help to raise the quality of analysis and also promote fairness in citation of published studies.[83]

Even a decade before the Bermuda principles, the editors of journals such as *Nucleic Acids Research* fostered the early development of GenBank and other genomic repositories by requiring researchers to deposit their data there as a precondition for publishing. Newer journals, such as the open access *Public Library of Science* journals, have made publication contingent on making the data 'freely available without restriction, provided that appropriate attribution is given and that suitable mechanisms exist for sharing the data used in a manuscript'.[84] *Nature* journals require authors to 'make materials, data and associated protocols promptly available to readers without preconditions'.[85]

Despite the increased call in many journals for data sharing, the ability of journals to compel such behavior is limited. In March 2009, for example, the journal *Epidemiology* declared that 'We invite our authors to share their data and computer code when the burden is minimal,' according to an editorial in that issue.[86] Miguel Hernán, an epidemiologist at Harvard University and a co-author of that editorial, acknowledged the trend toward data sharing but contended that forcing a sharing requirement on authors 'would be suicidal', especially with unresolved concerns over patient confidentiality, and would likely cause authors to submit their papers elsewhere.[87] Moreover, many journals have no written policy on the availability of either resources or primary data.[88]

---

82 Birney, E. and T. Hudson et al. (2009), 'Prepublication Data Sharing', *Nature*, **461** (7261), 168–70.

83 Ibid.

84 Nelson, B. (2009), 'Data Sharing: Empty Archives', *Nature*, **461** (7261), 160–63.

85 Ibid.

86 Ibid.

87 Ibid.

88 Schofield, P. et al. (2009), 'Post-publication Sharing of Data and Tools', *Nature*, **461** (7261), 171–3.

C The Complexity of Establishing a Suitable Infrastructure

The *Toronto Statement* also recommends that funding agencies offer infrastructural support by funding the creation and long-term maintenance of databases.[89] Among the costs associated with creating a suitable infrastructure to house and maintain the data and any associated physical specimens are: the maintenance of physical premises; the development of appropriate information technology; the preparation of data for storage or archiving, such as anonymising the data and documenting the variables; salaries for administrators, managers and staff; and the creation and maintenance of an accessible database. These costs are all ongoing ones.[90] Because granting agencies understandably focus primarily on research, they therefore frequently fail to invest in the infrastructural support necessary to support their archiving requirements, in a sort of 'tragedy of the commons'.[91]

The internationally collaborative nature of biomedical research exacerbates this problem by allowing each funding agency to leave the problem to another agency. Yet overcoming the infrastructure issue is crucial in order to optimize the benefits of technological progress in the life sciences. Taking human genome sequencing as one example, one source estimates that the cost of storing all known DNA sequence information in openly accessible databases is less than 1 percent of the sum necessary to generate such sequence data.[92]

V THE NEED FOR RESTRICTED ACCESS DATABASES

Notwithstanding the prescriptions in the *Toronto Statement* for encouraging data sharing, not all data are suitable for an open access approach. Indeed, samples and data that retain their link with an individual research participant must be protected with additional procedures. Examples of such data include personal information such as individual-level genotypic and phenotypic data; exposure to drugs and environmental factors; and

---

89 Birney, E. and T. Hudson et al. (2009), 'Prepublication Data Sharing', *Nature*, **461** (7261), 168–70.
90 Gitter, D. (2010), 'The Challenges of Achieving Open Source Sharing of Biobank Data', *Biotechnology Law Report*, **29** (6), 623–35.
91 Nelson, B. (2009), 'Data Sharing: Empty Archives', *Nature*, **461** (7261), 160–63.
92 European Life Science Infrastructure for Biological Information, *ELIXIR: Data for Life*, http://www.elixir-europe.org/bcms/elixir/Documents/Elixir_brochure.pdf.

pedigree data, including information about familial relationships, along with analyses of such data.[93] Experts have offered several technical and policy approaches to preserve patient privacy. They recommend, inter alia, the establishment of policies to assess the credentials of data users; the execution of clear contracts with data users that define the appropriate use of data; the formalization of liability rules for misuse of data; and the use of a technical data management approach that increases the number of research participants whose data will be aggregated where the data is considered more sensitive.[94] Experts note that subjecting data to access controls erects a barrier that limits the number of researchers who will reuse the resources. What is more, researchers may decline to abide by overly complicated data access agreements.[95]

Nevertheless, restricted access databases are also useful in terms of building upon the collaborations already established by scientific researchers, particularly those involved in small science.[96] One such biobank is the UK DNA Banking Network (UDBN), which is a secondary biobank, meaning that it aggregates and manages tissue samples and associated data gathered by clinicians who gather the samples in the course of studying particular diseases.[97]

In order to establish UDBN, the UK Medical Research Council offered grants to centers that housed DNA collections. These awards required the collections to be maintained as 'shared national resources' and 'made available to collaborators'. In addition, awardees were 'required to transfer a portion of each sample' to the network and 'to add any genotype data they obtain to the common database'.[98]

In granting access to its collections, the UDBN explicitly rejected the unrestricted 'open access' model, instead developing a 'fair access' regime, which derives inspiration from the 2003 United Nations Educational Scientific and Cultural Organization (UNESCO) International Declaration on

---

93 National Institutes of Health (2007), 'Policy for Sharing of Data Obtained in NIH Supported or Conducted Genome-Wide Association Studies (GWAS)', 28 August, http://grants.nih.gov/grants/guide/notice-files/NOT-OD-07–088.html.

94 Malin, B. et al. (2010), 'Technical and Policy Approaches to Balancing Patient Privacy and Data Sharing in Clinical and Translational Research', *Journal of Investigative Medicine*, **58** (1), 11–18.

95 Flicek, P. (2010), 'Challenges for the Data Management and Analysis of Large-Scale Human Genome Sequencing', Advances in Genomics conference, 29 January, http://www.advances-in-genomics.org/presentations/Flicek.pdf.

96 Yuille, M. et al. (2010), 'The UK DNA Banking Network: A "Fair Access" Biobank', *Cell Tissue Bank*, **11** (3), 241–51.

97 Ibid.

98 Ibid.

Human Genetic Data.[99] The UDBN distributes data via the project website to third party researchers, who apply online for registration. After verifying the researcher's credentials, the UDBN grants access to a restricted area of the website. The third party researchers can then communicate online with the data collectors in order to negotiate collaboration. If a collaborative relationship is successfully negotiated, the UDBN permits the data collector to grant the third party access to fuller data.[100]

The fair access model builds upon researchers' willingness to share data with collaborators, and concomitantly denies access to non-collaborators.[101] This approach also accepts a collector's right to 'exclusive access to his/her collection for the purposes of the investigational goals stated in the initial collection proposal', recognizing that granting a 'first mover' advantage is likely to motivate scientific discovery.[102] This model also acknowledges that tensions frequently arise when a potential collaborator requests data access from a data producer. Thus, the restricted access model provides a necessary complement to the open access databank, particularly in the realm of 'small science'.

---

99 Ibid.

100 Ibid.

101 Ibid.

102 Ibid.

7. Building with BioBricks: constructing a commons for synthetic biology research

Alison McLennan

On 21 May 2010, scientists at the J Craig Venter Institute in the United States announced that they had made a 'self-replicating synthetic bacterial cell'.[1] This achievement is the most famous outcome of synthetic biology, a burgeoning new field of biotechnology research. The achievement was made possible by our increasing ability to build DNA from its components and to stitch it together to form synthetic genomes.

The disciplines of molecular biotechnology, chemical engineering, genomics, and information technology converge in this new field of synthetic biology. With its promise of improved medical treatments and greener energy sources, along with its potential risks of bioterrorism and damage to the environment or human health, synthetic biology has been increasingly capturing the public imagination and attracting government and commercial attention. This excitement reached a crescendo with the announcement of the successful construction of the synthetic bacterial cell. Even the United States President Barack Obama joined the ensuing debate, asking his Presidential Commission for the Study of Bioethical Issues to investigate synthetic biology.[2]

This chapter will provide an overview of the field of synthetic biology, focusing in particular on the branch of synthetic biology that uses 'standard biological parts'. The chapter will then address the approach to intellectual property, sharing and commons development in this 'parts agenda'.

---

[1] Gibson, D. et al. (2010), 'Creation of a Bacterial Cell Controlled by a Chemically Synthesized Genome', *Science*, **329** (5987), 52–6.

[2] Their report was released in December 2010: The Presidential Commission for the Study of Bioethical Issues (2010), *New Directions: The Ethics of Synthetic Biology and Emerging Technologies*, Washington, DC: The Presidential Commission for the Study of Bioethical Issues, http://www.bioethics.gov/documents/synthetic-biology/PCSBI-Synthetic-Biology-Report-12.16.10.pdf.

I SYNTHETIC BIOLOGY

Synthetic biology refers to 'the design and fabrication of biological components and systems that do not already exist in the natural world', and 'the re-design and fabrication of existing biological systems'.[3] These new endeavours represent a departure from previous life sciences research in several respects. Firstly, molecular biologists have until now been focusing on seeking knowledge of living systems by breaking organisms and cells down into smaller and smaller entities, and understanding how these entities operate. This can be described as a reductionist approach to understanding biology. Some research in synthetic biology involves seeking to improve our understanding of biology by beginning to *build* life, taking these tiny parts and building new living systems from the bottom up.

Secondly, the preceding classical recombinant DNA technology aimed to manipulate the genes of various organisms, for example by adding a gene to confer a new property on a bacterium. Synthetic biology goes beyond this approach by attempting to build new organisms with functions that do not exist in nature.[4] While classical recombinant DNA technology can be seen as 'softening unpleasant edges' or adding value to existing organisms, 'synthetic biology creates life forms which are meant not to have any edges from the start', representing a shift from *manipulating* to actually *creating* organisms.[5]

Thirdly, synthetic biology has an emphasis on de-skilling. For example, one of the goals of the parts agenda is to make biological parts that can be easily assembled to make small machines that work in predictable ways.

Synthetic biology is a science driven by applications. Leading synthetic biologist Jay D. Keasling has commented that:

> Synthetic biology is the engineering of biology to solve important problems. It is basic research with a focus on application. This emerging field of fundamental science has great potential for developing solutions to large-scale societal challenges.[6]

---

3 Synthetic Biology Research Community, 'FAQ – What is Synthetic Biology?' http://syntheticbiology.org/FAQ.html.

4 Boldt, J. and O. Müller (2008), 'Newtons of the Leaves of Grass: Certain Ethical Implications of Synthetic Biology Research Go Beyond Those of Genetic Engineering', *Nature Biotechnology*, **26** (4), 387–9, 387.

5 Ibid., 387–8.

6 Keasling, J. (2010), 'Evidence to Committee on Energy and Commerce, United States House of Representatives, Effects of Developments in Synthetic Genomics and Implications for Energy and Health', Washington, DC, 27 May, 1.

The most high-profile of the societal challenges that synthetic biology hopes to address are those in medicine and energy. Potential medical applications include synthetic vaccine development, synthetic drug production and use of synthetic bacteria or viruses as 'pre-programmed cellular robots'[7] to target diseased cells. Several groups are working on using synthetic biology to produce biofuels, and significant private investment is occurring in this arena. Synthetic Genomics Inc., a private company formed to develop and commercialise synthetic biology research done at the J Craig Venter Institute, is working on a range of biofuels projects. These include a partnership with ExxonMobil Research and Engineering Company to develop biofuels from photosynthetic algae.[8]

The term 'synthetic biology' refers to several different approaches to biological engineering. These include approaches that can be described as the 'parts agenda'[9] and the 'construction approach'. Each of these angles on synthetic biology represents a different worldview as to the best way to move biological engineering forward.

In synthetic biology's parts agenda, traditional engineering principles are being combined with molecular and cell biology in order to reduce the complexity of biological engineering and increase its predictability and reliability.[10] Engineers are trying to apply engineering concepts such as standardisation, modularity and circuit design to DNA and genes. At the University of California Berkeley, Jay D. Keasling and colleagues have used these concepts in combination with metabolic engineering to engineer a strain of the bacteria *Escherichia Coli* to synthesise a precursor to the anti-malarial drug artemisinin, which is usually extracted from a plant.[11]

In the construction branch of synthetic biology, scientists are trying to build entirely synthetic organisms. Most famous among such projects is the endeavours of the J Craig Venter Institute in the United States to build a synthetic bacterial cell. The construction approach also includes efforts to generate a minimal organism 'that has been stripped down to a few basic

---

7 Mohan-Ram, Vid S. and J.M. Waxman (2008), 'Synthetic Biology Patent Applications Expected to Present New Challenges', *Life Sciences Law & Industry Report*, **2** (12), 1–3, 1.

8 ExxonMobil, 'Algae Biofuels', http://www.exxonmobil.com/Corporate/energy_climate_con_vehicle_algae.aspx.

9 This term has been used by Joachim Henkel and Stephen M. Maurer; see for example Henkel, J. and S. Maurer (2009), 'Parts, Property and Sharing', *Nature Biotechnology*, **27** (12), 1095–8.

10 Endy, D. (2005), 'Foundations for Engineering Biology', *Nature*, **438** (7067), 449–53.

11 Keasling, J. (2008), 'Synthetic Biology for Synthetic Chemistry', *ACS Chemical Biology*, **3** (1), 64–76.

evolutionary functions' and then adding on genes to express new functions or traits of interest.[12]

II THE SYNTHETIC BIOLOGY COMMONS

A The Parts Agenda

Synthetic biology's parts agenda was founded by MIT scientists Drew Endy, Randy Rettberg and Tom Knight.[13] These scientists had a focus on working with standard parts, as they all came from an engineering background. They moved into biology because they thought there was a gap in the field where engineers could contribute.[14] These scientists saw the 'engineering of useful synthetic biological systems' as 'still an expensive, unreliable and *ad hoc* research process'.[15] They blamed biology's lack of standards for biological function and other measures for costs and inefficiencies. Thus, they saw a need for 'the development of technologies and promulgation of standards that support the definition, description and characterization of the basic biological parts, as well as standard conditions that support the use of parts in combination'.[16] The goal was to transform biology from 'an artisanal process [...] to a true engineering discipline'.[17] They wished to reduce its unreliability and complexity in order to make 'routine the engineering of synthetic biological systems that behave as expected'.[18] Simply put, they wanted to 'make biology easier to engineer'.[19] This new approach has been eloquently summarised by Pleiss:

---

12 Boldt, J. and O. Müller (2008), 'Newtons of the Leaves of Grass: Certain Ethical Implications of Synthetic Biology Research Go Beyond Those of Genetic Engineering', *Nature Biotechnology*, **26** (4), 387–9, 387.

13 Rettberg, R. (2009), 'Evidence to the US National Academies Opportunities and Challenges in the Emerging Field of Synthetic Biology', Washington, DC, 10 July, 25.

14 Ibid., 26.

15 Endy, D. (2005), 'Foundations for Engineering Biology', *Nature*, **438** (7067), 449–53, 449.

16 Ibid., 450.

17 Boyle, James (2008), *The Public Domain: Enclosing the Commons of the Mind*, New Haven, CT and London: Yale University Press, 172.

18 Endy, D. (2005), 'Foundations for Engineering Biology', *Nature*, **438** (7067), 449–53, 452.

19 See for example Registry of Standard Biological Parts, 'Welcome to the Registry of Standard Biological Parts', http://partsregistry.org/Main_Page.

> traditional biological and biotechnological research aims at describing and modifying biological systems by analysing a large diversity of natural parts [...] while synthetic biology builds on a small inventory of robust, standardized parts to construct from them new, synthetic systems.[20]

In the parts agenda standardisation is achieved in two ways: through the development, definition and distribution of standard biological parts themselves; and through the development of technical standards for use in parts research. A 'standard biological part' is a sequence of DNA with well-characterised properties that acts as a 'functional unit'.[21] These parts are intended to operate much like LEGO^{R} bricks. As described by synthetic biologist Christina Smolke, these very simple functions can be added together to form larger devices. She explains that the aim is to have a 'very large library of refined parts that you can plug and play into the device platforms that you build'.[22] We could also draw an analogy with electrical circuits, as synthetic biologists are attempting to build 'genetic circuits' comprising functional units of DNA. Libraries of parts relating to different classes of molecules are needed for experiments in different areas, such as metabolites for biosynthesis, or disease biomarkers for therapeutics research.[23] The technical standards being developed relate to matters such as defining and describing parts, assembling them in the laboratory and measuring the activity of molecules encoded by parts (such as enzymes and other proteins).

B The BioBricks Foundation and the Registry of Standard Biological Parts

The BioBricks Foundation is a non-profit organisation founded in 2003 by scientists from MIT, Harvard and the University of California San Francisco.[24] The Foundation arranges seminars and conferences on synthetic biology as well as convening discussion on development of legal and technical standards.

---

[20] Pleiss, J. (2006), 'The Promise of Synthetic Biology', *Applied Microbiological Biotechnology*, **73**, 735–9, 736.

[21] Registry of Standard Biological Parts, 'Help: About the Registry', http://partsregistry.org/Help:About_the_Registry.

[22] Smolke, C. (2009), 'Evidence to the US National Academies, Opportunities and Challenges in the Emerging Field of Synthetic Biology', Washington, DC, 9 July, 142.

[23] Ibid., 143.

[24] The BioBricks Foundation, 'The BioBricks Foundation', http://www.biobricks.org/index.php.

A 'BioBrick™ standard biological part' is a particular type of biological part, one that 'meets the technical and legal standards set forth by the BioBricks Foundation'.[25] It is a 'nucleic acid encoded-molecular biological function (e.g. turn on/off gene expression), along with the associated information defining and describing the part'.[26] BioBrick parts are made available to synthetic biologists through the Registry of Standard Biological Parts, based at MIT. The overarching goal of the Foundation and the Registry is to support the parts agenda. More specifically, the technical and scientific goals are the 'development of well-specified, standard and interchangeable biological parts', 'to support development of open technical standards that define BioBricks™ standard biological parts', and 'to develop and provide educational and scientific materials to allow the public to use and improve existing BioBrick™ standard biological parts, and contribute new BioBrick™ standard biological parts'.[27]

The Registry is largely an online resource for researchers in synthetic biology, though physical samples of the DNA encoding parts are also provided. It is described as a 'continuously growing collection of genetic parts that can be mixed and matched to build synthetic biological devices and systems'.[28] This online resource now contains thousands of parts.

Users can search the Registry's online catalogue under various characteristics, such as biological function, to find parts that fit the specific requirements of their experiments. The catalogue provides information about each part including the individual who designed it, its DNA sequence, what it can be used for, its compatibility with BioBricks™ working standards and protocols, and whether it has been used by other Registry users. Designers of parts can record particular notes or instructions, and users can record their experience using parts so that this can be shared by other researchers.

Lab groups must register, and be approved, in order to participate in the Registry. It is available only to academic researchers and participants in the international Genetically Engineered Machine competition (iGEM) for undergraduates, discussed further below. In 2010, there were 90 academic labs registered as members of this 'Registry community'.[29] When a

---

25 The BioBricks Foundation, 'FAQ', http://www.biobricks.org/FAQ.php.

26 Registry of Standard Biological Parts, 'Help: About the Registry', http://partsregistry.org/Help:About_the_Registry; The BioBricks Foundation, 'FAQ', http://www.biobricks.org/FAQ.php.

27 The BioBricks Foundation, 'Our Goals', http://www.biobricks.org/Our_Goals.php.

28 Registry of Standard Biological Parts, 'Welcome to the Registry of Standard Biological Parts', http://partsregistry.org/Main_Page.

29 Registry of Standard Biological Parts, 'iGEM: Synthetic Biology Based on Standard Parts', http://ung.igem.org/Main_Page.

researcher has identified a part they wish to access, there are several ways to go about this. They can extract the part from one of the 'kits' of all parts that are distributed regularly, request it from the Registry, assemble it from small strips of natural DNA in their lab, or have the DNA commercially synthesised.[30] The synthesis option is still expensive, but the cost is ever-decreasing. Requesting a part from the Registry involves emailing the request to the Registry, receiving shipment of a piece of agar containing the part of interest (inside bacteria), and following a protocol to make up a solution of bacteria in the lab.[31] The researcher is then ready to use the part in experiments.

C Building a BioBricks Commons

As well as their technical scientific objectives, the BioBricks Foundation and Registry also aim to create an openly accessible resource for sharing of biological parts and associated information. In fact, it can be argued that they aim to create a 'commons' for synthetic biology researchers.[32] The BioBricks Foundation and Registry have created an online space for the sharing of experimental data, knowledge and ideas, and a system for the sharing of physical biological materials. They have also worked to build a community of synthetic biologists with a shared philosophy and common goals. Specific commons-related objectives of the BioBricks Foundation include 'the development of an open community of biological engineers', and 'to develop and implement legal strategies to ensure that parts remain freely available to the public'.[33]

The scientific and open access philosophy of the Foundation has been summarised in its BioBrick Public Agreement:

> The BioBricks Foundation, Inc. (the 'Foundation') was established to foster and advance innovation, research, standardization, and education in synthetic biology through the open design, construction, distribution, understanding, and use of BioBrick™ compatible parts, namely standardized genetic materials and

---

30 See for example Registry of Standard Biological Parts, 'Part:BBa_I742152: Get Part', http://partsregistry.org/partsdb/get_part.cgi?part=BBa_I742152.

31 See Registry of Standard Biological Parts, 'Help: Requesting Parts', http://partsregistry.org/Help:Requesting_Parts.

32 Boyle, James (2008), *The Public Domain: Enclosing the Commons of the Mind*, New Haven, CT and London: Yale University Press, 176–8.

33 Registry of Standard Biological Parts, 'Help: About the Registry', http://partsregistry.org/Help:About_the_Registry; The BioBricks Foundation, 'Our Goals', http://www.biobricks.org/Our_Goals.php.

> associated functional information, in ways that benefit the world. The Foundation believes that a free and easy-to-use legal framework for sharing and making use of engineered genetic materials underlies and serves these goals.[34]

The BioBricks Foundation and Directors of the Registry engage in a number of community building, leadership and education activities. The importance of this sharing and community building in synthetic biology is a theme in various published articles and testimonies to government committees by members of the parts agenda. For example, BioBricks Foundation President Drew Endy has stated that 'vibrant, open research communities and strategic leadership are necessary to ensure that the development and application of biological technologies remains overwhelmingly constructive'.[35]

The BioBricks Foundation acts as a focal point for community discussion and development of standards (including legal mechanisms) for research involving BioBrick parts. Any member of the synthetic biology community can access and comment on draft standards through the Foundation's website. The Foundation encourages the use of these openly developed BioBricks standards.

Since 2002, the Foundation has convened four large international conferences about synthetic biology, bringing the field's scientists together with lawyers, ethicists and government representatives. In addition to scientific achievements, some ethical and regulatory issues, such as the risk of bioterrorism, have been discussed at these interdisciplinary meetings.

The Registry also runs the annual international Genetically Engineered Machine (iGEM) competition, a synthetic biology competition for undergraduate students. iGEM has grown from a small competition featuring a handful of teams in 2004 to an international event participated in by more than 130 teams in 2010.[36] iGEM teams now come from universities, and even some high schools, from the United States, Canada, the European Union, Asia, South America, Australia and Africa. In fact, iGEM has grown so much that the annual jamboree is now too big for MIT's largest

---

34 Endy, D. et al. (2010), 'The BioBrick™ Public Agreement: DRAFT Version 1a', http://dspace.mit.edu/bitstream/handle/1721.1/50999/BPA_draft_v1a.pdf?sequence=1.

35 Endy, D. (2005), 'Foundations for Engineering Biology', *Nature*, **438** (7067), 449–53, 449.

36 Registry of Standard Biological Parts, 'iGEM 2011: Regions/iGEM Growth', http://2011.igem.org/Regions/iGEM_Growth.

auditorium.[37] From 2011 a regional structure will be put in place, with branches to be established in the Americas, Asia and Europe.[38]

Teams of undergraduates are led by experienced researchers at their university. They are given a set of biological parts, and over the summer they attempt to design and construct a functioning biological device. For example, in 2007 the University of California Berkeley's iGEM team attempted to construct a genetic device that would allow bacteria to make haemoglobin, generating 'Bactoblood'.[39]

In addition to lab-based experiments, teams can also include a 'human practices' element in their project, allowing them to investigate ethical, legal or social issues raised by synthetic biology. The Bactoblood team chose to investigate the patentability of their bacterial blood. Other teams have looked into issues such as public awareness and understanding of synthetic biology.

In November of each year all the participants come together for a jamboree held at MIT. They present their work, and judging of the competition takes place. Leading up to iGEM, there is a program of workshops and training, in which students and scientists can learn from leaders in the field. iGEM also allows students to learn about the broader issues surrounding synthetic biology, such as through outreach on biosecurity issues by members of the Federal Bureau of Investigation.[40]

Through the excitement and enthusiasm generated by iGEM, future synthetic biologists are created. iGEM has thus become a mechanism for enculturation of a whole new generation of synthetic biologists in the open, sharing philosophy of the BioBricks Foundation and the Registry, and in the understanding of public engagement, security, safety and intellectual property issues. Participants in iGEM have already gone on to become scientists in the field.[41] Drew Endy has commented that 'we have now, in a

37 Rettberg, R. (2009), 'Evidence to the US National Academies, Opportunities and Challenges in the Emerging Field of Synthetic Biology', Washington, DC, 10 July, 32.

38 Registry of Standard Biological Parts, 'iGEM 2011: Regions/Overview', http://2011.igem.org/Regions/Overview.

39 See University of California Berkeley 2007 iGEM Team (2007), 'Bactoblood', http://parts.mit.edu/igem07/index.php/Berkeley_UC.

40 You, E. (2010), 'Evidence to the Presidential Commission for the Study of Bioethical Issues, Synthetic Biology', Washington, DC, 9 July.

41 Registry of Standard Biological Parts, 'iGEM: Synthetic Biology Based on Standard Parts', http://ung.igem.org/Main_Page.

bottom-up, grass-roots fashion, de facto installed a genetic-engineering curriculum for the future of our field in 120 schools worldwide'.[42]

D Draft Legal Arrangements

In 2009, draft legal agreements titled 'BioBrick Public Agreement' were released for comment, and an updated version was released in January 2010. They were developed by a team including legal academics and interns at the Samuelson Clinic, University of California Berkeley Law School, and lawyers from a private firm specialising in intellectual property and technology.[43] The BioBrick Public Agreement contains a Contributor Agreement and a User Agreement. These are web-based click-through agreements. Records of the Contributor Agreements will be maintained by the BioBricks Foundation. The agreements focus on the contributed 'Materials', defined as 'the particular standardized genetic material(s), their uses, and any associated sequence or functional information' specified in the Contributor Agreement.

The Contributor is defined as the 'person, company, institution, or other entity submitting the Materials' to the Registry, and who enters into the agreement. They must provide information about all intellectual property relating to the contributed materials by disclosing any patents or patent applications, copyright or data rights. This includes rights relating to genetic sequences. The Contributor can indicate that to the best of their knowledge there are no intellectual property rights by stating 'none' in the relevant part of the agreement.

By entering into the Contributor Agreement, the Contributor agrees to permit Users to use the contributed Materials under the terms of the User Agreement. If the Contributor does have intellectual property rights in the contributed materials, they must 'irrevocably agree' not to assert or threaten to assert these patents or property rights (or future rights from patent applications), 'not to invite to license', and 'not to enforce the rights in any manner against or adverse' to the BioBricks Foundation or the Users.

The Contributor also agrees not to charge fees for use of the parts apart from nominal costs of manufacture and shipping. They can request that Users attribute the parts to the Contributor when they distribute or commercialise material arising from their use of the parts. The Agreement

---

42 Mooallem, J. (2010), 'Do-It-Yourself Genetic Engineering', *The New York Times Online*, 10 February.

43 See Endy, D. (2009), 'Open Biotechnology and the BioBrick Public Agreement', http://openwetware.org/images/f/fd/Why_the_BPAv1.pdf.

does not prevent Contributors from entering into a separate agreement with Users on different terms, so long as any such separate agreement does not 'diminish or derogate from' the rights of the User in the Materials provided under the User Agreement.

The User is defined as 'the person who may request and use the Materials, or the company, institution, or other entity on whose behalf User is authorized to act'. Once they click 'agree', Users 'have the right to use the Materials insofar as the Materials are within the public domain or the Contributors have promised not to assert any of the Contributors' proprietary rights against User by way of the applicable Contributor Agreements'. Users agree that if the Contributor has requested attribution, the User will make efforts to acknowledge the Contributor when they make available, commercialise or otherwise distribute the contributed Materials. If they undertake any of these distribution activities, including distributing the materials in a modified form, they must also acknowledge that the materials were used under the BioBrick Public Agreement by conspicuously including the BioBrick Public Agreement logo 'in all packaging or product inserts, publications, and grant related materials'.

From these agreements, it appears that a new Contributor Agreement would need to be executed each time a particular Contributor donates new materials to the Registry. However, an individual User would only need to agree to the User Agreement once for the Agreement will apply to all their use of material contributed under any Contributor Agreements.

E Why Create an Open Access Resource?

The reasons for creating a commons for synthetic biology research in this way are not expressly stated on the websites of the BioBricks Foundation or the Registry, or in articles published by their founders. The three scientists who conceived the parts agenda came from a computer engineering background. They were 'inspired' by open source software,[44] and may be trying to replicate the success of this approach in the field of synthetic biology research. Indeed, open biotechnology generally can be seen to have grown from the open source software movement.[45]

Concerns about intellectual property problems getting in the way of synthetic biology research are present in comments the founding scientists have made in various fora. In *The Public Domain*, intellectual property

---

44 Boyle, James (2008), *The Public Domain: Enclosing the Commons of the Mind*, New Haven, CT and London: Yale University Press, 176.

45 Joly, Y. (2010), 'Open Biotechnology: Licenses Needed', *Nature Biotechnology*, **28** (5), 417–19, 417.

expert James Boyle tells of synthetic biologists coming to him to ask how the categories of intellectual property would apply to their research, and how they could keep the science open.[46] Boyle found the scientists' desire to protect their research from being hindered by legal issues depressing, lamenting that 'Surely it would be better to have them doing, well, science?'[47] He suggested that an option would be to 'simply put all their work into the public domain immediately', and noted that this was occurring through the Registry of Standard Biological Parts.[48] This means that patents could not be sought for the publicly disclosed material.[49] He saw this as demonstrating a new development, the creation of a 'privately constructed commons where the public domain created by the state does not give you the freedom that you believe creativity needs in order to thrive'.[50]

Drew Endy has noted some practical problems in synthetic biology research, stating that we need legal standards to define the way in which parts from many different sources can be shared.[51] In his view, negotiating Material Transfer Agreements in order to get access to biological materials is time consuming. If a researcher can access the DNA sequence of the part they are interested in, it may be more efficient to have it synthesised rather than negotiate a Material Transfer Agreement.[52] Endy concluded that 'new low- or zero-cost legal strategies that facilitate and protect the exchange of the genetic information defining parts, and the use of parts in combination, are needed'.[53]

iGEM Director Randy Rettberg argues that patenting in biotechnology plays out differently from in the computer industry.[54] In the computer industry no particular individual patents seemed to be 'critical', and access to patented inventions could occur through cross licensing.[55] However in the biotechnology industry, Rettberg was concerned about the ability of

---

46 Boyle, James (2008), *The Public Domain: Enclosing the Commons of the Mind*, New Haven, CT and London: Yale University Press, 174.

47 Ibid.

48 Ibid., 177.

49 Ibid.

50 Ibid., 178.

51 Endy, D. (2005), 'Foundations for Engineering Biology', *Nature*, **438** (7067), 449–53, 450.

52 Ibid.

53 Ibid.

54 Rettberg, R. (2009), 'Evidence to the US National Academies Opportunities and Challenges in the Emerging Field of Synthetic Biology', Washington, DC, 10 July, 55–6.

55 Ibid., 56.

individual patents to impact upon the intellectual property environment. He cautioned that, in biotechnology:

> [...] individual patents could have extreme financial value. And that, I believe has produced a tendency in the academic community to use patents as lottery tickets to say [']I am doing an important piece of work. I am going to make sure it's patented. I don't expect that I'm going to make any money on that patent. But if anybody ever makes a billion dollars, I should get my share.['] And therefore a patent is put in place which acts as an intellectual block to future activity but for the most part doesn't produce a financial return.[56]

Rettberg is also clearly worried about the impact that patents could have on iGEM, commenting that 'There is not an actual research exemption. The courts struck that down. There is a de facto one, and will it work for iGEM in the future? I worry that some patent troll will send a letter to MIT and I will be shut down.'[57]

F Some Limitations of the Draft Legal Arrangements

i Patent issues

Some commentary on the optimal form of an open source mechanism for synthetic biology has assumed that individual parts are patentable.[58] This chapter will not test the actual patentability of parts as a legal issue. However, we can note that the United States Patent and Trademark Office databases do not currently contain any patents or patent applications containing the terms 'BioBrick' or 'standard biological part'.[59] This is to be expected, given the open access ethos of the parts agenda we would not expect to find patents on individual parts that have been put into the public domain. Theoretically, however, an invention meeting the definition of a standard biological part or composition of such parts could be patented.

The lack of current patents over parts does not mean that experiments using registered parts do not infringe patents. In his evidence to the Presidential Commission for the Study of Bioethical Issues, Drew Endy confirmed that 'the BioBricks Foundation is a public benefit organization.

---

56 Ibid.

57 Ibid., 38.

58 See for example Henkel, J. and S. Maurer (2007), 'The Economics of Synthetic Biology', *Molecular Systems Biology*, **3** (117), 1–4.

59 Search of United States Patent and Trademark Office Patents and Patent Applications databases, 26 November 2010.

It doesn't own the BioBrick parts. We haven't cleared freedom to operate on the uses of the genetic components.'[60] He went on to explain:

> In 2008, 1500 new BioBrick parts came into the competition, contributions from all over the world. If we wanted to be upstanding citizens in the world of property rights, we'd have to try and get claims on those and play within the patent-based system.[61]

Intellectual property expert Arti K. Rai also gave evidence to the Presidential Commission, stating that although a patent landscape study of the MIT Registry has not been done, there could be many relevant patents.[62] These patents or patent applications might cover DNA sequences, devices comprising multiple parts, or processes using parts. Rai stated that patent thickets have been a problem in the information technology, so this could also become problematic in synthetic biology research as there are currently many patents on gene sequences.[63]

In its current form, the Registry does not provide information about the intellectual property status of registered parts. Intellectual property status could include issues such as whether the sequence of the part overlaps with any patented DNA sequences or sequences for which patent applications are pending, and whether the contributors of that part have the authority to make the sequence of the part freely available or to allow the Registry to distribute the physical DNA comprising the part by registering it. The lack of intellectual property information means that if a researcher wanted to be sure they were not infringing any patents they would need to conduct patent searches for the sequences of each part they used and all experimental methods.

The iGEM website also does not contain information about patentability of parts and who would own any intellectual property that iGEM teams might generate. Some iGEM teams have filed patent applications on their work, for example the California Institute of Technology's 2010 team has a provisional patent on their '3D printing process using dual-wavelength light-induced lysis', though the biological parts used in this project were contributed to the Registry.[64]

---

60 Endy, D. (2010), 'Evidence to the Presidential Commission for the Study of Bioethical Issues, Synthetic Biology', Washington, DC, 8 July.

61 Ibid.

62 Rai, A. (2010), 'Evidence to the Presidential Commission for the Study of Bioethical Issues, Synthetic Biology', Philadelphia, Penn, 13 September.

63 Ibid.

64 California Institute of Technology 2010 iGEM Team (2010), 'Team:Caltech/Acknowledgments', http://2010.igem.org/Team:Caltech/Acknowledgments.

So, scientists will have to do some investigating if they wish to be sure that they are not infringing any patents when working with parts and devices. However, in her evidence to the Presidential Commission Rai argued that research shows academic life scientists are quite unconcerned about infringing patents in their work, and that the companies that own these patents tend not to sue them for infringement.[65] Another witness, Ashley Stevens of Boston University, concurred that 'infringement suits are highly expensive, probably a minimum of $5 million, and there's rarely sufficient economic value in basic research to justify enforcement actions'.[66] This suggests it would be valid to question how likely scientists working in synthetic biology are to be concerned about infringing patents when using material from the Registry.

In any case, if the BioBrick Public Agreement comes into effect, Contributors will have to provide information about the intellectual property status of their contributed material 'to the best of their knowledge'. If this information is then made easily available through the Registry, Users will know that the Contributor is not going to pursue them for patent infringement, and will be aware of any patents held on that material by other parties. However, this mechanism depends on accurate disclosure of intellectual property status by Contributors and will only apply prospectively, so that parts currently in the Registry will continue to have unclear intellectual property status.

ii Incentive to contribute

Although the Registry is said to be available only to academic labs, the agreements are expressly geared not only at academic researchers but also start-up companies and bioproducts firms. It could be queried why scientists, particularly in commercial enterprises, would wish to contribute parts or data to the Registry if they are not required to do so in order to access the information contained there. If disclosing information means its value is potentially reduced, what would be the incentive? In her evidence to the Presidential Commission Rai explained that various studies have demonstrated that scientists are very reluctant to share data, even *after* they have published a paper about their work.[67] According to economists Joachim Henkel and Stephen Maurer, the potential for patent infringement in

---

65 Rai, A. (2010), 'Evidence to the Presidential Commission for the Study of Bioethical Issues, Synthetic Biology', Philadelphia, Penn, 13 September.

66 Stevens, A. (2010), 'Evidence to the Presidential Commission for the Study of Bioethical Issues, Synthetic Biology', Philadelphia, Penn, 13 September.

67 Rai, A. (2010), 'Evidence to the Presidential Commission for the Study of Bioethical Issues, Synthetic Biology', Philadelphia, Penn, 13 September.

research using biological parts is high, and the need for scientists to disclose parts to the Registry is so pressing that an obligation to share in this way should be made a requirement of public funding grants.[68]

It has been suggested that synthetic biology can be described as a 'democratizing innovation effort'.[69] In his work *Democratizing Innovation*, technological innovation expert Eric von Hippel investigates the phenomenon of 'user innovation' and discusses the reasons why user innovators might freely release their innovations into a commons. He explains that many innovators have a dim view of the value of obtaining a patent (though noting that the pharmaceutical industry is an exception to this trend), and that the time and cost involved in patenting can make this avenue difficult for individuals and small or medium firms.[70] Also, trade secrets often cannot be kept secret for very long.[71] In these circumstances, it might make practical sense for a user to freely release their innovation rather than protect it with intellectual property rights.[72] Von Hippel also argues that 'innovators' active efforts to diffuse information about their innovations suggest that there are positive, private rewards to be obtained from free revealing'.[73] These include boosting the innovator's reputation amongst their peers, and benefits that flow from the fact that freely released innovations are more likely to be widely used by others, such as benefits from improvement or 'debugging' of the innovation by other users.[74]

A number of these influences can be seen at work in the BioBricks commons. Its formalised sharing and collaboration mechanisms enhance the benefits discussed above that scientists could expect from putting their work into the public domain individually. There may be limited incentive to patent biological parts, given that an open access resource already exists and could reduce the market for exploitation of the patent. The costs of patenting are also problematic, with Drew Endy stating that patenting the parts already contained in the Registry would cost far more than the project's entire budget:

---

68 Henkel, J. and S. Maurer (2009), 'Parts, Property and Sharing', *Nature Biotechnology*, **27** (12), 1095–8.

69 Johnson, R. (2009), 'Evidence to the US National Academies, Opportunities and Challenges in the Emerging Field of Synthetic Biology', Washington, DC, 10 July, 50.

70 Hippel, Eric von (2005), *Democratizing Innovation*, Cambridge, MA and London: The MIT Press, 84.

71 Ibid., 81–2.

72 Ibid., 85.

73 Ibid., 86.

74 Ibid., 86–9.

> I'm not for or against patents per se. But in this context, doing that for 1500 uses of genetic functions would cost about $25,000 per each one, which adds up to $37.5 million. The budget for this student and educational event distributed worldwide is $3.5 million. So the scale and pace of work in the technology is, in this case, outstripped and created not a governance challenge but I think an ownership, sharing, and innovation challenge.[75]

Contributing a biological part to the Registry could certainly be seen to enhance the reputation and credibility of the part's designers amongst their synthetic biologist peers by publicising the work, particularly if the part has been successfully used by the designers in experiments. The synthetic biology community is relatively small, and the scientists working in the field know who is 'doing good work'.[76] Similar to benefits the designers would obtain from publishing a paper about their work, contributing a part to the Registry allows the designers to be identified and contacted about their achievement by interested peers, who might offer suggestions about using and improving the part or opportunities to perform collaborative research. All Registry users are potentially able to resolve problems relating to a part or improve it, and they can add this knowledge to the Registry for the whole community's benefit. In this way the original designers benefit from putting their part into the public domain.

Use of a biological part increases its value, because this generates knowledge about how the part actually worked, reducing the effort and cost involved in future use of that part.[77] This experimental experience and data can then be shared through the Registry. Users would likely be more inclined to use parts that have already been successfully used by others.[78] In the Registry's online catalogue, parts are rated according to 'usefulness' and given Registry 'stars'. A part that is being successfully used and providing value to other Registry users could certainly boost the reputation of the original contributor.

Being involved in an open research community can also be a satisfying end in itself. For example, von Hippel quotes computer scientist Chris Hanson as explaining that

---

75 Endy, D. (2010), 'Evidence to the Presidential Commission for the Study of Bioethical Issues, Synthetic Biology', Washington, DC, 8 July.

76 Johnson, R.A. (2009), 'Evidence to the US National Academies, Opportunities and Challenges in the Emerging Field of Synthetic Biology', Washington, DC, 10 July, 50.

77 Henkel, J. and S. Maurer (2007), 'The Economics of Synthetic Biology', *Molecular Systems Biology*, **3** (117), 1–4, 1.

78 Ibid., 2.

> [...] creative programmers want to associate with each other: only their peers are able to truly appreciate their art. Part of this is that programmers want to earn respect by showing others their talents. But it's also important that people want to share the beauty of what they have found. This sharing is another act that helps build community and friendship.[79]

In the highly specialised field of synthetic biology using standard biological parts, it can certainly be imagined that belonging to a community of researchers with similar passions and goals, and sharing one's achievements with them, would in itself be a rewarding activity. A sense of wanting to 'give back' has also been noted in open source software projects, where some programmers wish to contribute something back to those who have allowed them to access software source code.[80] As synthetic biology progresses and young scientists or those who have participated in iGEM as students become more senior experts in the field, this is likely to become a more and more significant reason for making contributions to the Registry.

iii Obligations to funders

Apart from incentives to contribute parts and data to a commons, there is also the question of scientists' obligations to their institution or funders. For example, in the United States universities and other non-profit organisations or small businesses who receive federal funding for their research have obligations to the Federal Government under the *Bayh-Dole Act*.[81] This may affect their ability to release data into the public domain. The provisions of the *Bayh-Dole Act* appear in 35 USC § 200–212. § 200 states that it is 'the policy objective of the Congress to use the patent system to promote utilisation of inventions arising from federally supported research or development [...]'. Under these provisions, universities receiving federal funding can elect to keep the title to any inventions conceived in work carried out under the funding agreement.[82] If the organisation does not elect to retain rights, the federal agency can receive the title or the inventor may request that they be granted the rights.[83] If the funded organisation does retain the rights, they must agree to file a patent application before any

---

79 Hippel, Eric von (2005), *Democratizing Innovation*, Cambridge, MA and London: The MIT Press, 124.

80 Ibid., 87.

81 See 35 USC § 200–212. The inventions covered by the provisions of § 200–212 are referred to as 'subject inventions', defined in § 201 (e) as '[...] any invention of the contractor conceived or first actually reduced to practice in the performance of work under a funding agreement'. Universities are included in the definition of 'nonprofit organization' in § 201(i).

82 § 202 (a).

83 § 202 (c)(2) and § 202 (d).

statutory deadline owing to publication, sale or public use. They must also file corresponding patent applications in the other countries in which they wish to retain title, within reasonable times. If this is not done, the Federal Government may receive title to the inventions.[84] The federal agency funding the research can require periodic reporting on the utilisation of the invention.[85] Non-profit organisations cannot assign their rights to an invention without the approval of the funding agency.[86]

In the context of the BioBrick Public Agreement, this means that scientists whose research is federally funded could find that their organisation is in breach of its funding agreement if they do not patent their biological parts. Parts could still be disclosed to the Registry *after* the filing of a patent application, but agreeing to the BioBrick Public Agreement may be problematic as this requires promising not to assert the patents against users of the Registry (a statement of non-assertion). This will depend on the terms of the funding agreement. Similarly if the organisation does not retain title to the invention, and this is granted to the scientist who designed the part, contributing the part to the Registry may breach the terms of the grant of rights to the inventor. The Act only applies to *inventions*, so data that do not themselves comprise an invention could be released via the Registry, for example experimental experience using parts.

Restrictions on disclosure and exercise of intellectual property rights could also be imposed by a university itself, under its intellectual property policies. Freely sharing information that could potentially have commercial applications may not accord with a university's expectation of its scientists. In addition, it is worth noting that many synthetic biologists would be at least partially funded by commercial entities. Commercial funders may well impose conditions on disclosure of data and exercise of intellectual property rights such that the researcher would not have authority to disclose parts and other data to the Registry or agree to the BioBrick Public Agreement.

Currently parts *are* being contributed to the Registry, which had swelled to 3500 parts by mid 2009.[87] However, the growth is largely owing to new parts generated through the iGEM competition, in which teams are expected to submit any new parts they create.[88]

---

84 § 202 (c)(3).

85 § 202 (c)(5).

86 § 202 (c)(7)(A).

87 Rettberg, R. (2009), 'Evidence to the US National Academies, Opportunities and Challenges in the Emerging Field of Synthetic Biology', Washington, DC, 10 July, 36.

88 Ibid., 29.

III CHARACTERISING THE BIOBRICKS COMMONS

A Open Source Technology

This section considers whether BioBricks are properly described as an 'open source' resource for synthetic biology researchers. In the context of technology development, the term 'open source', originally used in software programming, means that the products or tools are 'free' to be used, modified and distributed, in the sense that these activities are not restricted owing to property rights.[89] Under the General Public License (GPL) used in many open source software projects, users of source code are free to copy, modify and distribute the software source code.[90] However, the GPL restricts the conditions on which users may distribute the accessed material. They must make the source code, or modified version, available under the same free access conditions on which they originally accessed the source code. This is known as the 'viral' nature of such licences.[91] Like viral particles, the open access terms are self-perpetuating. Users cannot impose any further conditions on the recipient. Users cannot incorporate the accessed code into products protected by intellectual property rights, or charge licence fees for products that modify or incorporate the accessed code.[92]

Some features of the BioBrick Public Agreement model distinguish it from open source projects such as those that use the GPL. Under the BioBrick Public Agreement, Users' future commercialisation and licensing strategies are not limited. Users are able to patent and commercialise any developments they make using BioBricks, as there is nothing in the agreements to prevent this. The BioBrick Public Agreement does not limit agreements the Users might make with other parties regarding the contributed materials, stating that nothing in the Agreement precludes Users (or Contributors) from voluntarily entering into a separate agreement with another party regarding the materials that may 'vary from or supersede' the Agreement. This means that, for example, Users are not limited in the conditions that they impose on access to any intellectual property they

---

89 This definition is adapted from Free Software Foundation Inc. (2010), 'The Free Software Definition', http://www.gnu.org/philosophy/free-sw.html.

90 Free Software Foundation Inc., 'The GNU General Public License (GPL) Version 2', http://www.opensource.org/licenses/gpl-2.0.php.

91 Hope, Janet (2008), *Biobazaar: The Open Source Revolution and Biotechnology*, Cambridge, MA and London: Harvard University Press, 11.

92 Free Software Foundation Inc., 'The GNU General Public License (GPL) Version 2', http://www.opensource.org/licenses/gpl-2.0.php.

develop using contributed material. If Users wish to distribute their work based on contributed materials, they do not have to make this available on the same conditions under which they accessed the materials originally. There is nothing to prevent the charging of fees for licences.

A further difference is that Users do not have to contribute any material or information to the Registry, such as any modifications or improvements they might make to parts accessed under the User Agreement. As Boyle has noted, in the absence of underlying property rights the strategic disclosure of synthetic biology information into the public domain does not provide a 'right to force subsequent innovators to release their code back into the commons'.[93] Unlike Users under the BioBrick Public Agreement, iGEM teams must use parts from the Registry and submit any new parts they create. Director Randy Rettberg has described the philosophy of iGEM as 'get and give', referring to the expectation that students put parts generated or improved back into the Registry.[94] Scientists using the Registry for their own purposes are not required to 'give back' in this way.

The open source software model described above requires underlying intellectual property rights, because the copyright in the source code allows the original developer to enforce the conditions of the open access licence.[95] By contrast, sharing under the BioBrick Public Agreement is not based on an intellectual property holder's decision about how to implement their rights. Instead, it can be argued that through the Registry the BioBricks Foundation is trying to be pre-emptive, to create 'sufficient prior art so basically patentability and enclosure would become impossible or at least very difficult'.[96] This is a deliberate and strategic disclosure of knowledge and data into the public domain. The strategic disclosure approach is revealed by the FAQs to the BioBrick Public Agreement (referred to as the BPA). In response to the question 'Why won't a patent "troll" file patents on all possible novel uses of BPA-contributed parts?', the FAQs state that 'contributors and others can readily disclose at the time of contribution

---

93 Boyle, James (2008), *The Public Domain: Enclosing the Commons of the Mind*, New Haven, CT and London: Yale University Press, 177.

94 Rettberg, R. (2009), 'Evidence to the US National Academies Opportunities and Challenges in the Emerging Field of Synthetic Biology', Washington, DC, 10 July, 29.

95 Boyle, James (2008), *The Public Domain: Enclosing the Commons of the Mind*, New Haven, CT and London: Yale University Press, 176.

96 Johnson, R. (2009), 'Evidence to the US National Academies, Opportunities and Challenges in the Emerging Field of Synthetic Biology', Washington, DC, 10 July, 59.

known and imaginable uses of so-contributed parts, making obvious many future uses'.[97]

An aspect of the model that does use intellectual property rights is the requirement that Contributors must agree not to assert their rights against Users. However, because it does not have the viral characteristics described above, this model is quite different from open source projects. It is more accurately described as an 'open biotechnology' project, as discussed below.

The fact that this access model is not truly open source may limit its capacity to operate as a commons for synthetic biology. The model ensures that scientists will have free access to biological parts as originally contributed. However, if they wish to access improved or modified biological parts, devices composed from multiple parts, or systems that use parts to achieve particular outcomes with practical applications, they would need to be aware of the possibility that property rights may exist and not be covered by the statement of non-assertion in the BioBrick Public Agreement. If the goal of the Agreement is to ensure free access to basic pre-competitive data and building blocks in synthetic biology – biological parts rather than applications using these parts – then the scope of the Agreements is probably sufficient to achieve this.

B Open Biotechnology

According to sociologist of science Robert K. Merton, communal ownership of scientific discoveries is an 'integral element of the scientific ethos'.[98] In 1973, he observed that:

> The substantive findings of science are a product of social collaboration and are assigned to the community. They constitute a common heritage. [...] The scientist's claim to 'his' intellectual 'property' is limited to that of recognition and esteem which, if the system functions with a modicum of efficiency, is roughly commensurate with the significance of the increments brought to the common fund of knowledge. [...] The institutional conception of science as part of the public domain is linked with the imperative for the communication of findings. Secrecy is the antithesis of this norm, full and open communication its enactment.[99]

---

97 The BioBricks Foundation, 'FAQ', http://www.biobricks.org/FAQ.php.

98 Merton, Robert K. (1973), *The Sociology of Science: Theoretical and Empirical Investigations*, Chicago: University of Chicago Press, 273.

99 Ibid., 273–4.

Merton called this 'communism'. He went on to identify the conflict between this norm and the concept of intellectual property.[100] This insightful observation was made well before the advent of gene patenting, patents on genetic tests and other controversies generated by a collision between biotechnology research and intellectual property rights.

More recently, this ethos of 'communism' has become known as 'openness', or 'sharing', in biological research. Openness has arisen partly for practical reasons relating to the limitations of the intellectual property system. For example, owing to the various uncertainties and costs associated with patenting in biotechnology research, scientists have begun looking beyond patenting to other strategies such as contractual arrangements.[101] These 'access agreements' do not require underlying intellectual property rights.[102] Current openness and sharing efforts in biological research are also underlain by a desire for improved outcomes of research efforts and better collaboration. This trend is particularly visible in medical research, in fields such as human genetic diseases and cancer.[103]

These influences can be seen at work in the BioBricks commons model. We can observe the ethos of communism in the proactive disclosure of scientific knowledge and the commons building activities of the participants in the parts agenda. By building a commons that exists outside the patent system, the model avoids the conflict between scientific norms and intellectual property identified by Merton. The practical problems and uncertainties of patenting in this field have been noted with concern by the founders of the parts agenda, as described above. These problems are avoided by designing a model of deliberate disclosure and non-assertion of intellectual property rights, based on contractual agreements.

Intellectual property and biotechnology expert Yann Joly has identified five defining characteristics of 'open biotechnology' projects. They must 'make use [...] of the internet and other information technologies', facilitate collaboration by scientists, 'include a strategy to ensure rapid public dissemination of information and research results', allow the use of results without restricting the 'freedom and integrity' of researchers, and 'not use

---

100 Ibid., 275.

101 Joly, Y. (2010), 'Open Biotechnology: Licenses Needed', *Nature Biotechnology*, **28** (5), 417–19, 418.

102 Ibid.

103 An example of a human genetics data-sharing project is the Sage Commons, see Sage Bionetworks, 'The Sage Commons', http://sagebase.org/commons/index.php.

intellectual property to limit access to the project' or its results.[104] Currently, the Registry has all of these characteristics. The Registry and the BioBricks Foundation both operate largely online, allowing members of the synthetic biology community to access and add information, contribute to discussions and participate in the development of technical standards. The open access nature of this resource facilitates collaboration within the community. Registered scientists in the field can contribute information, view information contributed by others, and use the online resources to identify those working on projects closely related to their own. The various community building activities, including Synbio conferences and iGEM, also facilitate collaboration. The BioBrick Public Agreement facilitates open dissemination and use of experimental results by providing a mechanism for the contribution of materials to the Registry on the condition that any property rights identified will not be asserted against other members of the community. This means that intellectual property does not limit access to the materials contained in the Registry. For these reasons, the model is more accurately described as 'open biotechnology' than 'open source'.

C Democratising Innovation and Innovation Communities

A number of aspects of Von Hippel's theory of democratising innovation are echoed in synthetic biology's parts agenda. Von Hippel argues that democratising innovation means a democratisation of the 'opportunity to create':

> Democratization of the opportunity to create is important beyond giving more users the ability to make exactly right products for themselves [...] the joy and the learning associated with creativity and membership in creative communities are also important, and these experiences too are made more widely available as innovation is democratized.[105]

This democratisation of the opportunity to create can be seen in the parts agenda, which aims to make biological engineering more predictable and easier to carry out by developing standardised methods and components. It is hoped that this will open up the field to a wider range of researchers. The agenda also extends the opportunity to create something new in biology to relatively inexperienced undergraduate and even high school students through iGEM.

---

104 Joly, Y. (2010), 'Open Biotechnology: Licenses Needed', *Nature Biotechnology*, **28** (5), 417–19.

105 Hippel, Eric von (2005), *Democratizing Innovation*, Cambridge, MA and London: The MIT Press, 123–4.

The distribution of sets of biological parts by the Registry to iGEM teams and other Registry users mirrors the concept of manufacturers making 'toolkits' available for user innovation.[106] Perhaps in the synthetic biology context this is more accurately characterised as lead users (those at the leading edge of their field, who would gain from solutions to problems within the field)[107] supplying other users with toolkits needed to innovate in the field.

The BioBricks commons, facilitated by the Registry and the BioBricks Foundation, has many of the features of an 'innovation community'. Hippel describes 'innovation communities' as:

> nodes consisting of individuals or firms interconnected by information transfer links which may involve face-to-face, electronic, or other communication. These can, but need not, exist within the boundaries of a membership group. They often do, but need not, incorporate the qualities of communities for participants, where 'communities' is defined as meaning 'networks of interpersonal ties that provide sociability, support, information, a sense of belonging, and social identity'.[108]

Scientists participating in the BioBricks commons are linked through both electronic communication and face-to-face interactions at the large Synbio conferences. Use of the Registry is limited to those laboratories that have been registered by its directors, creating a membership group. The BioBricks Foundation and the Registry provide various technical support and training for Registered Users, such as workshops, online tutorials and experimental protocols. The various community building and leadership activities of these organisations, including the Synbio conferences, open standards development and iGEM, facilitate a sense of the emergence of a new community within biological research.

CONCLUSION

Through the Registry of Standard Biological Parts, the BioBrick Public Agreement and other initiatives, leading synthetic biologists have created a commons in which their community can conduct its research. This commons exists outside the intellectual property system. Instead of pursuing private property rights, synthetic biologists are disclosing their work as a strategy to protect its future from patent thickets and other obstructive

[106] Ibid., 128.
[107] Ibid., 4.
[108] Ibid., 96.

intellectual property problems. The open access arrangements are not 'open source' in the viral sense, but are 'open' in the sense of creating a freely accessible resource that facilitates sharing and collaboration. The commons has attributes of both an 'open biotechnology initiative' and a user-driven 'innovation community'. It can be seen as part of a trend towards commons-based research in the biological sciences.

Currently, the operation of the commons is underpinned not by legal mechanisms but by a culture of sharing and goodwill between scientists working towards common goals. The philosophy of openness and sharing is being spread through iGEM and the Synbio conferences. The case study suggests that community building and leadership form a significant part of creating a successful commons for scientific research. As the field progresses it will need to move from trust towards contract based arrangements.[109] This is already occurring, with the release of draft legal arrangements for accessing and contributing materials to the Registry.

The draft BioBrick Public Agreement will facilitate open access to biological parts into the future. There are significant concerns relating to the draft Agreements, particularly in the interaction with industry, development of commercial applications using parts, and *Bayh-Dole* obligations for federally funded scientists. The lack of 'viral' character of these arrangements will mean that only biological parts or other materials as originally contributed will be within the BioBricks commons. Access to further innovations using these contributed materials will be at the discretion of individual innovators.

The BioBricks commons provides an example of scientists creating a space for their own research, right from the establishment of a new field. Here access to information and innovation is not dependent on government intervening in the intellectual property market, for example negotiating access arrangements by exercising retained rights in inventions produced with federal funds. Nor is access dependent on beneficent attitudes of key intellectual property holders. Instead, the people who are going to use the resource have taken steps to build their own commons and ensure access to what they need to do their research. The sense of community and the shared desire to see the field of synthetic biology move forward are likely to be significant contributors to the BioBricks commons' success.

---

109 Johnson, R. (2009), 'Evidence to the US National Academies, Opportunities and Challenges in the Emerging Field of Synthetic Biology', Washington, DC, 10 July, 50.

PART IV

Genetics, stem cells and nanotechnology

8. Regulating gene regulation: patenting small RNAs

Adam Bostanci, Jane Calvert and Pierre-Benoit Joly

Much has been written on the revolution in the life sciences that has taken place over the last 20 years. Perhaps the most important development has been an increasing recognition of complexity in biology, which has led to the emergence of new fields of research such as systems biology and epigenetics. In this chapter, we focus on one recent conceptual shift in particular: the shift away from gene-centric understandings of biological processes. This shift has been fed by – and has reinforced interest in – research on small RNAs, thanks to the recognition of the role of these molecules in many crucial biological processes. Small RNAs and the phenomenon of RNA interference (RNAi) have been seen as therapeutically and commercially promising, to the extent that some prominent commentators have spoken about a patenting 'gold rush' in this area.[1]

In the first section of the chapter, we characterise the patent system in general terms and highlight some recent developments with respect to inventions in the life sciences. We then give some background to the discovery that opened up the field of RNA interference and was eventually rewarded with the Nobel Prize to Andrew Fire and Craig Mello in 2006. We go on to briefly describe the mechanism of RNA interference, and show that it is now thought of as part of a more general and only partially understood RNA silencing system. This has implications for how we view some of the most important patents related to RNA interference, which are based on the research of leading scientists in the area (including the two Nobel Laureates, and other key scientists like Thomas Tuschl). This is in

[1] Angrist, M. and R. Cook-Deegan (2006), 'Who Owns the Genome?' *The New Atlantis*, **11** (Winter), 87–96.

sharp contrast to gene patenting, where the scientific establishment challenged the wholesale patenting of human genes and the associated policies of patent offices in the 1990s.

In our discussion, we consider some theoretical issues surrounding patenting in this area. Small RNAs take the focus away from genes as the most important molecules in biological processes. But we show that despite this demotion of the biological importance of the gene, we see the persistence of gene-centrism in patents on small RNAs. Several of the patents focus on one particular gene, and then attempt to claim any small RNA that targets it. In its persistent emphasis on the gene, the patent system works with an outdated understanding of the science. Arguably, the patent system is oblivious of recent shifts in the biological sciences.

We also show how the context-dependence and multifunctionality of some subclasses of small RNAs is not acknowledged in patenting, where the focus is on individual entities, rather than on relations between entities or processes.[2] It is not possible to predict the function of some small RNAs without first taking into account their particular environmental, cellular and developmental context. However, the legacy of 'the gene' and 'its function' is still given a central place in awarding product patents on RNAi. Finally, we suggest that it is possible to understand RNAi above all as a 'research tool' without gene-centric inferences. This, however, only raises further questions about the justification for patenting in this area. Ultimately, it seems likely that patenting has played an important part in encouraging the recent 'gold rush' associated with RNA interference.

I GENE PATENTING

A patent is a contract in which an inventor is given a temporary monopoly over an invention in return for providing a full description that allows other 'skilled artisans' to practise the invention. Patenting is associated with a political-legal doctrine holding that intellectual property rights are necessary to provide incentives for innovation and technological progress. Needless to say, the extent to which biological entities should be governed by the patent system has been heavily debated.[3] In this section, we characterise the patent system in general terms and highlight some recent developments with respect to inventions in the life sciences.

---

2 Simondon, G. (1989), *L'individuation Psychique et Collective*, Paris: Aubier.

3 See for example Kevles, Daniel (2002), *A History of Patenting Life in the United States with Comparative Attention to Europe and Canada*, Brussels: European Group on Ethics in Science and New Technology, European Commission.

The patent system is governed by well-known criteria for patentability. In order to merit a patent, an invention must be novel, inventive (or non-obvious in the USA) and useful. A patent must also fulfil the written description requirement; the so-called patent specification must be complete enough that a skilled practitioner can make and use the invention without undue experimentation. A less well-known but nevertheless important feature of patenting is that all inventions are examined with respect to the patent classification system, which defines the classes of subject matter that can be patented. From time to time, patent offices are confronted with new entities or phenomena, such as 'genes', 'genomes' and 'RNA interference'.[4] New classes of invention may then be created, or the new entities may otherwise be accommodated in the patent system. For example, patent applications for genes were initially challenged on the grounds that they were 'products of nature', but in a landmark case the Supreme Court of the United States ruled that natural substances, including genes, are patentable if they are 'isolated and purified'.[5] In the early days of biotechnology, a gene was usually patented as an isolated nucleic acid (DNA) used for the production of a protein, which would often also have a medical application. However, when sequencing DNA became routine in the mid 1990s, eminent scientists challenged patent applications to the United States Patent and Trademark Office (USPTO), which claimed DNA sequences that only had uses in research (as molecular probes), and whose biological function had only been identified in a speculative manner via searches in DNA databases.[6]

Following the gene patenting controversies, patent offices have revised their examination procedures for biotechnology-related inventions. Notably, in the United States, the Revised Guidelines require that the 'utility'

---

4 For analysis of genome patents, see Bostanci, A. and J. Calvert (2008), 'Invisible Genomes: The Genomics Revolution and Patenting Practice', *Studies in the History and Philosophy of the Biological and Biomedical Sciences*, **39**, 109–19.

5 Conley, J.M. and R. Makowski (2003), 'Back to the Future: Rethinking the Product of Nature Doctrine as a Barrier to Biotechnology Patents', *Journal of the Patent and Trademark Office Society*, **85**, 301–34 (Part I), 371–98 (Part II); and Demaine, L. and A. Fellmeth (2002), 'Reinventing the Double Helix: A Novel and Nonobvious Reconceptualization of the Biotechnology Patent', *Stanford Law Review*, **55** (2), 303–462.

6 Human Genome Organization (1995), 'HUGO Statement on Patenting of DNA Sequences', http://www.hugo-international.org/PDFs/Statement%20on%20Patenting%20of%20DNA%20Sequences%202000.pdf.

ascribed to a DNA patent must be specific, substantial and credible.[7] It is often assumed that one way of satisfying this requirement is by disclosing a function for the gene.[8] There have also been more subtle changes,[9] and the USPTO has been accused of using the written description requirement as a means of policing the scope of biotechnology patents.[10]

One basic tension with respect to inventions in biotechnology stems from the custom that patent claims can be for either a product or a process. Moreover, it is often assumed that the first patent on a particular product is more important than subsequent process patents involving this product.[11] This obviously has a bearing on nucleic acid patents where, on the one hand, the disclosure of a function has become a precondition of patentability and, on the other, developments in the life sciences have made it increasingly clear that the same nucleic acid sequence may have many different functions depending on the biological context.[12] In short, even if biotechnological inventions in the first instance were patented as products, considerable discussion has revolved around the need to associate a function with such a product.

---

7 United States Patent and Trademark Office (2001), 'Utility Examination Guidelines', *Federal Register*, Friday 5 January, **66** (4), 1092–9, http://www.uspto.gov/web/offices/com/sol/notices/utilexmguide.pdf.

8 Calvert, J. (2007), 'Patenting Genomic Objects: Genes, Genomes, Function and Information', *Science as Culture*, **16** (2), 207–23.

9 See for example Fitt, R. and E. Nodder (2009), 'The Industrial Applicability of Biotechnology Patents – A New Test for Europe', *Biotechnology Law Report*, **28** (2), 151–8.

10 Potter, K. (2009), 'Getting Written Description Right in the Biotechnology Arts: A Realist Approach to Patent Scope', *Biotechnology Law Report*, **28** (1), 1–17. We do not consider here the recent conflict over Myriad Genetics Inc.'s patents drawn to BRCA gene testing. United States District Court Judge Sweet ruled that DNA is 'the physical embodiment of biological information distinct in its essential characteristics from any other chemical found in nature' (*Association for Molecular Pathology et al.* v. *United States Patent and Trademark Office, Myriad Genetics, et al.* 702 F. Supp. 2d 181, (S.D.N.Y. 2010), *appeal docketed,* No. 2010–1406 (Fed. Cir. June 22, 2010). However, this ruling was largely overturned on appeal – see *Association for Molecular Pathology et al.* v. *United States Patent and Trademark Office, Myriad Genetics, et al.* 653 F.3d 1329, 2011 WL 3211513 (C.A. Fed. (N.Y.)), 99 USPQ 2d 1398.

11 Eisenberg, R. (1997), 'Structure and Function in Gene Patenting', *Nature Genetics*, **15**, 125–30.

12 Calvert, J. (2007), 'Patenting Genomic Objects: Genes, Genomes, Function and Information', *Science as Culture*, **16** (2), 207–23; and Ilag, L.L., L.M. Ilag and L.L. Ilag (2002), 'From Patenting Genes to Proteins: the Search for Utility via Function', *TRENDS in Biotechnology*, **20** (5), 197–9.

This discussion illustrates some of the issues that arise when the concepts and classifications that govern the patent system are adapted to accommodate new biological phenomena. In particular, the conceptualisation of products and processes, how patents in this area relate to genes and to biological functions, and what kind of evidence for anticipated therapeutic utility is required to patent a technique, all come to the fore when we discuss patents for small RNAs.

II RNA INTERFERENCE

For much of the history of molecular biology, RNA was understood as a mere messenger molecule between DNA and protein. The central dogma of molecular biology, which (in one version) states that there is a unidirectional flow of biological 'information' from DNA to RNA to protein, is often cited in this context.[13] In the 1980s, this understanding of RNA as a messenger molecule motivated attempts to inhibit protein synthesis by introducing so-called 'antisense' molecules into cells.[14] Antisense molecules are strands of As, Ts, Cs and Gs that are complementary to a particular messenger RNA (mRNA) sequence. The theory was that the 'sense' and the 'antisense' strand would 'zip up' together, thus blocking the production of the protein. Some of the companies that now carry the flag of RNA interference were previously dedicated to developing antisense drugs in this manner (Isis Pharmaceuticals, for instance). Only one antisense drug (Fomivirsen) has made it to market, however. One of the difficulties with these drugs is the delivery of the molecules into cells, as well as so-called unwanted 'off-target effects'. Despite the exuberant promises associated with the technology, it remains to be seen to what extent RNA interference can go beyond the legacy of antisense drug design.

Another important feature of the history of RNA interference are prior experimental findings that were retrospectively accounted for by RNA silencing. Most famously, when scientists attempted to deepen flower colour in Petunias by adding a DNA construct for producing extra pigment, they were surprised to find that the flowers actually lost their colour

---

13 Arguably the invocation of the 'central dogma' in this area is a red herring; as one of us (A.B.) will argue elsewhere.

14 Skoblov, M. (2009), 'Prospects of Antisense Therapy Technologies', *Molecular Biology*, **43** (6), 917–29.

when the gene was introduced.[15] Another set of anomalous results concerned control experiments in antisense research. Here it was observed that even if the 'sense' strand (complementary to the antisense strand and identical to the strand involved in protein production) was injected as a control, gene expression was reduced.

The key experiment, carried out by Fire and Mello, was to compare the inhibition of single stranded RNA to that of double stranded RNA. In February 1998, Fire and his colleagues published the now seminal paper 'Potent and specific genetic interference by double-stranded RNA in *Caenorhabditis elegans*' in *Nature.*[16] At the time there was no reason to expect that double-stranded RNA was more likely to block target mRNAs, but the paper reported that, surprisingly, 'double stranded RNA was substantially more effective at producing interference than was either strand individually'.[17] In fact, a few molecules of double-stranded RNA per cell were sufficient to 'silence' the corresponding genes to such an extent that the resulting phenotype resembled that of a knock-out organism where the gene in question had been deleted. The effect was sequence-specific and was observed with double-stranded RNA homologous to four different genes.[18]

These experiments formed the basis of the so-called 'Carnegie patent', as well as the 2006 Nobel Prize.[19] Experiments in Fire's laboratory also showed that the same effect could be achieved by feeding *C. elegans* with genetically engineered *E. coli* that produce double-stranded RNA, and even by soaking the worm in double-stranded RNA. This led to the conclusion that RNA interference provides a convenient research tool for studying the function of genes in *C. elegans.* RNAi was soon also applied successfully as a technique for exploring gene function in a wide range of experimental organisms.

Over the following years, the same effect was documented in plants and in other model organisms, such as *Drosophila*, and in developing mouse embryos.[20] However, initially it proved impossible to reproduce this RNA

---

15 Beal, J. (2005), 'Silence is Golden: Can RNA Interference Therapeutics Deliver?' *Drug Discovery Today*, **10** (3), 169–72.

16 Fire A., S. Xu, M. Montgomery, S. Kostas, S. Driver and C. Mello (1998), 'Potent and Specific Genetic Interference by Double-stranded RNA in Caenorhabditis elegans', *Nature*, **19,** 391(6669), 806–11.

17 Ibid. at 806.

18 Ibid.

19 Fire A. (2006), 'Gene Silencing by Double Stranded RNA', Lecture, Nobel Prize in Physiology or Medicine, 8 December, http://nobelprize.org/nobel_prizes/medicine/laureates/2006/fire_lecture.pdf.

20 Kennerdell, J. and R. Carthew (1998), 'Use of Double-stranded RNA-Mediated Genetic Interference to Demonstrate that frizzled and frizzled 2 Act

interference effect in mammalian cell lines. These difficulties were overcome when researchers focused on the length of the double-stranded RNA. Experiments in plants published in 1999 suggested that double-stranded RNA of approximately 25 nucleotides played a role in specific genetic interference.[21] Then Zamore and co-authors reported that long double-stranded RNAs introduced into *Drosophila* were processed to RNAs that were 21 to 23 nucleotides in length, and that it was these very short molecules that guided the destruction of the target mRNA.[22] This set of experiments formed the basis of the so-called Tuschl I patent application, which is part of an ongoing patent dispute.

In a paper published in *Nature* in 2001, Thomas Tuschl and his collaborators showed that the specific RNA interference effect could also be produced in mammalian cell lines with short RNAs.[23] However, the inhibition of gene expression was incomplete: it was reduced by between 3- and 25-fold. We mention this as a way of cautioning against giving too much weight to simplified accounts of RNA interference as a way of 'silencing' genes. The paper concluded that these short interfering RNAs 'provide a new tool for studying gene function in mammalian cells and may eventually be used as a gene-specific therapeutics'.[24] This conjecture sums up the biomedical promise associated with RNA interference, and the experiments reported in this paper formed the basis for the Tuschl II patent. Tuschl and several key scientists in this area went on to found Alnylam Pharmaceuticals in 2002, now considered to be one of the leading companies developing therapeutic strategies using RNA interference.

The focus on RNAs after 1998 also led to the articulation of a new class of RNAs that were dubbed microRNAs in *Science* in October 2001. Their history can be traced back to the early 1990s when researchers found a *C. elegans* gene that was not translated into a protein but instead encoded a small RNA molecule. Such RNAs were called small 'temporal RNAs' because they were regulated during the development of the organism. In

---

in the Wingless Pathway', *Cell*, **95**, 1017–26; and Wianny, F. and M. Zernicka-Goetz (2000), 'Specific Interference with Gene Function by Double-stranded RNA in Early Mouse Development', *Nature Cell Biology*, **2**, 70–75.

21 Hamilton A. and D. Baulcombe (1999), 'A Species of Small Antisense RNA in Posttranscriptional Gene Silencing in Plants', *Science*, **286**, 950–52.

22 Zamore, P., T. Tuschl, P. Sharp and D. Bartel (2000), 'RNAi: Double-Stranded RNA Directs the ATP-Dependent Cleavage of mRNA at 21 to 23 Nucleotide Intervals', *Cell*, **101**, 25–33.

23 Elbashir, S., J. Harborth, W. Lendeckel, A. Yalcin, K. Weber and T. Tuschl (2001), 'Duplexes of 21±Nucleotide RNAs mediate RNA Interference in Cultured Mammalian Cells', *Nature*, **411**, 494–8.

24 Ibid. at 494.

Science, three papers identified additional short RNAs in several organisms as well as in human cell lines and called them microRNAs (miRNAs). One of these papers concluded that 'sequence-specific, posttranscriptional regulatory mechanisms mediated by small RNAs are more general than anticipated'; and the 'challenge for the future is to define the function and find the potential targets of these novel miRNAs'.[25] It appears that this paper formed the basis for what is sometimes called the Tuschl III patent application.

Genetic research carried out in parallel showed that both RNAi and miRNA rely on essentially the same cellular machinery,[26] and that this machinery is conserved in different experimental organisms.[27] In the case of RNA interference, long double-stranded RNAs (which may be introduced by scientists or by natural processes like viral infections) are cleaved into the 21 to 27 nucleotide RNAs known as short-interfering RNAs (siRNAs). One strand of the siRNAs is subsequently incorporated into a piece of cellular machinery called an 'RNA-induced silencing complex', and this leads to the degradation of the target mRNA. In contrast, miRNAs are encoded in the genome and processed in several steps into mature molecules that are similarly incorporated into a RNA-induced silencing complex, whereupon they block protein synthesis. While siRNAs trigger the sequence-specific degradation of mRNA, miRNAs are generally thought to bind to regions of imperfect sequence complementarity in the mRNA and block translation.[28] This means that a single miRNA can affect the translation of a large number of mRNAs. Several miRNAs may also 'cooperate' in inhibiting the expression of a particular gene;[29] and it has been estimated that up to one-third of all human genes are controlled in

---

[25] Lagos-Quintana, M., R. Rauhut, W. Lendeckel and T. Tuschl (2001), 'Identification of Novel Genes Coding for Small Expressed RNAs', *Science*, **294**, 853–8 at 857.

[26] Grishok, A., A. Pasquinelli, D. Conte, N. Li, S. Parrish, I. Ha, D. Baillie, A. Fire, G. Ruvkun and C. Mello (2001), 'Genes and Mechanisms Related to RNA Interference Regulate Expression of the Small Temporal RNAs that Control *C. elegans* Developmental Timing', *Cell*, **106** (1), 23–34.

[27] As reviewed by Plasterk, R. (2002), 'RNA Silencing: The Genome's Immune System', *Science*, **296,** 1263–5.

[28] Zamore, P. and B. Haley (2005), 'Ribo-gnome: The Big World of Small RNAs', *Science*, **309,** 1519–24.

[29] Doench, J. and P. Sharp (2004), 'Specificity of microRNA Target Selection in Translational Repression', *Genes and Development*, **18**, 504–11.

this manner by miRNAs.[30] According to one estimate, there are in the order of 1000 human miRNAs.[31]

One hypothesis for the evolution of RNA silencing is that it acts as a type of immune system present in plants, animals and fungi (an analogy cited by Fire in his 2006 Nobel Lecture).[32] In plants, immunity against viruses can travel through the organism, with the implication that molecular machinery exists to support these processes. Similarly, RNA interference can spread through the body of the worm *C. elegans*.

Another important development that has brought RNA to the forefront of biological research has been the finding that a vast number of non-coding RNA transcripts are generated from genomic DNA sequences. Huge swathes of the human genome are expressed that were previously considered to be 'junk' DNA because they do not encode proteins.[33] Based on these findings, some scientists have proposed that small RNAs play a 'central role' in regulating development and physiology in animals.[34] RNA-mediated regulatory networks have also been cited in explanations for the evolution of complexity, and for the unexpected fact that complex organisms have relatively few genes compared to much simpler ones. Mattick argues that, as a system becomes more complex, more of the system has to be dedicated to regulation, and this function is increasingly ascribed to small RNAs.[35] The changing view of the importance of RNA is perhaps illustrated most emblematically in the description of the genome as an 'RNA Machine' in one recent review.[36]

---

30 Lewis, B., C. Burge and D. Bartel (2005), 'Conserved Seed Pairing, Often Flanked by Adenosines, Indicates that Thousands of Human Genes are MicroRNA Targets', *Cell*, **120**, 15–20.

31 Bentwich, I. et al. (2005), 'Identification of Hundreds of Conserved and Nonconserved Human microRNAs', *Nature Genetics*, **37**, 766–70.

32 Fire A. (2006), 'Gene Silencing by Double Stranded RNA', Lecture, Nobel Prize in Physiology or Medicine, 8 December, http://nobelprize.org/nobel_prizes/medicine/laureates/2006/fire_lecture.pdf. To the best of our knowledge, this broader relevance of this analogy has not been critically assessed.

33 Bertone, P. et al. (2004), 'Global Identification of Human Transcribed Sequences with Genome Tiling Arrays', *Science*, **306**, 2242–6; and Kapranov, P. et al. (2007), 'RNA Maps Reveal New RNA Classes and a Possible Function for Pervasive Transcription', *Science*, **316**, 1484–8.

34 Mattick, J. (2003), 'Challenging the Dogma: The Hidden Layer of Non-protein-coding RNAs in Complex Organisms', *Bioessays*, **10**, 930–39.

35 Mattick, J. (2004), 'RNA Regulation: A New Genetics?' *Nature Reviews Genetics*, **5**, 316–23.

36 Amaral P., M. Dinger, T. Mercer and J. Mattick (2008), 'The Eukaryotic Genome as an RNA Machine', *Science*, **319**, 1787–9.

So far, we have described two types of small RNA: siRNA and miRNA, a distinction also often drawn in the biotechnology-oriented literature.[37] A range of more recent findings complicates this picture, however. Most importantly, it has been shown that small RNAs can also interfere with the transcription of genomic DNA into RNA in the cell nucleus.[38] Additionally, while RNA interference was previously strictly an experimentally induced effect, 2008 saw the discovery of endogenous short-interfering RNAs (endo-siRNAs), as well as other classes of small RNAs, such as the so-called piRNAs, the longest of the short RNAs found so far.[39] Surprisingly, one recent paper has even reported that some small RNAs can stimulate the transcription of human genes.[40]

In summary, small RNAs are now thought to be involved in broader biological networks that are only partially understood. One recent review noted that although the RNAi, miRNA and piRNA pathways were thought to be distinct, 'the lines distinguishing them continue to fade'.[41] Even though the notion of 'gene silencing' genes has been used since the early 1990s (and is therefore presumably derivative from the notion of 'gene expression'), it is now used to refer to a more general and only partially understood set of biological processes that are mediated by small RNAs. A common metaphor is that current knowledge only represents the 'tip of an iceberg',[42] and many questions remain about what one paper has dubbed the 'RNAi code'.[43]

---

37 For example Mack, G. (2007), 'MicroRNA Gets Down to Business', *Nature Biotechnology*, **25** (6), 631–8.

38 Moazed, D. (2009), 'Small RNAs in Transcriptional Gene Silencing and Genome Defence', *Nature*, **457**, 413–20; and Bühler, M. (2009), 'RNA turnover and chromatin-dependent gene silencing', *Chromosoma*, **118**, 141–51.

39 Okamura, K. and E. Lai (2008), 'Endogenous Small Interfering RNAs in Animals', *Nature Reviews Molecular Cell Biology*, **9**, 673–8; and Nilsen, T. (2008), 'Endo-siRNAs: Yet Another Layer of Complexity in RNA Silencing', *Nature Structural & Molecular Biology*, **15**, 546–8.

40 Li, L.-C. et al. (2006), 'Small dsRNAs Induce Transcriptional Activation in Human Cells', *Proceedings of the National Academies of Sciences*, **103**, 17337–42.

41 Ghildiyal, M. and P. Zamore (2009), 'Small Silencing RNAs: An Expanding Universe', *Nature Reviews Genetics*, **10** (2), 94–108 at 103.

42 For example, Taft, R. et al. (2010), 'Non-coding RNAs: Regulators of Disease', *Journal of Pathology*, **220**, 126–39.

43 Ghildiyal, M. and P. Zamore (2009), 'Small Silencing RNAs: An Expanding Universe', *Nature Reviews Genetics*, **10** (2), 94–108.

III PATENTING SMALL RNAS

As we have seen above, now seminal scientific experiments have also been the basis of a series of important patent applications connected with RNA interference, which we discuss in this section. Patent applications have also been filed on gene-specific RNAi targets and on modifications of RNA molecules that enhance their pharmaceutical properties, a legacy of anti-sense approaches to drug development.[44] There are also patent applications on miRNAs, as well as methods for inhibiting miRNAs,[45] which we will only characterise very briefly.[46]

A The 'Carnegie Patent'

The first patent application concerning RNA interference, based on the Nobel Prize-winning experiments by Fire and Mello, was filed in December 1997 on behalf of the Carnegie Institution of Washington and the University of Massachusetts. The patent application initially sought to claim a general method for inhibiting the expression of a target gene in a cell by introducing double-stranded RNA with a nucleotide sequence identical to a portion of the gene: the very principle of RNAi. As a matter of fact, the application claimed *any* double-stranded RNA targeting *any* gene in *any* organism. The patent specification explained that the invention could be used for the treatment or prevention of disease, for example by targeting cancer genes or pathogen genes. The specification went on to show that the invention had been demonstrated by experiments in *C. elegans* and that it was 100-fold more effective than conventional antisense approaches.

Examiners at the USPTO objected that the invention had to be restricted to either animals or plants, on the grounds that different methods are used for delivering nucleic acids in each case, and because success in animals is not predictive of success in plants. The applicants responded that their invention claims were generic, and argued that recent scientific publications showed that 'the success of the claimed invention in animals *is* predictive of success in plants'.[47] These arguments did not persuade the examiner,

---

44 Schmidt, C. (2007), 'Negotiating the RNAi Patent Thicket', *Nature Biotechnology*, **25** (3), 273–5.

45 Mack, G. (2007), 'MicroRNA Gets Down to Business', *Nature Biotechnology*, **25** (6), 631–8.

46 A detailed characterisation of patents and biotechnology activity with respect to miRNAs would be a subject for another paper.

47 Fire, A. et al. (1998), 'Genetic Inhibition by Double-stranded RNA', United States Patent No. 6,506,559.

however, and the claims were finally restricted to RNA interference *in vitro* in plant, animal and invertebrate cells, and RNA interference *in vivo* in invertebrates, as demonstrated in *C. elegans*. The examiner also objected to broad claims referring to any portion of a gene and requested that the nucleotide sequences to be claimed were specified. However, after several arguments the generic nature of the invention in this respect was accepted. Approximately 50 non-exclusive licences of the resulting US patent, known as the 'Carnegie patent', have been granted.[48]

B Tuschl I

A second patent application that sought to claim fundamental aspects of RNA interference is called Tuschl I after Thomas Tuschl, a German scientist then in Göttingen but now at the Rockefeller University in New York. This patent application was based on experiments with a strain of *Drosophila* that first suggested that short RNAs mediate the degradation of mRNA. Although the initial patent application sought to cover gene knock-down cells, organisms and even disease models, the claims were narrowed to isolated RNA molecules 'from about 21 to about 23 nucleotides' in length during examination. However, providing a description of the very large number of claimed RNAs in general terms and in a way that did not include prior art (such as the Carnegie patent) has been a difficulty for this application. In their correspondence with the examiners, the applicants mentioned the wide adoption of RNAi in research, in order to argue that it was unreasonable for the USPTO to demand a written description of each RNA molecule to be claimed:

> The commercial success of this discovery lies in its seemingly simplistic elegance. [...] The testament to this is an entire discipline in the academic scientific community that practices the invention. [...] In addition, a thriving sector of the biological industry has developed due to the RNAi technologies.[49]

In later amendments, the claims have been narrowed to RNAs that are 21 to 23 nucleotides in length and perfectly complementary to mammalian and human mRNAs. Pharmaceutical uses of such RNA molecules have also been added to the claims. However, the examiners have continued to find

---

48 Duffy, A. and G. Kowalczyk (2008), 'U. Mass Medical School and Carnegie announce licensing agreements with Oxford BioMedica', Press Release, 7 January, http://www.eurekalert.org/pub_releases/2008–01/ci-um010708.php#.

49 Tuschl, T., P. Zamore, P. Sharp and D. Bartel (2001), 'RNA Sequence-specific Mediators of RNA Interference', US Patent Application No. 20020086356.

difficulties with these broad claims and objected that experiments at the time of filing did not enable all the applications claimed in the patent application, particularly the pharmaceutical applications.

Meanwhile, a dispute has broken out between the assignees of the patent. In June 2009, the Max Planck Society, joined by Alnylam, claimed that some inventions from the Tuschl II patent family have inappropriately been included in the Tuschl I application (in particular a feature referred to as the so-called '3' overhang') by the Whitehead Institute, which pursued this patent application also on behalf of the University of Massachusetts and the Massachusetts Institute of Technology. Significantly, the University of Massachusetts has licensed the Tuschl I patent to Sirna Therapeutics, a company acquired by Merck for $1.1 billion in 2006. Haussecker suggests that the likely loss of support for the controversial parts of the Tuschl I patent application by all assignees except for the University of Massachusetts will reinforce the predominant position of Alnylam Pharmaceutical on the strength of the Tuschl II patent.[50] One patent ostensibly from the Tuschl I patent family has now been granted in Europe. However, this does not lay claim to the most disputed parts of the invention. In fact, the claims cover RNA interference for *in vitro* modification of cells (which may be returned to the organism in question) and as a way of 'examining the function of a gene'.[51]

C Tuschl II

The so-called Tuschl II patent claims a method for preparing double-stranded RNA molecules (of 19 to 23 nucleotides in length with 3' overhang) that mediate cleavage of an mRNA in a mammalian cell. The examination history of this patent application was relatively uneventful. As with the Tuschl I patent, the utilities put forward in the patent include determining the function of a gene, and modulating or inhibiting that function, in particular for disease-related genes, which, it states, may lead to 'valuable information and therapeutic benefits' in agriculture, medicine and veterinary science.[52] The patent emphasises mammalian and human applications.

---

[50] Haussecker, D. (2010) 'Tuschl Litigation Decidedly Shifting into Max Planck-Alnylam's Favor', RNAi Therapeutics, http://rnaitherapeutics.blogspot.com/2010/10/tuschl-litigation-decidedly-shifting.html.

[51] Bartel, D., P. Sharp, T. Tuschl and P. Zamore (2010), 'RNA Sequence-specific Mediators of RNA Interference', European Patent No. 1,309,726.

[52] Tuschl, T., S. Elbashir and W. Lendeckel (2004), 'RNA Interference Mediating Small RNA Molecules', US Patent No. 7,056,704, claiming EPO priority of 1 December 2000.

D siRNA Patents

The patents discussed so far concern the general phenomenon of RNA interference. In contrast, many patent applications and some issued patents cover the use of short RNAs to inhibit the expression of specific disease-related genes. Sirna Therapeutics was granted the first target-specific RNAi patent in 2006 (No. 7,022,828).[53] This patent specifies which gene is targeted (I Kappa B kinase-gamma), but covers *any* modified short RNA targeting the gene (as well as any other nucleic acid molecule that targets the gene). As we discuss in more detail below, this may be a case of persistence of gene-centrism in the patent system, because the gene is still the target of the patent.

E miRNA Patents – Tuschl III

The patent application sometimes known as Tuschl III has miRNAs rather than siRNAs as its focus. It originally covered more than 120 miRNAs (as well as the corresponding antisense molecules, precursors and chemically modified variants), but, as one might expect, the USPTO issued a restriction requirement. In other words, it was held that each isolated miRNA sequence potentially constitutes a separate invention, and hence requires a separate patent application. At this point miRNA-122 was elected (a product claim) and a patent was quickly granted. The specification argues that such molecules may have 'diagnostic or therapeutic applications' and can be used as a 'modulator of the expression of genes'.[54]

IV THE 'GOLD RUSH' FOR SMALL RNAs

As some commentators have noted, there has been a veritable 'gold rush' of patent applications connected with RNA interference.[55] More than 2000 patent applications have been submitted to the USPTO, but very few of them have been issued.[56] Below, we further characterise the claims that have been made about RNA interference and patents in this area.

---

[53] McSwiggen, J. (2002), 'siRNA Treatment of Diseases or Conditions related to Levels of IKK-gamma', United States Patent No. 7,022,828.

[54] Tuschl, T., M. Lagos-Quintana, W. Lendeckel, J. Meyer and R. Rauhut (2005) 'MicroRNA Molecules', United States Patent Application 20050059005.

[55] Angrist, M. and R. Cook-Deegan (2006), 'Who Owns the Genome?' *The New Atlantis*, **11** (Winter), 87–96.

[56] Schmidt, C. (2007), 'Negotiating the RNAi Patent Thicket', *Nature Biotechnology*, **25** (3), 273–5.

A siRNAs

Intuitively, the idea of using short interfering RNAs to 'knock down' genes that have functions that are relevant to disease may seem to be an attractive therapeutic strategy, especially when many diseases have been associated with changes in gene expression in previous research. Moreover, the sequence-specificity of RNA interference suggests that it might be possible to design siRNA drugs against any chosen target, perhaps even to distinguish between disease-specific versions and normal versions of genes.[57] Haussecker further suggests that synthetic siRNAs as therapeutic agents were also attractive because they 'bear semblance to traditional drugs', and thus made it possible for companies to put clear water between themselves and gene therapy, which is regarded as a failed technology by investors.[58] Perhaps another factor was that much of the intellectual property developed for antisense approaches to drug development remained relevant to RNA interference, and several companies in this area (Isis Pharmaceuticals being the prime example) have jumped on what one might call the 'RNAi bandwagon'.

In addition to the patents we have described above, companies have been busy assembling patent portfolios with which they hope to control drug development in this area. Isis Pharmaceuticals claims to have more than 1600 patents in the field, and 'some observers say that it's difficult to work with RNA technology without licensing from Isis'.[59] Sirna Therapeutics has applications on over 250 mammalian and viral genes that are targeted by siRNA.[60] Alnylam Pharmaceuticals 'claims over 150 issued patents of relevance to RNAi, including exclusive access to seven that it believes are fundamental to all therapeutic uses of the technology'.[61] These companies have also been courted by 'big pharma', as illustrated by the acquisition of

---

57 For example Twyman, R. (2003), 'Silencing is Golden: RNA Interference in Research and Medicine', *Wellcome News*, **36** (Q3) 10–11.

58 Of course, RNAi can also be produced by approaches that are reliant on techniques typical of gene therapy, as discussed for example by Haussecker: Haussecker, D. (2008), 'The Business of RNAi Therapeutics', *Human Gene Therapy*, **19** (5), 451–62 at 456. The effects of short RNAs are often explored by introducing genetic modifications into experimental systems.

59 Mack, G. (2007), 'MicroRNA Gets Down to Business', *Nature Biotechnology*, **25** (6), 631–8 at 633.

60 Schmidt, C. (2007), 'Negotiating the RNAi Patent Thicket', *Nature Biotechnology*, **25** (3), 273–5.

61 Ibid. at 473.

Sirna by Merck.[62] In other deals several large pharmaceutical companies have entered agreements with RNAi-focused companies to gain access to their patent portfolio. As Haussecker notes, the field of RNA interference has been 'exceptionally adept in monetizing IP and attracting early-stage funding',[63] although there are signs that the enthusiasm for medical applications of RNA interference has begun to ebb more recently.[64]

Companies in this area continue to sing the praises of RNA interference. In addition to the intuitive arguments cited above, Alnylam Pharmaceuticals, for example, emphasises on its website that RNA interference is a 'natural pathway involved in regulation of gene expression', which can be 'harnessed to create a new class of innovative medicines'. Moreover, visitors to the company's website are told that RNA interference operates upstream from the production of defective proteins that might cause disease. Hence, RNA interference is said to be akin to 'stopping a flood by turning off the faucet' as compared with conventional medicines that simply 'mop up the floor'.[65]

But there are well-known problems with using small RNAs in drug development. Unless they are chemically modified, they break down in seconds in the bloodstream and they cannot easily cross cell membranes because they are negatively charged. Hence, much of the activity in biotechnology has revolved around so-called 'delivery' systems for synthetic siRNAs. Another difficulty associated with siRNAs are 'off-target effects', which can occur because any given short RNA may be complementary to several genomic DNA sequences and can thus interfere with the expression of multiple genes.

Currently, the most advanced short-interfering RNA therapy is being developed by Alnylam Pharmaceuticals to treat a serious lung infection. This siRNA therapeutic, which can be delivered directly into the lung, targets a particular viral gene that is required for its replication. Prior to this, one of the earliest ideas was to use siRNA to treat age-related macular degeneration, a condition where there is excessive growth of blood vessels in the eye. Several companies pursued studies that involved injecting siRNA targeting directly into the eye, thus avoiding issues associated with delivery. However, a study published in 2008 found that other short RNAs

---

62 Osborne, R. (2007), 'Companies Jostle for Lead in RNAi, Despite Uncertainties', *Nature Biotechnology*, **25**, 1191–2.

63 Haussecker, D. (2008), 'The Business of RNAi Therapeutics', *Human Gene Therapy*, **19** (5), 451–62 at 456.

64 Ledford, H. (2010), 'Drug Giants turn their backs on RNA Interference', *Nature*, **468**, 487.

65 Alnylam Pharmaceuticals, www.alnylam.com.

that were not specific for vascular endothelial growth factors also had a therapeutic effect.[66] Ongoing trials in this area were subsequently discontinued.

The complexity of RNA silencing has also highlighted the possibility of unpredictable effects. One recent study has shown that a siRNA has a strong toxic effect in mice, possibly because it interferes with endogenous miRNA processing.[67] Other studies indicate that the various RNA constructs have the potential to trigger the immune system in ways that are not well understood.[68]

B miRNAs

Similarly, miRNAs have been associated with therapeutic and commercial promise. The limited number of miRNAs implies a pool of potentially valuable intellectual property. As miRNAs affect the translation of multiple mRNAs, it has been speculated that miRNAs could be effective drug targets. The hope is that interfering with miRNAs 'could create cascading effects of major importance in complex disease situations'.[69] Rather than targeting a single gene product with siRNA, miRNAs might facilitate a 'one-hit-multiple-target' approach, which could perhaps influence an entire disease pathway.[70] MicroRNAs have also been shown to be involved in many different diseases. Cancers, for example, have been found to have a particular 'miRNA expression signature', and miRNAs have quickly been assimilated into the genetic model of cancer.[71] Hence, miRNAs have been described as tumour suppressors because a low level of the miRNA is associated with cancer. It is thought that therapy might involve administration of the missing miRNAs to block cancer cell growth (Hammond

66 Kleinman, M. et al. (2008), 'Sequence- and target-independent Angiogenesis Suppression by siRNA via TLR3', *Nature*, **452**, 591–7.

67 Grimm, D. et al. (2006), 'Fatality in Mice due to Oversaturation of Cellular microRNAs/short Hairpin RNA Pathways', *Nature*, **441**, 537–41.

68 Marques, J. and R. Williams (2005), 'Activation of the Mammalian Immune System by siRNAs', *Nature Biotechnology*, **23,** 1399–405. For a general review of difficulties associated with RNAi see Castanotto, D. and J. Rossi (2009), 'The Promises and Pitfalls of RNA-interference-based Therapeutics', *Nature*, **457**, 426–33.

69 Mack, G. (2007), 'MicroRNA Gets Down to Business', *Nature Biotechnology*, **25** (6), 631–8 at 632.

70 Osborne, R. (2007), 'Companies jostle for lead in RNAi, despite uncertainties', *Nature Biotechnology*, **25**, 1191–2.

71 Pecorino, Lauren (2008), *Molecular Biology of Cancer: Mechanisms, Targets, and Therapeutics*, Oxford: Oxford University Press.

2005).[72] Another frequently mentioned application concerns the targeting of a specific miRNA (miRNA-122) as a therapy for the hepatitis C virus (Lanford et al. 2010).[73] But as drug targets the complexity and multifunctionality of miRNAs could be problematic, because of unexpected side effects and downstream consequences. According to one appraisal, this means that 'the medical utility of miRNAs is still entirely up in the air'.[74]

V DISCUSSION: SMALL RNAs AND THE LEGACY OF GENE-CENTRISM

Recent biological research has been characterised by a shift away from the gene as the most important biological molecule and by an increasing recognition of biological complexity. Small RNAs have contributed to this shift because an unexpectedly large number of them have been identified in recent research, and because these molecules have been found to be involved in the regulation of gene expression (although the general biological significance of RNA silencing has yet to be worked out). This shift to small RNAs even challenges standard definitions of a 'gene' as a DNA sequence that codes for a protein.[75] Instead, it has become clear that much of the genome 'codes' for a considerable number of small but functional RNA molecules. And in the realm of RNA silencing the significance of sequence-specific interactions between nucleic acids lies in their roles in regulatory networks, rather than in specifying a particular protein sequence.

Despite these important changes in biological understanding, RNAi-related patents continue to be drawn to genes. In the context of drug development, for example, patents have been granted for any siRNAs that target a particular gene (such as I Kappa B kinase-gamma in the Sirna

---

72 Mack, G. (2007), 'MicroRNA Gets Down to Business', *Nature Biotechnology*, **25** (6), 631–8.

73 Jopling, C., M. Yi, A. Lancaster, S. Lemon and P. Sarnow (2005), 'Modulation of Hepatitis C Virus RNA Abundance by a Liver-Specific MicroRNA', *Science*, **309** (5740), 1577–81; and Lanford, R., E. Hildebrandt-Eriksen, A. Petri, R. Persson, M. Lindow, M. Munk, S. Kauppinen and H. Ørum (2010), 'Therapeutic Silencing of MicroRNA-122 in Primates with Chronic Hepatitis C Virus Infection', *Science*, **327** (5962), 198–201.

74 Fambrough in Mack, G. (2007), 'MicroRNA Gets Down to Business', *Nature Biotechnology*, **25** (6), 631–8 at 634.

75 Pearson, H. (2006), 'What is a Gene?' *Nature*, **441**, 399–401; and Barnes, Barry and John Dupré (2008), *Genomes and What to Make of Them*, Chicago: University of Chicago Press.

Therapeutics case).[76] Following the USPTO's revised utility guidelines, discussed above, it is often necessary to identify 'the function' of 'the gene' in order to get a patent. With respect to patents on siRNAs, the USPTO has in fact issued similar guidance to the effect that it is not sufficient to patent short-interfering RNAs simply as probes or tools for blocking a gene. Instead, it is necessary to also have some evidence of the function of the gene and why its inhibition is warranted as an experimental therapeutic strategy.[77] As a consequence, the gene remains the focus of the patent, and information about gene function is taken to entail therapeutic potential. But why should it be sufficient to provide information about the function of a gene, itself a problematic notion, in order to be allowed to patent an siRNA that targets this gene?

The regulatory molecules themselves, miRNAs in particular, are being patented as compositions of matter. Perhaps this can in part be explained in terms of the enthusiasm some biologists have shown for these molecules. Nevertheless, the patenting of such regulatory molecules as compositions of matter raises much the same issues as traditional patents on genes as DNA molecules. Our previous work has highlighted the problems associated with the assumption that there is a one-to-one relationship between a particular gene and a particular biological function.[78] And the multifunctionality of miRNAs in particular, where one miRNA can target hundreds of different genes, and several miRNAs can target the same gene, is even more striking than the multifunctionality of protein-coding genes. MicroRNAs also behave differently in different contexts, depending on the environmental, cellular and developmental milieu, meaning that an analysis of the DNA sequence coding for a particular miRNA will be insufficient to understand its action. In fact, according to Burian, 'it is virtually impossible to determine whether or not a particular sequence of ~22 nucleotides is a microRNA', without knowing its cellular context, because a sequence of 22 nucleotides that does behave as an miRNA in one context will not necessarily do so in another.[79] Also, miRNA targets 'can be

---

76 McSwiggen, J. (2002) 'SiRNA Treatment of Diseases or Conditions Related to Levels of IKK-Gamma', United States Patent No. 7,022,828.

77 LeGuyader, J. (2004), 'Patenting Interfering RNA, 'USPTO Presentation', www.cabic.com/bcp/020304/JLeGuyader_PIRNA.ppt.

78 See Calvert, J. (2007), 'Patenting Genomic Objects: Genes, Genomes, Function and Information', *Science as Culture*, **16** (2), 207–23; and Calvert, J. and P. Joly (2011), 'How Did the Gene Become a Chemical Compound? Shifting Ontologies of the Gene and the Patenting of DNA', *Social Science Information*, **50** (2) (forthcoming).

79 Burian, R. (2009), 'Is Molecular Genetics Becoming Less Reductionistic? Notes from recent case studies on mapping *C. elegans* and the discovery of

scattered over the entire genome and respond differentially in different cell types, at different developmental stages, or in the presence of different exogenous chemicals'.[80] In this light it is perhaps a mistake to assume that the isolation of the miRNA is the fundamental invention. Why then should an initial patent for an miRNA be more important than subsequent ones that develop different uses for the same miRNA in other contexts?

A Gene-centrism

The medical and commercial promise of this area can be put down to three interrelated features of RNA interference: First, the idea of using short-interfering RNAs to 'knock down' genes that have functions that are relevant to disease is intuitive and has been attractive to broader audiences. Second, RNAi has been extremely successful as a tool for exploring gene function in a wide range of experimental systems in contemporary biological research. Third, patents have been awarded, and further patent applications are pending, that mention both these features, sometimes even in the patent claims. Our argument here is that the intuitive appeal of RNA interference as a therapeutic strategy is an expression of what Ken Waters has characterised as 'gene-centrism' in molecular biology.

According to Waters, gene-centrism is the idea that 'genes direct all the important developmental and metabolic processes in living systems'.[81] This idea often appears to be borne out by empirical studies that show, for example, that 'genes' (i.e. genetic differences) can make a specific difference between sets of healthy and diseased organisms in a particular environment. However, as Waters argues, this does not imply that the gene in question is *the* cause of the disease. Many other causes will be involved and many other intervention strategies are conceivable (such as modifying the environment). Similarly, if we find differences in the gene expression profiles of healthy and diseased organisms, this does not imply that targeting over-expressed genes will necessarily be a viable therapeutic strategy. It

---

microRNA', Invited Lecture, Institut d'histoire et de Philosophie des Sciences et des Techniques, Paris, December 4, 27.

80 Ibid.

81 Kellert, Stephen, Helen Longino and C. Kenneth Waters (eds) (2006), *Scientific Pluralism: Minnesota Studies in the Philosophy of Science*, vol. 19, Minneapolis: University of Minnesota Press, 190–214 at 190.

is possible that 'alternate, less intuitive targets may be more effective, or correction of the deviated parameter may itself have harmful system-wide effects'.[82]

B Reassessing Patents for RNA Interference

Companies emphasise that RNA interference is a 'natural pathway' and that the patents they have assembled are 'fundamental' to drug development in this area. However, our discussion leads us to a different assessment of patents related to RNA interference. Above all, it is important to note that the patents mentioned above are based on the initial discoveries in the area of RNA silencing. The 'initial' patents are by no means necessarily the 'fundamental' patents. Moreover, if we set aside the ideas associated with gene-centrism, there seems to be no general reason for expecting that knocking down over-expressed genes should be a successful therapeutic strategy. In light of the complexity of RNA silencing, the contributions that these patents will make to the eventual formulation of useful products, such as therapeutics, may be relatively small.

C RNAi as a Research Tool

What genes *can* do is provide an entry point for experimental manipulation, and RNA interference is perhaps best understood in this tradition.[83] Knocking down genes and creating a difference between organisms that are 'normal' and those where the expression of a gene has been disabled by means of RNA interference can give us insights into the processes in which the gene participates. In this sense it is perhaps possible to understand RNA interference as a *research tool* without gene-centric inferences. However, this does not of course imply that such experiments elucidate *the* function of a gene, perhaps as a basis for a patent application. It is possible that in some contexts a gene appears to have no function because it is simply not expressed, and consequently there is no difference between the normal organism and the one that has been subjected to RNA interference.[84] As we

---

82 Ahn A., M. Tewari, C.-S. Poon and R. Phillips (2006), 'The Limits of Reductionism in Medicine: Could Systems Biology Offer an Alternative?' *Public Library of Science Medicine*, **3** (6), 709–13 at 710.

83 Kellert, Stephen, Helen Longino and C. Kenneth Waters (eds) (2006), *Scientific Pluralism: Minnesota Studies in the Philosophy of Science*, vol. 19, Minneapolis: University of Minnesota Press, 190–214.

84 If we ignore off-target effects and the broader silencing effects small RNAs can have.

have argued above, any small RNA may have several functions, depending on biological context.

The widespread adoption of RNAi in the various experimental systems of contemporary biology is a testament to the success of RNAi as a research tool. Perhaps surprisingly this is also cited as a key justification for some of the patents on the RNAi phenomenon, sometimes even in the claims. However, there is no good reason to expect that research tools will necessarily double as therapeutic agents. We also do well to remember that the RNA interference effect that we access may be the tip of the iceberg of complex interconnecting networks of gene regulation and control, which is still incompletely understood. If it is above all a research tool, then do we actually need patents on RNA interference?[85]

CONCLUSION

RNAi has been widely taken up in biological research, because it is a simple technique for exploring gene function in a wide range of experimental systems. The discovery of RNA interference has also contributed to a partial regeneration of the biotechnology industry. In this chapter, we have examined the growing scientific and patent literatures in this area of biotechnology, and we have shown that some familiar but problematic conceptual schemes are still at work in patenting.

In particular, 'the gene' and 'its function' are still given a central place in awarding patents related to RNA interference. We have seen that much of the patenting in this area continues to be focused on the patenting of products, when justified by something that pretends to be a biological function. In contrast to Angrist and Cook-Deegan, who point out that the backlog of RNAi-related patent applications at USPTO will take years to sort out,[86] we have characterised these patents in some detail. We would also point out that for some of these patent applications (like Tuschl I)

---

85 The patenting of research tools has been discussed in general in the following recent articles: Lee, P. (2005), 'Inverting the Logic of Scientific Discovery: Applying Common Law Patentable Subject Matter Doctrine to Constrain Patents on Biotechnology Research Tools', *Harvard Journal of Law & Technology*, **19** (1); and Pressman, L., R. Burgess and R. Cook-Deegan (2006), 'The Licensing of DNA Patents by US Academic Institutions: An Empirical Survey', *Nature Biotechnology*, **24**, 31–9.

86 Angrist, M. and R. Cook-Deegan (2006), 'Who Owns the Genome?' *The New Atlantis*, **11** (Winter 2006), 87–96.

there is considerable uncertainty over the precise nature of the invention for which protection is being sought.[87]

Research into small RNAs can be understood as part of a more general shift in the life sciences towards an appreciation of complexity and regulation. But instead of undermining existing patenting paradigms, we see an attempt to patent every molecular entity that is involved in gene expression.[88] The range of biological entities that can be patented is simply being extended, even though each biological molecule may participate in many different processes. All this makes the recent 'gold rush' in the patenting of small RNAs more worrisome. If only a few small RNAs are central to several important regulatory pathways, then their ownership could lead to patents with very broad scope. The potential dangers of patent thickets and the 'tragedy of the anti-commons'[89] are perhaps exacerbated in the case of miRNAs, which operate in complexly integrated regulatory networks.

Our parallel surveys of the patenting and scientific literature also show that we should not overestimate the promise associated with RNA interference, although we have seen that many of the highly speculative claims associated with RNA interference are not only trumpeted by biotechnology companies but have also found their way into patent documents. Hence, we must ask whether the 'gold rush' for RNA interference was in part precipitated by the ability to obtain patents, which at least tacitly endorsed these therapeutic strategies. Perhaps this is what Haussecker meant when he wrote that patents are 'the current life blood of RNA therapeutics'.[90]

---

[87] Tuschl, T., P. Zamore, P. Sharp and D. Bartel, (2001), 'RNA Sequence-specific Mediators of RNA Interference', US Patent Application No. 20020086356.

[88] This resonates with Simondon's discussion of individuation: an invention is taken to apply to an individuated entity as opposed to a relational one: Simondon, G. (1989), *L'individuation psychique et collective*, Paris: Aubier.

[89] This is the argument that the existence of patents leads to the neglect of large areas of research: see Heller, M. and R. Eisenberg (1998), 'Can Patents Deter Innovation? The Anticommons in Biomedical Research', *Science*, **280**, 698–701.

[90] Haussecker, D. (2008), 'The Business of RNAi Therapeutics', *Human Gene Therapy*, **19** (5), 451–62 at 460.

9. Stem cell patents: looking for serenity

Amina Agovic

Human embryonic stem cell research is the subject of great hope and controversy. International and regional policies on stem cell research are complex and lack harmonisation. Recently, however, the policy battles in respect of innovation in stem cell research have begun to shift from controversies within research to skirmishes over patent law and related intellectual property rights.

Human embryonic stem cells are generally harvested from the inner cell mass of a blastocyst-stage embryo and this necessitates that an embryo is destroyed. This inevitable destruction of the human embryo is a major cause of controversy in relation to not only the research itself, but also the patenting of research results. Moreover, this embryo destruction was a primary cause for the European Patent Office (EPO) to reject patent applications such as the Edinburgh (T1079/03), the Wisconsin Alumni Research Foundation (WARF) (T1374/04) and the CIT (T522/04).

In the United States, ever since the isolation of the first human embryonic stem cell line in 1998, many scientists have increasingly complained about the hindering effects of the WARF stem cell patents on possibly life-saving research involving human embryonic stem cells. Now, the latest ruling on the American WARF patent has the potential to alter the patentability criteria in life sciences.

Both the European and the American WARF cases illustrate the ethical conundrums surrounding patenting human embryonic stem cells. As a result, ethical tensions surrounding patentability of human embryonic stem cells require a better understanding of patentability of human embryonic stem cells in a wider legal and ethical context.

This chapter aims to draw a comparison between ethical concerns surrounding patentability of human embryonic stem cells in Europe and the United States. It begins with a brief introduction to the stem cell sciences and the research controversy. A description of the patentability requirements for stem cell technologies, including a reflection on the moral or ethical exclusions in patent laws follows. Further, the chapter evaluates ethical tensions that arise with patenting such inventions. The latest developments in respect of the WARF stem cell patents in both Europe and the United States are examined, as well as the European CIT patent application and Oliver Brüstle's patent. Europe and America continue to express a significant lack of enthusiasm to engage in resolving ethical tensions that surface with patentability of human embryonic stem cells. This chapter, by examining the risks of maintaining the status quo and the repercussions such may have on stem cell innovation in both nations, emphasises a need for a patent system in line with its original goal of balance between the inventor and society at large.

I STEM CELL SCIENCE AND CONTROVERSY

Embryonic stem cells offer groundbreaking insights into human health and diseases and have the potential to provide for therapeutic cures to many ailments in the future. Stem cells represent, figuratively speaking, an 'elixir of life' in a sense that they 'can only reproduce what the body is or should have naturally (i.e., it is replacing cells that should already be there)'.[1] This remarkable feature allows them to proliferate indefinitely *in vitro* (in a test tube), branding these cells a miracle cure.

Embryonic stem cells were first independently extracted from mouse embryos in 1981, and in 1998 Professor James Thomson's group at the University of Wisconsin successfully derived embryonic stem cell lines from human blastocysts in 1998.[2] Despite being a relatively new field of scientific exploration, many promising research results suggest that stem cells are ever closer to the prospect of damaged tissue regeneration in clinical applications.[3] The past decade of stem cell research has increased scientific understanding of specific fatal and debilitating diseases like diabetes, cancer, Parkinson's disease and other neurodegenerative diseases.[4]

---

1. Burns, L. (2009), 'You Are Our Only Hope: Trading Metaphorical "Magic Bullets" for Stem Cell Superheroes', *Theoretical Medicine and Bioethics*, **30**, 427–42 at 432.
2. Thomson, J. et al. (1998), 'Embryonic Stem Cell Lines Derived from Human Blastocysts', *Science*, **282** (5391), 1145–7.
3. Ibid.; Orlic, D. et al. (2001), 'Bone Marrow Cells Regenerate Infarcted Myocardium', *Nature*, **410** (6829) 701–5; and Burd, A. (2007), 'Stem Cell Strategies in Burns Care', *Burns*, **33** (3), 282–91.
4. Martinezserrano, A. and A. Bjorklund (1996), 'Protection of the Neostriatum Against Excitotoxic Damage by Neurotrophin-producing, Genetically Modified Neural Stem Cells', *Journal of Neuroscience*, **16** (15), 4604–16; Cao, Q., R. Benton and S. Whittemore, 'Stem Cell Repair of Central Nervous System

Naturally, human embryonic stem cell research has attracted great interest in the United States, Australia, Europe and internationally – both for its scientific and medical potential, and for its commercial promise. For example, market forecasts directed at stem cell technologies range from a modest USD 100 million to a rather optimistic USD 10 billion by 2010.[5]

However, to obtain an embryonic stem cell, an embryo is grown in the laboratory for five to seven days at which point the stem cells are harvested. Once these cells are extracted from the embryo, the embryo is automatically destroyed.[6] The destruction of the human embryo is a significant cause for controversy facing this type of scientific exploration. As such, increasing international attention is focused on a technology called somatic cell nuclear transfer (SCNT) or 'therapeutic cloning', a process of creating an embryo through a cloning procedure so as to derive human embryonic stem cells for purposes of research and potentially for tissue transplantation.[7]

Therapeutic cloning involves using a patient's cells to create an embryo, which is allowed to develop only to the blastocyst stage and solely for the purpose of research into developing cures and tissue for transplantation. There appears to be no intention of implanting the embryo into a uterus for development of a cloned human being.[8] In contrast, reproductive cloning results in a cloned human being. To date, scientists are yet to create human

---

Injury', *Journal of Neuroscience Research*, **68** (5), 501–10; and Di Giorgio, F. et al. (2007), 'Non-cell autonomous effect of glia on motor neurons in an embryonic stem cell-based ALS model', *Nature Neuroscience*, **10** (5), 608–14.

5 Korobkin, Russel and Stephen Munzer (2007), *Stem Cell Century: Law and Policy for a Breakthrough Technology*, New Haven, CT: Yale University Press, 4.

6 Burns, L. (2009), 'You Are Our Only Hope: Trading Metaphorical "Magic Bullets" for Stem Cell Superheroes', *Theoretical Medicine and Bioethics*, **30**, 427–42.

7 The President's Council on Bioethics (2004), *Monitoring Stem Cell Research*, Washington, DC: The President's Council on Bioethics, 5–6, 22–23, 385, http://bioethics.georgetown.edu/pcbe/reports/stemcell/; Robertson, J. (1999), 'Two Models of Human Cloning', *Hofstra Law Review*, **27**, 616; and Schechter, J. (2010), 'Promoting Human Embryonic Stem Cell Research: A Comparison of Policies in the United States and the United Kingdom and Factors Encouraging Advancement', *Texas International Law Journal*, **45** (3), 603–29.

8 The President's Council on Bioethics (2004), *Monitoring Stem Cell Research*, Washington, DC: The President's Council on Bioethics, 5–6, 22–23 at 385; Robertson, J. (1999), 'Two Models of Human Cloning', *Hofstra Law Review*, **27**, 616; and Schechter, J. (2010), 'Promoting Human Embryonic Stem Cell Research: A Comparison of Policies in the United States and the United Kingdom and Factors Encouraging Advancement', *Texas International Law Journal*, **45** (3), 603–29 at 609, 611.

embryonic stem cell lines from cloned blastocyst[9] but the technology is likely to substantiate[10] in the near future.

SCNT does not just involve the destruction of human embryos for purposes of human embryonic stem cell research, but also the creation of embryos for destruction and research, which is even more morally contentious for some.[11] There are significant concerns that the distinction between therapeutic cloning and reproductive cloning will eventually disappear – 'that once cloned embryos exist, it will be difficult to prevent scientists or physicians from using them for reproduction'.[12]

Further controversy arises in respect of chimeras, trans-species organisms created by mixing stem cells or embryos from one species with early embryos of another species.[13] This technology carries an inherent risk of creating creatures with a significant number of human nerve and brain cells, making them potentially useful subjects for biomedical research.[14] Moreover, the idea is to utilise these techniques in order to develop organs 'genetically compatible with a human stem cell donor and harvest them in chimeras in preparation for human organ transplant' thereby minimising

---

9 Schechter, J. (2010), 'Promoting Human Embryonic Stem Cell Research: A Comparison of Policies in the United States and the United Kingdom and Factors Encouraging Advancement', *Texas International Law Journal*, **45** (3), 603–29 at 607.

10 Ritter, M. (2008), 'Scientists Make Human Embryo Clones', *Huffington Post*, 17 January, http://www.huffingtonpost.com/huff-wires/20080117/cloned-embryos; and Smith, D. (2008), 'Scientists Attempt Stem Cell Breakthrough', *Sydney Morning Herald* (Sydney), 17 September.

11 Rewerski, P. (2007), 'The Need for a New US Stem Cell Research Policy: A Comparative Look at International Stem Cell Research Laws', *University of Illinois Journal of Law, Technology and Policy Review*, **1**, 415–31 at 416; and Luk, E. (2006), 'The United Kingdom and Germany: Differing Views on Therapeutic Cloning and How the Belgian Resolution Brings Them Together', *Michigan State University Journal of Medicine and Law*, **10**, 523–56 at 538.

12 Robertson, J. (1999), 'Two Models of Human Cloning', *Hofstra Law Review*, **27**, 609–38 at 616; and Schechter, J. (2010), 'Promoting Human Embryonic Stem Cell Research: A Comparison of Policies in the United States and the United Kingdom and Factors Encouraging Advancement', *Texas International Law Journal*, **45** (3), 603–29 at 609.

13 Kopinski, N. (2004), 'Human-Nonhuman Chimeras: A Regulatory Proposal on the Blurring of Species Lines', *Boston College Law Review*, **45**, 619–66 at 624–5; and Robert, J.S. and F. Baylis (2003), 'Crossing Species Boundaries', *American Journal of Bioethics*, **3**, 1–13 at 8.

14 Winickoff, D. et al. (2009), 'Opening Stem Cell Research and Development: A Policy Proposal for the Management of Data, Intellectual Property, and Ethics', *Yale Journal Health Policy, Law and Ethics*, **9**, 52–127 at 79.

the risk of transplantation rejection.[15] The major issue with this method is that this may blur species lines.[16]

As a result, the notion that not all forms of research are ethically desirable is escalating in itself. For instance, for the biotechnology industry and stem cell technologies, the availability of patent protection is crucial in order to succeed in the industry. However, this need to patent almost all biotechnology innovation has fallen under heavy criticism.[17] This extreme dependency[18] on patent protection has begun to expose fundamental uncertainties that afflict the very nature of the patent protection. As a result, the patent offices are more and more under attack for issuing patents on stem cell related innovation and thereby stimulating and encouraging a form of research otherwise deemed ethically undesirable. While the biotechnology industry has a positive influence on the economy and public health, scholars and society have begun questioning what role, if any, patent law should play in dealing with morally controversial inventions.

II PATENTING STEM CELLS AND ETHICAL CHALLENGES

Prior to the American landmark decision of *Diamond* v. *Chakrabarty*,[19] life forms were deemed 'products of nature' rather than patentable human inventions. Although the decision confirmed that phenomena of nature in their natural states continue to be unpatentable, the Supreme Court of the United States ruled that a major exemption exists should the products have been isolated from their natural state through human intervention. Shortly

15 Kopinski, N. (2004), 'Human-Nonhuman Chimeras: A Regulatory Proposal on the Blurring of Species Lines', *Boston College Law Review*, **45**, 619–66 at 630; and Robert, J.S. and F. Baylis (2003), 'Crossing Species Boundaries', *American Journal of Bioethics*, **3**, 1–13 at 8.

16 Winickoff, D. et al. (2009), 'Opening Stem Cell Research and Development: A Policy Proposal for the Management of Data, Intellectual Property, and Ethics', *Yale Journal Health Policy, Law and Ethics*, **9**, 52–127 at 79.

17 See for example International Expert Group on Biotechnology, Innovation and Intellectual Property (2008), *Towards a New Era of Intellectual Property: From Confrontation to Negotiation*, Montreal: International Expert Group on Biotechnology, Innovation and Intellectual Property, 35.

18 Lerner, J. (1995), 'Patenting in the Shadow of Competitors', *Journal of Law and Economics*, **38**, 463–95 at 472.

19 *Diamond* v. *Chakrabarty*, 477 US 303 (1980).

afterwards, the European and international patent offices began issuing patents on living things.[20]

At present, patenting requirements for mechanical innovations are the same as for material of human origin. The *TRIPS Agreement* 1994 prescribes minimum standards[21] for patentability of inventions to be complied with by each of the WTO member state. Article 27.1 of the *TRIPS Agreement* 1994 sums up standard patentability requirements:

> patents shall be available for any inventions, whether products or processes, in all fields of technology, provided that they are new, involve an inventive step and are capable of industrial application.

However, despite this clear-cut definition, the patenting of human embryonic stem cells and material of human origin remains contested, more so owing to the general nature of bio-patenting and specific ethical and social confines it carries along.[22]

The *TRIPS Agreement* 1994 allows for certain inventions to be excluded from patentability should they go against the public order or morality.[23] Most international and national patent laws include a sort of 'morality' clause in their legislations so as to 'safeguard' social welfare, but the scope and interpretation differ between nations.[24] In Europe, for example, concepts of morality and public order are unique to the *European Patent Convention* (EPC)[25] where Article 53(a) (the 'morality clause') aims to

---

20 Chapman, A. (2009), 'The Ethics of Patenting Human Embryonic Stem Cells', *Kennedy Institute of Ethics Journal*, **19** (3), 261–88, 267.

21 *Marrakesh Agreement Establishing the World Trade Organization*, opened for signature 15 April 1994, 1867 UNTS 3 (entered into force 1 January 1995) annex 1C ('*Agreement on Trade-related Aspects of Intellectual Property Rights*' – *TRIPS Agreement* 1994).

22 Hermeren, Göran (2000), 'Patents and Licensing, Ethics, International Controversies', in Thomas Murray and Maxwell Mehlman (eds) (2000), *Encyclopedia of Ethical, Legal and Policy Issues in Biotechnology*, New York: Wiley, 817–25.

23 Article 27.2 of the *TRIPS Agreement* 1994.

24 See for example, *Patent Law of the People's Republic of China* 2000 (China) art 5 where Chinese patent law forbids patents for 'any invention-creation that is contrary to the laws of the State or social morality or that is detrimental to public interest'; *Patents Act* 1970 (India) s 3 where Indian patent legislation defines non-patentable inventions as those that 'include an invention the primary or intended use of which would be contrary to law or morality or injurious to public health'; *Intellectual Property Code of the Philippines* (Republic Act 8293) (Philippines), s 22.6 where the Philippines bars 'anything which is contrary to public order or morality' from patentability.

25 *Convention on the Grant of European Patents* ('*European Patent Convention*') of 5 Oct 1973 [2000] art 1 Official Journal Special Edition 1/2007. The *European*

safeguard the interests of the public. Article 53(a) EPC clearly states that 'European patents shall not be granted in respect of a) inventions the commercial exploitation of which would be contrary to 'ordre public' or morality [...]' Article 53(a) EPC is reflected in Article 6 (1) of the European Parliament's *Directive on the Legal Protection of Biological Inventions* (*European Biotechnology Directive*).[26]

The core conflict over stem cell patents relates to fears that patents on material of human origin risk upsetting the delicate balance between the private and public benefits of patent law in favour of economic gain. Although developments in biotechnology and stem cell research have brought about specific ethical challenges and complexities, many wonder whether these challenges should be addressed by the patent system. As a result, there have been anguished deliberations over the morality and ethics of patenting human embryonic stem cell research, and the scope of traditional patentable subject matter.

Thus, when it comes to human embryonic stem cell innovation, the key concerns include uncertainty about how to apply patent law to stem cell innovation; ethical objections to patenting such inventions; and the research involving human embryos as well as the post-grant commercial exploitation.[27] It is important to distinguish between ethical concerns related to the stem cell research itself and those that arise with the patenting of stem cell inventions.[28]

In Europe, patentability of human embryonic stem cells instantly translates into the interpretation of the European morality clause. To some, the morality clause is seen as 'a veiled attempt to regulate innovation'[29] where a failure to obtain patent protection would act as a disincentive to engage in stem cell research activities. To others, granting exclusive monopoly rights

---

Patent Convention establishes 'a system of law, common to the Contracting States, for the grant of patents for invention'. A single application may lead to grant of a patent in each country designated, but, once issued, the patents are treated in the same way as a set of national applications.

26 *European Parliament Directive 98/44/EC of the European Parliament and of the Council on the Legal Protection of Biotechnological Inventions* [1998] OJ (L 213) 13 ('*European Biotechnology Directive*').

27 Australian Law Reform Commission (2004), *Gene Patenting and Human Health, Discussion Paper 68*. Sydney: Australian Commonwealth, http://www.austlii.edu.au/au/other/alrc/publications/dp/68/m [15.60]–[16.67].

28 Thomson, P. (2007), 'Ethical Aspects and Patents in Lifescience', Paper for the WIPO symposium on IP and Bioethics, Geneva, 4 September, http://www.wipo.int/meetings/en/doc_details.jsp?doc_id=85513.

29 Mills, Oliver (2005), *Biotechnological Inventions: Moral Restraints and Patent Law*, Aldershot: Ashgate, 34.

over morally contentious inventions becomes particularly troublesome in the stem cell area because there may be some inventions that no one should 'own' or be encouraged by governments to invent.[30] This has resulted in overwhelming anxiety amidst the public, governments and even scientists about the impact of patenting stem cell inventions. Additionally, issuance of human embryonic stem cell patents is said to have put the patent system on a slippery slope to commodification of human life, diminishing of human dignity and failure to provide shared ownership of key social goods.[31]

By contrast, present American patent law does not contain an express provision permitting patent applications to be refused or challenged on ethical or moral grounds. There is no prohibition on patents claiming human beings, or the processes for their generation, equivalent to the morality clause in Europe. Yet American case law provides for inclusion of 'public policy' and 'well being of the society' in the patenting procedure. It allows for inventions deemed 'frivolous or injurious to the well-being, good policy, or sound morals of a society' to be excluded from patentable subject matter, but this patent moral utility doctrine is all but futile.[32] Ambiguity as to the nature of what can be patented remains, but so far the United States Patent and Trademark Office (USPTO) and the courts agree that the isolation and purification of naturally occurring products is patentable.[33] Any questions directed to a patent's morality will only arise post patent issue.

The American legislature's attempt at rectifying the situation resulted in the Weldon amendment, which forbids patenting human beings at any stage of development.[34] Nevertheless, it appears that the Congress negotiators agreed that the bar on patenting genetically engineered human embryos, foetuses and human beings would not extend to human stem cells, genes, tissue, cells and 'other biological products'.[35]

---

[30] Gordon, W. (2002), 'Authors, Publishers, and Public Goods: Trading Gold for Dross', *Loyola of Los Angeles Law Review*, **36**, 159–98 at 164.

[31] Chapman, A. (2009), 'The Ethics of Patenting Human Embryonic Stem Cells', *Kennedy Institute of Ethics Journal*, **19** (3), 261–88 at 267.

[32] *Lowell* v. *Lewis*, 15 Fed Cas 1018 (1817) (Story J) quoted in *Tol-O-Matic Inc* v. *Promo Produkt-und Marketing Gesellschaft MbH*, 945 F 2d 1546, 1553 (1991).

[33] Chapman, A. (2009), 'The Ethics of Patenting Human Embryonic Stem Cells', *Kennedy Institute of Ethics Journal*, **19** (3), 261–88, 267.

[34] Rogan, J. (2003), 'Letter to the United States Senate Committee and House Appropriations Committees', United States Patent and Trademark Office, 20 November.

[35] The Henry J. Kaiser Family Foundation (2003), 'National Politics and Policy, Lawmakers Reach Agreement on Appropriations Bill Language Barring Patents on Human Organisms', Kaisernetwork.org, 25 November, http://www.kaisernetwork.org/daily_reports/rep_index.cfm?%20hint=2&DR_ID=21029.

Nevertheless, the battle for stem cell patents has placed a heavy burden on patent laws and put patent agencies at the forefront of dealing with economic progress and social concerns. Thus far, both the United States Congress and the European Parliament refuse to engage in the discourse concerning finer points of their respective patent laws, ethics and human embryonic stem cells. But absent clear guidelines on human embryonic stem cell research, scientists and patent attorneys get to determine the limits of patent eligibility by describing the subject matter in their patent applications.[36] Following the *Chakrabarty* decision, the American government passed over many opportunities to rectify or modify the traditional 'mechanical' patent system and make it more suited to the newer technologies such as biotechnology and research involving material of human origin.[37] In addition, the Supreme Court of the United States only recently passed on yet another opportunity to clarify the issue of patents on living things.[38]

By comparison, over the course of only a few years, Europe has experienced its fair share of opposition to patenting life forms ranging from the OncoMouse[39] patent to the Edinburgh[40] patent to the WARF[41] patents. The WARF patents have caused many to argue that material of human origin should not be patented because the power afforded to the patent holder is out of proportion with the benefit provided to society, or that it is immoral for patent owners to have a certain degree of 'ownership' over other life forms.[42]

---

36 Bagley, M. (2004), 'Patent First, Ask Questions Later: Morality and Biotechnology in Patent Law', *William and Mary Law Review*, **45**, 469–547.

37 Brody, B. (2006), 'Intellectual Property and Biotechnology: The U.S. Internal Experience – Part I', *Kennedy Institute of Ethics Journal*, **16**, 1–37; and Brody, B. (2006), 'Intellectual Property and Biotechnology: The U.S. Internal Experience – Part II', *Kennedy Institute of Ethics Journal*, **16** (2), 105–28.

38 Klein, R. and M. Mahoney (2008), 'LabCorp v. Metaboite Laboratories: The Supreme Court Listens but Declines to Speak', *Journal of Law, Medicine and Ethics*, **36** (1), 141–9.

39 Leder, P. and T. Stewart (1986), 'Method for Producing Transgenic Animals', European Patent No: EP 0169672.

40 Smith, A. (1996), 'Isolation, Selection and Propagation of Animal Transgenic Stem Cells', European Patent No. EP695351 ('Edinburgh patent').

41 Thomson, J. (1997), 'Primate Embryonic Stem Cells', European Patent No. EP 0770125.

42 For example, Coulter, J. (1990), 'Parliamentary Debates', Australian Parliament, Senate, 22 August; and White, A. (2004), 'The Ethics – Gene Patenting and Human Health', *Australian Intellectual Property Journal*, **15** (1), 6–17.

III HUMAN EMBRYONIC STEM CELL PATENTS IN EUROPE AND AMERICA

While Europe deems the mere notion of patents on human embryonic stem cells reprehensible and persists with debates on their appropriateness, the United States continues to issue patents on privately funded human embryonic stem cell research and wreak havoc in respect of federally funded human embryonic stem cell research. On 23 August 2010, a federal district court judge blocked President Obama's 2009 executive order on expanding federal funding to cover stem cell research. The judge claimed that the order infringed a ban on federal money being used to destroy embryos.[43] Yet, on 28 September 2010, the United States Court of Appeals in Washington 'granted the Obama administration's request to allow the funding from the National Institutes of Health while it appeals a judge's order blocking the research'.[44]

To date, the USPTO has issued over 8000 stem cell patents, hundreds of patents claiming methods of isolating stem cells, methods of differentiating stem cell lines, and methods of using stem cells in treatment.[45] Some of these patents include claims to totipotent stem cells. For example, Geron Corporation obtained a patent on 'methods and materials for the growth of primate-derived primordial stem cells in feeder-free culture', which claims a cellular composition comprising undifferentiated primate primordial cells. That is, the patent claims both pluripotent and totipotent primate stem cells, and does not exclude human primate stem cells.[46] According to the USPTO, methods directed to cloning of human beings seemingly do not include methods directed to or including a human organism. Despite the ban on patenting human beings at any stage of development, human totipotent cells are patentable in the United States.

As the United States has faced great political disputes over whether to fund human embryonic stem cell research, Europe has also suffered from divisions and disagreements between nation states, and institutions. In Europe, the morality clause is now being called on to perhaps cover either

---

43 Gardiner, H. (2010), 'US Judge Rules Against Obama's Stem Cell Policy', *New York Times*, 23 August.

44 Pickler, N. (2010), 'Stem Cell Research Funding Appeal: Court OK's Federal Support While Appeal Continues', *The Huffington Post*, September.

45 Konski, A. and D. Spielthenner (2009), 'Stem Cell Patents: A Landscape Analysis', *Nature Biotechnology*, **27** (8), 722–6 at 723.

46 Bodnar, A. (2004), 'Methods and Materials for the Growth of Primate-derived Primordial Stem Cells in Feeder-free Culture', United States Patent No. 6,800,480, 5 October.

more than intended or to be utilised as intended from the very origins of the modern patent system. The trouble is that the framers of the EPC behaved very oddly: they required the EPO to take the morality clause into account, but failed to offer any guidance as to which and whose morality norms one should rely on. Such lack of support has all but made the patent examiners look the other way when dealing with morally controversial inventions. While the EPO can reject a patent application should it fall foul of the morality clause, Europe's current lack of clarity on the morality clause further adds to the already complex discourse on the relationship between stem cells and patents.

Thus, for example, the EPO bars patenting human embryonic stem cells if the invention or making of it has resulted in an indispensable use of or destruction of the human embryo.[47] The EPO appears to use 'the destruction of the human embryo' as the primary criterion in determining whether a particular human embryonic stem cell invention is patentable or not. Surprisingly, the EPO has equated this determining factor with the fundamental principles enshrined within the *European Convention for the Protection of Human Rights and Fundamental Freedoms* of 1950 (ECHR), for example in recitals 16, 38 and 43 of the *European Biotechnology Directive*.[48]

Nonetheless, it appears that neither the European nor the American patent system has managed to avoid controversy in respect of patentability of human embryonic stem cells. Likewise, neither institution has offered any adequate and functioning resolve on dealing with the ethical issues encompassing this type of scientific exploration. One prime example of such inadequacy is evident in legal actions concerning the WARF patents in the United States and their application in Europe.

A European Pre-Historical Approach to Stem Cells

The past few years of the EPO's rulings[49] have failed to provide any clear-cut answer to the patentability of human embryonic stem cells, which

---

47 *Wisconsin Alumni Research Foundation* (G 0002/06) EPO OJ 2009, 306, Order [2].

48 Herrmann, J.R. and M. Rowlandson (2008), 'The Role of Ethics and Morality in EU Law', *Journal of International Biotechnology Law*, **5** (6), 241–51 at 246.

49 For example see Thomson, J. (2007), 'The Example of Stem Cells', *Hastings Centre Report*, **37** (1), 12–13; Smith, A. (1996), 'Isolation, Selection And Propagation Of Animal Transgenic Stem Cells', European Patent No. EP695351 ('Edinburgh patent'); *Transgenic animals/Harvard* (T 315/03) EPO OJ 2005, 246 ('*Transgenic animals/Harvard*'); Anderson, D. and D. Stemple (1995), 'Mammalian Multipotent Neural Stem Cells', European Patent No. EP 93921175 (21 June 1995) ('Caltech Patent'). See also for example German

resulted in many hopes being pinned on the so called WARF appeal[50] before the EPO's Enlarged Board of Appeal. With the WARF appeal, the EPO was given an opportunity to set the record straight and provide guidance on the issue of patent morality. On 25 November 2008, the Enlarged Board of Appeal delivered a long awaited WARF ruling[51] that, disappointingly, failed to answer key questions on the patent protection available to stem cell research in Europe. In the WARF appeal, the Enlarged Board of Appeal demonstrated the EPO's lack of will to tackle key questions concerning patent protection available to stem cell research in Europe. The EPO asserted its established policy that it will not grant a patent if the invention relies exclusively on a method that necessitates the destruction of a human embryo from which the claimed products are derived. The WARF decision further confirms that this ban stands 'even if the said method is not part of the claims'.[52] The Enlarged Board of Appeal stated that 'what needs to be looked at is not just the explicit wording of the claims but the technical teaching of the application as a whole as to how the invention is to be performed'. It further added that 'to restrict the application of Rule 28(c)[53] [...] to what an applicant chooses explicitly to put in his claim would have the undesirable consequence of making avoidance of the patenting prohibition merely a matter of clever and skilful drafting of such claim'.[54] The Enlarged Board of Appeal made it clear that the WARF decision does not concern the general question of human stem cell patentability. In addition, the Enlarged Board of Appeal rejected the request for a

---

courts' approach to patentability of stem cells, Brüstle, O. (2010), 'Neural Precursor Cells, Method for the Production and Use Thereof in Neural Defect Therapy', European Patent No. EP1040185 ('Brüstle patent application'); and *Greenpeace* v. *Oliver Brüstle* Decision of the German Bundespatentgericht (BPatG) of 5 Dec 2006, 3 Ni 42/04.

50 *Stem Cells/WARF* (T 1374/04) EPO OJ 2007, 313 ('*Stem Cells/WARF*') under G0002/06.

51 *Wisconsin Alumni Research Foundation* (G 0002/06) EPO OJ 2009, 306 ('*Use of embryos/WARF*').

52 *Wisconsin Alumni Research Foundation* (G 0002/06) EPO OJ 2009, 306 ('*Use of embryos/WARF*').Order [2].

53 *European Patent Convention* 2000 r 28(c): 'Under Article 53(a) European patents shall not be granted in respect of biotechnological inventions which, in particular, concern the following: (c) uses of human embryos for industrial or commercial purposes.'

54 *Wisconsin Alumni Research Foundation* (G 0002/06) EPO OJ 2009, 306, Reasons for the Decision [22].

preliminary ruling on the matter by the European Court of Justice owing to lack of any legal and institutional link between the EPO and the European Union.[55]

Overall, the WARF decision is disappointing in many respects. Regardless of the end result of the patentability of human embryonic stem cells, the problem with the WARF decision is that it does not go into any analysis of legal principles and guidelines that surround the morality clause in Article 53(a) of the *European Patent Convention*. For now, the EPO in this decision has closed off any legal avenues that may further explore the specific exclusion of industrial and commercial uses of human embryos. However, the legal boundaries of the specific moral exclusions from the perspective of EU law have yet to be examined by the European Court of Justice.[56]

B CIT Patent Application (T522/04)

The patent application claims a population of mammalian neural crest cells from other embryo cells and uses thereof by Caltech.[57] The application was originally refused by the EPO Examining Division during the examining process under Article 53(a) and Rule 28c (former Rule 23d(c))[58] EPC. This decision[59] was appealed by the Applicant but the appeal proceedings were stayed until the WARF referral (G2/06) was dealt with by the Enlarged Board of Appeal.

The contentious claim 1 involved method of proliferating *in vitro* a clonal population of neural crest stem cells of mammalian origin. This clearly includes cells of human origin. The Appellant's argument that 'the cells specified as subject of the claimed method can be obtained not only from

55 *Wisconsin Alumni Research Foundation* (G 0002/06) EPO OJ 2009, 306, Reasons for the Decision [22]; and Nurton, J. (2008), 'EPO Rulings Clarify Biotech Protection', *Managing Intellectual Property*, **185**, 14.

56 Plomer, Aurora (2006), *Stem Cell Patents: European Patent Law and Ethics Report* (FP6), 'Life Sciences, Genomics and Biotechnology for Health', SSA LSSB-CT-2004- 005251, 50, http://www.nottingham.ac.uk/~llzwww/StemCellProject/project.report.pdf. See also Plomer, Aurora and Paul Torremans (eds) (2009), *Embryonic Stem Cell Patents: European Patent Law and Ethics*, Oxford: Oxford University Press.

57 Anderson D. and D. Stemple (1995), 'Mammalian Multipotent Neural Stem Cells', European Patent No. EP0658194.

58 Rule 28 (c) of the *European Patent Convention* 2000 provides that European patents shall not be granted in respect of biotechnological inventions which, in particular, concern the following '(c) uses of human embryos for industrial or commercial purposes'.

59 *California/Stem Cells* (T 522/04) 2009 EPOR 45.

embryos, but also from the adult peripheral nervous system and central nervous system' was found not to be supported by the application as filed (WO 94/02593). Given that at the time of filing 'the only teaching of how to prepare human neural crest stem cell cultures is the use (involving their destruction) of human embryos', the EPO reached the conclusion that, at the filing date, the method claimed necessarily involved the destruction of human embryos. Claim 1 was held to fall under the prohibition of the morality clause (Article 53(a)) EPC taken in combination with Rule 28(c) EPC and the appeal was dismissed in May 2009.[60] The Board of Appeal cited WARF decisions G 0002/06 and T 1374/04 in its reasoning.

C Brüstle's Patent

Europe and its competitors have now set their sights on the pending Brüstle[61] appeal before the Bundesgerichtshof (German Federal Supreme Court) regarding a German patent. Oliver Brüstle of the University of Bonn sought a patent directed to a method to differentiate neural cells from mammalian embryonic stem cells, which was rejected, appealed and subsequently granted in an amended form. Brüstle applied for a patent for a method of converting embryonic stem cells into nerve cells that could potentially be used to treat neurological trauma and disease, with claims directed to neural precursor cells derived from embryonic stem cells and methods for producing the neural precursor cells. However, the claims did not mention the use of embryos for producing embryonic stem cells. Brüstle was granted an amended version of the patents containing claims directed to neural precursor cells derived from embryonic stem cells, but – crucially – excluding cells that are derived from embryonic stem cells prepared from human embryos on the grounds of morality. This indicates that only cells derived from human embryonic stem cells obtained from other sources such as human oocytes or human embryonic germ cells may be patented in Germany. Nonetheless, in 2004 Greenpeace filed the opposition to the patent.

The German Federal Patent Court revoked Brüstle's patent claims, which referred to embryo-derived human embryonic stem cells and cell lines. Brüstle subsequently appealed to the Bundesgerichtshof claiming that 'the patented methodology, based on already available ESC lines, does not destroy any embryos'. Interestingly, Brüstle's research on human

---

60 *California/Stem Cells* (T 522/04) 2009 EPOR 45 [B(1)].

61 Brüstle, O. (2010), 'Neural Precursor Cells, Method for the Production and Use Thereof in Neural Defect Therapy', EP1040185; and *Greenpeace* v. *Oliver Brüstle* Decision of the German Bundespatentgericht (BPatG) of 5 Dec 2006, 3 Ni 42/04.

embryo-derived embryonic stem cell lines was legal, which at the time did not provide adequate reasons to ban his research results from being patented.[62] The Brüstle appeal was postponed in view of the then pending WARF appeal. Given that the EPO rejected the WARF patent application and failed to clarify the issue of determining the morality clause as found in patent law, the Bundesgerichtshof chose to forward a number of key questions to the Court of Justice of the European Union concerning the interpretation of 'human embryo' in the sense of Article 6 of the *European Biotechnology Directive* (equivalent to Article 53(a) EPC). Article 234 of the Treaty establishing the European Community (EC Treaty) grants the Court of Justice of the European Union the jurisdiction to give preliminary rulings with respect to 'inter alia the validity and interpretation of acts of the institutions of the European Community' such as the *European Biotechnology Directive* in this instance. The first question concerns the meaning of the term 'human embryos' and what it includes under Article 6(2)(c) (equivalent to Rule 28c).[63] Secondly, the Bundesgerichtshof asks about the meaning of the expression 'uses of human embryos for industrial or commercial purposes' and whether this includes any commercial exploitation within the meaning of Article 6(1) of the *European Biotechnology Directive*, especially use for the purposes of scientific research. The third question relates to the patentability of technical teaching under Article 6(2)(c) even if the use of human embryos does not form part of the technical teaching claimed, but is a necessary precondition for the application of that teaching.

The Brüstle case will likely provide one of the first insights into the ramifications of the WARF ruling for national patents. The European Court of Justice is expected to deliver its decision within one to two years. How the European Court of Justice rules may break or make not only the patent applications, but the progress of the life sciences as an industry in Europe.

i Observations before the European Court of Justice

Written observations are submitted after notification of the request for a preliminary ruling to the parties and Member States of the European Union. According to the European Court of Justice Registry, written observations were submitted by Portugal (which has a restrictive view on stem cell research) and Sweden and the United Kingdom (both of which

---

62 Editor (2010), 'Lagging Laws', *Nature Neuroscience*, **13** (1), 1.

63 *European Biotechnology Directive* art 6(2)(c). On the basis of paragraph 1, the following, in particular, shall be considered unpatentable: '(c) uses of human embryos for industrial or commercial purposes'.

have a fairly liberal view on patentability of human embryonic stem cell-related inventions and research).

Thus, for instance, Swedish Observations depict the Swedish Government's fundamental position that the concept of human embryos has not been harmonised by the *European Biotechnology Directive* and that the authorities of each Member State should be the ones making this assessment. In addition, Sweden believes that it is still early to harmonise the definition of the human embryo at the level of the European Union.[64]

Further, Sweden adds that if the stem cells can be derived from the already established cell lines, then this should be taken into account in view of the *European Biotechnology Directive*, which is concerned with the grant of patent rights, not what happens before or after this. As such, the patentability cannot depend on whether a fertilised embryo has been used at an early stage of the development of the invention.

ii Advocate General's opinion

On 17 March 2011, the European Court of Justice Advocate General Yves Bot delivered an opinion, which recognises that the Oliver Brüstle case is the first time that the court has had to consider the concept of 'use of embryos for industrial or commercial purposes' as found in Article 6(2) of the *European Biotechnology Directive*.

In the opinion, the Advocate General suggested that 'totipotent cells carrying within them the capacity to evolve into a complete human being must be legally classified as human embryos' and as such cannot be patented. However, the Advocate General concurred that a pluripotent stem cell in isolation cannot be seen to constitute a human embryo in itself. In fact, most Member States of the European Union agreed that pluripotent stem cells are not human embryos.[65]

Yet it is puzzling that the Advocate General would proclaim that 'a procedure using other embryonic stem cells, known as pluripotent cells' cannot form part of the patentable subject matter either, if the first step 'requires the destruction or modification of the embryo'.[66] The Advocate General believes that the origin of the pluripotent cell cannot be ignored.[67]

---

64 Utrikesdepartementet, Skriftligt yttrande (6 May 2010) (free translation).

65 *Oliver Brüstle* v. *Greenpeace eV* (Case C-34/10) (Opinion of Advocate General Bot) (10 March 2011) at [98]–[99].

66 European Court of Justice (2011), 'Opinion of the Advocate General in Case C-34/10', Press Release, No. 18/11, 10 March, 1.

67 *Oliver Brüstle* v *Greenpeace eV* (Case C-34/10) (Opinion of Advocate General Bot) (10 March 2011) at [103]-[104].

The Advocate General continues by stating that an industrial application using embryonic stem cells amounts to the use of human embryos as 'simple base material', which in itself would be contrary to ordre public and ethics.[68] Consequently, the Advocate General concludes that Article 6(2)(c) of the *European Biotechnology Directive* 'must be interpreted to the effect that the concept of a human embryo applies from the fertilisation stage to the initial totipotent cells and to the entire ensuing process of the development and formation of the human body'. Such interpretation includes the blastocyst as well.[69]

However, it is difficult to see how the Advocate General's statement on the patentability of pluripotent stem cells can be reconciled with the view that patentability is excluded where 'the application of the technical process for which the patent is filed necessitates the prior destruction of human embryos or their use as base material, even if the description of that process does not contain any reference to the use of human embryos'.[70] While the European Court of Justice usually follows the opinion of the Advocate General, it does not necessarily do so in every instance. Overall, this will be a difficult task for the European Court of Justice to decide on.

D United States WARF Patent Revoked

The five most dominant human embryonic stem cell patents[71] in the United States belong to WARF,[72] Geron,[73] and North Carolina State University.[74]

---

68 Ibid. [110].

69 Ibid. [115].

70 Ibid. [117].

71 Radin, Margaret (1996), *Contested Commodities: The Trouble with Trade in Sex, Children, Body Parts and Other Things*, Cambridge, MA: Harvard University Press, 92; European Group on Ethics in Science and New Technologies (2010), *Ethical Aspects of Patenting Inventions Involving Human Stem Cells*, Opinion no. 16 EGE, 13, http://ec.europa.eu/europan_group_ethics/docs/avis16_en.pdf; and Konski, A. and D. Spielthenner (2009), 'Stem Cell Patents: A Landscape Analysis', *Nature Biotechnology*, **27** (8), 722–6 at 723.

72 Thomson, J. (1998), 'Primate Embryonic Stem Cells', United States Patent No. 5,843,780; Thomson, J. (2001), 'Primate Embryonic Stem Cells', United States Patent No. 6,200,806; and Kaufman, D. and J. Thomson (2001), 'Hematopoietic Differentiation of Human Pluripotent Embryonic Stem Cells', United States Patent No. 6,280,718.

73 Mandalam, R. and C. Xu, Geron Corporation, 'Medium for Growing Human Embryonic Stem Cells', United States Patent No. 7,297,539 ('Geron patent').

74 Petitte, J. and Z. Yang (2004), 'Method of Producing an Avian Embryonic Stem Cell Culture and the Avian Embryonic Stem Cell Culture Produced by the Process', United States Patent No. 5,340,740 ('North Carolina State University').

However, WARF's embryonic stem cell patents alone are said to represent groundbreaking technologies in this field.[75] WARF holds around 64 patents on stem cell lines, including extensive rights to five unmodified stem cell lines themselves.[76] Its stem cell patents cover both the method of isolating human embryonic stem cells (the process) and the product in terms of the five unmodified stem cell lines. Both patents cover the specific processes used to isolate and purify stem cells and the specific stem cell lines that the group created, and also, with qualifications, all primate and human embryonic stem cells capable of stable proliferation in a certain *in vitro* culture for one year with the potential to differentiate into different types of mature cell.[77] These patents are set to expire by 2015. A third stem cell patent issued to WARF claims unmodified human embryonic stem cell lines per se, regardless of their 'creator'.[78]

In 2004, Proposition 71, also known as the *Stem Cell Research and Cures Act* 2004 (California), was endorsed by the California electorate. This provided USD 3 billion for stem cell research.[79] Yet, WARF general counsel Elisabeth Donley said that WARF would require licence fees and royalties to be paid on all embryonic stem cell research in California funded under Proposition 71. As a result, in 2006 the Los Angeles based Foundation for Taxpayer and Consumer Rights and New York based Public Patent Foundation sought re-examination of the claims of all three WARF patents.

Both foundations expressed their concern over the breadth of those claims and said that:

---

75 The Danish Council of Ethics (2004), 'The Ethics of Patenting Human Genes and Stem Cells', Copenhagen, 28 September; and Nuffield Council on Bioethics (2004), 'The Ethics of Patenting DNA', Discussion Paper, Nuffield Council on Bioethics, 12 August; and Konski, A. and D. Spielthenner (2009), 'Stem Cell Patents: A Landscape Analysis', *Nature Biotechnology*, **27** (8), 722–6.

76 Taft, K. and S. Webb (2005), 'United States: Stem Cells: Their Promise, Their Problems', *Mondaq*, 9 March.

77 Hazuka, C. (2002), 'Supporting the Work of Lesser Geniuses: An Argument for Removing Obstructions to HESC Research', *University of Miami Law Review*, **57**, 157–220 at 172. See also Thomson, J. (1998), 'Primate Embryonic Stem Cells', United States Patent No. 5,843,780; and Thomson, J. (2001), 'Primate Embryonic Stem Cells', United States Patent No. 6,200,806.

78 Thomson, J. (2001), 'Primate Embryonic Stem Cells', United States Patent No. 6,200,806.

79 Legislative Analysts' Office (2004), *Proposition 71 Stem Cell Research. Funding. Bonds. Initiative Constitutional Amendment and Statute*, http://www.lao.ca.gov/ballot/2004/71_11_2004.htm.

> By demanding significant financial consideration before allowing research to be performed, the owner of 5,843,780, 6,200,806 and 7,029,913 patents[80] is impeding, and in some cases literally stopping, domestic human embryonic stem cell research in its infancy. This not only harms scientific advance here in the United States, it also has a harmful economic impact on Americans by diverting taxpayer dollars meant for research to pay for licensing fees. In the words of one industry insider, this aggressive patent assertion is 'stifling industrial research and investment'.[81]

The argument for invalidity is based on prior art and three scientific publications that demonstrate the ability to derive embryonic stem cells in various animals, including mice, pigs and sheep. The prior art patent was granted to an Australian inventor Robert Lindsay Williams. The Foundations argued that, with a priority date of 20 January 1995, this anticipated both Thomson patents 5,843,780 and 6,200,806.[82]

On 29 February 2008, a patent examiner with the USPTO upheld the '913 patent. On 11 March 2008, the USPTO upheld the validity of the modified claims of the remaining two WARF patents with no possibility for appeal.[83] The USPTO has been heavily criticised for awarding broad patents in respect of stem cell research per se. Seth Shulman said that 'perhaps the biggest lesson of all, though, surrounds the chronic myopia of the [USPTO] in awarding such needlessly all-encompassing patents as it has in this field'.[84] In line with current United States patent law, only one of the three above outlined WARF patent rulings could be appealed by the requester and appellant. The patent appealed was WARF patent '913.

---

80 Thomson, J. (1998), 'Primate Embryonic Stem Cells', United States Patent No. 5,843,780; Thomson, J. (2001), 'Primate Embryonic Stem Cells', United States Patent No. 6,200,806; and Thomson, T. (2006), 'Primate Embryonic Stem Cells', United States Patent No. 7,029,913.

81 The Foundation for Taxpayer and Consumer Rights and the Public Patent Foundation (2006), 'Request for *Ex Parte* Re-examination' in respect of US Patent No. 5,843,780, 2, http://www.pubpat.org/assets/files/warfstemcell/780Request.pdf.

82 Rimmer, Matthew (2008), *Intellectual Property and Biotechnology: Biological Inventions*, Cheltenham and Northampton, MA: Edward Elgar, 261.

83 Plas, J.V. (2008), 'Patent Office Upholds Remaining WARF Stem Cell Patents', *Wisconsin Technology Network*, 11 March, http://wistechnology.com/articles/4601/.

84 Shulman, S. (2001), 'The Morphing Patent Problem', Owning the Future, *Technology Review*, November.

On 28 April 2010, the USPTO Board of Appeals and Interferences rejected the controversial WARF '913 patent (issued 18 April 2006) claiming pluripotent human embryonic stem cells.[85] Claim 1 is the only independent claim on appeal and it was amended several times during the re-examination proceedings (amendments occurred on 30 May 2007, 2 October 2007 and 4 October 2007).[86] The USPTO invalidated the patent on the grounds of anticipation by an earlier patent and obviousness in light of the 'significant guideposts' in the literature for deriving the cells at the time of invention.[87]

IV THE NEED FOR BALANCE AND STABILITY

In essence, the conflict over human embryonic stem cell patents may be seen as a disturbance within the patent system itself, where 'the tension between the economic and cultural dynamics' is responsible for governing the politics of patenting.[88] Patent law has been depicted as a 'stressful if fertile union between certain contradictory principles: self-interest and the common good; monopoly rights and liberty; the ownership of ideas and public disclosure of knowledge'.[89] This union is a result of the dual goals of patent law, to benefit society by encouraging the provision of new and useful goods, and to encourage and reward inventiveness.[90] These goals are achieved by rewarding an inventor for innovation via a limited monopoly and knowledge-sharing via disclosure of the invention which in turn encourages investment and provides benefit to society. The underlying idea is that without this investment, inventors in stem cell research, for example,

---

85 *Foundation for Taxpayer and Consumer Rights* v. *Patent of Wisconsin Alumni Research Foundation* (Appeal 2010–001854), the Board of Patent Appeals and Interference http://www.consumerwatchdog.org/resources/WARFDecision042910.pdf.

86 Ibid.

87 *Foundation for Taxpayer and Consumer Rights* v. *Wisconsin Alumni Research Foundation*, BPAI No. 2010–001854 (28 April 2010).

88 Salter, B. (2007), 'Patenting, Morality and Human Embryonic Stem Cell Science: Bioethics and Cultural Politics in Europe', *Regenerative Medicine*, **2** (3), 301–11 at 301.

89 Kass, L. (1981), 'Patenting Life', *Journal of the Patent Office Society*, **63**, 570 at 580.

90 MacQueen, Hector, Charlotte Waelde and Graeme Laurie (2008), *Contemporary Intellectual Property: Law and Policy*, Oxford: Oxford University Press, 365. See also art. 7 of the *TRIPS Agreement* 1994.

would not be encouraged to invest.[91] In fact, patents are viewed as a necessity for human embryonic stem cell innovation to occur and provide crucial incentives for investment in research and development.[92] David Kappos, the Director of the USPTO, explained, 'as goes the patent office, so goes the economy. The USPTO either launches or buries innovation [...].'[93] Patent availability is said to be a precondition to gaining a competitive advantage in the field. But how, if at all, patent law should be applied to human embryonic stem cell innovation is a question that still lacks an adequate answer both in Europe and in the United States.

Dramatic changes in the modern economy over the last few decades have resulted in unprecedented changes in patentable subject matter. Until recently, the patent system dealt with manufacturing methods and mechanical inventions, whereas nowadays patent protection is required for materials and in new areas such as the life sciences. Patent law has begun moving into more delicate areas 'which are the realm of society, and it needs to be in balance, fixed in a new way so society has more influence over the process'.[94] The *United Nations Covenant on Economic, Social and Cultural Rights* 1966 reflects this in Article 15, which mentions the need for balance between the interests of society and intellectual property rights. The morality clause attempts to do exactly that, but the application of the morality provision is yet to be worked out adequately.

So far, the EPO refuses to interpret the morality clause and even takes a narrow approach in examining stem cell inventions.[95] On the other hand, the United States is increasingly finding it difficult to continue its patent policy of 'everything made by man under the sun'. Doubts and complaints have almost overwhelmed the American patent system, which has reached a point of such uncertainty that 'the only patent that is valid is one which this (the Supreme) Court has not yet been able to get its hands on'.[96] Despite a

---

91 See for example Organisation for Economic Co-operation and Development (OECD) (2004), *Patents and Innovation: Trends and Policy Changes*, Paris: OECD, 9–10.

92 de Beer, J. (2007), 'Biotrespass', *Bulletin of Science Technology and Society*, **27** (4), 287–99.

93 Carter, T. (2010), 'A Patent on Problems', *ABA Journal*, **96** (3), 7.

94 European Patent Office (2007), 'Interview with Patent Expert Greenpeace International Dr Christoph Then', Open Ended Interviews, http://documents.epo.org/projects/babylon/eponet.nsf/0/106FF770E1FA4E02C12572D8003A49FD/$File/Interview_Then.pdf.

95 Drahos, P. (1999), 'Biotechnology Patents, Markets and Morality', *European Intellectual Property Review*, **21** (9), 441–9.

96 *Jungerson* v. *Ostby and Barton Co*, 335 US 560, 80 USPQ 32 (dissent by Jackson J) (1948), [32].

significant need to tackle the issues, both the USPTO and the American judiciary have retreated from the important matters of addressing the patentability of human embryonic stem cells and dealing with ethical issues surrounding new and, at times, morally contentious technologies. Both the European and American current approaches to patentability tend to ignore the wider social and economic implications that raise 'questions of morality and ordre public by the very nature of patent law itself'.[97]

In Europe, for instance, under current circumstances, it is insufficiently clear how the prohibition of the destruction of the human embryo can form a primary and indisputable factor in determining patentability of human embryonic stem cells. It does not help that the *European Biotechnology Directive* is vague regarding patentability of human embryonic stem cells and that it contains a significant loophole in the context of the use of human embryos. The ongoing debate among experts and the public concerning stem cell patentability has now focused on trying to find a clear ethical balance in this case. The European Parliament did not want any patents to be issued claiming human embryos, but it appears that the *European Biotechnology Directive* failed to clearly present this bar from a legal point of view. As a result, the EPO began taking the initiative by developing its own case law on the issue of patenting human embryonic stem cells with inconsistent, confusing and at times unpredictable outcomes.[98] This has resulted in a sort of chaos at the EPO, whose rulings indicate a clear lack of any substantive definition of what constitutes morality or ethics within the European context[99] and a lack of procedural guidance on interpreting the morality clause.

It is crucial to reiterate that modern patent law has a strong historical connection to morality, whereas nowadays the patent community believes that morality has little or nothing to say on the issuing of patents. Or, on the other hand, if morality has anything to do with patents, the patent system is not the appropriate place to consider such issues.[100] However, to suggest that ethics are somehow outside the scope of patent law or that there is no place for ethics within the patent system implies that the patent system is in

---

97 European Patent Office (2007), 'Interview with Patent Expert Greenpeace International Dr Christoph Then', Open Ended Interviews, http://documents.epo.org/projects/babylon/eponet.nsf/0/106FF770E1FA4E02C12572D8003A49FD/$File/Interview_Then.pdf.

98 Ibid.

99 The Opposition Division seems to agree in its decision in the *Edinburgh case* T1079/03.

100 Drahos, P. (1999), 'Biotechnology Patents, Markets and Morality', *European Intellectual Property Review*, **21** (9), 441–9.

some way amoral or unethical, outside of the society's common moral norms. It is difficult to see why patent law would be a special area of law that would allow it to be separated from moral standards that other areas of law and society are subjected to. In issuing morally or socially controversial, or even unacceptable patents – human clones, material of human origins, and so on – the USPTO is exercising ethics, albeit bad ethics. Yet, bad ethics is still ethics of a sort.[101] On the other hand, the EPO's attempts to incorporate ethics through the decision-making process have left behind a trail of confusing and inconsistent precedents, adding to the uncertainty and anxiety among the patent community, public and governments.

In addition, if all innovation is allowed to be patented then nothing is disallowed. Then nothing is wrong. Then nothing can ultimately be deemed against the social welfare. If patent law is segregated from morality, then any patent morality is relative. But, if moral relativism exists within the patent system, then negative repercussions on the public resulting from exploitation of patent rights cannot exist because negative effects on the social welfare become a relative matter.

In their attempts to rectify the problem of patenting biotechnology and stem cell inventions, both Europe and America have offered words of caution. So, for example, recommendation 10(ix) asks that the Committee of Ministers call on governments of member states and the Commission of the European Communities 'to adopt a cautious policy with regard to the granting of patents for biotechnological inventions and applications so as to take due account of ethical considerations and environmental safety concerns'.[102] The dissenting opinion of four justices, written by Justice William Brennan in *Diamond* v. *Chakrabarty*, agreed that Congressional guidance is imperative because 'the composition sought to be patented uniquely implicated matters of public concern'.[103]

Interestingly, following an analytical study of patent law, Geertrui Van Overwalle and Esther van Zimmeren concluded that patent law is essentially incapable of performing its major objectives and functions within the

---

101 Crane, Andrew and Dirk Matten (2004), *Business Ethics*, Oxford: Oxford University Press, 8. See also Agovic, Amina (2010), 'Patent Morality – The Little Battler', in Niklas Bruun (ed.), *In Search of New IP Regimes*, Oy Nord Print Ab: IPR University Center.

102 *Council of Europe Recommendation 1213 on Developments in Biotechnology and the Consequences for Agriculture* [1993]. Assembly debate on 12 May 1993 (34th sitting) (see Doc. 6780, report of the Committee on Agriculture, Rapporteur: Mr Gonzalez Laxe). Text adopted by the Assembly on 13 May 1993 (36th Sitting).

103 *Diamond* v. *Chakrabarty*, 477 US 303 (1980) (dissent by Brennan J, with whom White, Marshall, and Powell JJ join dissenting).

current social and political context.[104] This inability to fulfil its objectives and functions highlights some significant limits of patent law. This is especially relevant in areas of biotechnology where the major challenge centres on ethics and morality within the patent system.

In any case, the pursuit of an ethical balance in patent systems in respect of patentability of human embryonic stem cells should be conducted in a way that not only keeps up with rapid advances in life sciences but that also takes into account the underlying utilitarian philosophy of the modern patent system. That is, patenting of human embryonic stem cells should be in line with the original aims of the patent system: to reward the inventor with the monopoly, for the benefit of society.

CONCLUSION

There is a need to strike a balance between what is best for the public (i.e. patients) and a financial incentive (i.e. patents) for the private sector of stem cell research to actually develop stem cell therapies. While the current patent system has essentially remained unchanged for centuries, questions arise as to its flexibility to adapt to technological advances of today.

This moral uniqueness found in patent law places the patent system at the crossroads of science and technology and at the centre of the stem cell debate. The breathtaking pace of advances in stem cell research and emerging technologies today is crucial to global prosperity. This necessitates new approaches to developing a policy framework that would be appropriate to the new realities of our technology driven world. While no one wants to sacrifice the potential benefits of stem cell research, no one wants to ignore the potential perils of the new technology.

However, the problem of patenting human embryonic stem cells is as acute as ever. Nowhere is this more apparent than in an array of confusing and puzzling EPO precedents. Instead of addressing the core of the problem of how to interpret the morality clause in an efficient and legally certain manner, the EPO uses the 'human embryo destruction' test as undoubtedly a 'real' moral test. However, to the extent that moral considerations may conceivably play a 'background' role in helping to clarify the scope of exclusion under specific provisions such as Rule 28(c), the relevant moral principle(s) to be applied here cannot be the principle of 'embryo destruction', primarily owing to a lack of moral consensus in Europe

104 Van Overwalle, Geertrui and Esther van Zimmeren (2009), 'Chapter 22 – Functions and Limits of Patent Law', in Erik Claes, Wouter Devroe and Bert Keirsbilck (eds), *Facing the Limits of the Law*, Berlin: Springer.

precluding uses of human embryos that necessarily involve 'embryo destruction'.[105]

Overall, it appears that the moral principle of 'embryo destruction' is a prevalent test in the EPO decisions on the Edinburgh patent, the WARF applications and more recently the CIT application.

Far from settling the question of patent morality, the morality clause and the divergent opinion among the patent community, society, industry and all involved has led to confusion and a very unsatisfactory situation for all concerned. The result is that discourse continues and there is little consensus. Meanwhile, patent examiners at both the EPO and the USPTO face complex issues of patent morality and no tools are provided for dealing with stem cell innovation. In Europe, the WARF decision failed to clarify the matter, and high expectations are now placed on the Brüstle patent questions forwarded to the European Court of Justice.

As prescribed in the *TRIPS Agreement* 1994, there should be scope for dealing with ethical concerns in patent law. Furthermore, there are particular instances where it might be contrary to morality to issue a patent because such would facilitate the commercial development of certain technologies. Yet, the present European morality clause does not adequately deal with this matter.

By passing on a chance for reform, the patent system in Europe is not simply going to maintain its status quo. In the long term, the position will not hold. Europe's neighbours will be surging forward and enjoy access to foreign investment, development funds, numerous services and legal certainty, while the EPO fades into the distance. Patent morality is a key way for the EPO to provide for the patent system's long-term stability and economic success. The EU has almost half a billion customers and EU global trade is approximately a fifth of the world's imports and exports.[106] If the EPO is to take complete advantage of morally controversial technologies and stem cell research, it needs to reform its approach to morally contentious patent applications. If nothing is done, the patent system will remain stuck in current thinking, shackled to present technology and unable to keep pace in a changing twenty-first century. This should be unacceptable not only to the inventors and investors but also to the public who relies on innovation to improve life.

---

[105] Plomer, Aurora (2006), *Stem Cell Patents: European Patent Law and Ethics Report* (FP6), 'Life Sciences, Genomics and Biotechnology for Health', SSA LSSB-CT-2004-005251, http://www.nottingham.ac.uk/~llzwww/StemCell Project/project.report.pdf.

[106] Europa, (2005), 'A Major Trading Power', Europa, http://europa.eu/abc/keyfigures/tradeandeconomy/tradingpower/index_en.htm.

In contrast to the EPO practice of trying to deal with ethical issues concerning patenting human embryonic stem cells, the American approach of allowing patents on 'all under the sun' fails to show the existence of any checks and balances. However checks and balances have been a part of the USPTO since its establishment.[107] The landmark decision of *Diamond* v. *Chakrabarty* shifted responsibility regarding moral concerns within the patent system to Congress. Many bills have sought to limit the scope of the patentable subject matter in biotechnology and life sciences, but almost all have failed to be passed. The Congress and the patent institution fail to realise that it is by far more difficult to comply with ethical and social concerns once the patent has been issued, especially since American law is built on prospective rather than retroactive acts.[108] In addition, the lack of clear guidelines on patenting human embryonic stem cells paves the way for scientists and patent attorneys to determine the limits of patent eligibility by describing the subject matter in their patent applications without taking social welfare into account. To let the patent community self-regulate and set limits on the patentable subject matter could allow for a dangerous expansion of biotechnology patents, even into areas such as human cloning. But, to deprive biotechnology and life sciences companies of patent protection will only result in competition moving to a more stem-cell-patent friendly nation. However, to remove considerations of social welfare and patent morality from the patent system may have a far greater impact on society than the competition moving off shore.

In addition, the American judicial system is too passive and the Parliamentarians are unable to perceive and fix patenting problems before it becomes too late. Further, the problem of patent system overhaul is very real and this demands that a solution be found swiftly and in line with present legal principles. More than anything, the USPTO should act proactively to avoid chaos and stalling of technological progress.

Criticism directed at the patent system can no longer be left unattended. At some point in the near future, the patent office will need to react and adapt to the ethical dilemma of patenting human embryonic stem cells. Delay in addressing the interpretation of the morality clause and patenting of human embryonic stem cells is not professional by any standards. As such, the patent office needs to evaluate alternatives and decide on the best approach to assessing human embryonic stem cell inventions and

---

107 Khan, B. Zorina (2005) *The Democratisation of Inventions: Patents and Copyrights in American Economic Development, 1790–1920*, Cambridge: Cambridge University Press, ch. 2.

108 See for example *Dash* v. *Van Kleeck*, 7 Jonhs 477 (Sup Ct of Judicature, 1811); and *Terrett* v. *Taylor*, 9 Cr 43 (1815).

interpreting the morality clause. This is ever more important as the worldwide economy is moving at an accelerated speed, and no country can afford to be left behind. In order to support the development of the biotechnology and life sciences industry on either European or American soil, the patent system therefore needs to be updated. It needs to be rejuvenated and to begin contributing to social welfare. The patent system 'has to be based on the premise that invention will have merit for all the people of the world' thereby striking out nationalistic arguments.[109] Additionally, the European and American patent systems should promote how European and American innovation, including their patent systems, can contribute to humankind.

This means that science, politics and patent law should engage in a dialogue. The only way to truly change the current attitudes at the EPO or the USPTO and to tackle the issue of human embryonic stem cell inventions and other morally controversial inventions is by changing the current policy on dealing with patent morality.

CODA

As this book went to press, the European Court of Justice delivered its ruling on the patentability of stem cells: *Oliver Brüstle* v. *Greenpeace eV* (Case C-34/10), Judgment of the European Court of Justice (Grand Chamber) (18 October 2011). It found:

1. Article 6(2)(c) of Directive 98/44/EC ... must be interpreted as meaning that: ... it is for the referring court to ascertain, in the light of scientific developments, whether a stem cell obtained from a human embryo at the blastocyst stage constitutes a 'human embryo' within the meaning of Article 6(2)(c) of Directive 98/44.
2. The exclusion from patentability concerning the use of human embryos for industrial or commercial purposes set out in Article 6(2)(c) of Directive 98/44 also covers the use of human embryos for purposes of scientific research, only use for therapeutic or diagnostic purposes which are applied to the human embryo and are useful to it being patentable.
3. Article 6(2)(c) of Directive 98/44 excludes an invention from patentability where the technical teaching which is the subject-matter of the patent application requires the prior destruction of human embryos or their use as base material.

---

109 Plomer, Aurora (2006), *Stem Cell Patents: European Patent Law and Ethics Report* (FP6), 'Life Sciences, Genomics and Biotechnology for Health', SSA LSSB-CT-2004-005251, http://www.nottingham.ac.uk/~llzwww/StemCell Project/project.report.pdf.

10. Cosmo, Cosmolino: patent law and nanotechnology

Alison McLennan and Matthew Rimmer

'She would sing to it sweetly, Cosmo, Cosmolino, world, little world'

Helen Garner, *Cosmo, Cosmolino*[1]

Gary Marchant and his colleagues observe that nanotechnology is an exemplar of an emerging technology:

> Nanotechnology is the latest in a growing list of emerging technologies that includes nuclear technologies, genetics, reproductive biology, biotechnology, information technology, robotics, communication technologies, surveillance technologies, synthetic biology, and neuroscience. As was the case for many of the technologies that came before, a key question facing nanotechnology is what type of regulatory oversight is appropriate for this emerging technology.[2]

Ian Kerr and Goldie Bassie pithily put the question, 'How ought today's policy-makers to address such concerns about a technology "so new that, in truth, it barely exists"?'[3]

Nanotechnology is the field of science and technology that is focused around the hundred nanometer scale downwards.[4] It refers to devices

1 Garner, Helen (1992), *Cosmo, Cosmolino*, Melbourne: McPhee Gribble and Penguin Books, 144.

2 Marchant, G., D. Sylvester and K. Abbott (2009), 'What Does the History of Technology Regulation Teach Us About Nano Oversight?' *Journal of Law, Medicine, and Ethics*, **37**, 724–31 at 724.

3 Kerr, I. and G. Bassie (2004), 'Not *That* Much Room? Nanotechnology, Networks and the Politics of Dancing', *Health Law Journal*, **12**, 103–23.

4 For a discussion of the historical development of nanotechnology, see Drexler, K. Eric (1986), *Engines of Creation: The Coming Era of Nanotechnology*, Garden City: Anchor Press/DoubleDay; Peterson, C. (2004), 'Nanotechnology: From Feynman to the Grand Challenge of Molecular Manufacturing', *IEEE Technology and Society Magazine*, **23** (4), 9–15; Kulinowski, K. (2004), 'Nanotechnology: From "Wow" to "Yuck"?' *Bulletin of Science, Technology & Society*, **24** (1), 13–20; and Drexler, E. (2004). 'Nanotechnology: From Feynman to Funding', *Bulletin*

whose parts can be measured in nanometres, or billionths of a metre. Nanotechnology often involves bio-mimicry.[5] Scientists in this field often study nature's models and then imitate or take inspiration from these designs and processes to solve human problems.[6] The National Nanotechnology Initiative in the United States provides the following definition of nanotechnology:

> Nanotechnology is the understanding and control of matter at dimensions between approximately 1 and 100 nanometers, where unique phenomena enable novel applications. Encompassing nanoscale science, engineering, and technology, nanotechnology involves imaging, measuring, modelling, and manipulating matter at this length scale. A nanometer is one-billionth of a meter. A sheet of paper is about 100,000 nanometers thick; a single gold atom is about a third of a nanometer in diameter. Dimensions between approximately 1 and 100 nanometers are known as the nanoscale. These properties may differ in important ways from the properties of bulk materials and single atoms or molecules.[7]

Ian Kerr and Goldie Bassie reflect that consensus over the definition of nanotechnology is elusive: 'Because both root words are broad, there is a risk that proposed regulatory mechanisms could be over- or under-inclusive.'[8] The question of definition and classification is certainly a pressing issue with respect to the application of patent law to the field of nanotechnology.

Nanotechnology is a multidisciplinary field of research.[9] This area of science and technology cuts across the traditional disciplinary fields of

---

of Science, Technology & Society, **24** (1), 21–7; Hornyak, Gabor Louis, H.F. Tibbals and Joydeep Dutta (2008), *Introduction to Nanoscience and Nanotechnology*, London: Taylor and Francis; Roco, Mihail and William Sims Bainbridge (eds) (2010), *Societal Implications of Nanoscience and Nanotechnology*, London: Springer; and Rogers, Ben, Sumita Pannathur and Jesse Adams (2011), *Nanotechnology: Understanding Small Systems*, Taylor and Francis.

5 Benyus, Janine (1997), *Biomimicry: Innovation Inspired by Nature*, New York: Harper Perennial.

6 Smith, D. (2000) 'Adventures in Lilliput', *The Sydney Morning Herald*, 11 December.

7 National Science and Technology Council (2011), *The National Nanotechnology Initiative: Strategic Plan*, Washington, DC: Executive Office of the President, The White House, 4 February, 3.

8 Kerr, I. and G. Bassie (2004), 'Not *That* Much Room? Nanotechnology, Networks and the Politics of Dancing', *Health Law Journal*, **12**, 103–23.

9 Roco, Mihail and William Sims Bainbridge (2003), *Converging Technologies for Improving Human Performance: Nanotechnology, Biotechnology, Information Technology and Cognitive Science*, AA Dordrecht: Kluwer Academic.

biology, chemistry, physics and engineering. Through analysing publication and citation data, Alan Porter and Jan Youtie have situated nanotechnology in the context of other scientific disciplines and fields.[10] The researchers noted: 'It is commonplace to read that nanoscience and nanotechnology are highly multidisciplinary, and that the convergence of nanotechnology, biotechnology, information technology and cognitive sciences (NBIC) will lead to big changes in many aspects of our lives and modern society.'[11] The researchers concluded: 'Six subject categories dominate both the original nanopapers and the cited references: "materials science, multidisciplinary" contained the most nanopapers, followed by "physics, applied", "chemistry, physical", "physics, condensed matter", "nanoscience and nanotechnology", and "chemistry, multidisciplinary".'[12] To a lesser extent, there are also connections with biomedical sciences, environmental sciences and agricultural sciences.

John Holdren, the Director of the Office of Science and Technology Policy, emphasizes that nanotechnology enables 'a broad spectrum of applications in electronics, medicine, energy, manufacturing, advanced materials, and other fields'.[13]

Gyorgy Scrinis and Kristen Lyons comment that nanotechnology will have a significant impact upon the fields of agriculture and food:

> Nanotechnology is rapidly moving from the laboratory and onto the farm, supermarket shelves and the kitchen table. For example, a new range of 'smart' agricultural inputs and products are being developed, such as nano-seed varieties with in-built pesticides that will release under certain environmental conditions; nano-encapsulation techniques may make it possible to alter the nutritional composition, flavour and other attributes of food to match consumers' personal tastes and physiological requirements; and 'smart' food packaging able to detect the presence of pathogens.[14]

---

10 Porter, A. and J. Youtie (2009), 'Where Does Nanotechnology Belong in the Map of Science?' *Nature Nanotechnology*, **4**, 534–6.

11 Ibid., 534.

12 Ibid., 535.

13 National Science and Technology Council (2011), *The National Nanotechnology Initiative: Strategic Plan*, Washington, DC: Executive Office of the President, The White House, 4 February, iii.

14 Scrinis, G. and K. Lyons (2007), 'The Emerging Nano-Corporate Paradigm: Nanotechnology and the Transformation of Nature, Food and Agri-Food Systems', *International Journal of Sociology of Food and Agriculture*, **15** (2), 22–44.

Scrinis and Lyons wonder whether nano-foods will face a hostile reaction comparable with that which greeted genetically modified crops.[15] Indian nanotechnology professor T. Pradeep makes the case that nanotechnology could be applied 'in agriculture for breeding crops with higher levels of micro-nutrients, enhancing pest detection and control, and improving food processing techniques'.[16]

The field of nanomedicines is well-established. Research is occurring into applications such as vaccine and drug delivery, dentistry, diagnostic imaging and tissue engineering. Magnetic nanoparticles are being developed for use in medical imaging and fluorescent nanoparticles are being developed for use in diagnosis.[17] Nano-capsules are being developed to transport drugs and even genes into cells.[18] It is hoped that using various 'nanovehicles' will improve drug absorption, stability in the body and various other factors currently limiting drug efficacy.[19]

Wendy Sanhai from the United States Food and Drug Administration and her colleagues discuss the potential and challenges of nanomedicines, observing: 'Nanomedicine promises a transfigured portrait of better health care, health economics and personalized medicine'.[20] Sanhai and her collaborators suggest that there is a need to bring together an array of interests – 'pharma and bio, academia and industry, technologist and toxicologist, patient advocate, clinician and ethicist, federal, state and private' – in a true public-private partnership to address the challenges of making nanomedicines safe.[21]

It has been argued that nanotechnology can also make an important contribution to address environmental protection, sustainable development and climate change. Richard Jones quips: 'Green nanotechnology, then, is not just a convenient slogan but an inescapable necessity for our time.'[22] He reflects that the technology offers 'the prospects of nano-enabled clean energy and clean water, with some vaguer and more general

---

15 Ibid., 37.

16 Pradeep, T. (2008), *Nano: The Essentials, Understanding Nanoscience and Nanotechnology*, New York: McGraw-Hill, 365.

17 Mohanty, C. et al. (2009), 'Nanobiotechnology: Application of Nanotechnology in Therapeutics and Diagnosis', *International Journal of Green Nanotechnology: Biomedicine*, **1** (1), B24–B38, B28–B29.

18 Ibid., B31–B35.

19 Ibid., B30–B31.

20 Sanhai, W., J. Sakamoto, R. Canady and M. Ferrari (2008), 'Seven Challenges for Nanomedicine', *Nature Nanotechnology*, **3**, 242–4.

21 Ibid.

22 Jones, R. (2007), 'Can Nanotechnology Ever Prove That It Is Green?', *Nature Nanotechnology*, **2**, 71–2 at 72.

notions of nanotechnology facilitating cleaner and more sustainable modes of production sitting in the background'.[23] Ron Pernick and Clint Wilder comment: 'Nanotech could end up enabling many next-generation clean technologies, such as advanced batteries, water-salination and water-filtration membranes, building insulation, and of course, solar power.'[24] Geoffrey Hunt comments: 'The cleaner energy technologies of wind and solar power are now taking off, and well-directed nanotechnology research and development programmes could speed up the transfer to such energy technologies.'[25] Moreover, he suggests that 'carbon capture and storage techniques may well be greatly improved by a range of nano-efficiencies'.[26]

Patent law has a significant instrumental and symbolic role in regulating nanotechnology. A 2011 report of the United States Federal Trade Commission noted that 'the patent system plays a critical role in promoting innovation across industries from biotechnology to nanotechnology, and by entities from large corporations to independent inventors'.[27] This chapter considers the much contested legal, ethical and social issues involved with regulating the patenting of nanotechnology. Section I considers the efforts of patent offices to classify nanotechnology and the empirical evidence about patent filing rates. Section II examines whether there is a 'tragedy of the anticommons' emerging in respect of nanotechnology. It contemplates access mechanisms – such as the defence of experimental use, patent pools, open innovation models and technology transfer. Section III explores ethical and social concerns associated with nanotechnology – in particular, issues about the impact upon human health and the environment.

---

23 Ibid., 71.

24 Pernick, Ron and Clint Wilder (2008), *The Clean Tech Revolution: Discover the Top Trends, Technologies and Companies to Watch*, New York: Harper Collins Business, 46–7.

25 Hunt, Geoffrey (2006), 'The Global Ethics of Nanotechnology', in Geoffrey Hunt and Michael Mehta (eds), *Nanotechnology: Risks, Ethics and Law*, London: Earthscan, 189.

26 Ibid., 190.

27 Federal Trade Commission (2011), *The Evolving IP Marketplace: Aligning Patent Notice and Remedies with Competition*, Washington DC: Federal Trade Commission, 1.

I CLASSIFYING NANOTECHNOLOGY

A United States Patent and Trademark Office

The United States Patent and Trademark Office (USPTO) appears to have assumed that inventions in the field of nanotechnology are eligible for patent protection as patentable subject matter. Some commentators have expressed consternation over the impact of the 'machine-or-transformation' test in the case of *in re Bilski* upon research in the field of nanoscience.[28] Eric Laury was relieved by the ruling of the Supreme Court of the United States in *Bilski* v. *Kappos*,[29] observing:

> If the Supreme Court had required strict adherence to the machine or transformation test, it would have had seismic implications for patents already issued, and certainly would have caused trouble for pending applications relating to biotechnology and nanotechnology. The trouble arises from the precarious position that biotech and nanotech patents are not theoretically involved with either a machine nor do they transform matter. Generally speaking, they are processes involved with analysis, diagnosis, treatment or some type of function. Notable examples are those nanoparticles involved in drug delivery. Even though the particles behave differently at the nanoscale level, they essentially are not transformed or part of a machine and would thus fall outside patent protection under the rigid Federal Circuit test. Now, there is a least the possibility for these types of inventions to move forward in the process of patentability.[30]

Of course, the eligibility of nanotechnology inventions for patent protection depends upon its precise sub-discipline. Nano-materials could be analogized to the protection of mechanical inventions and engineering under patent law. Nano-foods could be protected like agricultural products, as in *JEM Ag Supply Inc* v. *Pioneer Hi-Bred International Inc*.[31] Nano-biotechnology could raise interesting questions about the distinction between 'products of nature' and inventions.[32] Nano-medicine and nano-therapeutics could be protected, following *Laboratory Corporation of*

---

28 *In re Bilski*, 545 F.3d 943 (Fed. Cir., 2008).

29 *Bilski* v. *Kappos*, 130 S. Ct. 3218 (2010).

30 Laury, E. (2010), 'Patentable Subject Matter Paradigm Evolves: Bilski's Impact on Nanotechnology', NanoLaw Blog, 1 July, http://blogs.law.widener.edu/nanolaw/2010/07/01/patentable-subject-matter-paradigm-evolves-bilskis-impact-on-nanotechnology/.

31 *JEM Ag Supply Inc* v. *Pioneer Hi-Bred International Inc* 534 US 124 (2001).

32 *Diamond* v. *Chakrabarty*, 447 U.S. 303 (1980); *Association for Molecular Pathology et al.* v. *United States Patent and Trademark Office, Myriad Genetics, et al.* 702 F. Supp. 2d 181, (S.D.N.Y. 2010), *appeal docketed*, No. 2010–1406 (Fed. Cir. June 22, 2010); and *Association for Molecular Pathology et al.* v. *United States*

American Holdings v. *Metabolite Laboratories, Inc*[33] and *Prometheus Laboratories Inc.* v. *Mayo Collaborative Services*.[34] There has been a debate at an international level as to the extent that clean technologies could be excluded from the scope of patent protection.[35]

In response to such concerns expressed by research and industry groups about the expertise of patent offices in dealing with emerging technologies,[36] the USPTO has created a specialist unit to examine nanotechnology patents. As part of its efforts to improve the ability to search and examine nanotechnology-related patents, USPTO established a new cross-reference digest for nanotechnology designated Class 977/Dig.1, entitled 'Nanotechnology'.[37] The USPTO defines the term 'nanostructure' to mean 'an atomic, molecular, or macromolecular structure that: (a) has at least one physical dimension of approximately 1–100 nanometers; and (b) Possesses a special property, provides a special function, or produces a special effect that is uniquely attributable to the structure's nanoscale physical size'.[38] The nanotechnology art collection provides for disclosures related to 'nanostructure and chemical compositions of nanostructure', 'device that include at least one nanostructure', 'mathematical algorithms, e.g., computer software, etc., specifically adapted for modeling configurations or properties of nanostructure', 'Methods or apparatus for making, detecting, analyzing, or treating nanostructure'; and 'specified particular uses of nanostructure'.[39] Class 977 has a number of sub-classifications.

Patent and Trademark Office, Myriad Genetics, et al. 653 F.3d 1329, 2011 WL 3211513 (C.A. Fed. (N.Y.)), 99 USPQ 2d 1398.

33 *Laboratory Corporation of American Holdings* v. *Metabolite Laboratories, Inc.* 548 U.S. 124 (2006).

34 *Prometheus Laboratories Inc.* v. *Mayo Collaborative Services* No. 2008–1403, 2010 WL 5175124 (December 17, 2010); and *Mayo Collaborative Services* v. *Prometheus Laboratories, Inc., cert granted*, – U.S. –, 2011 WL 973139 (U.S. June 20, 2011) (No. 10–1150).

35 See Rimmer, Matthew (2011), *Intellectual Property and Climate Change: Inventing Clean Technologies*, Cheltenham and Northampton, MA: Edward Elgar.

36 NanoBusiness Alliance (2002), 'US Patent Examiners May Not Know Enough About Nanotech', *Small Times*, http://www.smalltimes.com/articles/stm_print_screen.cfm?ARTICLE_ID=267691.

37 United States Patent and Trademark Office, 'New Cross-Reference Digest for Nanotechnology', http://www.uspto.gov/web/patents/biochempharm/crossref.htm.

38 Ibid.

39 Ibid.

Table 10.1 USPTO sub-classifications of nanotechnology for Class 977

700	NANOSTRUCTURES
701	*Integrated with dissimilar structures on a common substrate*
724	*Devices having flexible or movable element*
734	Fullerenes (i.e., graphene-based structures, such as nanohorns, nanococoons, nanoscrolls, etc.) or fullerene-like structures (e.g., WS2 or MoS2 chalcogenide nanotubes, planar C3N4, etc.)
754	Dendrimer (i.e., serially branching or 'tree-like' structure)
755	Nanosheet or quantum barrier/well (i.e., layer structure having one dimension or thickness of 100 nm or less)
762	Nanowire or quantum wire (axially elongated structure having two dimensions of 100 nm or less)
773	Nanoparticle (structure having three dimensions of 100 nm or less)
778	Within specified host or matrix material (e.g., nanocomposite films, etc.)
788	Of specified organic or carbon-based composition
810	Of specified metal or metal alloy composition
811	Of specified metal oxide composition (e.g., conducting or semiconducting compositions such as ITO, ZnOx, etc.)
813	Of specified inorganic semiconductor composition (e.g., periodic table group IV-VI compositions, etc.)
827	Formed from hybrid organic/inorganic semiconductor compositions
831	Of specified ceramic or electrically insulating compositions
832	Having specified property (e.g., lattice-constant, thermal expansion coefficient, etc.)
839	MATHEMATICAL ALGORITHMS, E.G., COMPUTER SOFTWARE, ETC., SPECIFICALLY ADAPTED FOR MODELING CONFIGURATIONS OR PROPERTIES OF NANOSTRUCTURE
840	MANUFACTURE, TREATMENT, OR DETECTION OF NANOSTRUCTURE
841	Environmental containment or disposal of nanostructure material
842	For carbon nanotubes or fullerenes
849	With scanning probe
880	With arrangement, process, or apparatus for testing

882	Assembling of separate components (e.g., by attaching, etc.)
887	Nanoimprint lithography (i.e., nanostamp)
888	Shaping or removal of materials (e.g., etching, etc.)
890	Deposition of materials (e.g., coating, CVD, or ALD, etc.)
894	Having step or means utilizing biological growth
895	Having step or means utilizing chemical property
900	Having step or means utilizing mechanical or thermal property (e.g., pressure, heat, etc.)
901	Having step or means utilizing electromagnetic property (e.g., optical, X-ray, electron beamm, etc.)
902	SPECIFIED USE OF NANOSTRUCTURE
903	For conversion, containment, or destruction of hazardous material
904	For medical, immunological, body treatment, or diagnosis
932	For electronic or optoelectronic application
961	For textile or fabric treatment
962	For carrying or transporting
963	MISCELLANEOUS

The patent attorney, Raj Bawa, has argued that the USPTO classification for nanotechnology is flawed 'because the USPTO has copied the NNI's narrow definition of nanotechnology for classification purposes, which has resulted in a skewed classification system, especially with respect to nanomedicine and bio-nanotechnology inventions'.[40] He observed that 'this classification scheme is neither sufficiently descriptive to accommodate many of the unique properties that nanotechnology inventions exhibit nor does it address the interdisciplinary nature and range of technologies encompassed by nanotechnology'.[41] He contends that 'reforms are urgently needed at the USPTO to address problems ranging from poor patent quality and questionable examination practices to inadequate search capabilities [...] and a sky-rocketing patent application backlog'.[42]

There has been an exponential increase in the number of patent applications filed by both public and private entities in respect of nanotechnology

[40] Bawa, R. (2007), 'Nanotechnology Patent Proliferation and the Crisis at the U.S. Patent Office', *Albany Law Journal of Science & Technology*, **17** (3), 699–736 at 707.

[41] Ibid., 707.

[42] Ibid., 701.

inventions. A USPTO patent examiner, Vivek Koppikar, gives a sense of the early history of this field:

> In many cases, technology which was first conceived in the 1970s and early 1980s, often as academic curiosities, have now become a major area of commercial development under the nanotechnology rubric [...] Quantum dots and dendrimers, similarly, are examples of nanomaterials first patented in the mid-1980s. By 1994, over ten patents issued per year, and by 2003, over 100 patents were issuing per year referring to each of these materials.[43]

The 2008 Chen study analysed the nanotechnology patents granted by the USPTO between 1976 and 2006.[44] During this time, 7406 patents were granted – amongst 12 885 inventors, and 2196 institutions from 46 different countries.[45] Leading patent owners included IBM (209 patents); the University of California (184), the US Navy (99), Eastman Kodak (90), Massachusetts Institute of Technology (76), Micron Technology (75), Hewlett-Packard (67), the Xerox Corporation (63), the 3M Company (59) and Rice University (51).[46]

In 2010, J. Steven Rutt observed that the USPTO had classified 5909 patents as being under Class 977 as nanotechnology patents.[47] He observed that 'typically, the annual number of granted nano patents is only 250 – 500 patents per year'.[48] Rutt noted that the vast majority of the patents appear to be owned by US entities. Of other patent applicants, leading non-US entities included: '(1) Japan, 1,004 (18 percent), (2) Germany, 188 (three percent), and (3) South Korea, 182 (three percent).'[49]

In addition to patent filings in respect of nanotechnology, there has been significant activity with respect to trade marks in relation to nanotechnology. In a study of the use and the misuse of nanotechnology in trade mark applications, Du Mont is alarmed that the term 'Nano' has been deployed,

---

43 Koppikar, V., S. Maebius and J.S. Rutt (2004), 'Current Trends in Nanotech Patents: A View From Inside the Patent Office', *Nanotechnology Law and Business*, **1**, 24–30 at 25.

44 Chen, H., M. Roco, X. Lin and Y. Lin (2008), 'Trends in Nanotechnology Patents', *Nature Nanotechnology*, **3**, 123–5.

45 Ibid.

46 Ibid.

47 Rutt, J.S. (2010), 'Recent Nanotech Patent Trends: Top Ten Observations', Nanocleantech Blog, http://www.nanocleantechblog.com/tags/cleantech-patent-landscape-rep/.

48 Ibid.

49 Ibid.

even on products and services that are unrelated to the science of nanotechnology.[50] Most notably, Apple, the information technology and consumer electronics company, has deployed the term 'nano' in relation to the MP3 player, the iPod Nano.[51] Du Mont contends that the USPTO and courts 'can help clear up some of the confusion caused by NANO-imposters'.[52]

B European Patent Office

Sensitive to the impact of emerging technologies,[53] the European Patent Office (EPO) has also sought to consider how best to reform its administrative practices and classifications to better address nanotechnology. The EPO predicts that nanotechnology will be a pioneering field: 'Nanotechnology is considered by many to be one of the key technologies of this century, with an expected market volume of EUR 1 trillion in 2015.'[54] Nonetheless, the EPO reflects that '[a]s with any new technology, there is some anxiety over how the patenting process will adapt'.[55]

Christian Kallinger, patent examiner and team leader of the nanotechnology working group of the EPO, and his colleagues have discussed the challenges posed to patent offices by new technologies.[56] He reflects that nanotechnology is a field of convergence between a number of scientific fields and disciplines:

> Patent offices worldwide are constantly confronted with new and emerging technologies of which nanotechnology is the most recent example. What makes nanotechnology a special case compared to other booming technologies (e.g., biotechnology or information technology), is the fact that it does not originate from a single area of science or technology. Rather, nanotechnology is a very interdisciplinary field. Because nanotechnology is defined purely by size, it may be part of almost any area of science or technology; it is just as relevant to

---

50 Du Mont, J. (2008), 'Trademarking Nanotechnology: Nano-Lies and Federal Trademark Registration', *AIPLA Quarterly Journal*, **36** (2), 148–78.

51 Apple Retail Store, http://www.apple.com/au/ipodnano/.

52 Du Mont, J. (2008), 'Trademarking Nanotechnology: Nano-Lies and Federal Trademark Registration', *AIPLA Quarterly Journal*, **36** (2), 148–78 at 178.

53 See European Patent Office (2007), *Scenarios for the Future: How might IP Regimes Evolve by 2025? What Global Legitimacy Might Such Regimes Have?* Munich: European Patent Office, http://documents.epo.org/projects/babylon/eponet.nsf/0/63A726D28B589B5BC12572DB00597683/$File/EPO_scenarios_bookmarked.pdf.

54 European Patent Office, 'Nanotechnology in European Patents: Opportunity and Challenges', http://www.epo.org/news-issues/issues/nanotechnology.html.

55 Ibid.

56 Kallinger, C. et al. (2008), 'Patenting Nanotechnology: A European Patent Office Perspective', *Nanotechnology Law and Business*, **5**, 95–105.

> biotechnologists and physicists as it is to electrical and mechanical engineers or materials scientists.[57]

Kallinger maintains that the EPO is well-prepared for nanotechnology: 'Guidelines for Examination as well as a number of nanotechnology related Board of Appeal decisions, do provide examiners with guidance on how to answer questions relating to the novelty of selection inventions, the assessment of an inventive step in case of miniaturization or the definition of common general knowledge in young technologies'.[58] In his opinion, 'Given this existing framework in the EPO, there exists no need for a new patent regime in nanotechnology.'[59]

The EPO has settled upon the following definition of nanotechnology: 'The term nanotechnology covers entities with a controlled geometrical size of at least one functional component below 100 nanometres in one or more dimensions susceptible of making physical, chemical or biological effects available which are intrinsic to that size. It covers equipment and methods for controlled analysis, manipulation, processing, fabrication or measurement with a precision below 100 nanometres.'[60] The EPO has sought to develop special classifications for nanotechnology to ensure that patent documents can be properly identified by introducing the 'Y01N' tags.[61] The EPO has developed half a dozen sub-categories dealing with varieties or species of nanotechnology.

Table 10.2 European Patent Office sub-classifications of nanotechnology

Code	**Nanotechnology field**
Y01N2	Nanobiotechnology
Y01N4	Nanotechnology for information processing, storage and transmission
Y01N6	Nanotechnology for materials and surface science
Y01N8	Nanotechnology for interacting, sensing or actuating
Y01N10	Nano optics
Y01N12	Nanomagnetics

57 Ibid., 95–6.
58 Ibid., 105.
59 Ibid., 105.
60 European Patent Office, 'Nanotechnology in European Patents: Opportunity and Challenges', http://www.epo.org/news-issues/issues/nanotechnology.html.
61 Ibid.

At present, the filing rates in respect of nanotechnology patents are modest. The 2008 Chen Study highlighted that the EPO had granted 3596 nanotechnology patents between 1978 and 2006.[62] Leading patent owners included the Japan Science and Technology Agency (78); L'Oreal (60); IBM (50); Rohm & Haas (47); Samsung (45); Eastman Kodak (40); CEA (39); CNRS (37); Matsushita Electric Industrial (32); and BASF (31).[63]

The EPO is hopeful, though, that the field will grow: 'Although the number of nanotechnology patent applications filed with the EPO is still relatively small – in the lower single-digit percentages of all European patent applications – filing figures have been slowly rising, and many experts are now predicting that nanotechnology will become the next big growth field after biotechnology.'[64] The EPO has emphasized that there is a need to ensure that the patents granted have a high quality:

> Experts have observed that there is a growing trend to file for patents earlier, and file more of them. That could potentially hamper innovation by acting as a disincentive for other institutions to embark on similar research. Although this is a general trend in patenting, nanotechnology-related inventions could be especially affected since it is a young and growing field. To avoid an inflation of low-quality patent applications that could clog up the EPO and create a backlog, the EPO has introduced a quality policy to bring certainty to the market, for both the applicant and the public.[65]

The EPO has collaborated with the USPTO and the Japan Patent Office on questions of patent practice with respect to nanotechnology applications: 'Collaborating on the nanotechnology issue could help all three offices tackle common problems that have arisen with the emergence of this exciting new technology.'[66]

The EPO's system of classification of nanotechnology has been adopted and adapted under the International Patent Classification system: 'From now on all patent offices worldwide will classify nanotechnology under the new symbol B82Y, building on the EPO's internal Y01N coding system which was specifically created by the Office as no scheme under the IPC was

---

62 Chen, H., M. Roco, X. Lin and Y. Lin (2008), 'Trends in Nanotechnology Patents', *Nature Nanotechnology*, **3**, 123–5.

63 Ibid.

64 European Patent Office, 'Nanotechnology in European Patents: Opportunity and Challenges', http://www.epo.org/news-issues/issues/nanotechnology.html.

65 Ibid.

66 Ibid.

available to capture all nanotechnology-related technologies and applications.'[67] EPO President Benoît Battistelli has observed: 'The introduction of a new international classification subclass to capture the diverse aspects and applications of nanotechnology is recognition of the EPO's pro-active efforts to align the worldwide classification system for patents with the information needs of researchers, innovators and patent experts in promising emerging technologies.'[68]

There have been tensions between the approach of the EPO and the European Parliament. In 2009, the European Parliament passed a resolution on the regulation of nanotechnology, expressing concerns about the breadth of patents granted in respect of nano-materials.[69] The resolution noted that 'the European Union is lagging behind its current main competitors – the USA, Japan and South Korea – who account for over half of the investment and two-thirds of the patents filed worldwide'.[70] The resolution feared that the 'broad application of patents to nanomaterials, as well as the excessive cost of patenting and the absence of patent access facilities for very small businesses and small and medium-sized enterprises (SMEs), could stifle further innovation'.[71] The resolution recognised 'that it is essential to remove the obstacles preventing very small businesses and SMEs in particular from accessing patents and calls at the same time for patent rights to be limited to specific applications or production methods of nanomaterials, and only to be extended to nanomaterials themselves on an exceptional basis, to avoid stifling innovation'.[72]

There has been criticism from civil society groups and non-governmental organizations that patent offices such as the EPO have been technocratic and bureaucratic in their approach to patents and nanotechnology, and failed to address larger questions about the ethics of such inventions, and potential risks to human health and environmental safety. The Canadian based ETC Group succinctly complains: 'Patent examiners have managed to negotiate their way around the absence of global definitions of nano, as well as the characterization and standardization methodologies that would support them, and have largely ignored its cross-sectoral, multidisciplinary

---

67 European Patent Office (2011), 'EPO Classification Work Leads to New International Standard', Press Release, 28 January, http://www.epo.org/news-issues/news/2011/20110128.html.

68 Ibid.

69 *European Parliament Resolution of 24 April 2009 on Regulatory Aspects of Nanomaterials*, http://www.europarl.europa.eu/sides/getDoc.do?type=TA&reference=P6-TA-2009–0328&language=EN.

70 Ibid.

71 Ibid.

72 Ibid.

nature as well as the biggest risk tiny tech poses: the potential reach of exclusive monopoly to the fundamental building blocks of all of nature.'[73]

C The Japan Patent Office

The Japan Patent Office (JPO) has also developed a special classification system for nanotechnology.[74] In 2007, the JPO noted in its annual report: 'The content of patent applications is becoming more complex and advanced in line with the innovative progress in technology, centering on such cutting-edge fields as nanotechnology and biotechnology.'[75] Shinjiro Ono, former Deputy Commissioner of the JPO, noted that even the strongest patent offices struggled to cope with emerging technologies: 'Even the large offices cannot recruit experienced examiners in sufficiently large numbers to cope with the speed and complexity of developments in emerging technologies, as well as the increased volume of applications.'[76]

The 2008 Chen study found that there were 1150 nanotechnology patent publications by the Japanese Patent Office between 1976 and 2006.[77] The leading patent applicants included Nippon Electric (109), the Japan Science and Technology Agency (70), the National Institute for Materials Science (52), the National Institute of Advanced Industrial Science and Technology (48), the Matsushita Electric Industrial (45), the Agency of Industrial Science and Technology (43), Tokyo Shibaura Electric (43), Sony (32), Canon (31), and Seiko Instruments (27).[78]

II SHARING NANOTECHNOLOGY

There have already been legal skirmishes over patent rights in respect of nanotechnology. The NanoBusiness Alliance has vociferously supported

---

73 ETC Group (2010), *The Big Downturn: Nanogeopolitics*, Ottawa: ETC Group, 36, http://www.etcgroup.org/upload/publication/pdf_file/nano_big4web.pdf.

74 Japan Patent Office, http://www.jpo.go.jp/.

75 Japan Patent Office (2007), *Annual Report*, 8, http://www.jpo.go.jp/shiryou_e/toushin_e/kenkyukai_e/pdf/annual_report2007/part1.pdf.

76 See European Patent Office (2007), *Scenarios for the Future: How might IP Regimes Evolve by 2025? What Global Legitimacy Might Such Regimes Have?*, Munich: European Patent Office, http://documents.epo.org/projects/babylon/eponet.nsf/0/63A726D28B589B5BC12572DB00597683/$File/EPO_scenarios_bookmarked.pdf.

77 Chen, H., M. Roco, X. Lin and Y. Lin (2008), 'Trends in Nanotechnology Patents', *Nature Nanotechnology*, **3**, 123–5.

78 Ibid.

and defended the enforcement of patent rights by nanotechnology companies. The Alliance was critical of the *Patent Reform Act* of 2007 (US), complaining that 'this pending legislation contains provisions that will create uncertainty and weaken the enforceability of validly issued patents' and 'pose serious negative consequences for continued innovation and American technological leadership in a competitive global economy'.[79] The NanoBusiness Alliance complained: 'In effect the bill increases the cost of filing a patent while reducing the cost of infringing on one.'[80] In a submission to the Federal Trade Commission's inquiry into patent law and competition, the Alliance argued: 'Changes which reduce our ability to receive adequate compensation for infringement of those patents will make it difficult to protect our intellectual property, and therefore will discourage investment in our field.'[81] Notwithstanding such negativity from the nanotechnology industry, the *America Invents Act* of 2011 (US) passed United States Congress.[82]

In *Nano-Proprietary Inc.* v *Canon Inc*, the Court of Appeals for the Fifth Circuit considered an action brought by a nanotechnology company which holds field emission display patents against its licensee for breach of contract – through allegedly sublicensing patented technology to a corporation.[83] The judge dismissed the call by Nano Propietary Inc. that it was entitled to damages.

In *Nanoexa Corp.* v. *University of Chicago*, Nanoexa Corporation alleged that the University of Chicago and UChicago Argonne LLC were in breach of contract and liable for patent misuse by impermissibly restricting the scope of a patent licence agreement between the parties.[84] The matter is still under review.

In July 2006, Elan Pharmaceutical International, Ltd. filed a complaint in the United States District Court for the District of Delaware alleging that the cancer treatment, 'Abraxane,' manufactured by Abraxis Bioscience,

---

79 Federal Trade Commission (2011), *The Evolving IP Marketplace: Aligning Patent Notice and Remedies with Competition*, 1.

80 Ibid.

81 Nanobusiness Alliance (2009), 'Federal Trade Commission Hearing on Evolving IP Marketplace – Evolution of Remedies: Comment Number 540872–00017', http://www.ftc.gov/os/comments/iphearings/540872–00017.htm.

82 *Leahy-Smith America Invents Act*, 2011 (US) Pub. L. No. 112–29.

83 *Nano-Proprietary Inc.* v *Canon Inc*. 537 F. 3d 394 (2008).

84 *Nanoexa Corp.* v. *University of Chicago*, 2010 WL 4236855 (N.D.Cal.).

Inc. infringed two of Elan's patents.[85] In June 2008, a Delaware jury delivered a $55.2 million jury verdict to Elan.[86]

In July 2009, Nanosys, the United States producer of nanotech quantum dots, settled a patent infringement lawsuit against Nanoco Technologies Ltd of the United Kingdom.[87] The lawsuit claimed that the Nanoco's quantum dot technology infringed upon five seminal quantum dot patents held by Nanosys. In the settlement, Nanoco agreed to terminate its current United States business for its core-shell quantum dots. Jason Hartlove, chief executive officer of Nanosys, commented: 'By enforcing ownership of our intellectual property, the manufacturers remain the real winners in having access to proven, trusted advanced material architecture, including quantum dot applications.'[88] The company, Nanosys, has a portfolio of over 700 patents covering its quantum dot technology and its process-ready components for solid state lighting, solar power and electronic display systems.

There has been some concern that the field of nanotechnology will suffer from 'a tragedy of the anticommons' as a result of patent thickets in the field.[89] Reflecting upon such emerging trends, Raj Bawa laments: 'Currently, there are too many players holding too many nanotech-related patents; this has created the current fragmented, messy patent landscape.'[90] Terry Tullis fears that 'the active enforcement by nanotechnology patent holders of their exclusivity rights ultimately could result in the creation of a

---

85 Milford, P. (2008), 'Elan Wins $55.2 Million Jury Award From Abraxis', Bloomberg, 13 June.

86 Ibid.

87 *Nanosys Inc.* v. *Nanoco Technologies Ltd.*, case number 09-cv-00258, in the U.S. District Court for the Western District of Wisconsin. Nanosys (2009), 'Nanosys Reaches Settlement of Patent Infringement Lawsuit Against Nano Technologies for Quantum Dot Technology', Press Release, 23 July.

88 Nanosys (2009), 'Nanosys Reaches Settlement of Patent Infringement Lawsuit Against Nano Technologies for Quantum Dot Technology', Press Release, 23 July.

89 Heller, M. (1998), 'The Tragedy of the Anticommons: Property in the Transition from Marx to Markets', *Harvard Law Review*, **111** (3), 621–88; Heller, M. and R. Eisenberg (1998), 'Can Patents Deter Innovation? The Anticommons in Biomedical Research', *Science*, **280**, 698–701; Heller, Michael (2008), *The Gridlock Economy: How Too Much Ownership Wrecks Markets, Stops Innovation and Costs Lives*, New York: Basic Books; and Heller, Michael (ed.) (2010), *Commons and Anticommons*, vols I and II, Cheltenham and Northampton, MA: Edward Elgar.

90 Bawa, R. (2007), 'Nanotechnology Patent Proliferation and the Crisis at the U.S. Patent Office', *Albany Law Journal of Science & Technology*, **17** (3), 699–736 at 728.

nanotechnology anticommons – a situation in which a scarce resource becomes prone to underuse because there are too many owners holding the right to exclude others from that resource, and no one has an effective privilege of use'.[91] Sean O'Neill and collaborators urged: 'In order to reduce uncertainty regarding the rights to nanotechnology inventions, patent applicants, and patent owners must resist the urge to claim overly broadly in the hope of getting a windfall of nanotechnology rights.'[92]

A Experimental Use

Historically, the defence of experimental use in patent law evolved from cases dealing with mechanical inventions. The defence was, in effect, a 'freedom to tinker'. The research exemption is particularly important in the field of nanotechnology, given the patent thickets and litigation in this area. There has been much debate as to whether there is a 'freedom to tinker' with respect to nano-machines.

There has been much disquiet about the narrow definition of experimental use in the United States. In the case of *Madey* v. *Duke University,* Gajarsa J considered the defence of experimental use in the context of a dispute between the inventor of a patented laser and his former university.[93] His Honour observed of the limited operation of the defence of experimental use: 'In short, regardless of whether a particular institution or entity is engaged in an endeavor for commercial gain, so long as the act is in furtherance of the alleged infringer's legitimate business and is not solely for amusement, to satisfy idle curiosity, or for strictly philosophical inquiry, the act does not qualify for the very narrow and strictly limited experimental use defense.'[94]

Terry Tullis was alarmed by the narrow interpretation given to the defence of experimental use in the United States courts:

> The lack of an experimental use defense is deeply troubling for nanotechnology innovation. Basic tenets of scientific research require the duplication of tests to validate past procedures. If an exclusive licensee of a nanotechnology patent

91 Tullis, T. (2005–2006), 'Application of the Government License Defense to Federally Funded Nanotechnology Research: The Case for a Limited Patent Compulsory Licensing Regime', *University of California, Los Angeles Law Review*, **53**, 279–314 at 283.

92 O'Neill, S., K. Hermann, M. Klein, J. Landes and R. Bawa (2007), 'Broad Claiming in Nanotechnology Patents: Is Litigation Inevitable', *Nanotechnology Law and Business*, **4**, 595–606.

93 *Madey* v. *Duke University* 307 F.3d 1351 at 1362 (2002).

94 Ibid.

> decides to not allow competitors even to repeat a patented fundamental method or technology of nanoscience, then any attempt to validate or even improve upon that patent will result in liability for infringement.[95]

Nicholas Zovko contends: 'In light of the Federal Circuit's decision in *Madey* v. *Duke University*, the current narrow scope of the experimental use defense to patent infringement will likely stifle nanotechnology innovation.'[96] He concludes: 'The courts or Congress should listen to the arguments that were made in the biotechnology arena and broaden the scope of the experimental use defense to permit non-commercial experimental use.'[97]

In the European Union, nation states have taken a broad, flexible approach to the defence of experimental use. Article 27(b) of the *Community Patent Convention* 1975 has provided the basis for an experimental use exception that exempts 'acts done for experimental purposes relating to the subject mater of the patented invention'.[98] While there remain variations in the interpretation of the defence of experimental use in different jurisdictions in the European Union, there is scope for the defence of experimental use to apply to potentially both commercial and non-commercial uses.[99]

Wim Helwegen has considered the defence of experimental use, from the perspective of nanotechnology: 'Economies that want to rely on knowledge-intensive products should pay attention to the clarity of their research exemptions.'[100] He suggests that the approach of the European Union is preferable:

> From a legal point of view, entering the nanotech patent landscape is certainly more attractive in Europe than it is in the United States. The provisions implemented at Member-State level allow experiments relating to the subject

---

95 Tullis, T. (2005–2006), 'Application of the Government License Defense to Federally Funded Nanotechnology Research: The Case for a Limited Patent Compulsory Licensing Regime', *University of California, Los Angeles Law Review*, **53**, 279–314 at 301.

96 Zovko, N. (2006), 'Nanotechnology and the Experimental Use Defense to Patent Infringement', *McGeorge Law Review*, **37** (1), 129–56 at 156.

97 Ibid.

98 *Convention for the European Patent for the Common Market* 1975 (*Community Patent Convention*).

99 Holzapfel, H. and J. Sarnoff (2008), 'A Cross-Atlantic Dialog on Experimental Use and Research Tools', *IDEA: Intellectual Property Law Review*, **48**, 123–4.

100 Helwegen, W. (2010), 'The Research Exemption from a Nanotechnology Perspective', *European Intellectual Property Review*, **32** (7), 341–51 at 348.

> matter of the invention, independent of commercial intent. The system of research exemptions seems to be working rather well in terms of harmonisation.[101]

He commented: 'In emerging technologies like nanotechnology that are practised by a broad variety of entities, from multinationals to start-ups and universities, a research exemption as restrictive as the current US exemption does not represent a balanced trade-off between the monopoly right of the patentee and the public interest.'[102]

B Compulsory Licensing

Compulsory licensing could be deployed to provide access to nanotechnology inventions in return for reasonable compensation to the patent owner. The *TRIPS Agreement* 1994 recognizes that members of the World Trade Organization can authorize the use of patented inventions, subject to a number of procedural conditions. Terry Tullis has made the case for a limited compulsory licensing regime in respect of nanotechnology.[103] He comments:

> In the area of patents, Article 31 of the *TRIPS Agreement* 1994 already permits World Trade Organization members to grant compulsory patent licenses under the limited circumstances of national emergency, antitrust violations, and public noncommercial use. Future intellectual property harmonization or nanotechnology treaties could also expand existing government license rights. Other countries have little cause to object to this proposed augmentation of the government license defense because there is no impact on foreign intellectual property.[104]

Tullis concludes: 'A well-formulated government license defense would provide a means for overcoming the innovation-impeding effects of absolute exclusion rights by assessing infringement along a spectrum of use.'[105]

---

101 Ibid.

102 Ibid., 344.

103 Tullis, T. (2005–2006), 'Application of the Government License Defense to Federally Funded Nanotechnology Research: The Case for a Limited Patent Compulsory Licensing Regime', *University of California, Los Angeles Law Review*, **53,** 279–314.

104 Ibid., 311.

105 Ibid., 308.

C Patent Pools

Patent pools have been presented as a solution to patent thickets in an area of technological fields – ranging from motion pictures and DVDS to essential medicines and genetics, and even clean technologies. It is not surprising that some commentators have suggested the option of patent pools to address patent thickets in nanotechnology. Alexander Lee contends that the crowded nanotechnology patent landscape could result in a decreased rate of innovation, widespread intellectual property disputes and a reduced rate of consumer acceptance.[106] He comments upon the possible impacts of patent thickets in the area of nanotechnology: 'The potentially large number of poor quality patents may actually slow down this field's pace of advancement.'[107] Nevertheless, Lee recognizes that there are a number of factors to consider before a patent pool could be formed – including questions of standards, the operation of the marketplace, the nature of the patent holdings and larger matters of competition. In a case study on the dendritic nanotechnology market, Lee concludes that a patent pool will not be necessary for the continued advancement of this application.

Clarkson and Dekorte agree that 'there is a growing concern that future innovation and commercialization will be inhibited by the explosive rate of nanotechnology patenting and the potential for the formation of patent thickets'.[108] The pair note that such problems have bedevilled both the biotechnology and the information technology sectors: 'In other industries characterized by cumulative innovations and multiple blocking patents, the existence of such densely concentrated patent rights can have the perverse effect of stifling innovation rather than encouraging it.'[109] Clarkson and Dekorte observe: 'An alternative strategy would be for nanotechnology patent holders to form a patent pool, both clearing the nanothicket for that technology and facilitating "one-stop technology platform licensing".'[110] Such arrangements could be based upon the standards established by the American National Standards Institute's Nanotechnology Standards Panel. Nonetheless, the pair are conscious that 'The standards-setting

---

[106] Lee, A. (2006), 'Examining the Viability of Patent Pools for the Growing Nanotechnology Patent Thicket', *Nanotechnology Law and Business*, **3**, 317–27.

[107] Ibid.

[108] Clarkson, G. and D. Dekorte (2006), 'The Problem of Patent Thickets in Convergent Technologies', *Annals of the New York Academy of Sciences*, **1093**, 180–200.

[109] Ibid., 197.

[110] Ibid., 197–8.

process for nanotechnology is in its infancy, however, and the prospects for the development of standards remain uncertain.'[111]

D Public Sector Licensing

There has been great public and private support for research into nanotechnology. In the United States, the *21st Century Nanotechnology Research and Development Act* 2003 (US) provided significant funding for federal nanotechnology research and development. This has been reinforced by the National Nanotechnology Initiative.[112] There has been much government and industry investment in nanotechnology in the members of the European Union[113] and Japan. In January 2007, the Australian Government announced that the Australian Research Council announced new funding for a National Nanotechnology Strategy and a Niche Manufacturing Flagship.[114]

Mark Lemley identifies a number of features of nanotechnology, which will pose particular issues for patent law.[115] He is particularly intrigued that universities and public research institutions hold patents in surprisingly large proportions:

> The technology is still in its infancy, and many of the patents that have issued so far – certainly many of the basic building-block patents that are most relevant here – issue to research labs doing fundamental science, rather than to specific product implementations. It is not too surprising that most of those basic research labs are located in universities. Indeed, this may be a pattern with enabling technologies.[116]

Lemley suggests that there are a number of possible solutions if the field of nanotechnology becomes cluttered with strong and broad patents. He notes that the law could promote access to research tools: 'If there is a significant risk that nanotechnology innovation will be retarded by broad upstream patents, we can replicate by law the result we got by accident in the biotech, software, hardware, and Internet industries – freedom to use basic tools and

---

111 Ibid., 198.

112 National Nanotechnology Initiative, http://www.nano.gov/.

113 European Commission (2007), 'Nanotechnology', http://cordis.europa.eu/nanotechnology/.

114 Australian Government (2007), *Australian Nanotechnology: Capability and Commercial Potential*, 3rd edn, Invest Australia, http://www.investaustralia.gov.au/media/IR_Nano_NanotechReport.pdf.

115 Lemley, M. (2005), 'Patenting Nanotechnology', *Stanford Law Review*, **58,** 601–30.

116 Ibid., 615.

processes with patents only on downstream implementations.'[117] Moreover, he suggests that there could be scope for government funding agencies to encourage non-exclusive licensing in respect of nanotechnology inventions developed by the public sector.

An example of public sector licensing is the partnership between the Cambridge start-up, Nano-Terra,[118] and Harvard University, involving the licensing of rights to more than 50 patents. Kevin Bullis commented: 'The wide-ranging set of patents – the result of research done in the Harvard chemistry lab of George Whitesides – covers everything from techniques for designing materials that assemble themselves into microscopic lenses and data storage devices, to tools for patterning complex nanoscale circuits over large, irregularly shaped surfaces.'[119] A representative of Nano-Terra, observed that the patent filing and maintenance costs alone top $2 million.[120] The company plans to use its intellectual property portfolio as the basis for a business strategy that will market its ability to quickly adapt the tools covered by the patents to create products for its clients.

No doubt, there will be legal battles over the ownership of nanoscience and nanotechnology, involving difficult questions as to whether a scientist's 'duty to research' involves a 'duty to invent'.[121]

E Technology Transfer

In analysing how intellectual property in nanotechnology can avoid 'the mistakes of biotech', Richard Gold identifies a shift that is occurring from 'Old IP' to 'New IP' in the biotechnology industry.[122] One of the themes of New IP is recognition of the need to facilitate research capacity in less developed countries.[123]

---

117 Ibid., 628.

118 Nano-Terra Inc., http://www.nanoterra.com/.

119 Bullis, K. (2007), 'Key Nanotech Patents Licensed', *Technology Review*, 6 June, http://www.technologyreview.com/biomedicine/18829/.

120 Nanotechnology Development Blog (2007), 'Harvard Nanotechnology Patents to Nano-Terra', 6 June, http://www.nanotechnologydevelopment.com/news/harvard-nanotechnology-patents-to-nano-terra.html.

121 See Monotti, Anne and Sam Ricketson (2003), *Universities and Intellectual Property: Ownership and Exploitation*, Oxford: Oxford University Press; and *University of Western Australia* v. *Gray* [No. 20] (2008) FCA 498; *University of Western Australia* v *Gray* [2009] FCAFC 116; and *University of Western Australia* v *Gray* [2010] HCATrans 11 (12 February 2010).

122 Gold, Richard (2009), 'Avoiding the Mistakes of Biotech: How Intellectual Property Can Be Better Managed to Advance Nanotechnology Research', *Studies in Ethics, Law, and Technology*, **3** (3), Article 4.

123 Ibid., 6–7.

Nanotechnology raises ethical issues regarding the use of the technology to address problems that confront developing countries and transfer of nanotechnology applications to developing countries. As Pradeep notes, 'Nanoscience and its deriving technologies have the potential to improve the state of the developing world, if the applications are designed and tailored to best fit the needs of the people.'[124] Fields in which nanotechnology has relevance to sustainable development include improvements to drinking water, food yields and medicine – diagnosis, screening, drug delivery – energy production and storage, and environmental remediation.[125]

Fabio Salamanca-Buentello and his colleagues have argued that 'nanotechnology can be harnessed to address some of the world's most critical development problems'.[126] First, the group observes that 'developing countries are already harnessing nanotechnology to address some of their most pressing needs'.[127] Second, the team highlights 'ten applications of nanotechnology most likely to benefit developing countries, and demonstrate that these applications can contribute to the attainment of the United Nations Millennium Development Goals'.[128] Third, Salamanca-Buentello and his colleagues call upon 'the international community to accelerate the use of these top nanotechnologies by less industrialized countries to meet critical sustainable development challenges'.[129] However, Mexican researchers, Guillermo Foladori and Noela Invernizzi, questioned the suitability of nanotechnology to address development concerns.[130] The pair noted: 'It is not necessarily true that the technology that is best and meets our needs will be the one to survive.'[131]

Intellectual property regimes will play a role in nanotechnology's success in addressing the problems of the developing world. Pradeep argues that nanotechnology offers a 'unique chance to bridge the technological gap between the industrialised and developing world', but that 'joint and

---

124 Pradeep, T. (2008), *Nano: The Essentials*, New York: McGraw-Hill, 368.
125 Ibid., 363–7.
126 Salamanca-Buentello, F., D.L. Persad, E.B. Court, D.K. Martin, A.S. Daar and P.A. Singer (2005), 'Nanotechnology and the Developing World', *Public Library of Science Medicine*, **2** (5), e97, 302.
127 Ibid.
128 Ibid.
129 Ibid.
130 Foladori, G. and N. Invernizzi (2005), 'Nanotechnology for the Poor?' *Public Library of Science Medicine*, **2** (8), e280. doi:10.1371/journal.pmed.0020280.
131 Ibid.

concerted efforts by all concerned are needed to rule out future factions and new defiance'.[132]

III REGULATING NANOTECHNOLOGY

Ethical objections to nanotechnology fall into several categories including concern about inherent risks of the technology (such as damage to human health or the environment), fear of unethical use of the technology (for example in weapons delivery systems) and concerns about unethical consequences of ownership and commercial exploitation of nanotechnology (such as access to technology in developing countries, and impacts upon traditional farmers). In fact the ETC Group goes so far as to warn that together, converging technologies such as nanotechnology 'will allow human security and health – even cultural and genetic diversity – to be firmly in the hands of a convergent technocracy'.[133]

In a 2006 report entitled the *Ethics and Politics of Nanotechnology*, UNESCO provides a thoughtful exploration of ethical issues associated with nanotechnology.[134] In its introduction, the report notes the welter of hopes, fears and anxieties expressed about the discipline in popular media and discourse:

> Nanotechnology could become the most influential force to take hold of the technology industry since the rise of the Internet. Nanotechnology could increase the speed of memory chips, remove pollution particles in water and air and find cancer cells quicker. Nanotechnology could prove beyond our control, and spell the end of our very existence as human beings. Nanotechnology could alleviate world hunger, clean the environment, cure cancer, guarantee biblical life spans or concoct super-weapons of untold horror [...] Nanotechnology could lead to the next industrial revolution. Nanotechnology could transform the food industry. Nanotechnology could repair the ozone layer. Nanotechnology could change everything.[135]

---

132 Pradeep, T. (2008), *Nano: The Essentials*, New York: McGraw-Hill, 368.

133 ETC Group (2004), 'Down on the Farm: The Impact of Nano-scale Technologies on Food and Agriculture', Ottawa: ETC Group, 5, http://www.etcgroup.org/upload/publication/80/02/etc_dotfarm2004.pdf.

134 United Nations Educational, Scientific and Cultural Organization (UNESCO) (2006), *The Ethics and Politics of Nanotechnology*, Paris: UNESCO, http://unesdoc.unesco.org/images/0014/001459/145951e.pdf.

135 Ibid., 3.

Neutrally, the report concludes: 'The emergence of consensus concerning the direction, safety, desirability and funding of nanotechnology will depend on how it is defined, and on who will be included as a result.'[136]

In July 2007, a group of civil society organizations issued a set of eight agreed principles for regulation of nanotechnology which they proposed that government and industry take up.[137] These principles provide a snapshot of common concerns about nanotechnology. The principles advocate a precautionary approach to regulation and emphasize transparency and public participation:

> I. A Precautionary Foundation: Product manufacturers and distributors must bear the burden of proof to demonstrate the safety of their products: if no independent health and safety data review, then no market approval.
>
> II. Mandatory Nano-specific Regulations: Nanomaterials should be classified as new substances and subject to nano-specific oversight. Voluntary initiatives are not sufficient.
>
> III. Health and Safety of the Public and Workers: The prevention of exposure to nanomaterials that have not been proven safe must be undertaken to protect the public and workers.
>
> IV. Environmental Protection: A full lifecycle analysis of environmental impacts must be completed prior to commercialization.
>
> V. Transparency: All nano-products must be labeled and safety data made publicly available.
>
> VI. Public Participation: There must be open, meaningful, and full public participation at every level.
>
> VII. Inclusion of Broader Impacts: Nanotechnology's wide-ranging effects, including ethical and social impacts, must be considered.
>
> VIII. Manufacturer Liability: Nano-industries must be accountable for liabilities incurred from their products.[138]

These principles argue that nanotechnology requires specific regulation rather than being subject to existing regulation of technology.

---

136 Ibid., 21.

137 ETC Group (2007), 'Press Release: Broad International Coalition Issues Urgent Call for Strong Oversight of Nanotechnology', 31 July, http://www.etcgroup.org/en/node/651.

138 Ibid.

One of these organizations, the ETC Group, objects strongly to nanotechnology for a variety of reasons. Overall, its position can be summarized as concern about possible harmful effects of nanotechnology research and products using nanotechnology, concern that these risks are not being adequately investigated, and that therefore nanotechnology is proceeding in an under regulated and potentially dangerous fashion. The science of nanotechnology is progressing very rapidly, and assessing the risks associated with the technology is only likely to become more difficult.[139] In this context, 'Government is the bureaucratic tortoise chasing the technological hare.'[140] In relation to under regulation of nanotechnology, the ETC Group stated in 2004 that:

> [N]o government has developed a regulatory regime that addresses the nanoscale or the societal impact of the invisibly small [...] By allowing nanotech products to come to market in the absence of public debate and regulatory oversight, governments, agribusiness and scientific institutions have already jeopardised the potential benefits of nano-scale technologies.[141]

In 2010 the ETC Group released an updated nanotechnology report and noted some action that had been taken to improve regulation of nanotechnology in the European Union and within the remit of the Environmental Protection Agency in the United States.[142] However, they described the approach of the United States Government as 'toothless'.[143] Overall, the report gave the impression of persisting alarm about under regulation in the nanotechnology area.

---

139 Rejeski, D. (2008), 'Comment on A Framework Convention for Nanotechnology', *Environmental Law Reporter*, **38**, 10518–9 at 10518; see also Hunt, Geoffrey and Michael Mehta (eds) (2006), *Nanotechnology: Risks, Ethics and Law*, London: Earthscan.

140 Rejeski, D. (2008), 'Comment on A Framework Convention for Nanotechnology', *Environmental Law Reporter*, **38**, 10518–9 at 10518. See also Abbott, K., G. Marchant and D. Sylvester (2008), 'A Framework Convention for Nanotechnology', *Environmental Law Reporter*, **38**, 10507–17.

141 ETC Group (2004), 'Down on the Farm: The Impact of Nano-scale Technologies on Food and Agriculture', Ottawa: ETC Group, 1, http://www.etcgroup.org/upload/publication/80/02/etc_dotfarm2004.pdf.

142 ETC Group (2010), *The Big Downturn: Nanogeopolitics*, Ottawa: ETC Group, 13–17, http://www.etcgroup.org/upload/publication/pdf_file/nano_big4web.pdf.

143 Ibid., 17.

A Risks to Human Health and the Environment

There has been ongoing concern about potential risks of exposure to nanoparticles for both human health and the environment. In relation to health, human exposure to nanoparticles could be through products such as foods and cosmetics or via particles released into the environment in products such as pesticides. Concerning characteristics of nanoparticles include their high surface reactivity[144] and 'high mobility within the body'.[145] There is evidence, for example, that they can penetrate cells and can also pass into the brain.[146] Studies on the toxic effects of other particles indicate that length and diameter of particles and their surface area are relevant to whether they cause damage within the body, along with other factors such as the dose received.[147] There is evidence that some substances can become toxic when their particles are nanoscale.[148]

Concerns about health risks relate not only to potential products of nanotechnology, but also to products that are already on the market, for example, sunscreens. There is concern that some nanoparticles contained in sunscreens can be toxic to cells, as well as some evidence that they can be absorbed through the skin and may 'interact with the immune system'.[149] In 2007–2008 there was a state parliamentary inquiry in Australia regarding, among other terms of reference, 'the health, safety and environmental risks and benefits of nanotechnology' and 'the appropriateness of the current regulatory frameworks in operation for the management of nanomaterials over their life-cycle'.[150] The inquiry recommended that the 'ingredient

---

144 The Royal Society and the Royal Academy of Engineering (2004), *Nanoscience and Nanotechnologies: Opportunities and Uncertainties*, London: The Royal Society, 35, http://www.nanotec.org.uk/finalReport.htm.

145 Faunce, Thomas (2010), 'Exploring the Safety of Nanoparticles in Australian Sunscreens', *International Journal of Biomedical Nanoscience and Nanotechnology*, **1** (10), 87–94 at 87.

146 The Royal Society and the Royal Academy of Engineering (2004), *Nanoscience and Nanotechnologies: Opportunities and Uncertainties*, London: The Royal Society, 35 and 41, http://www.nanotec.org.uk/finalReport.htm.

147 Ibid., 37–41.

148 Ibid., 40.

149 Faunce, Thomas (2010), 'Exploring the Safety of Nanoparticles in Australian Sunscreens', *International Journal of Biomedical Nanoscience and Nanotechnology*, **1** (10), 87–94 at 89.

150 New South Wales Legislative Standing Committee on State Development (2008), *Nanotechnology in New South Wales*, Sydney: New South Wales Parliament at iv.

labelling requirements for sunscreens and cosmetics include the identification of nanoscale materials'.[151]

Lack of information and data about the effects of nanoparticles is a theme in literature in this area. There are gaps in our understanding of their impacts on the human body.[152] We do not know how long nanoparticles can remain in the environment and how they might degrade over time.[153] More information is needed about the behaviour of nanoparticles, such as their interactions with cells and cellular components, and their tendency to bioaccumulate.[154] In addition to the lack of information about impacts on human health, there is also a dearth of information about how other organisms could be affected by the presence of nanoparticles in the environment.[155] Because of these uncertainties, the United Kingdom's Royal Society and Royal Academy of Engineering's 2004 report recommended interdisciplinary research in this area,[156] and that release of 'manufactured nanoparticles and nanotubes into the environment be avoided as far as possible'.[157] In 2008, the Expert Panel on Nanotechnology of the Council of Canadian Academies recommended: 'Given the current limited state of scientific knowledge regarding many nanomaterials, the panel identifies the need to give priority to the development and resourcing of a strategic research agenda to improve our understanding of the risks associated with each specific class of nanomaterials.'[158]

B Unethical Use of Nanotechnology

Many of the applications of nanotechnology research in fields such as aerodynamics, materials science and electronics also have potential military

---

151 Ibid., xxi.

152 Faunce, Thomas (2010), 'Exploring the Safety of Nanoparticles in Australian Sunscreens', *International Journal of Biomedical Nanoscience and Nanotechnology*, **1** (10), 87–94.

153 Pradeep, T. (2008), *Nano: The Essentials*, New York: McGraw-Hill, 353.

154 The Royal Society and the Royal Academy of Engineering (2004), *Nanoscience and Nanotechnologies: Opportunities and Uncertainties*, London: The Royal Society, 49, http://www.nanotec.org.uk/finalReport.htm.

155 The Royal Society and the Royal Academy of Engineering (2004), *Nanoscience and Nanotechnologies: Opportunities and Uncertainties*, London: The Royal Society, 45, http://www.nanotec.org.uk/finalReport.htm.

156 Ibid., 48–9.

157 Ibid., 50.

158 Council of Canadian Academies (2008), *Small is Different: A Science Perspective on the Regulatory Challenges of the Nanoscale,* the Expert Panel on Nanotechnology, Ottawa: Council of Canadian Academies, 1, http://www.scienceadvice.ca/en/assessments/completed/nanotechnology.aspx.

or security applications. Nanotechnology has potential military applications in sensors, detection of weapons and explosives, ballistics, electronics and information and communications technologies.[159] For example, a research group at Massachusetts Institute of Technology is investigating ways to 'develop and exploit nanotechnology to dramatically improve the survivability of Soldiers' in areas such as 'Light Weight, Multifunctional Nanostructured Fibers and Materials' and 'Blast and Ballistic Protection'.[160] Significant investment in military applications of nanotechnology is occurring around the world. For example, in 2011 the proposed nanotechnology research and development budget for the United States Department of Defense is US $348.5 million.[161] Smaller investments in this field are also occurring in the United Kingdom, India and Sweden.[162]

The military applications of nanotechnology are seen by opponents of the technology as a significant ethical concern. Nanotechnology could have both defensive and offensive applications.[163] There is also concern that some nanotechnology research has a 'dual use' character as it could be used both for weapons development and non-weapons uses. For example, the ETC Group has argued that nanocapsules, which are being researched for use as medical tools, could also be used as mechanisms to deliver chemical and biological weapons.[164]

C Objections to Patenting of Nanotechnology

The ETC Group has expressed ethical and political objections to the patenting of nanotechnology: 'Nano-scale manipulation in all its forms offers unprecedented potential for sweeping monopoly control of elements and processes that are fundamental to biological function and material

159 Nasu, H. and T. Faunce (2009/2010), 'Nanotechnology and the International Law of Weaponry: Towards International Regulation of Nano-Weapons', *Journal of Law, Information and Science*, **20**, 21–54 at 25–7.

160 Massachusetts Institute of Technology Institute for Soldier Nanotechnologies, 'Institute for Soldier Nanotechnologies', http://web.mit.edu/isn/; and Massachusetts Institute of Technology Institute for Soldier Nanotechnologies, 'Research', http://web.mit.edu/isn/research/index.html.

161 National Nanotechnology Initiative, 'Funding', http://www.nano.gov/html/about/funding.html.

162 Nasu, H. and T. Faunce (2009/2010), 'Nanotechnology and the International Law of Weaponry: Towards International Regulation of Nano-Weapons', *Journal of Law, Information and Science*, **20**, 21–54 at 22.

163 Ibid., 28–9.

164 ETC Group (2004), 'Down on the Farm: The Impact of Nano-scale Technologies on Food and Agriculture', Ottawa: ETC Group, 15, http://www.etcgroup.org/upload/publication/80/02/etc_dotfarm2004.pdf.

resources.'[165] The ETC Group fears that 'Atomtech will eclipse genetic engineering because it involves *all* matter – both living and non-living'.[166] The ETC Group expounds on the topic:

> Whereas the biotech industry had to overcome a number of legislative prohibitions (that living material was non-patentable, for example) to secure intellectual property over organisms and their component parts, Atomtechies face far fewer constraints, giving matter moguls a substantial edge over genetic engineers [...] The prohibition on product of nature patents was rendered vacuous by the 1980 Supreme Court decision. Today, with the world's largest corporations gearing up to work down at the nano-level, it is only a matter of time before industry convinces patent examiners that the genetically-engineered microbe of twenty-two years ago is no different from the atomically-engineered elements of today. Between nuclear colliders, atomic force microscopes, and cameras that can photograph light as it meanders through a retina, the nanotech industry will be in a political position to argue that any tinkering with the elemental products of nature is patent-worthy.[167]

The ETC Group objected to patents being granted in respect of nanotechnology, suggesting that it was akin to patenting elements of the Periodic Table: 'Patenting at the nano-scale can mean monopolizing the basic elements that make life possible.'[168] The civil society organization maintained that nanotechnological devices should not be conceived of as patentable subject matter because they are products of nature and scientific discoveries. The ETC group believes that a moratorium should be placed on research involving molecular self-assembly and self-replication.

The ETC Group is concerned about patenting and commercialization practices in nanotechnology, including control of the products of

---

165 ETC Group (2003) *The Big Down: From Genomes to Atoms; Atomtech: Technologies Converging at the Nano-Scale,* Ottawa: ETC Group, 8, http://www.etcgroup.org/upload/publication/171/01/thebigdown.pdf.

166 Ibid.

167 Ibid., 41.

168 ETC Group (2002), 'Patenting Elements of Nature', 25 March, http://www.etcgroup.org/upload/publication/220/01/nanopatentsgeno.rtf.pdf. In the chemical age, there was controversy as to whether patents should be issued in respect of chemical elements. Famously, Marie and Pierre Curie, the joint winners of the Nobel Prize in Physics in 1903, refused to file patent applications in relation to the isolation of the elements of radium #88 and polonium #84 because of her belief that the chemical elements belonged in the public domain. By contrast, the United States Nobel Laureate Glenn Seaborg acquired United States patent No. 3,156,523 in respect of Americium #95 on November 10, 1964. He also named his second 'invented' element, Curium #96, and received U.S. patent No. 3,161,462, which was granted December 15, 1964.

nanotechnology and access to these products. They have drawn an analogy with the perceived unethical behaviour of agricultural biotechnology companies in relation to patented genetically modified crops.[169] The ETC Group comments that 'based on current trends, atom scale technologies will further concentrate economic power in the hands of giant multinational corporations. How likely is it that the poor will benefit from the technology that is outside their control?'[170]

D Patent Law and Ethics

The relationship between patent law and ethics has been described as an 'uncomfortable' one that involves 'uncertainty and ambiguity'.[171] The role of the patent system and the proper function of patent examining bodies in dealing with issues of ethics and morality are controversial. The ethical issues discussed above could create questions over the patent validity of patent applications involving nanotechnology. For example, in Europe Article 53 (a) of the *European Patent Convention* provides that patents shall not be granted for:

> (a) inventions the commercial exploitation of which would be contrary to 'ordre public' or morality; such exploitation shall not be deemed to be so contrary merely because it is prohibited by law or regulation in some or all of the Contracting States [...].[172]

Both issues of risk to human health or the environment and potential weapons use of nanotechnology may be relevant in considering whether a claimed invention falls within this morality exclusion. The European case law that can shed light on the question of how ethical and risk issues relating to nanotechnology are likely to affect its patentability is limited.[173]

---

169 ETC Group (2004), 'Down on the Farm: The Impact of Nano-scale Technologies on Food and Agriculture', Ottawa: ETC Group, 8, http://www.etcgroup.org/upload/publication/80/02/etc_dotfarm2004.pdf.

170 ETC Group (2003) *The Big Down: From Genomes to Atoms; Atomtech: Technologies Converging at the Nano-Scale,* Ottawa: ETC Group, 3, http://www.etcgroup.org/upload/publication/171/01/thebigdown.pdf.

171 Bently, Lionel and Brad Sherman (2009), *Intellectual Property Law*, 3rd edn, Oxford: Oxford University Press, 453, 457.

172 *Convention on the Grant of European Patents* ('*European Patent Convention*') of 5 Oct 1973 Official Journal Special Edition 1/2007 (entered into force 7 October 2000).

173 See, for instance, *Harvard/OncoMouse* (1990) OJ EPO 476; *Plant Genetic Systems* (1995) EPOR 357; *Howard Florey/H2 Relaxin* (1995) EPOR 541; *Harvard/Onco-mouse* [2003] OJ EPO 473; *Harvard/Transgenic Animals*,

Applying the case law to nanotechnology inventions could, for example, involve the weighing up of the benefits of the invention to mankind with the potential risks to human health and the environment. However, as noted above, the information about these risks is limited and so the balancing exercise would be difficult.

Some commentators maintain that it is not the proper role of the patent system to address ethical, social and environmental concerns. For instance, Gregory Mandel argues that 'the struggles of the patent system to handle its core competency of reviewing patent applications for nanotechnology inventions should raise significant questions about the system's ability to handle any additional mission involving the regulation of health, safety, and environmental impacts of nanotechnology'.[174] He asserts: 'Even if the risk of nanotechnology to human health, safety, and the environment were perceived to be greater than it is currently generally considered, there are far better ways to regulate such risk than a blanket slowing and secretizing of nanotechnology development.'[175]

By contrast, Ian Kerr and Goldie Bassie contend that there is a need for a broader public policy debate over the regulation of nanotechnology: 'In the face of rapid change, it is not good enough to simply debate about what we think is and is not scientifically possible today.'[176] The pair call for an inclusive model of governance of nanotechnology: 'An approach that creates a broader nanonetwork, involving other social actors in discussions and decision-making about the future regulation of nanotechnology, would enhance legitimacy and foster public trust.'[177]

E *Sui Generis* Regimes

In the past, *sui generis* regimes have been minted to address certain emerging technologies – such as plant breeders' rights, access to genetic resources and circuit layouts. Special legislation was mooted to address

---

T315/03 [2006] OJ EPO 15; *Euthanasia Compositions/Michigan State University* (2005), T 0866/01; http://www.epo.org/law-practice/case-law-appeals/recent/t010866eu1.html; and *Stem Cells/WARF* (T 1374/04) EPO OJ 2007, 313 ('*Stem Cells/WARF*').

174 Mandel, Gregory (2011), 'Regulating Nanotechnology through Intellectual Property Rights', in Graeme Hodge, Diana Bowman and Andrew Maynard (eds), *International Handbook on Regulating Nanotechnologies*, Cheltenham and Northampton, MA: Edward Elgar, 388–407 at 393.

175 Ibid., 403.

176 Kerr, I. and G. Bassie (2004), 'Not *That* Much Room? Nanotechnology, Networks and the Politics of Dancing', *Health Law Journal*, **12**, 103–23 at 117.

177 Ibid., 122.

computer software and hardware – but, in the end, information technology received comprehensive intellectual property coverage from a combination of copyright law, patent law, trade secrets, trademarks and Internet domain names. The possibility of developing a *sui generis* regime to address nanotechnology has been discussed in academic debate. Douglas Sylvester and Diana Megan Bowman wonder whether nanotechnology will precipitate a reformulation of intellectual property law: '[Although] emerging technologies, such as biotechnology and information communication technology, have challenged the traditional patent landscape, the overarching framework and design has largely remained intact – But will this always be the case?'[178]

Siva Vaidhyanathan is somewhat more sceptical as to whether the patent regime is appropriately adapted for the field of nanotechnology: 'The dream of nanotechnology – engineering substances at the scale of one nanometer – reveals many of the dangers of an overprotective patent system.'[179] He comments: 'Paradoxically, an overprotective patent system threatens the potential benefits of a fully realized nanotechnology industry.'[180] Vaidhyanathan wonders whether a *sui generis* system of protection would be more appropriate for nanotechnology:

> Regardless, it is worth considering whether a special set of rules should apply to nanotechnology. Theoretically, the patent system is supposed to be non-discriminatory. It should operate the same way under the same principles regardless of the type of technology at hand. But in practice, different fields do work differently in the patent system. Perhaps nanotechnology would grow more equitably, efficiently, and predictably if its patents worked for a shorter time, perhaps 10 years instead of 20. Perhaps there should be a global nanotechnology patent database run through the United Nations. And perhaps the United Nations should sponsor regular conferences and reports to consider the ethical, economic, and environmental effects of nanotechnology.[181]

Vaidhyanathan expressed concerns that a 'tragedy of the anti-commons' would afflict the nanotechnology field, much as Heller and Eisenberg predicted of biomedical research.[182] He submitted that there is a need to

---

178 Sylvester, D. and D.M. Bowman (2010), 'English Garden or Tangled Grounds? Navigating the Nanotechnology Patent Landscape', 12 January, http://ssrn.com/abstract=1535650.

179 Vaidhyanathan, Siva (2006), 'Nanotechnology and the Law of Patents: A Collision Course', in Geoffrey Hunt and Michael Mehta (eds), *Nanotechnology: Risk, Ethics, Law*, London: Earthscan, 225–36.

180 Ibid., 225.

181 Ibid., 234.

182 Ibid., 232–3.

encourage greater openness in this field of science: 'The proprietary and competitive nature of the current nanotechnology community does not bode well for transparency and equity.'[183]

Dennis Karjala contends that copyright law is an inappropriate template for emerging technologies – such as nanotechnology: 'Just as the ease of copying software led us to use copyright as an anti-misappropriation scheme for computer programs, the eventual ease of copying even hardware will require a similar reconsideration of how we protect nanotech innovation under intellectual property law.'[184] He concludes: 'When biotech and nanotech reach the stage at which new functional products will be, like complex computer programs, designed by application of well understood engineering principles in a straightforward manner – costing time, skill, energy, and money but not involving much, if any, patentable invention – we should resist calls for protection under copyright and devise a scheme that more aptly addresses the underlying problem.'[185]

Mark Lemley suggests that trade secrets protection could be attractive for nanotechnology developers because 'even when nanotechnology products are released on the open market reverse engineering them may be significantly more difficult for competitors than it would be in other fields'.[186]

Gary Marchant and his colleagues are hostile to the suggestion that there should be *sui generis* regulation of nanotechnology:

> Proposals for sui generis regulation of nanotechnology products create a similar risk of irrational discrimination. If we have two products with the same functionality – one produced using nanotechnology and the other not – it is not certain or obvious that the nanotechnology version will necessarily be the riskier of the two. They may have the same risks or in some cases the nanotechnology product might even be safer.[187]

The group concludes that 'automatically treating nanotechnology as more dangerous and thus needing additional regulation will be putting a thumb

---

183 Ibid., 233.

184 Karjala, D. (2011), 'Protecting Innovation in Computer Software, Biotechnology, and Nanotechnology', *Virginia Journal of Law and Technology*, **16** (1), 42–65.

185 Ibid.

186 Lemley, M. (2005), 'Patenting Nanotechnology', *Stanford Law Review*, **58**, 601–30.

187 Marchant, G., D. Sylvester and K. Abbott (2009), 'What Does the History of Technology Regulation Teach Us About Nano Oversight?' *Journal of Law, Medicine, and Ethics*, **37**, 724–31 at 726.

on the scale against nanotechnology, deterring companies from using nanotechnology except when no other alternative is available'.[188]

In extra-judicial comments, Gajarsa J of the Court of Appeals for the Federal Circuit, and the author of the decision in *Madey* v. *Duke University*, has expressed confidence that the patent system will foster and promote nanotechnological inventions: 'I submit that these developments will come about because of our dynamic patent system that has been able over 200 years to absorb new technologies and has responded vigorously to technical advances.'[189]

CONCLUSION

This chapter has considered the manifold challenges posed by nanotechnology to patent law, policy and practice. Mark Lemley reflects that there are multiple reasons to monitor this field of technology.[190] He observes:

> Nanotechnology patents bear watching. They have characteristics that may well turn out to be fundamentally different than patents in any other industry in the last eighty years. How the market responds to these characteristics will determine whether and how the law must step in and tailor the rules of patent law to the needs of this nascent industry. It will also give us broader insight into the role of patents in enabling technologies.[191]

This chapter has considered various efforts by the USPTO, EPO and JPO to classify nanotechnology and include the field in the schema of patent administration. It has highlighted the importance of the patent offices engaging in rigorous, cross-disciplinary examination of nanotechnology patent applications. This chapter has argued that there is a need to promote access to nanotechnology – through a broad defence of experimental use, flexible compulsory licensing provisions, the deployment of patent pools, public sector licensing and meaningful technology transfer provisions. There is a need to consider the regulation of nanotechnology in a larger context of policy discussions over public health, environmental protection, and military peace.

---

188 Ibid., 726.

189 Gajarsa, A. (2002) 'Quo Vadis? The Fifth Annual Honorable Helen Wilson Nies Memorial Lecture in Intellectual Property Law', *Marquette Intellectual Property Law Review*, **6**, 1–9 at 8–9.

190 Lemley, M. (2005), 'Patenting Nanotechnology', *Stanford Law Review*, **58,** 601–30.

191 Ibid., 630.

PART V

Biodiversity, food security and climate change

11. Patenting the Kakadu plum and the Marjarla tree: biodiscovery, intellectual property and Indigenous knowledge

Sarah Holcombe and Terri Janke

The Australian Government's 2009 endorsement of the *United Nations Declaration on the Rights of Indigenous Peoples* 2007 brings a renewed recognition and focus on Indigenous knowledge ownership in the area of Indigenous cultural and intellectual property.[1] However, embedding this rights discourse into the action and language of natural resource management has not been taken up by the Australian Government, as yet. Australia's federal system, under the *Environment Protection and Biodiversity Conservation Act* 1999 (Cth) is ad hoc, with several states and territories creating their own biological resources acts and regional groups their own management processes. The Government-commissioned Hawke Inquiry into the *Environment Protection and Biodiversity Conservation Act* 1999 (Cth) found that 'each jurisdiction has different rules and requirements for accessing biological resources' and that the 'Nationally Consistent Approach' policy developed in 2000 'should be reinvigorated'.[2]

Prior to this Inquiry, in 2008, both of the authors were engaged by the Natural Resource Management Board of the Northern Territory to develop resources and tools to ensure that when ethnobiological knowledge is integrated into resource management programs and in research generally in that jurisdiction it is done ethically and equitably. This chapter will discuss several of the challenges we encountered, which accord with the

---

1 *United Nations Declaration on the Rights of Indigenous Peoples* 2007, adopted by General Assembly Resolution 61/295 on 13 September 2007

2 Hawke, Allan et al. (2009), *The Australian Environment Act – A Report of the Independent Review of the Environment Protection and Biodiversity Conservation Act 1999*, Canberra: Commonwealth of Australia, 236, http://www.environment.gov.au/epbc/review/publications/final-report.html.

findings of the recent Hawke Inquiry, and some of the recommendations we made for change, which went beyond the Inquiry's findings. To shed light on the issues we provide several case studies of the management of traditional knowledge in relation to biological resources, such as the Mary Kay Kakadu Plum Patent Opposition[3] and the Marjarla patent.[4] Finally we also provide some critical comment on the recent *Nagoya Protocol* 2010 and find common ground with the growing interest in the development of bio-cultural community protocols to assert local Indigenous priorities using a rights-based approach.

First, we will briefly provide some background. Australia, along with 192 other nation states, is a signatory to the *Convention on Biological Diversity* 1992.[5] To satisfy its obligations under the *Convention on Biological Diversity* 1992, the Australian Government implemented the *Environment Protection and Biodiversity Conservation Act* 1999 (Cth), mentioned earlier, and associated Regulations (2000 and 2005). However, this Act only applies to Commonwealth areas, such as national parks and reserves such as Ramsar wetlands and world heritage areas[6] and to issues of 'national environmental significance'.[7] So in terms of land mass it has extremely limited application. Critics of the Act also maintain that the regime represents a minimalist implementation of the *Convention on Biological Diversity* 1992. For instance, that the laws and regulations do not cover biological resources that are publicly available on a commercial basis is a significant legal loophole.[8] Likewise, as Robinson notes, many plant tissues,

---

3 Mary Kay Inc. et al. (2006), 'Compositions Comprising Kakadu Plum Extract or Acai Berry Extract', World Intellectual Property Organization Patent No. WO/2007/084998 filed on 19 January 2007, priority date 19 January 2006 possible owing to a United States patent; and Jones, B., D. Gan, J. Aravena and M. Hines (2007), 'Compositions Comprising Kakadu Plum Extract or Acai Berry Extract', Australian Patent Application No. 2007205838.

4 Griffith University and Jarlmadangah Burru Aboriginal Corporation (2008), 'Novel Compounds', Australian Patent Application No. 2004293125.

5 *Convention on Biological Diversity* 1992, opened for signature 5 June 1992, 1760 UNTS 143 (entered into force 29 December 1993).

6 Janke, Terri (2009), *Indigenous Ecological Knowledge and Natural Resources in the Northern Territory: Report on the Current Status of Indigenous Intellectual Property*, Darwin: Natural Resource Management Board NT, 82.

7 Australian Department of Sustainability, Environment, Water, Population and Communities, *Environment Protection and Biodiversity Conservation Act 1999: Frequently Asked Questions*, http://www.environment.gov.au/epbc/publications/epbc-act-fact-sheet.html.

8 We are also not the first to critique the Australian *Environment Protection and Biodiversity Conservation Act* 1999 (Cth): a Commonwealth Public Inquiry that was held in relation to 'Access to Biological Resources in Commonwealth

seeds or genetic materials are commercially available from nurseries. From a biodiscovery perspective this is problematic, as researchers can purchase biological materials without obtaining permits and then take them away for further research. If they are taken offshore it is likely to be very difficult to establish retrospective benefit-sharing arrangements.[9] This is likely to have been the case in the United States company Mary Kay's use of the Kakadu plum in cosmetics in their current pursuit of a patent, which will be discussed further as a case study.

Under the Australian federal system, the Northern Territory and Queensland have developed their own biological resources acts, while Western Australia and Tasmania are in the process of doing so. Queensland passed the *Biodiscovery Act* 2004 (Qld) and the Northern Territory the *Biological Resources Act* 2006 (NT). The Queensland Act makes no specific mention of Indigenous people or traditional knowledge in relation to biodiscovery. The decision-making process in sections 33–41 sets out a process whereby the state enters into a benefit-sharing agreement with the biodiscovery entity. The decision-making process provides no opportunities for community participation and consultation (including for Indigenous communities). Whilst the Queensland Biotechnology Code of Ethics encourages consultation and benefit sharing with Indigenous knowledge holders, the Code applies only where the biodiscovery entity gets funding or

---

Areas'. In Voumard, John (2000), *Commonwealth Public Inquiry Into Access To Biological Resources In Commonwealth Areas*, Canberra: Environment Australia, http://www.ea.gov.au/biodiversity/science/access/inquiry/index.html, Fourmile noted that it is of direct relevance to the Hawke Interim Report and is important to revisit. Of particular value is 'While the EPBC Act addresses the important provisions contained in Articles 8(j), 10(c) and 18.4 of the *Convention on Biological Diversity*, it falls short of providing intellectual property-style protection for community-held traditional knowledge. While prior informed consent procedures and contractual provisions can give a degree of legal certainty to protecting traditional knowledge, recognition of such knowledge as intellectual property will provide a higher degree of certainty for all parties and attract greater recognition in court proceedings.' See Fourmile, Henrietta (2000), 'Indigenous Interests in Biological Resources in Commonwealth Areas', in John Voumard, *Commonwealth Public Inquiry Into Access To BiologicalResources In Commonwealth Areas*, Canberra: Environment Australia http://www.ea.gov.au/biodiversity/science/access/inquiry/index.html, 24.

9 Robinson, D.F. (2010), 'Traditional Knowledge and Biological Product Derivative Patents: Benefit Sharing and Patent Issues relating to Camu Camu, Kakadu Plum and Acai Plant Extracts', Traditional Knowledge Bulletin – Topical Issues Series, United Nations University, Institute of Advanced Studies, 9, http://www.unutki.org/news.php?doc_id=174.

support from the Queensland Government.[10] The *Biological Resources Act* 2006 (NT) does acknowledge the rights of Indigenous knowledge holders and the requirement for benefit-sharing agreements to be negotiated over the commercial use of this knowledge. However, one of the shortfalls of the Act is that knowledge obtained from a published source, such as a scientific journal, rather than directly from an Indigenous person, is not considered 'Indigenous knowledge' under Section 29(2)(b) of the Act.[11]

These state and territory Acts, as with the *Environment Protection and Biodiversity Conservation Act* 1999 (Cth), demand that access and benefit-sharing agreements be negotiated between what they refer to as the access provider and the access party. As at late 2009 there were no permits on the Commonwealth register in respect of commercial benefit sharing in Australia. In the Northern Territory, under the *Biological Resources Act* 2006 (NT), of the approximately 45 commercial benefit sharing deeds entered between the Northern Territory and biodiscovery groups, only 3 have been completed with Indigenous parties. This is surprising, given that almost half of the Northern Territory is Aboriginal Freehold Title. This would suggest a significant level of non-compliance with the regime by the large number of public and private bioprospecting entities in Australia. In practice it also means that the regime is not generating commercial benefit to be shared with Indigenous communities and groups.

The Australian Government approach taken to genetic resources is encapsulated in 'cooperative federalism' and the principle of subsidiarity; in other words a hands-off approach. In 2002, a ministerial council agreed to the non-binding *Nationally Consistent Approach for Access to and Utilisation of Australia's Native Genetic and Biochemical Resources*, referred to as the *Nationally Consistent Approach*.[12] As an intergovernmental (Commonwealth with states and territories) approach, the agreement requires that all parties 'recognise the need to ensure the use of traditional knowledge is undertaken with the cooperation and approval of the holders of that knowledge and on mutually agreed terms'.[13] However, the difficulty

---

10 Janke, Terri (2009), *Indigenous Ecological Knowledge and Natural Resources in the Northern Territory: Report on the Current Status of Indigenous Intellectual Property*, Darwin: Natural Resource Management Board NT, 90.

11 Ibid., 89.

12 Natural Resource Management Ministerial Council (2002), *Nationally Consistent Approach for the Access to and Utilisation of Australia's Native Genetic and Biochemical Resources*, Canberra: Department of the Environment and Heritage, http://www.environment.gov.au/biodiversity/publications/access/nca/pubs/nca.pdf.

13 Ibid., 5.

in implementing this approach was noted by the government in a submission to a working group of the *Convention on Biological Diversity* 1992 in relation to access and benefit sharing, when they commented that 'countries with federal structures of government like Australia face very specific challenges when introducing national access laws'.[14] These 'challenges' are highlighted by the different approaches taken to managing Indigenous knowledge in the various state and territory Acts.

I THE INDIGENOUS ECOLOGICAL KNOWLEDGE MANAGEMENT RESOURCES

And so in this context the authors were engaged by the Northern Territory Natural Resources Management Board in 2008/2009 to develop three resources for three different audiences,[15] under their program brief: 'Indigenous Cultural and Intellectual Property Protocols for Indigenous Ecological Knowledge Management'. The purpose of this program was to

> develop guiding principles, practical advice and a clear statement of the rights and obligations of natural resource management practitioners in respect of Indigenous Intellectual Property rights in a Northern Territory context.

Unlike other states and territories, there is only 1 Natural Resources Management body for the whole of the Northern Territory, while there are almost 60 different regions and associated Natural Resources Management bodies across the country (16 in Queensland alone). So the resources we developed have *potentially* a territory wide application. Aboriginal people in the Northern Territory are also fortunate as approximately two-thirds of the land is now Aboriginal inalienable freehold title under the *Land Rights Act* 1976 (NT) and permits are required to access this land. The permit process for research requires that the proponents outline in detail the potential value of the research for Indigenous land holders and any benefit-sharing arrangements negotiated.[16] The principle of gaining free

---

14 The Australian Government (2005), 'Submission to the *Ad Hoc Open-ended Working Group on Access and Benefit-sharing* under the *Convention on Biological Diversity*', UNEP/CBD/WG-ABS/3/INF/1/Add.1, 25 January.

15 Michael Davis and Peter Veth were also on the project team.

16 Central Land Council, Project Principles and Special Purpose Permits, http://www.clc.org.au/onlinepermits/specialpurposeprinciples.aspx elaborates on the project principles for special purpose permits that should be followed when working with Aboriginal people throughout the central land council region, whether or not legally ALRA land. See also Central Land Council, Special

prior and informed consent is embedded in the process.[17] So the process of gaining consent and negotiating benefit-sharing agreements for accessing biological resources is ideally stronger in this jurisdiction than it would be in Commonwealth areas under the *Environment Protection and Biodiversity Conservation Act* 1999 (Cth).

The three resources developed included:

1) *Guidelines for Indigenous Ecological Knowledge Management: including repatriation and archiving*[18]

2) *Maintain and Strengthen Your Culture: Handbook for Working with Indigenous Ecological Knowledge and Intellectual Property*[19]

3) *Report on the current status of Indigenous Ecological Knowledge and Intellectual Property in the Northern Territory*[20]

We will not discuss the content of these resources in any detail. However, a central component of them was that they included the language of the *United Nations Declaration on the Rights of Indigenous Peoples* 2007 and a central plank of them was ensuring recognition of Indigenous cultural and intellectual property through practical advice.

Purpose Permit Application Form, http://www.clc.org.au/OnlinePermits/ApplicantDetails.aspx?0=Special.

17 For several Northern Territory case studies, see Holcombe, S. and N. Gould (2010), 'A Preliminary Review of Ethics Resources, with a Particular Focus on those Available on-line from Indigenous Organisations in WA, NT and Qld', *Australian Aboriginal Studies Journal*, No. 2., 107–25.

18 Holcombe, Sarah (2009), *Indigenous Ecological Knowledge and Natural Resources in the Northern Territory: Guidelines for Indigenous Ecological Knowledge Management (including Archiving and Repatriation)*, Darwin: Natural Resource Management Board NT, 12, www.nrmbnt.org.au/files/iek/IEK%20&%20NRM%20NT%20Guidelines%20final.pdf.

19 Davis, Michael (2009), *Maintain and Strengthen your Culture: Handbook for Working with Indigenous Ecological Knowledge and Intellectual Property*, commissioned by the Natural Resource Management Board NT, Darwin, www.nrmbnt.org.au/files/iek/Handbook%20for%20Working%20with%20IEK%20&%20IP.pdf.

20 Janke, Terri (2009), *Indigenous Ecological Knowledge and Natural Resources in the Northern Territory: Report on the Current Status of Indigenous Intellectual Property*, Darwin: Natural Resource Management Board NT.

A What is Indigenous Cultural and Intellectual Property?

This concept, as we utilised it within the resources above, refers to all the elements that make up Indigenous people's cultural heritage. It includes genetic materials, cultural objects and languages, songs, ceremonies, photographs, sound recordings, films and written reports (including archival and historical documents) as well as Aboriginal knowledge systems. In other words, it comprises all the tangible *and* intangible elements of Aboriginal culture.[21] Because it specifically incorporates the intellectual property regime it attempts to capture the interface of these different knowledge management systems; of Indigenous knowledge management paradigms (local, particular) as well as the global intellectual property system. The concept of Indigenous cultural and intellectual property recognises that this is the space within which all research operates; commercial and non-commercial. We cannot avoid the intellectual property system.

B Northern Territory Indigenous Ecological Knowledge Resources (2008): Principles and Issues

Although we sought transferability of principles and approach across all land tenures in the Northern Territory, we did not want the resources to appear overly proscriptive. Nevertheless, a key issue in ensuring that these resources have legitimacy is their enforceability, as well as their general take-up. Part of the brief was 'analysis of any required legislation, research and capacity needed to protect Indigenous Intellectual Property', to 'identify challenges and barriers to the development of Intellectual Property policy' and to 'determine the capacity and will of government and institutions to enforce protection of Indigenous Intellectual Property'. This was a significant task and we took a multi-pronged approach to answering it.

In relation to the protection and recognition of Indigenous cultural and intellectual property, we found, as anticipated, that like elsewhere in Australia there were no regulatory mechanisms, laws or policies that specifically provide rights to Indigenous peoples over their Indigenous cultural and intellectual property. The *Biological Resources Act* 2006 (NT) goes part way to recognising Indigenous cultural and intellectual property rights. However, it is likely to be applied unevenly, depending on whether or not the

---

[21] Holcombe, Sarah (2009), *Indigenous Ecological Knowledge and Natural Resources in the Northern Territory: Guidelines for Indigenous Ecological Knowledge Management (including Archiving and Repatriation)*, Darwin: Natural Resource Management Board NT, 12, www.nrmbnt.org.au/files/iek/IEK%20&%20NRM%20NT%20Guidelines%20final.pdf.

resource is on Aboriginal land. Moreover, it is contingent upon whether expert advice (including legal advice) is available to negotiate a benefit-sharing agreement at the time that access is granted.[22]

There was also very little awareness amongst the groups that we worked with that there was a specific *Biological Resources Act* 2006 (NT). And at the time of our research to develop the resources, no such benefit-sharing agreements had been entered into.[23] However, ultimately, we noted that the strongest protection for Indigenous cultural and intellectual property lies in the development of national *sui generis* law (that is, law 'of its own kind') that recognises the unique qualities of Indigenous cultural and intellectual property. It seems to us that the Northern Territory Natural Resource Management Board along with the other state and territory Natural Resource Management Boards could lobby the Federal Government to this effect.

However, we note that in a 2009 submission on 'Development of elements of *sui generis* systems for the Protection of Traditional Knowledge, innovations and practices', the Australian Government stated that it:

> Support[s] cooperative, non-binding measures to enhance national mechanisms for the respect, preservation and maintenance of traditional knowledge, innovations and practices of Indigenous and local communities in accordance with Article 8(j) [and …] encourage[s] the Working Group to further discuss and explore non-intellectual property based elements of *sui generis* systems.[24]

The issues of concern here are the 'non-binding measures' and 'non-intellectual property based elements of *sui generis* systems'. It seems to us that this stance, on this international level, also reflects the Australian Government's ad hoc and weak domestic approach. We return to these obvious limitations later in a discussion of the new *Nagoya Protocol* 2010.[25]

Nevertheless, in this same submission, the Australian Government did indicate that it was supportive of the 'development of guidelines for

---

22 Janke, Terri (2009), *Indigenous Ecological Knowledge and Natural Resources in the Northern Territory: Report on the Current Status of Indigenous Intellectual Property*, Darwin: Natural Resource Management Board NT, 10.

23 Ibid., 89.

24 The Australian Government (2009), 'Submission on Development of Elements of sui generis Systems of the Protection of Traditional Knowledge, Innovations and Practices', Ad Hoc Open-ended Inter-sessional Working Group on Article 8(j) and Related Provisions of the *Convention on Biological Diversity*, Sixth meeting, Montreal 2–6 November, 2.

25 *Nagoya Protocol on Access to Genetic Resources and the Fair and Equitable Sharing of Benefits Arising From Their Utilization to the Convention on Biological Diversity*, COP 10 Decision ('*Nagoya Protocol* 2010').

national measures for the respect, preservation, maintenance of traditional knowledge, innovations and practices of Indigenous and local communities relevant to the conservation and sustainable use of biodiversity' [per Article 8(j)].[26] They then proceeded to outline the range of federal government programs which 'support the recording, storage and transfer of traditional ecological and cultural knowledge'.[27] Although supportive of these programs, it seems to us that without a systematic approach to ethically managing the knowledge that emerges from these program activities – as this also affords protection of Indigenous cultural and intellectual property – and as this ensures an Australia-wide approach – such program outcomes are not robust. This is not only because they tend to be limited to those groups fortunate enough to reside on biodiverse rich Aboriginal land, but such programs are usually time-limited.[28]

Likewise, in the Indigenous Ecological Knowledge Project findings, in terms of the management of Indigenous cultural and intellectual property, we found in the funding agreements between the Commonwealth and regional Indigenous Natural Resources Management bodies, that although the intellectual property was vested in the funding recipient (which was not always the traditional knowledge holders), the Government required an automatic non-exclusive, worldwide, perpetual and free licence to any material. This included the restructured nationwide Caring for Our Country Program (from where most of all Indigenous Natural Resource Management funding derives), as well as the previous National Heritage Trust funding. Thus, we noted that it was crucial for mutually negotiated and agreed contracts to recognise the Indigenous cultural and intellectual property that may emerge as a result of the Natural Resource Management work carried out under the contract. It seemed ironic that in the 2008 contracts between Indigenous groups and the Northern Territory Natural Resources Management Board to undertake work under the Indigenous Ecological Knowledge Program banner, this issue of the funding body requiring an automatic licence to use any resultant intellectual property still remained as a standard clause.

---

26 The Australian Government (2009), 'Submission on Development of Elements of sui generis Systems of the Protection of Traditional Knowledge, Innovations and Practices', Ad Hoc Open-ended Inter-sessional Working Group on Article 8(j) and Related Provisions of the *Convention on Biological Diversity*, Sixth meeting, Montreal 2–6 November, 2.

27 Ibid.

28 See Altman, J. and P. Whitehead (2003), 'Caring for Country and Sustainable Indigenous Economic Development: Opportunities, Constraints and Innovation', *CAEPR Working Paper* 20/2003, http://caepr.anu.edu.au/Publications/WP/2003WP20.php.

We noted that currently the strongest mechanisms for protecting Indigenous cultural and intellectual property in commercial and non-commercial research are the use of regional or local protocols – such as those we developed and as these include Indigenous cultural and intellectual property rights – *and* their enforcement by contract. In developing the resources, we were conscious of the need to respect the range of local protocols and knowledge management tools that already exists within some Indigenous organisations; such as representative bodies and Indigenous land management corporations. Enabling the principles within the *United Nations Declaration on the Rights of Indigenous Peoples* 2007 can only be achieved at a fundamentally local level, as ultimately the most important principle around which all of the others revolve is that of free and prior informed consent. Effective free and prior informed consent necessitates a two-way collaborative and ongoing dialogue that should enhance local self-determination.

One of the resources developed specifically targeted this issue of free and prior informed consent: the plain English Handbook *Maintain and Strengthen your Culture: Handbook for Working with Indigenous Ecological Knowledge and Intellectual Property* was developed for Indigenous Natural Resource Management practitioners – such as rangers.[29] This resource is useful for ensuring that local Indigenous knowledge holders can know what their rights and responsibilities are, to then enable assertion of them. Such an assertion has to start locally and the growing international movement toward developing bio-cultural community protocols that define customary rights and interests in biological resources is, we argue, a potentially valuable path forward.[30] One of the reasons such an approach is even more compelling for Indigenous peoples currently, in Australia, as for Africa or Asia, is the adoption in 2010 by the parties to the *Convention on Biological Diversity* 1992 of the *Nagoya Protocol* 2010.

---

[29] Davis, Michael (2009), *Maintain and Strengthen your Culture: Handbook for Working with Indigenous Ecological Knowledge and Intellectual Property*, commissioned by the Natural Resource Management Board NT, Darwin, www.nrmbnt.org.au/files/iek/Handbook%20for%20Working%20with%20IEK%20&%20IP.pdf.

[30] See also: Natural Justice (2009), *Bio-cultural Community Protocols: A Community Approach to Ensuring the Integrity of Environmental Law and Policy.* Nairobi: United Nations Environment Programme (UNEP) and Natural Justice; and Bavikatte, K. and H. Jonas (2010), 'Bio-cultural Community Protocols Enforce Biodiversity Benefits: A Selection of Cases and Experiences', *Endogenous Development Magazine*, **6**, 4–6, http://www.compasnet.org/afbeeldingen/Magazines/ED%20Magazine%206/Magazine_ED6.html.

The *Nagoya Protocol* 2010 has recently been adopted by the members of the *Convention on Biological Diversity* 1992 and all signatory countries including Australia have a year until February 2012 to respond and sign up.[31] However, the status of the Protocol is somewhat contested and it requires the ratification of 50 states to enter into force.[32] But, given that it is part of the *Convention on Biological Diversity* 1992, it would be binding, unlike the *Bonn Guidelines* for instance. However, the obligations do not go much further than the *Convention on Biological Diversity* 1992 – so it is not going to bind countries to much higher standards. The access to genetic resources and benefit-sharing framework within the *Convention on Biological Diversity* 1992 – including the recent *Nagoya Protocol* 2010 – recognises (the language is 'reaffirms') the sovereign rights of states over natural resources and the authority of states to decide over the use of genetic resources. Although the principle of national sovereignty is important in promoting equitable benefit sharing between countries, as Swiderska notes, it is generally interpreted as government *ownership*, with the rights of other actors, notably Indigenous and local communities, often unclear or unrecognised.[33]

Likewise, the emphasis in the *Nagoya Protocol* 2010 is largely on protecting only the intangible or intellectual aspects of traditional knowledge systems, abstracted from the cultural, biological and customary law context that sustains them.[34] The language of 'genetic resources' in the *Nagoya Protocol* 2010, rather than the broader concept of biological resources, speaks to this focus on tangible and isolatable 'products'. For our purposes here the term 'biological resources' is a 'broader category that also includes genetic resources, but also organisms or parts thereof, populations, or any other biotic component of ecosystems with actual or potential use or value for humanity'.[35]

---

31 *Nagoya Protocol on Access to Genetic Resources and the Fair and Equitable Sharing of Benefits Arising From Their Utilization to the Convention on Biological Diversity*, COP 10 Decision ('*Nagoya Protocol* 2010').

32 United Nations (2011), 'UN Protocol on Sharing the World's Genetic Wealth Opens for Signing', Press Release, UN News Centre, 2 February, http://www.un.org/apps/news/story.asp?NewsID=37453&Cr=&Cr1.

33 Swiderska, Krystyna (2007), 'Protecting Traditional Knowledge: A Framework Based on Customary Laws and Bio-cultural Heritage', in Bertus Haverkort and Stephen Rist (2007), *Endogenous Development and Bio-Cultural Diversity: The Interplay of Worldviews, Globalisation and Locality*, Leusden, Netherlands and Bern, Switzerland: Compas and Centre for Development and Environment, 358–65 at 361.

34 Ibid., 359.

35 Secretariat of the Convention on Biological Diversity (2008), *Access and Benefit-sharing in Practice: Trends in Partnerships Across Sectors,* Montreal: CBD Technical series No. 38, 8.

Access to genetic resources and benefit sharing is being driven by intellectual property regimes such as patents and plant breeders' rights or plant variety protection. As Swiderska notes, intellectual property regimes are becoming increasingly ubiquitous as a result of trade agreements of the World Trade Organization and the proliferation of bilateral free trade agreements.[36] This is accelerating the commercial use and privatisation of Indigenous knowledge and resources. Yet, the *Convention on Biological Diversity* 1992 itself in Article 8(j) – in relation to *in situ conservation* – takes a more holistic approach in its emphasis on respecting, preserving and maintaining knowledge innovations and practices of Indigenous and local communities embodying traditional lifestyles relevant for conservation and sustainable use. Thus, it seems to us that the access to genetic resources and benefit-sharing systems – with its focus on intellectual property and commercialisation – downgrades this more complex reality of Indigenous knowledge as it is embedded in and guided by the customary knowledge economy.

In a submission to the secretariat of the *Convention on Biological Diversity* 1992, a group of African countries noted that formal 'monetary and non-monetary benefits to Indigenous Local Communities (ILCs) in exchange for the use of their traditional knowledge, within which benefit-sharing agreements must be negotiated should not be the sole aim of benefit-sharing agreements'.[37] The process and the outcome of a benefit-sharing agreement between Indigenous local communities and the relevant stakeholders must affirm aspects of their traditional lifestyles that conserve and sustainably use biological diversity. They go on to outline the way in which a community protocol would operate in this space:

> a community protocol is an outlining of ecological values on which prior informed consent (PIC), mutually agreed terms (MAT) and benefit sharing would be based. A useful analogy for a community protocol would be the 'bill of rights' in the constitution of a country that lists the core values of a people. It enunciates a community's core values and while it remains a flexible instrument,

---

[36] Swiderska, Krystyna (2007), 'Protecting Traditional Knowledge: A Framework Based on Customary Laws and Bio-cultural Heritage', in Bertus Haverkort and Stephen Rist (eds), *Endogenous Development and Bio-Cultural Diversity: the Interplay of Worldviews, Globalisation and Locality*, Leusden, Netherlands and Bern, Switzerland: Compas and Centre for Development and Environment, 358–65 at 358.

[37] Natural Justice (2009), 'Submission Information and Views in preparation for the meeting of the Expert Group on Traditional Knowledge associated with Genetic Resources. Answers to Questions posed to the Expert Group on Traditional Knowledge associated with GR as specified in COP decision IX/12', 9.

> it provides community members and outside interests a level of certainty about the principles about which an ABS agreement will be negotiated. Community protocols are perhaps the best chance for ILCs to ensure that their way of life and values are respected and promoted. Merely relying on the benefits of an ABS agreement without affirming their 'ecological values' would reduce ILCs to sellers of TK who warm themselves on the embers of a lifestyle that is fast dying out.[38]

It seems to us that this concept builds upon that put forward by Posey and Dutfield in 1996, of 'Traditional Resource Rights', which also incorporated the concept of 'community Intellectual Property Rights'.[39] Indeed, they state that 'Traditional Resource Rights go beyond other sui generis models in that they seek not only to protect knowledge relating to biological resources but also to assert the right of peoples to self determination and the right to safeguard "culture" in its broadest sense'.[40] Bio-cultural community protocols and the affiliated concept of the 'Traditional Knowledge Commons' – a model for regulating non-commercial use of traditional knowledge – further defines the range of local possibilities for developing frameworks of control over access to genetic resources and benefit sharing.[41]

II MARY KAY KAKADU PLUM OPPOSITION CASE

In order to highlight the issues associated with intellectual property and biodiscovery – particularly those around in situ conservation – we will now discuss the Mary Kay Kakadu plum patent case, which has caused some concerns for Indigenous people in the north of Australia.

---

38 Ibid.

39 Posey, Darrell and Graham Dutfield (1996), *Beyond Intellectual Property: Toward Traditional Resource Rights for Indigenous Peoples and Local Communities*, Ottawa: International Development Research Centre, 95–100.

40 Ibid., 95.

41 Natural Justice (2009), *Bio-cultural Community Protocols: A Community Approach to Ensuring the Integrity of Environmental Law and Policy*, Nairobi: United Nations Environment Programme (UNEP) and Natural Justice; and Natural Justice (2009), *Imagining a Traditional Knowledge Commons: A Community Approach to Sharing Knowledge for Non-commercial Research*, Rome: International Development Law Organisation.

A About the Fruit

The 'Kakadu plum' (*Terminalia ferdinandiana*) is a small fruit-bearing tree found in the north of Australia, in the Northern Territory and Western Australia. It is known by Indigenous people in the Northern Territory as the 'Billy Goat plum'.[42] Western Australian Indigenous groups refer to it as the 'gubinge'.[43] The fruit has long been used by Northern Territory and Western Australian Indigenous groups for food and medicine, including use as an antiseptic.

The plum has recently come under the spotlight because of its remarkably high concentration of vitamin C and ascorbic acid. The plum has been used in a range of products, particularly food products such as jams and teas but also health drinks. The commercial benefits for Indigenous communities was recognised in the 2006 report *Small-scale Commercial Plant Harvests by Indigenous Communities,* where the demand for the fruit was calculated at 10 to 12 tonnes per annum, with an estimated return of roughly $10 per kilo.[44] That report noted that, although in its infancy, the Kakadu plum industry has the potential to develop into a sustainable business opportunity for certain well positioned Indigenous communities. Development of such industries should be encouraged as a means of generating economic and social benefits for Indigenous communities. The connection of Indigenous people and Indigenous ecological knowledge to bush products of this nature is in fact a selling point and value adding within the community can significantly increase returns.[45] The fact that the plant is spread across a significant region of the Top End in reality can mean that there is likely to be a range of communities that are potentially in strategic locations for market access.

A number of Indigenous businesses have developed a link into the supply chain. For example, in Western Australia, Bruno 'Winnawal' Dann harvests gubinge in the bush and supplies 'Loving Earth' with the raw fruit in order

---

42 Janke, Terri (2009), *Indigenous Ecological Knowledge and Natural Resources in the Northern Territory: Report on the Current Status of Indigenous Intellectual Property*, Darwin: Natural Resource Management Board NT, 149: Northern Territory Natural Resource Management Board consultations, 2008.

43 Pure Well-being, www.purewellbeing.com, a United Kingdom company, attributes 'gubinge' as a word from the Nyul Nyul language name for the Kimberley version of the Kakadu Plum, which is also called the Billygoat Plum or Murunga.

44 Gorman, Julian and Peter Whitehead (2006), *Small-scale Commercial Plant Harvests by Indigenous Communities*, Canberra: Rural Industries Research and Development Corporation, 17.

45 Ibid.

for it to be dried and processed into a powder for sale.[46] According to the website, Dann is returning the traditional harvesting and growing practices to his community, and all profits from the sale of the gubinge are sent back to the Nyul Nyul community. How the monies are distributed is, itself, an interesting ethnographic question, which has been explored elsewhere for other non-timber forest products.[47] The 'bush harvest' of gubinge allows Indigenous people to undertake economic exploitation, once a licence is obtained from the Department of Environment and Conservation in Western Australia.[48] It is reported that the fruit sells for $25 to $30 a kilo.[49]

B About the Patent

In 2007, the US cosmetics company Mary Kay Inc applied for an international patent with the World Intellectual Property Organization (WIPO) under the *Patent Cooperation Treaty* 1970 for 'compositions comprising Kakadu plum extract or acai berry extract'.[50] The international application status report prepared in July 2008 considered that a great deal of the patents' claims were obvious, and could be argued to lack an inventive step. The applicant was asked to come back with further clarifications; and now each country patent office (listed in the world patent application) is considering the obviousness of the patent.

The application is currently being considered by the Australian Patents Office at IP Australia.[51] To be patentable, the Mary Kay application patent

---

46 Pure Wellbeing, http://www.purewellbeing.com/contents/en-uk/d180.html.

47 Wynberg, Rachel, Doris Schroeder, Samantha Williams and Saskia Vermeylen (2009), 'Sharing Benefits Fairly: Decision Making and Governance', in Rachel Wynberg, Doris Schroder and Roger Chennells (eds), *Indigenous Peoples, Consent and Benefit Sharing: Lessons from the San Hoodia Case*, Dordrecht: Springer, 231–60; and Wynberg, R. and S. Laird (2007), '"Less is Often More" Governance of a Non-timber Forest Product, Marula (Sclerocarya birrea subsp. Caffra) in Southern Africa', *International Forestry Review*, **9** (1), 475–89.

48 Department of Agriculture and Food (2009), 'Picking Gubinge – the Right Way', http://wn.com/Kakadu_plum.

49 Bran, M. (2010), 'Gas Hub Threatens Gubinge: Traditional Owner', ABC Rural, 24 December, http://www.abc.net.au/rural/nt/content/201012/s3101215.htm?site=kimberley.

50 Mary Kay Inc. et al. (2006), 'Compositions Comprising Kakadu Plum Extract or Acai Berry Extract', World Intellectual Property Organization Patent No. WO/2007/084998 filed on 19 January 2007, priority date 19 January 2006 possible owing to a United States patent.

51 Jones, B., D. Gan, J. Aravena and M. Hines (2007), 'Compositions Comprising Kakadu Plum Extract or Acai Berry Extract', Australian Patent Application No. 2007205838.

must show novelty and involve an inventive step. To assess this question, the invention is compared with the prior art base.[52] The prior art base is likely to contain traditional remedies, and the healing properties of the Kakadu plum are already published and publicly available.[53] But the question for the patent examiner is whether the application of the plum extract to cosmetics is obvious. According to a spokesperson of Mary Kay, the plum has never been used for cosmetic formulation before.[54] Robinson notes that there appears to be some use of the Kakadu plum in cosmetics such as by the company Outback Earth in the 1990s and Lilypilly Australia, but these products are not the subject of a patent.[55] Robinson has filed his article as a Section 27 submission to the Patents Office, which allows third parties to make statements regarding a patent's obviousness claim. Further, the Gundjeihmi Aboriginal Corporation, representative organisation of the Mirrar people of Kakadu National Park, says that the plum has been used by the Mirrar as a food and medicine for 'longer than anyone can remember'.[56] The Australian Patents Office of IP Australia issued its first examination report in December 2010. The carefully considered response poses questions of novelty that will require clarification before the patent will be accepted.

---

52 Section 7(1) of the *Patents Act* 1990 (Cth).

53 Robinson notes that the plant has been documented by Gorman, Woods and Brand. Robinson, D.F. (2010), 'Traditional Knowledge and Biological Product Derivative Patents: Benefit Sharing and Patent Issues relating to Camu Camu, Kakadu Plum and Acai Plant Extracts', Traditional Knowledge Bulletin – Topical Issues Series, United Nations University, Institute of Advanced Studies, http://www.unutki.org/news.php?doc_id=174; Gorman, J.T., A.D. Griffiths and P.J. Whitehead (2006), 'An Analysis of the Use of Plant Products for Commerce in Remote Aboriginal Communities of Northern Australia', *Economic Botany*, **60** (4), 362–73; Woods, B. (1995), 'A Study of the Intra-specific Variations and Commercial Potential of *Terminalia ferdinandiana (Excell)* (the Kakadu Plum)', M.Sc. Thesis, Northern Territory University, Darwin; and Brand, J.C., C. Rae, J. McDonnell, A. Lee, V. Cherikoff and A.S. Truswell (1982), 'The Nutritional Composition of Australian Aboriginal Bushfoods', *Food Technology in Australia*, **35**, 293–8.

54 A spokesperson of Mary Kay as cited in Powell, R. and L. Murdoch (2010), 'Patent Fight Erupts over Kakadu Plum', *The Sydney Morning Herald*, 4 December, 11.

55 Robinson, D.F. (2010), 'Traditional Knowledge and Biological Product Derivative Patents: Benefit Sharing and Patent Issues relating to Camu Camu, Kakadu Plum and Acai Plant Extracts', Traditional Knowledge Bulletin – Topical Issues Series, United Nations University, Institute of Advanced Studies, 4, http://www.unutki.org/news.php?doc_id=174.

56 Powell, R. and L. Murdoch (2010), 'Patent fight erupts over Kakadu plum', *The Sydney Morning Herald*, 4 December, 11.

Last year, IP Australia released an information kit for Indigenous people called 'Dream Shield' providing advice on the intellectual property system, including the use of patents for Indigenous people to protect their intellectual property.[57] However, in reality there are limited opportunities for Indigenous people to patent traditional knowledge, partly because of the costs involved and expertise required. The technicalities involved in isolating the molecular structure of natural resources to patent new medicines and drugs is more comfortably the domain of pharmaceutical companies than remote communities.

C How Did Mary Kay Get Access to the Plum?

According to an article in the *Sydney Morning Herald*, Mary Kay claims that the plums were sourced from 'a local supplier, who has sourced and harvested the Kakadu plum with a licence, under government regulations'.[58] This may be the case and the company may well have been within its rights if it sourced the plum from commercial harvesters and those who are supplied by these harvesters, particularly outside of the Northern Territory. This could also happen to the gubinge that Bruno Dann supplies to wholesalers and retailers.

In the Northern Territory, the *Biological Resources Act* 2006 (NT) requires bioprospectors to enter into benefit-sharing agreements with owners of the land they want to access to obtain the raw resources. This means that a benefit-sharing agreement would be required for biological resources accessed from Aboriginal land held under the *Aboriginal Land Rights Act* 1974 (NT). As Morse notes, Mary Kay may have obtained samples of the Kakadu plum on a commercial basis through a nursery or a commercial supplier, in which case it would be under no obligation to share benefits with Indigenous communities for access to the genetic resources in question.[59]

---

57 IP Australia (2010), *Dream Shield: A Guide to Protecting Designs, Brands and Inventions for Aboriginal and Torres Strait Islanders*, http://www.ipaustralia.gov.au/resources/dream_shield.shtml; and IP Australia (2010), Nanga Mai Arung – Dream Shield: A Guide to Protecting Designs, Brands and Inventions for Aboriginal and Torres Strait Islanders, Canberra: Commonwealth of Australia, http://www.ipaustralia.gov.au/pdfs/news/dream_shield/print/dreamshield.pdf.

58 Powell, R. and L. Murdoch (2010), 'Patent Fight Erupts over Kakadu Plum', *The Sydney Morning Herald*, 4 December, 11.

59 Morse, J. (2011), 'Nurturing Nature, Nurturing Knowledge: The Nagoya Protocols on Access and Benefit Sharing', Terri Janke and Company, January (in press).

This example shows the shortfall in the law. Mary Kay is unlikely to have had direct contact with living Indigenous people in gaining access to the Kakadu plum resources. Mary Kay is unlikely to have followed community protocols about respecting culture or sharing benefits with Indigenous people in the same way that Loving Earth returns profits to Bruno Dann and the Nyul Nyul community, and other Indigenous people involved in bush harvest.

The right of Indigenous people to collect the Kakadu plum and other food sources for their own 'traditional use' is recognised by a number of pieces of Commonwealth and Northern Territory legislation (for example, section 122 of the *Territory Parks and Wildlife Act* 2010 (NT)). The right has also been recognised in relation to land granted under the *Aboriginal Land Rights (Northern Territory) Act* 1976 (Cth) (section 71) and the *Native Title Act* 1993 (Cth). Thus, for most areas in the Northern Territory, the basic starting point is that Aboriginal people are permitted to harvest the Kakadu plum, as long as they are doing so 'in accordance with Aboriginal tradition'. But sale to commercial entities like Loving Earth may require a permit. Aboriginal people wanting to harvest and *sell* the Kakadu plum in the Northern Territory must apply to the Territory Parks and Wildlife Commission for a permit to do so, as must other Territorians. Aboriginal people collecting the Kakadu plum for commercial purposes, *even when on Aboriginal land*, are required to apply for such a permit. They are, however, exempt from paying the usual royalty rate of $1 per kilo to the Northern Territory Government. Given the widespread occurrence of the Kakadu plum across the Northern Territory, permits for commercial harvest of the Kakadu plum have to date been freely issued. So it is likely that the harvested crop could have been sold to the company supplying Mary Kay without infringement of these laws. However, the access to genetic resources regime for traditional knowledge is aimed at allowing Indigenous people to benefit from the use of their knowledge. In terms of the traditional knowledge relating to the genetic resources, the *Nagoya Protocol* 2010 states that the party accessing the knowledge would be obliged to obtain the prior informed consent of the Indigenous communities that hold the knowledge, and ensure their involvement on mutually agreed terms.[60] However, as knowledge of the Kakadu plum's high beneficial properties is now publicly available, it is unlikely that Mary Kay would

---

60 *Nagoya Protocol on Access to Genetic Resources and the Fair and Equitable Sharing of Benefits Arising From Their Utilization to the Convention on Biological Diversity*, COP 10 Decision ('Nagoya Protocol'), art 7.

be legally obliged to obtain the prior informed consent of Indigenous people to use this knowledge.

III MARJARLA PATENT APPLICATION: CASE STUDY

The example of the development of the Marjarla patent provides insight into a collaborative partnership model between a university and Indigenous people in the case of the use of the Jarlmadangah community traditional knowledge and genetic resources.

Despite the absence of traditional knowledge rights recognised in Australian patent legislation, the Marjarla patent is the result of a benefit-sharing relationship between the Jarlmadangah Community Inc[61] and Griffith University. The subject of this patent is the Marjarla tree, known to Aboriginal communities in the Kimberley region of Western Australia to have healing powers, and ability to sooth pain. Hence, its potential use as an analgesic.[62]

A senior Nyikina-Mangala elder John Watson had half his finger bitten off by a crocodile when hunting in his country, hours away from medical assistance.[63] He had used the bark of the Marjarla tree to soothe the wound. He then related this story to Paul Marshall, the community's CEO, who subsequently got in touch with Professor Ron Quinn, a scientist with Griffith University, in Brisbane. A sample of the plant was delivered to Professor Quinn, and the elder Watson explained how the bark was chewed and also applied to the finger wound.[64]

---

61 Jarlmadangah Community Inc, http://www.jarlmadangah.com/1.htm.

62 Miles, J. (2008), 'Aborigine, Scientist find Pain Relief in Marjarla Tree', *Courier Mail*, 21 October, http://www.couriermail.com.au/news/queensland/native-tree-bark-eases-pain/story-e6freoof-1111117814741.

63 John Watson as interviewed in 'Traditional Knowledge' video, 2010, http://www.ipaustralia.gov.au/resources/dream_shield_vid1.shtml. The Dream Shield IP Australia website also uses this Marjarla example as a case study; see IP Australia (2010), *Dream Shield: A Guide to Protecting Designs, Brands and Inventions for Aboriginal and Torres Strait Islanders*, http://www.ipaustralia.gov.au/resources/dream_shield.shtml and IP Australia (2010), *Nanga Mai Arung – Dream Shield: A Guide to Protecting Designs, Brands and Inventions for Aboriginal and Torres Strait Islanders*, Canberra: Commonwealth of Australia, http://www.ipaustralia.gov.au/pdfs/news/dream_shield/print/dreamshield.pdf.

64 Miles, J. (2008), 'Aborigine, Scientist find Pain Relief in Marjarla tree', *Courier Mail*, 21 October, http://www.couriermail.com.au/news/queensland/native-tree-bark-eases-pain/story-e6freoof-1111117814741.

Professor Quinn, Director of the Eskitis Institute for Cell and Molecular Therapies,[65] commenced research on the plant, isolating several compounds that were active in the relief of pain. From 2003 to 2009, four patents were filed relating to the healing properties of the plant by Griffith University and Jarlmadangah Burru Aboriginal Corporation as joint applicants. Two have lapsed, and one filed in 2009 is still under examination.[66] The patent 2004293125 is for 'novel analgesic compounds, extracts containing same and methods of preparation' and was sealed in 2010.[67]

However, as noted above, the patent system has limited value for protecting traditional knowledge; technically the patent's life is limited to 20 years, after which time it can be copied and used by all.[68] Yet, the Jarlmadangah Burru Community, as co-applicant of the patent, will always be the traditional knowledge owners of this knowledge and, although this is not likely to be contested locally, the global circulation of their knowledge 'product' will be out of their control post-patent.

---

65 This centre at Griffith University has become one of the leading natural product discovery units in the world through the partnership with AstraZeneca: Secretariat of the Convention on Biological Diversity (2008), *Access and Benefit-sharing in Practice: Trends in Partnerships Across Sectors,* Montreal: CBD Technical series No. 38, 9. This report provides a case study of this partnership – but prior to the Marjarla patent. The report notes that 'traditional knowledge was not collected as part of the AstraZeneca-Griffith University partnership' (47). Surprisingly the report is uncritical of the Australian Government's access to genetic resources and benefit-sharing legal frameworks – stating for instance that 'All states remain committed to the implementation of the *Bonn Guidelines*', (43). There is scant evidence for this given the lack of consistency nationwide with implementing or enforcing any sort of ABS frameworks in relation to TK and indeed the lack of interest in Indigenous knowledge shown for instance by the Queensland Government in their *Biodiscovery Act* 2004 (Qld) – the state where Griffith University is located.

66 Jarlmadangah Burru Aboriginal Corporation and Griffith University (2008), 'Novel Compounds', Australian Patent Application No. 2008905126 (patent has since lapsed); Jarlmadangah Burru Aboriginal Corporation and Griffith University (2008), 'Novel Compounds', Australian Patent Application No. 2004293125; Jarlmadangah Burru Aboriginal Corporation and Griffith University (2008), 'Novel Compounds', Australian Patent Application No. 2003906558 (lapsed); and Jarlmadangah Burru Aboriginal Corporation and Griffith University (2009), Australian Patent Application No. 2009904829 (filed and under examination).

67 Jarlmadangah Burru Aboriginal Corporation and Griffith University (2008), 'Novel Compounds', Australian Patent Application No. 2008905126 (patent has since lapsed).

68 The *Patents Act* 1990 (Cth) allows for patents owners to apply for an extra five years' protection for pharmaceutical patents.

In 2008, it was reported that Jarlmadangah Burru Aboriginal Corporation and Griffith University have entered into an agreement with Avexis (a biotechnology company), which allows Avexis to commercialise the plant in return for a share of the income. The agreement also provides Jarlmadangah Burru Aboriginal Corporation with the opportunity to cultivate and supply the plant.[69] This may not be an ideal outcome for the community, as horticulture is not conducive to the intergenerational knowledge transfer of Indigenous ecological knowledge, as bush harvest is for instance. Questions also remain as to whether the horticulture would be local or interstate and whether community members would benefit from learning new skills and/or benefit financially. In other contexts of attempted horticulture of bush products in central Australia, there has been local resistance and a marked preference for bush harvest.[70]

The outcome in this case study is different from the Mary Kay example, where there was no direct contact with local Indigenous knowledge holders, and subsequently no opportunity for benefit sharing.

The Marjarla plant bark remedy constitutes secret use or confidential traditional knowledge, because it does not appear to have been well known or used as a remedy prior to the 'discovery' by Professor Quinn. Unlike the Kakadu plum, where the knowledge was widely publicly available, this access point for the knowledge was direct information from the traditional knowledge owners and not textbooks, or already marketed raw materials.[71]

The delivery of the sample was given by the community council representative to the researcher – so it was not an independent discovery of the scientist.

---

69 Lyon, Z. (2008), 'Bartier Perry Give Community Helping Hand', *Lawyers Weekly*, 21 November.

70 Yates, P. (2009), 'The Bush Foods Industry and Poverty Alleviation in Central Australia', *Dialogue*, **28** (2), 47–56; Walsh, F. and J. Douglas (2009), 'Aboriginal Harvesters who Sell Bush Foods and Seeds from Central Australia for Food and Mining Revegetation Markets: Their Practices, Priorities and Opinions', Alice Springs: Desert Knowledge Co-operative Research Centre Report, 72; and Holcombe, S., P. Yates and F. Walsh (2011), 'Reinforcing Alternative Economies: Self-motivated Work by Central Anmatyerr People to Sell Akatyerr (Desert Raisin) in Central Australia', *Rangelands Journal* (forthcoming).

71 Morse, Jeremy (2011), *Know your Rights to your Indigenous Plant Knowledge – A Guide for Indigenous Knowledge Holders on Recording and Commercializing Indigenous Plant Knowledge*, Darwin: Aboriginal Bush Traders, http://www.aboriginalbushtraders.com/.

One of the inventors, Quinn, is the Director of the Eskitis Institute at Griffith University, which has an intellectual property policy and an intellectual property procedure.[72] This enables the university to own or be assigned the intellectual property of its employee research scientists such as Quinn. However, it would not own the intellectual property of third parties, including any confidential knowledge given to researchers identifying plants and their healing properties. The policy is silent on Indigenous cultural and intellectual property, and the use of traditional knowledge but, clearly, the patent details recognise the co-ownership of the patent by the Jarlmadangah Aboriginal community.

The Marjarla case study illustrates the scope for mutually agreed benefits and sharing of intellectual property to be included as part of the biotechnological model. Here, the bringing together of traditional knowledge with the natural product expertise of the Eskitis Institute led to a new partnership between Griffith University and the Jarlmadangah Burru Aboriginal Community to develop novel pain-killing drugs from traditional Aboriginal medicine. Whilst yet to develop into a saleable drug with royalty returns, there is a good foundation to develop a community supply chain that can link in at a range of levels for local economic empowerment, but also cultural reinforcement through harvesting. The long-term results are yet to be seen.

IV ABORIGINAL BUSH TRADERS

Aboriginal Bush Traders is a not-for-profit organisation set up to foster Indigenous people of the Northern Territory into entering into businesses that supply the bushfoods industries. An initiative of Darwin Regional Indigenous Advancement & Community Development Employment Program Inc, Aboriginal Bush Traders aims to assist in the commercialisation and trade of 'authentic products and services based on Aboriginal culture, art and experiences'.[73] The organisation does this by providing services that relate to bush harvest, art, product development, and tourism and retail activities. An example of a service they currently provide is the Bush Harvest Business Development Officer who works with Aboriginal communities, in the Darwin region, on the identification and collection of bush harvest products using appropriate resource management principles. Jeremy Morse prepared a guide for Indigenous knowledge holders on

---

72 Griffith University, 'Intellectual Property Policy', http://www62.gu.edu.au/policylibrary.nsf/binders/8a581b3be85b69e34a2570530063fdcf?opendocument.

73 See Aboriginal Bush Traders, http://www.aboriginalbushtraders.com/.

recording and commercialising plant knowledge entitled *Know Your Rights to your Indigenous Plant Knowledge*.[74]

This guide provides information to Indigenous people and their advisors on the laws relating to Indigenous cultural and intellectual property rights, and preparing for the commercial potential at the early research phase. Obviously this applies to knowledge and species that are to be newly researched, where the knowledge and potential commercial extent are not the subject of the prior art base. The document provides a systematic guide to the essential protocols for each stage of the commercialisation of Indigenous plant knowledge from identification of the plant/knowledge, through to research and development, and production and marketing.

Know Your Rights to your Indigenous Plant Knowledge demonstrates how owners of Indigenous ecological knowledge can best protect their knowledge through use of legal tools such as trademarks, and written agreements related to prior informed consent and access and benefit-sharing agreements. The guide notes such things as the recording of knowledge as a protective measure to safeguard against patent applications based on knowledge and the use of confidentiality agreements so researchers do not record knowledge and use it in ways not agreed to. With the help of resources such as this owners of traditional knowledge can ensure that the misuse of traditional knowledge leading to a patent is less likely to occur and that any benefits that arise from the use of traditional knowledge can be shared with the knowledge holders and relevant communities.

V AD HOC EFFORTS TO DEVELOP A SYSTEMATIC NATIONWIDE APPROACH

There have been some attempts by the Federal Government to develop Australia wide resources in this area of traditional knowledge, intellectual property and natural resource management. In 2003 the then Federal Department of the Environment and Heritage and the Natural Heritage Trust department[75] hosted a workshop in central Australia at Alice Springs, which addressed issues surrounding the loss, protection, documentation, and transfer of Indigenous knowledge. Participants included representatives from Commonwealth and state agencies, and a diverse range of

---

[74] Morse, Jeremy (2010), *Know your Rights to your Indigenous Plant Knowledge – A Guide for Indigenous Knowledge Holders on Recording and Commercializing Indigenous Plant Knowledge*, Sydney: Terri Janke and Company Pty Ltd.

[75] Now the Federal Department of Sustainability, Environment, Water, Population and Communities.

stakeholder groups.[76] Including the Workshop Outcomes report, four short publications were released, directed at regional Natural Resource Management bodies and Commonwealth agencies. They outlined issues and recommendations in engaging with Indigenous knowledge.[77] These resources and guidelines are referred to (albeit infrequently) on Natural Resource Management body websites, but they are in no way comprehensive. For instance, none of the resources mention the relevant principles from the *Convention on Biological Diversity* 1992, such as article 8 (j), to which Australia is a signatory and the associated 'Nationally Consistent Approach'.[78]

These resources – now under the auspices of the Department of Sustainability, Environment, Water, Population and Communities resources – are also outdated, now over eight years old. This is partly owing to the raft of new government policies and the departmental restructuring in the land management area. The *Guidelines for Regional Bodies*,[79] for instance, purports to 'provide guidelines, processes and protocols for regional bodies to work effectively with Indigenous Knowledge in regional natural resource management'. However, in four pages it is difficult to offer any guidance, except at a generalised aspirational level.[80] Likewise, this lack of uniformity between the newly developed state and territory biological resources acts, the Nationally Consistent Approach and the international obligations under the *Convention on Biological Diversity* 1992 mean that it has been left up to local and regional bodies to develop their own processes for ensuring

---

76 National Heritage Trust (2003), *Indigenous Knowledge Forum, Alice Springs 28–29 May 2003: Workshop Outcomes*, Canberra: Commonwealth of Australia, www.environment.gov.au/indigenous/publications/pubs/workshop.pdf.

77 The four publications can be found at: http://www.environment.gov.au/indigenous/publications/workshop.html.

78 Natural Resource Management Ministerial Council (2002), *Nationally Consistent Approach for the Access to and Utilisation of Australia's Native Genetic and Biochemical Resources*, Canberra: Department of the Environment and Heritage, http://www.environment.gov.au/biodiversity/publications/access/nca/pubs/nca.pdf.

79 Natural Heritage Trust (2004), *Working with Indigenous Knowledge in Natural Resource Management: Guidelines for Regional Bodies*, Canberra: Commonwealth of Australia, http://www.environment.gov.au/indigenous/publications/pubs/guidelines.pdf.

80 A further workshop was held in June 2008, also hosted by the Federal Department of the Environment, Water, Heritage and the Arts, on Indigenous Information Management. This workshop produced the document 'Indigenous Knowledge Management Systems (databases) Guide for Indigenous Communities'.

that Indigenous cultural and intellectual property is recognised, protected and managed accordingly.

The emerging importance placed by, then, the Department of the Environment, Water, Heritage and the Arts on the ethical management of engagement with Indigenous peoples and their local knowledge is also illustrated by the range of aspirational documents indicating the need to develop such resources within Indigenous land management organisations.

A further workshop was held in June 2008, also hosted by then Department of the Environment, Water, Heritage and the Arts, on Indigenous Information Management. This workshop produced the document *Indigenous Knowledge Management Systems (databases) Guide for Indigenous Communities*.[81] Again, this document is not a 'Guide', but rather a grab bag of information and ideas that unfortunately does not reference key resources in this area, most notably the United Nations University and the Institute of Advanced Studies report *The Role of Registers and Databases in the Protection of Traditional Knowledge*.[82]

CONCLUSION

Although the Australian Government has not yet introduced the language and principles of the *United Nations Declaration on the Rights of Indigenous Peoples* 2007[83] into their Indigenous policy programs this does not prevent Indigenous groups from incorporating it into their own programs and policies.[84] For instance, there is no reference to it at all, in the recent Hawke Inquiry into the *Environment Protection and Biodiversity Conservation Act* 1999 (Cth). So it would seem that it is up to Indigenous interests to activate

---

81 The Department of the Environment, Water, Heritage and the Arts (2008), 'Indigenous Knowledge Management Systems (Databases): Guide for Indigenous Communities', http://www.environment.gov.au/indigenous/publications/pubs/knowledge-management-guide.pdf.

82 United Nations University and the Institute of Advanced Studies (2004), *The Role of Registers and Databases in the Protection of Traditional Knowledge: A Comparative Analysis*, Japan: United Nations University and the Institute of Advanced Studies, http://www.ias.unu.edu/binaries/UNUIAS_TKRegistersReport.pdf.

83 *United Nations Declaration on the Rights of Indigenous Peoples* 2007, adopted by General Assembly Resolution 61/295 on 13 September 2007.

84 In late 2010, the Australian Human Rights Commission produced a detailed community guide to the Declaration to encourage use in the way by indigenous people and their representative organisations: Australian Human Rights Commission (2010), *The Community Guide to the UN Declaration on the Rights of Indigenous Peoples*, Sydney: The Australian Human Rights Commission.

and enable the *United Nations Declaration on the Rights of Indigenous Peoples* 2007. The articles 31.1 and 31.2 in relation to Indigenous cultural and intellectual property and articles in the preamble are immediately relevant here. The preamble to the *United Nations Declaration on the Rights of Indigenous Peoples* 2007 reinforces the need to 'recognise that respect for Indigenous knowledge, cultures and traditional practices contributes to sustainable and equitable development and proper management of the environment'.[85]

It is relevant to recall that three of the reasons the previous Howard Government gave when they refused to endorse the *United Nations Declaration on the Rights of Indigenous Peoples* 2007 were 'concerns over the extension of Indigenous Intellectual Property rights under the Declaration as unnecessary under current international and Australian law',[86] concerns about references to self-determination and their potential to be misconstrued. Marise Payne of the Liberal Party complained of 'the potential abuse of the right under the Declaration for Indigenous peoples to "unqualified consent on matters affecting them" or free and prior informed consent'.[87]

These crucial planks of Indigenous cultural and intellectual property, self-determination and free prior informed consent seem to concern the current government just as much – and they are central components in the Indigenous ecological knowledge resources developed for the Northern Territory.

The focus of the *Nagoya Protocol* 2010 on 'genetic resources' is disappointing, as it limits the parameters of any access to genetic resources regime to working within the intellectual property regime and the subsequent focus on the commercialisation of traditional knowledge.[88] If we are to learn anything from the Australia Government's ad hoc approach to the implementation of the *Convention on Biological Diversity* 1992, the success of the access to genetic resources framework to deliver traditional knowledge rights of Indigenous people in relation to rights over genetic resources

85 *United Nations Declaration on the Rights of Indigenous Peoples* 2007, adopted by General Assembly Resolution 61/295 on 13 September 2007.

86 Payne, M. (2007), 'Matters of Urgency: United Nations Declaration on the Rights of Indigenous Peoples', Senate Hansard, Australian Parliament, 10 September.

87 Ibid.

88 *Nagoya Protocol on Access to Genetic Resources and the Fair and Equitable Sharing of Benefits Arising From Their Utilization to the Convention on Biological Diversity*, COP 10 Decision ('*Nagoya Protocol* 2010').

is also likely to be limited. The current bioprospecting regimes have loopholes, as discussed above.

Ensuring that there is a minimum standard across all jurisdictions can only be facilitated and, ultimately, enforced by the Federal Government; notably in relation to access and benefit-sharing agreements. Whilst there are some further clarifications needed to the *Nagoya Protocol* 2010, the framework potentially provides Indigenous people with a chance to control the use of their traditional knowledge associated with genetic resources. However, without government intervention and guidance at the state and territory level, we have no reason to believe that even the current minimalist access to genetic resources regime will provide real leverage for Indigenous peoples over access to genetic resources. It seems to us that extending the scope of the *Nagoya Protocol* 2010 to also incorporate biological resources would not only be more consistent with the intent of the *Convention on Biological Diversity* 1992 and article 8 (j), but also allow more scope for the development of localised bio-cultural community protocols.

It seems to us that greater recognition of Indigenous community bio-cultural protocols, as these manage local knowledge, has a range of values. The utilisation of known materials saves research time in already being humanly tested over years of Indigenous use (saving the need for human trials on drug testing) and also recognises that, in many of these cosmetic applications, Indigenous cultural practices are used to market the product indirectly. Local economic opportunities may also, ideally, be facilitated.

The two case studies cited show two very different results that have occurred in the absence of a national regulatory framework or *sui generis* laws. The Mary Kay Kakadu plum case study illustrated how traditional knowledge and a traditional resource cultivated can be taken for commercial use without any recourse under current intellectual property laws or the access to a genetic resources regime. On the other hand, the Marjarla patent shows an approach of working together and willingness to enter into good faith relationships over time between Indigenous knowledge holders and their community and a university. The joint ownership of a biotechnological patent between an Indigenous community and a university sets new ground in Australia and it will no doubt be instructive to follow the development of the product and the associated benefit sharing as it unfolds on the ground.

12. Climate-ready crops: intellectual property, agriculture and climate change

Matthew Rimmer

In *The Climate Change Review*, Ross Garnaut emphasised that 'Climate change and climate change mitigation will bring about major structural change in the agriculture, forestry and other land use sectors'.[1] He provides this overview of the effects of climate change on food demand and supply: 'Domestic food production in many developing countries will be at immediate risk of reductions in agricultural productivity due to crop failure, livestock loss, severe weather events and new patterns of pests and diseases.'[2] He observes that 'Changes to local climate and water availability will be key determinants of where agricultural production occurs and what is produced.'[3]

Gert Würtenberger has commented that modern plant breeding is particularly concerned with addressing larger issues about nutrition, food security and climate change: 'Modern plant breeding has an increasing importance with regard to the continuously growing demand for plants for nutritional and feeding purposes as well as with regard to renewable energy sources and the challenges caused by climate changes.'[4] Moreover, he notes that there is a wide array of scientific and technological means of breeding new plant varieties: 'Apart from classical breeding, technologies have an

1 Garnaut, Ross (2008), *The Garnaut Climate Change Review: Final Report*. Cambridge: Cambridge University Press, 531, http://www.garnaut review.org.au.

2 Ibid., 534.

3 Ibid., 535.

4 Würtenberger, G. (2009), 'An Analysis of the European and International Plant Variety Protection Systems', *Journal of Intellectual Property Law and Practice*, 4, 914.

important role in the development of plants that satisfy the various requirements that industrial and agricultural challenges expect to be fulfilled.'[5] He comments: 'Plant variety rights, as well as patents which protect such results, are of increasingly high importance to the breeders and enterprises involved in plant development programmes.'[6] There has been larger interest in the intersections between sustainable agriculture, environmental protection and food security.[7]

The debate over agricultural intellectual property is a polarised one, particularly between plant breeders, agricultural biotechnology companies and a range of environmentalist groups. Susan Sell comments that there are complex intellectual property battles surrounding agriculture:

> Seeds are at the centre of a complex political dynamic between stakeholders. Access to seeds concerns the balance between private rights and public obligations, private ownership and the public domain, and commercial versus humanitarian objectives.[8]

Part I of this chapter considers debates in respect of plant breeders' rights, food security and climate change in relation to the *UPOV Convention* 1991.[9] Part II explores efforts by agricultural biotechnology companies to patent climate-ready crops. Part III considers the report of the Special Rapporteur for Food, Olivier De Schutter. It looks at a variety of options to encourage access to plant varieties with climate adaptive or mitigating properties.

---

5 Ibid.

6 Ibid.

7 Hahlbrock, Klaus (2009), *Feeding the Planet: Environmental Protection Through Sustainable Agriculture*, London: Haus Publishing Ltd.

8 Sell, Susan (2009), 'Corporations, Seeds, and Intellectual Property Rights Governance', in Clapp, Jennifer and Doris Fuchs (eds), *Corporate Power in Global Agrifood Governance*, Cambridge, MA: MIT Press, 187–224.

9 *International Convention for the Protection of New Varieties of Plants*, opened for signature 2 December 1961, 815 UNTS 89 (entered into force 24 April 1968), as revised on 10 November 1972, 23 October 1978, and 19 March 1991 ('*UPOV Convention* 1991').

I PLANT BREEDERS' RIGHTS

Historically, *sui generis* regimes were established to provide for intellectual property protection for new plant varieties.[10] Margaret Llewelyn and Mick Adcock describe the key features of this unique regime:

> In contrast to the patent system, the plant variety rights system is not technology neutral but is overtly technology and sector specific. The right is granted over plant groupings which can be shown to be new, distinct, uniform and stable – these criteria are subject specific and not capable of more general application as is the case with the patent granting criteria.[11]

Typically, plant breeders' rights provide protection for 20 years for plants, and for trees and vines for up to 25 years. The international lawyer Frederick Abbott comments: 'Plant variety protection is relevant to [alternative energy resources]/[climate change mitigation technologies] because new varieties of plant may be developed for use in generating energy, and new varieties of plant may be developed to mitigate the impact of climate change (such as plants that exhibit improved drought-resistant characteristics).'[12] It is argued, in this section, that there is a need to rethink plant breeders' rights in light of their role in respect of food security, climate change and development. Such a position aligns with the future plans of the Director-General of the World Intellectual Property Organization, Francis Gurry. The plant breeders' rights cannot remain static; they need to evolve in order to better accommodate developments in law, science, global needs and the marketplace.

---

10 Kloppenburg, Jack (1988), *First the Seed: The Political Economy of Plant Biotechnology 1492–2000*, Cambridge: Cambridge University Press; Llewelyn, Margaret and Mike Adcock (2006), *European Plant Intellectual Property*, Oxford and Portland: Hart Publishing; Aoki, Keith (2008), *Seed Wars: Controversies and Cases on Plant Genetic Resources and Intellectual Property*, Durham, NC: Duke University Press; and Bently, Lionel and Brad Sherman (2009), *Intellectual Property*, 3rd edn, Oxford: Oxford University Press, 591–601.

11 Llewelyn, Margaret and Mike Adcock (2006), *European Plant Intellectual Property*, Oxford and Portland: Hart Publishing, 19.

12 Abbott, F. (2009), 'Innovation and Technology Transfer to Address Climate Change: Lessons from the Global Debate on Intellectual Property and Public Health', International Centre for Trade and Sustainable Development Platform on Climate Change, Trade Policies and Sustainable Energy, Issue Paper No. 24, June, 6.

A The UPOV Convention 1991

Francis Gurry is not only the Director-General of the World Intellectual Property Organization, but he is also the Secretary-General of the UPOV (the Union for the Protection of New Plant Varieties).[13] In his speech at the 2008 session of the Union, he emphasised the relevance of plant breeding to global issues, such as food security, climate change and energy security:

> I think there are a number of recent developments in the world, and in our environment, which make the traditional task of plant breeding all the more relevant today, much more relevant and more important than ever before. Of course, in one sense the fundamental task, biological task, of human beings has always been to feed themselves and socially, of course, that translates itself into a task of ensuring food security and sustainability and affordability [...] There is the challenge that we all face in the area of climate change, the number of climate incidents or accidents or natural disasters, whether geophysical or hydrological or meteorological is constantly on the rise, this presents great challenges for agriculture in particular; our own lifestyle and behaviour presents a great many challenges – we have increasing numbers of settlements in exposed regions of the world, a problem of urbanization, we have soil erosion, salinity difficulties and the extreme challenge of the diminishing availability of water. We also face an environment in which energy prices are extremely volatile and has enormous impact on our behavior and our responses to the volatility of energy prices and, of course, we are also in an environment of increasing globalization of markets which has a very important impact on the distribution of food.[14]

Gurry argues that '[Plant breeders' rights...] is a very powerful tool to support both plant breeding and the production and distribution of its products in the form of high-quality seeds and propagating materials to farmers and other users.'[15] He stressed: 'So there is, I think, constant mission on our part to [...] demonstrate the practical ways in which the plant breeders' rights system actually assists the central missions of humanity in ensuring food security and sustainability around the world.'[16] Much like the World Intellectual Property Organization, the UPOV focuses on reconciling the need for harmonisation, and development concerns.

The *International Convention for the Protection of New Varieties of Plants* (*UPOV Convention* 1961) was designed to ensure that the member

---

13 Gurry, F. (2008), 'Statement to the Forty-Second Ordinary Session of the International Union for the Protection of New Varieties of Plants', 30 October, http://www.upov.int/en/documents/c/42/c_42_21.pdf.

14 Ibid.

15 Ibid.

16 Ibid.

states party to the Convention acknowledge the achievements of breeders of new plant varieties by making available to them an exclusive property right, on the basis of a set of uniform and clearly defined principles.[17] The *UPOV Convention* 1961 was revised in Geneva in 1972, 1978 and 1991.[18] Both the 1978 and the 1991 Acts set out a minimum scope of protection and offer member states the possibility of taking national circumstances into account in their legislation.

The *UPOV Convention* 1991 is guided by the mission statement 'to provide and promote an effective system of plant variety protection, with the aim of encouraging the development of new varieties of plants, for the benefit of society'.[19] The defenders of the *UPOV Convention* 1991 maintain that plant breeders' rights will produce a range of benefits including: 'economic benefits, for example through varieties with improved yield leading to reductions in the price of end-products for consumers, or improved quality, leading to higher value products with increased marketability; health benefits, for example through varieties with improved nutritional content; environmental benefits, for example through varieties with improved disease resistance or stress tolerance; and even pure pleasure, for example with ornamental plants'.[20]

The *UPOV Convention* 1991 emphasises that 'Each Contracting Party shall grant and protect breeders' rights.'[21] Article 5 provides that 'the breeder's right shall be granted where the variety is (i) new, (ii) distinct, (iii) uniform and (iv) stable'. Article 6 elaborates upon the criteria of novelty. Article 7 defines distinctiveness. Article 8 considers uniformity. Article 9 examines stability. Article 14 provides more detailed provisions defining the exclusive acts concerning propagating material in relation to which the holder's authorisation is required. In exceptional circumstances authorisation may be required in relation to any of the specified acts done with harvested material of the variety, but only where the holder has had no

---

17 *International Convention for the Protection of New Varieties of Plants*, opened for signature 2 December 1961, 815 UNTS 89 (entered into force 24 April 1968), as revised on 10 November 1972, 23 October 1978, and 19 March 1991 ('*UPOV Convention*').

18 Ibid.

19 Ibid.

20 International Union for the Protection of New Varieties of Plants (2005), *UPOV Report on the Impact of Plant Variety Protection*, Geneva: International Union for the Protection of New Varieties of Plants, http://www.upov.int/en/publications/pdf/353_upov_report.pdf.

21 Article 2 of the *UPOV Convention* 1991.

reasonable opportunity to exercise his right in relation to the propagating material.

Article 15 highlights the compulsory exceptions to plant breeders' rights for acts done privately and for non-commercial purposes; experimental purposes; and for the purpose of breeding other varieties. Article 15 also has an optional exception for farmers' rights to 'restrict the breeder's right in relation to any variety in order to permit farmers to use for propagating purposes, on their own holdings, the product of the harvest which they have obtained by planting, on their own holdings, the protected variety'. Article 16 considers the exhaustion of the breeders' right. Article 17 provides scope for compulsory licensing, subject to equitable remuneration. Article 19 sets out the minimum periods of duration for plant breeders' rights – 20 years for plants, and 25 years for trees and vines.

Margaret Llewelyn and Mick Adcock consider the current debate over the value of the UPOV system:

> It is true that whilst UPOV continues to grow in membership and the number of rights granted has increased, it is in many respects a child of its time and area. It developed out of the needs of European agricultural plant breeders in the post-war industrialisation of plant breeding.[22]

The pair nonetheless reflect that 'whilst there may be problems with the system which exists, attempts to introduce new rights, or significantly to revise the existing ones (with all the political ramifications involved), might only serve to make the position worse than better'.[23]

Notwithstanding such aspirations to address matters of public policy, the UPOV has been criticised at times by both other United Nations agencies and by non-government organisations. In 2009, a number of farmer advocacy groups – the Association of Plant Breeding for the Benefit of Society and the European Coordination Via Campesina – were denied observer status at the UPOV on the grounds that they lacked technical competence in plant breeders' rights. In response, the Third World Network organised an open letter, protesting the exclusion of these groups: 'The decision [...] shows that UPOV is not inclusive or supportive of farmer and civil society organizations that work in the interest of peasants, small and

---

22 Llewelyn, Margaret and Mike Adcock (2006), *European Plant Intellectual Property*, Oxford and Portland: Hart Publishing, 198.

23 Ibid., 199.

medium scale farmers and the broader public.'[24] In October 2010, the UPOV relented, and admitted the two new civil society observers.[25]

In an event entitled the 'Future of UPOV in a Changing World', there was debate as to whether UPOV could accommodate future challenges such as climate change and the ongoing problem of food security.[26]

B National Regimes

There has been significant debate as to whether the plant breeders' rights regime has become outdated or anachronistic in light of new scientific developments and the expansion of the patent system. In a paper, Jay Sanderson and Kathryn Adams summarise such concerns: 'A recurring theme in these criticisms is that advances in science, as well as the availability of alternative forms of protection (notably, patents and trade marks), mean that a sui generis scheme of plant breeder's rights is no longer effective.'[27] Sanderson and Adams note that there has been a decrease in applications since 2003, speculating: 'These figures might suggest that changing environmental conditions such as drought and increased salinity have had an effect on plant breeding investment either by reducing the level of plant breeding, or by focusing breeding programs on developing particular traits (for example, drought resistance and salinity tolerance) in new plant varieties.'[28] Furthermore, the commentators suggest that the 'changed climatic conditions may have affected broader commercial considerations such as a grower's ability to purchase new plant varieties and market share, making both breeders and growers more selective about the varieties that they are either protecting or purchasing'.[29] The authors conclude by saying that there is 'a complex interplay between changing

---

24 The Third World Network (2010), 'Call to Support Farmer and Civil Society Organisations' Application to UPOV For Observer Status', 11 March, http://www.twnside.org.sg/title2/intellectual_property/info.service/2010/ipr.info.100305.htm.

25 Mara, K. (2010), 'Change Coming to Quiet UN Plant Variety Protection Agency', *Intellectual Property Watch*, 26 October, http://www.ip-watch.org/weblog/2010/10/26/change-coming-for-quiet-un-plant-variety-protection-agency/.

26 International Centre for Trade and Sustainable Development and the Quaker United Nations Office (2010), 'The Future of UPOV In A Changing World: Issues and Challenges', 25 March, http://ictsd.org/i/events/dialogues/72761/.

27 Sanderson, J. and K. Adams (2008), 'Are Plant Breeder's Rights Outdated? A Descriptive and Empirical Assessment of Plant Breeder's Rights in Australia, 1987–2007', *Melbourne University Law Review*, **32** (3), 980–1006 at 981–2.

28 Ibid., 995.

29 Ibid., 995.

environmental conditions, emerging science and technology and the transformation of the funding of plant breeding programs'.[30]

Doug Waterhouse, the Registrar for Plant Breeders' Rights in Australia, observed: 'In response or in anticipation to climate change, new varieties are being developed without a chill factor requirement, they have low or no chill requirement that means that the area available for growing those plants can be moved perhaps north from where they are now.'[31]

The Australian Department of Agriculture, Fisheries, and Forestry has contended: 'The Plant Breeder's Rights Intellectual Property Protection program promotes innovation in new plant varieties, including some that specifically address salinity and erosion control. Other varieties, such as those that use less herbicide, water or fertilizer, and natives, have beneficial conservation side effects.'[32]

The Australian Plant Breeders' Rights Office provides an example of a plant with resistance to environmental stresses.[33] The Office notes that 'a pioneering wheat-breeding program gives farmers a chance to fight back against the big dry'.[34] The Office elaborates:

> Scientists at the CSIRO have responded to water shortage issues with pioneering research that has led to the development of Drysdale, a wheat variety with superior water use efficiency. Dr Richard Richards heads the research team that developed Drysdale. The group worked with the Australian National University throughout the 1980s to identify a selection method based on plant respiration to improve water use efficiency in wheat-breeding programs. The method was a world first and, when applied to a cross of two high-quality and disease-resistant varieties, produced Drysdale. The variety yields at least 10 per cent more under dry conditions than ordinary wheats.[35]

The Office is hopeful that plant breeders' rights will encourage further innovation in respect of climate-ready crops.

It is hard to judge to what extent plant breeders' rights are used to support plant breeding with a view to certain environmental benefits. There

---

30 Ibid.

31 Dakis, S. (2008), 'PBR Helping with Climate Change', *ABC News*, 27 August, http://www.abc.net.au/rural/sa/content/2006/s2348159.htm.

32 Department of Agriculture, Fisheries and Forestry (2002–3), 'Ecologically Sustainable Development and Environmental Performance', in Annual Report, Canberra: Department of Agriculture, Fisheries and Forestry, 297, http://www.daff.gov.au/__data/assets/pdf_file/0012/5700/annual_report0203.pdf.

33 Plant Breeders Rights Office, 'Drysdale', http://www.ipaustralia.gov.au/pbr/drysdale.shtml.

34 Ibid.

35 Ibid.

has been some commentary that plant breeders' rights have been dominated by ornamental plants and trees. One civil society organisation – the ETC Group – comments:

> The industrial food chain focuses on far fewer than 100 breeds of five livestock species. Corporate plant breeders work with 150 crops but focus on barely a dozen. Of the 80,000 commercial plant varieties in the market today, well over half are ornamentals.[36]

The ETC Group complains: 'In the face of climate chaos, the industrial chain is imposing a patent regime that prizes uniformity over diversity and enforces a technological model that costs more and takes more time to breed one genetically-engineered variety than it does to breed hundreds of conventional varieties.'[37]

At present, it is difficult to read registers for plant breeders' rights in Australia and elsewhere, and single out applications that have a particular significance for climate change. Instead, one has to search the sites of individual plant breeders – for instance, Fleming's Nurseries highlights its drought tolerant plants.[38] Arguably, the *sui generis* regime needs 'green databases' much like the European Patent Office initiative.

Consideration should also be given to a fast-track mechanism in relation to plant breeders' rights for accelerating 'green' applications. If green patents can be 'fast-tracked', why not plant breeders' rights?

In relation to the Australian regime of plant breeders' rights, Paul Brennan of CropGen International contends that there will be 'substantial changes in the outcomes of plant breeding programs to accommodate the different growing environments and stresses that can be anticipated' and 'changes in Australia's portfolio of commercial crop species'.[39] He contends that there may be a need to reconfigure plant breeders' rights to boost investment in plant breeding: 'Clearly, substantial increases in investment in Australia's plant breeding industry will be required for the public and private good of Australia.'[40] Paul Brennan contends that there could be

---

36 ETC Group (2009), 'Who Will Feed Us?: Questions for the Food and Climate Crises', *ETC Group Communique*, November, Issue 102, 1, http://www.etcgroup.org/upload/publication/pdf_file/ETC_Who_Will_Feed_Us.pdf.

37 Ibid.

38 Flemings Nurseries, http://www.flemings.com.au/topten.asp?type=7.

39 Brennan, P. (2007), 'Plant Variety Intellectual Property Rights in a Changing and Challenging Environment', 20 Years of Plant Breeder's Rights Symposium, Canberra, 8 June, http://www.cropgeninternational.com/media/PBR_Future_070506_v2.pdf.

40 Ibid.

scope for broadening the protection afforded by plant breeders' rights; lengthening the duration of protection; and enhancing the enforcement of plant breeders' rights. Given the significant amount of research on agricultural biotechnology, food and energy, there may well be a need to revisit the requirement of 'essential derivation'.[41] At present, the holder of plant breeders' rights also enjoys protection in relation to any essentially derived varieties. That policy setting may help make the patent regime a more attractive legal regime to genetic engineers.

In the wake of the litigation between the Grain Pool and Cultivaust in the Federal Court of Australia,[42] and the Full Federal Court of Australia,[43] the Australian Government made amendments to the plant breeders' rights regime, in particular removing the broad food and fuel exception. The *Plant Breeder's Rights Amendment Act* 2002 (Cth) repealed the food and fuel exception in section 18 – which provided a defence to claims of infringement of plant breeders' rights where plants were used for the purposes of food or fuel. Introducing the legislation, the Minister for Agriculture, Fisheries, and Forestry, Warren Truss, emphasised that plant breeding was essential in providing 'the building blocks for a diverse range of industries; competitive, high quality food and fibre products to meet increasing world demand and; practical solutions to health and environmental concerns'.[44] Truss observed: 'Section 18 was designed to protect public interest concerns regarding limitations that might be placed on the use of the propagating material, for example, reflecting the concern that the [plant breeders' rights] owner should not be able to prevent, say, the making of bread or the brewing of beer from legitimately purchased grain.'[45] He was concerned that the provision 'should not deny the holder of PBR the opportunity to exercise their right of exclusion regarding normal commercial activities, including, for example, the first point of sale in the trading of food commodities'.[46] Truss said that the food and fuel exception is 'a clear disincentive to investment in plant breeding and is a likely explanation why innovation in plant breeding in the field crops sector is lagging that of other

---

41 Sanderson, J. (2006), 'Essential Derivation, Law and the Limits of Science', *Law in Context*, **24** (1), 34–53.

42 *Cultivaust* v. *Grain Pool Pty Ltd* [2004] FCA 638.

43 *Cultivaust* v. *Grain Pool Pty Ltd* [2005] FCAFC 223.

44 Truss, W. (2002), 'The *Plant Breeder's Rights Amendment Bill* 2002 (Cth)', the House of Representatives, Hansard, 12 December, 10589.

45 Ibid., 10590.

46 Ibid., 10590.

sectors'.[47] The *Plant Breeder's Rights Amendment Act* 2002 (Cth) instead set up a new regime under section 18 dealing with equitable remuneration.

In light of larger concerns about food security and climate change, one wonders whether a plant breeders' rights regime, particularly for a developing country or a least developed country, might benefit from a food or fuel exception.

The Advisory Council on Intellectual Property handed down a lengthy report on enforcement of plant breeders' rights.[48]

C *Sui Generis* Regimes

Some critics argue that the *UPOV Convention* 1991 is ill-suited to the needs of developing countries, least developed countries, and small island states. As such, some commentators promote *sui generis* regimes for providing protection for plant varieties such as the *Plant Variety Protection and Farmers' Rights Act* 2001 (India).[49] The Indian regime is an innovative piece of legislation, which brings together plant breeders' rights, farmers' rights and access to genetic resources. The preamble emphasises that the legislation is designed to 'recognise and protect the rights of the farmers in respect of their contribution made at any time in conserving, improving and making available plant genetic resources for the development of new plant varieties' and 'to protect plant breeders' rights to stimulate investment for research and development, both in the public and private sector, for the development of new plant varieties'.[50] The African Union has also developed a *sui generis* model law – the *African Model Legislation for the Protection of the Rights of Local Communities, Farmers and Breeders, and for the Regulation of Access to Genetic Resources*.[51]

Graham Dutfield comments that 'concern has been raised that the UPOV system was drawn up mainly by European countries and is designed

---

47 Ibid., 10591.

48 Advisory Council on Intellectual Property (2010), *A Review of Enforcement of Plant Breeder's Rights*, Canberra: Commonwealth of Australia, http://www.acip.gov.au/library/ACIP_PBR_Enforcement_Final_Report.pdf.

49 Sahai, Suman (2002), 'India's Plant Variety Protection and Farmers' Rights Legislation', in Drahos, Peter and Ruth Mayne (eds), *Global Intellectual Property Rights: Knowledge, Access, and Development*, New York: Palgrave Macmillan, 214–23.

50 Ibid.

51 *African Model Legislation for the Protection of the Rights of Local Communities, Farmers and Breeders, and for the Regulation of Access to Genetic Resources* 2000 http://www.cbd.int/doc/measures/abs/msr-abs-oau-en.pdf.

to accommodate the specific characteristics of the capital-intensive large-scale commercial agricultural systems that generally prevail on the continent'.[52] He observed: 'As a result, it is often argued, the system is unsuitable for most developing countries.'[53] Dutfield observes that 'among such critics, the current system of [intellectual property rights] protection for plants has raised concerns over their impact on food security in three areas: (i) [plant variety protection] and research priorities; (ii) the interests of poor farmers; and (iii) the availability of genetic resources for further breeding'.[54]

First, Dutfield comments that there is a problem in that the *UPOV Convention* 1991 has skewed research priorities: 'Empirical evidence casts doubt on whether [plant variety protection] (as well as patents) does much to encourage investment in plant breeding except in just a few commercially important crop species such as wheat and soya bean and ornamentals.'[55] He notes that there has been empirical research into jurisdictions, such as Kenya, where a large number of protected varieties of plants were not even food crops: 'In Kenya, for example, from 1997 to 2003, out of a total of 611 [plant variety protection] applications, 247 were for foreign-bred roses.'[56] Dutfield concludes: 'It is conceivable, then, that PVP may contribute to a trend, whereby traditional diverse agro-ecosystems, containing a wide range of traditional crop varieties, are replaced with monocultures of single agrochemical-dependent varieties, with the result that the range of nutritious foods available in local markets becomes narrower.'[57] This complaint is often made by civil society groups. GRAIN, for instance, complains: 'The loss of genetic diversity in agriculture is destroying farmers' capacities to adapt production to new pressures, such as population growth and climate change.'[58] The group alleges: 'The UPOV system encourages this process by providing powerful commercial breeders with the right to [intellectual property rights'] sanctioned monopolies.'[59]

---

[52] Dutfield, Graham (2008), 'Turning Plant Varieties into Intellectual Property: The UPOV Convention', in Tansey, Geoff and Tamsin Rajotte (eds), *The Future Control of Food: A Guide to International Negotiations and Rules on Intellectual Property, Biodiversity, and Food Security*, London: Earthscan, 27 at 40.

[53] Ibid.

[54] Ibid., 41.

[55] Ibid., 41.

[56] Ibid., 41.

[57] Ibid., 41.

[58] GRAIN (1998), 'Ten Reasons Not to Join UPOV: Global Trade and Biodiversity in Conflict', *Seedling*, http://www.grain.org/seedling/?id=10.

[59] Ibid.

Second, Dutfield comments that the restrictions on farm-saved seed under the *UPOV Convention* may adversely affect the freedom of smallholder farmers to buy seed from sources other than the original breeders or their licensees. He observes: 'Many of these farmers [in developing countries] are smallholders from whom the saving and across-the-fence and inter-community exchanging of seed are common practices.'[60] Dutfield notes, however: 'While finding non-[Plant Variety Protection] seed may not yet be a serious difficulty for developing country farmers, this situation may change.'[61] He observes: 'In some developed countries, it is becoming difficult for farmers to find non-PVP varieties of some crops.'[62] He also comments that plant breeders' rights are further supplemented by an array of other forms of legal protection – contract law, genetic use restriction technologies and other seed laws.

Graham Dutfield suggests that genetic use restriction technologies could have some environmental benefits: 'For one thing, it could allay one of the concerns of some opponents of genetically modified crops, which is the risk that genes from these plants may cross over to other species, a phenomenon called horizontal gene transfer which, ironically, many advocates of GM agriculture dismiss as being nothing to worry about anyway.'[63] Stephen Hubicki and Brad Sherman note that certain forms of genetic use restriction technologies could be deployed in relation to specific traits – such as resistance to environmental stresses.[64] Thus, 'a specific genetically-engineered trait, such as drought or disease resistance, will not be expressed in a plant unless the crop is exposed to an external agent, such as a chemical'.[65]

Third, Dutfield emphasises that the plant breeders' rights regime has a broad breeders' exemption, whereas patent law has a limited experimental use defence: 'Plant breeders and other supporters of UPOV tend to stress the necessity of being able to freely access genetic material including that

---

60 Dutfield, Graham (2008), 'Turning Plant Varieties into Intellectual Property: The UPOV Convention', in Tansey, Geoff and Tamsin Rajotte (eds), *The Future Control of Food: A Guide to International Negotiations and Rules on Intellectual Property, Biodiversity, and Food Security*, London: Earthscan, 27 at 41.

61 Ibid., 42.

62 Ibid., 42.

63 Dutfield, G. (2003), 'Should We Terminate Terminator Technology?' *European Intellectual Property Review*, **25** (11), 491–5 at 493.

64 Hubicki, S. and B. Sherman (2005), 'The Killing Fields: Intellectual Property and Genetic Use Restriction Technologies', *The University of New South Wales Law Journal*, **28** (3), 740.

65 Ibid., 741.

which is IPR protected.'[66] He laments that there is a need for further empirical research in this area: 'This discussion on how PVP affects food security and nutrition in developing countries leads one to consider in more general terms the applicability of such an IPR to these countries.'[67]

Arguably, the *UPOV Convention* 1991 needs to be updated and modernised, so that it addresses global issues, such as climate change, food security and nutrition.

II PATENT LAW AND AGRICULTURAL BIOTECHNOLOGY

In the field of patent law, technology developers have argued that genetically modified crops will be better adapted to the environmental stresses caused by climate change. It is worthwhile examining the patent portfolios and strategies of some of the key companies in the area (Table 12.1). In its amicus brief in the matter of in *J.E.M. Ag Supply, Inc.* v. *Pioneer Hi-Bred International Inc.*, the Biotechnology Industry Organization explicitly argues that agricultural biotechnology has the potential to have a positive environmental impact, particularly by enhancing the resistance of crops to climate stressors:

> The biotechnology industry in the United States has the potential to radically increase current levels of agricultural production, dramatically improve the quality of health care, and measurably contribute to a cleaner environment. Agricultural biotechnology, for example, promises to improve food production by enhancing crops' resistance to natural enemies and harsh environmental conditions, thereby lessening reliance on pesticides and other crop inputs, reducing farmers' costs (including fuel costs), and controlling soil erosion and compaction.[68]

The Council for Biotechnology – a peak body that represents biotechnology companies such as BASF, Bayer CropScience, Dow AgroSciences LLC, DuPont, Monsanto Company, and Syngenta – has argued that agricultural biotechnology has benefits for energy security, climate

---

66 Dutfield, Graham (2008), 'Turning Plant Varieties into Intellectual Property: The UPOV Convention', in Tansey, Geoff and Tamsin Rajotte (eds), *The Future Control of Food: A Guide to International Negotiations and Rules on Intellectual Property, Biodiversity, and Food Security*, London: Earthscan, 27 at 42–3.

67 Ibid., 43.

68 Biotechnology Industry Organization, 'Brief in Support of the Respondent in *J.E.M. Ag Supply, Inc.* v. *Pioneer Hi-Bred International Inc.*', 2001 WL 689273 (U.S.) (Appellate Brief), 2.

change[69] and water conservation.[70] The position of such agricultural biotechnology companies has received support from some politicians. For instance, Jorgo Chatzimarkakis, a German Member of the European Parliament, has contended: 'Agricultural biotechnology can help by reducing the production of greenhouse gases, helping crops adapt to varied and often adverse environments, and by helping to increase yields while using fewer hectares of land and other inputs.'[71]

It should be noted that patent law provides broader protection than plant breeders' rights. Patent law, rather than being limited to merely plants, can potentially protect a wide range of agricultural inventions using environmental or climactic applications. It should also be noted that patents relating to agriculture and climate are not merely limited to climate-ready crops. The United States Patent and Trademark Office's classification of clean technologies included the subcategory 'environmentally friendly farming', which includes alternative irrigation technique (USPC 405/36–51), animal waste disposal or recycling (USPC 210/610–611; 71/11–30), fertilizer alternative, e.g., composting (USPC 71/8–30), pollution abatement, soil conservation (USPC 405/15), water conservation (USPC 137/78.2–78.3; 137/115.01–115.28) and yield enhancement (USPC 504).[72] The subcategory of 'alternative energy production' includes Agricultural Waste (USPC 44/589), Biofuel (USPC 44/605; 44/589), and Fuel from animal waste and crop residues (USPC 44/605). The subcategory of 'environmental purification, protection, or remediation' includes plants and plant breeding (USPC 800/260–323.3).[73] Thus, an array of agricultural inventions could be considered to be patentable subject matter.

---

69 The Council for Biotechnology (2008), 'Agricultural Biotechnology: Benefits for Biofuels and U.S. Energy Security', http://www.whybiotech.com/resources/briefs_benefitsforbiofuels.asp.

70 The Council for Biotechnology (2008), 'Agricultural Biotechnology: Benefits for Water Scarcity in the United States', http://www.whybiotech.com/resources/briefs_benefitsforwaterscarcity.asp.

71 Chatzimarkakis, J. (2009), 'Getting an Appetite for Biotechnology', Viewpoint, *BBC News*, 24 February, http://news.bbc.co.uk/2/hi/sci/tech/7905567.stm.

72 United States Patent and Trademark Office (2009), 'Pilot Program for Green Technologies Including Greenhouse Gas Reduction', *Federal Register*, **74** (234), 8 December, 64666–9.

73 Ibid.

Table 12.1 Climate ready patent claims (30 June 2008 to 30 June 2010)

Assignee	# Patent Families	% of Total	Total Patents and Applications	Total Issued Patents
DuPont	114	44%	240	104
BASF (including CropDesign and Metanomics)	48	18%	522	53
Monsanto	11	4%	122	3
Mendel Biotechnology	4	2%	232	21
Syngenta	6	2%	39	2
Evogene	8	3%	64	1
Bayer	7	3%	43	2
Dow	3	1%	18	1
No Assignee	17	7%	99	5
Others	43	16%	272	28
Total	262	100%	1663	221

Source: ETC Group 2010[74]

A Pioneer Hi-Bred International and DuPont

In its Statement on Global Climate Change, the conglomerate DuPont emphasises: 'DuPont believes that the science is sufficient to compel prudent action to address global climate change.'[75] DuPont emphasises: 'The role of science and technology development and diffusion, and governmental incentives and support for such efforts, are critical.'[76] The conglomerate

[74] ETC Group (2010), 'Gene Giants Stockpile Patents on "Climate-Ready" Crops in Bid to Become "Biomassters": Patent Grab Threatens Biodiversity, Food Sovereignty', *ETC Group Communique*, November, Issue 102, http://www.etcgroup.org/en/node/5220.

[75] DuPont, 'Statement on Climate Change', http://www2.dupont.com/Media_Center/en_US/position_statements/global_climate.html.

[76] Ibid.

intended to bring a range of products to market: 'These sustainable solutions include alternative energy sources such as photovoltaic solar cells and next generation biofuels, value-adding materials produced from agricultural feedstocks rather than petroleum like our DuPont™ Sorona® polymer fiber, and energy efficiency aids such as next generation refrigerants and DuPont™ Tyvek® HomeWrap®.'[77]

Pioneer Hi-Bred International is a business of the chemical and science-based conglomerate DuPont, which aspires to be 'the world's leading source of customized solutions for farmers, livestock producers and grain and oilseed processors'.[78] DuPont Executive Vice President James C. Borel has contended that innovations in biotechnology can both address the global needs for food and fibre, and can reduce dependence on fossil fuels: 'Through biotechnology, we have the opportunity to transform how we produce and distribute food and how we source energy and materials.'[79] The company has also promoted public-private partnerships as a means of addressing global food security. Vice President of DuPont Crop Genetics Research and Development William S. Niebur told attendees at the Crawford Fund Annual Development Conference: 'The need to increase agricultural productivity amid the uncertainties of global climate change, political instabilities, shrinking availability of arable land and increasing world populations present challenges no single company, government or research institution can solve alone.'[80]

Pioneer Hi-Bred International Inc. holds a 2004 United States patent in respect of 'Transcriptional activators involved in abiotic stress tolerance'.[81] In the description, the patent application identifies the field of invention as the 'field of plant molecular biology, more particularly to regulation of gene expression in plants'.[82] The background to the invention is particularly illuminating and worth quoting extensively from:

---

77 Ibid.

78 DuPont Australia (2006), 'Drought Tolerance a Target of New International Crop Research Agreement', Press Release, 6 April, http://www2.dupont.com/Media_Center/en_AU/australia_news/2006/article20060406.html.

79 DuPont (2009), 'Biotechnology Can Address Global Mega Trends, DuPont Leader Says: Biotechnology Stewardship is Critical to Meeting Global Needs for Increased Food and Energy', BioJapan 2009, 7 October.

80 Pioneer HiBred International (2009), 'Public-Private Partnerships Key to Achieving Global Food Security, Says DuPont Leader', 2009 Crawford Annual Fund Conference, Canberra, Australia, 27 October.

81 Sivasankar, S., T. Helentjaris and D. Xu (2004), 'Transcriptional activators involved in abiotic stress tolerance', United States Patent No. 7,253,000.

82 Ibid.

> Insufficient water for growth and development of crop plants is a major obstacle to consistent or increased food production worldwide. Population growth, climate change, irrigation-induced soil salinity, and loss of productive agricultural land to development are among the factors contributing to a need for crop plants which can tolerate drought. Drought stress often results in reduced yield. In maize, this yield loss results in large part from plant failure to set and fill seed in the apical portion of the ear, a phenomenon known as tip kernel abortion.[83]

Pioneer Hi-Bred International Inc. also holds a companion 2006 United States patent on 'Transcriptional activators involved in abiotic stress tolerance'.[84] The company has also lodged applications under the *Patent Cooperation Treaty* 1970 to facilitate the international exploitation of the patents.[85]

B BASF

In its submission to the Supreme Court of the United States in *JEM Ag Supply* v. *Pioneer Hi-Bred International*, BASF expressed its preference for patent protection as opposed to plant variety protection or plant patent protection: 'Utility patent protection for plants and plant varieties is vital to BASF's agricultural businesses, and to seed producers and farmers, whose business depends upon novel and improved agricultural products, including plants.'[86]

In its 2009 *Annual Report*, BASF emphasised that it was 'committed to ambitious global climate protection that rapidly implements existing, economically-viable solutions'.[87] Moreover, the company comments: 'Products and processes that improve energy and raw material efficiency or contribute towards the use of renewable energies represent an important area of innovation for BASF.'[88] The company observed of its work in respect of agricultural biotechnology:

---

83 Ibid.

84 Sivasankar S., T. Helentjaris, D. Xu and A. Lyznik (2006), 'Transcriptional Activators Involved in Abiotic Stress Tolerance', United States Patent No. 7,317,141.

85 Sivasankar, S., D. Selinger and N. Brugiere (2009), 'Transcriptional Activators Involved in Abiotic Stress Tolerance', WIPO Publication No: WO/2009/094527.

86 BASF Corporation, 'Brief in Support of the Respondent in *J.E.M. Ag Supply, Inc.* v. *Pioneer Hi-Bred International, Inc.*', 2001 WL 689271 (U.S.) (Appellate Brief).

87 BASF Corporation (2009), *Annual Report: Economic, Environmental and Social Performance*, 91, http://www.report.basf.com/2009/en/servicepages/downloads/files/BASF_Report_2009.pdf?cat=b.

88 Ibid., 92.

> As a partner to leading seed companies, BASF Plant Science develops traits, or genetic properties, such as drought and disease resistance or higher yields. We offer seed companies solutions ranging from development to end product, depending on the characteristics and the crop. Together with our partners, we bring the seeds to market. In addition to crops that offer farmers higher and more secure yields, we also develop plants that are more nutritious or can be used as renewable raw materials.[89]

The company highlighted: 'Together with Monsanto, we develop higher-yielding and stress-tolerant corn (maize), soybean, canola (oil-seed rape) and cotton plants.'[90]

BASF holds a large portfolio of patents relating to plants and environmental stress. Notably, the company holds a patent entitled 'Transcription Factor Stress-Related Proteins and Methods of Use in Plants'.[91] This invention relates generally to nucleic acid sequences encoding proteins that are associated with abiotic stress responses and abiotic stress tolerance in plants. In response to criticism by the ETC Group, Jonathan Bryant, managing director of BASF's United States division, contended that the company was motivated by the common good: 'We're all looking to bring our technology together for a common good.'[92]

C Monsanto

The agricultural and chemical company Monsanto has argued: 'Through plant breeding, Monsanto is producing more resilient, better-adapted crops by continuous selection as local conditions change.'[93] The firm observes that 'Monsanto's pipeline of biotech crops is focused on environmental-stress adaptation' – including 'drought-tolerant crops and nitrogen-efficiency genes'.[94] Monsanto notes: 'Drought-tolerant crops are designed to provide greater yield stability in years when crops would otherwise suffer due to drought conditions.'[95] The company adds: 'Nitrogen-use efficiency has the potential to contribute to a significant reduction of agricultural

89 Ibid., 73.
90 Ibid., 73.
91 Da Costa, S. (2007), 'Transcription Factor Stress-Related Proteins and Methods of Use in Plants', United States Patent No. 7,619,137.
92 Weiss, R. (2008), 'Firms Seek Patents on "Climate Ready" Altered Crops', *The Washington Post*, 13 May, A04.
93 Monsanto (2008), 'Agriculture Can Help Keep Carbon in Balance', http://www.monsanto.com/responsibility/our_pledge/healthier_environment/climate_change.asp (available on the Internet Archive's Wayback Machine).
94 Ibid.
95 Ibid.

[greenhouse gas] emissions while maintaining high grain yield at lower nitrogen levels.'[96]

Monsanto has obtained a portfolio of patents in this field. They have filed for a patent in respect of 'drought tolerant corn with reduced mycotoxin'.[97] In response to criticism by the ETC Group, Ranjana Smetacek, a spokeswoman for Monsanto, said companies deserve praise for developing crop varieties that will survive climate change:

> I think everyone recognizes that the old traditional ways just aren't able to address these new challenges. The problems in Africa are pretty severe. We aim to be at once generous and also cognizant of our obligation to shareholders who have paid for our research.[98]

Smetacek noted that Monsanto and BASF are participating in a project, funded by the Bill and Melinda Gates Foundation, to develop drought-resistant corn that would be made available to farmers in four southern African countries royalty-free.

In June 2009, Monsanto and BASF announced that they had discovered that 'a naturally-occurring gene can help corn plants combat drought conditions and confer yield stability during periods of inadequate water supplies'.[99] The companies observe that 'they will use the gene in their first-generation drought-tolerant corn product which is designed to provide yield stability to their farmer customers' and that the product 'will be the first biotechnology-derived drought-tolerant crop in the world'.[100] The companies said that the drought-tolerant corn contains the *cspB* gene, from *Bacillus subtilis*. Robert Fraley, Chief Technology Officer for Monsanto, commented:

> The development of this trait demonstrates the strength of our robust discovery engine which is fueled by our ongoing investment in R&D. It also reflects our commitment to our farmer customers and a recognition of the investment they

---

96 Ibid.

97 Monsanto (2009), 'Drought Tolerant Corn with Reduced Mycotoxin', World Intellectual Property Organization Patent Cooperation Treaty Application, WO/2009/049110.

98 Weiss, R. (2008), 'Firms Seek Patents on "Climate Ready" Altered Crops', *The Washington Post*, 13 May, A04.

99 Monsanto and BASF (2009), 'Monsanto, BASF Scientists Disclose Discovery of Gene Conferring Drought Tolerance in Corn Plants', Press Release, 9 June, http://www.basf.com/group/pressrelease/P-09–274.

100 Ibid.

> make in our products. Drought-tolerant corn will be another tool with which we can help them meet the challenges facing agriculture today.[101]

Jürgen Schweden, Senior Vice President of Research and Development at BASF, observed: 'With our complementing technologies, we are able to bring more and better products to farmers faster.'[102] The two companies hope to release the drought-tolerant corn product for commercialisation in 2012. The two companies have sought regulatory submissions for cultivation in the United States and Canada, and for import to Mexico, the European Union and Colombia.

Monsanto scientists have published the findings of their research in the journal *Plant Physiology*.[103] Michael Edgerton of Monsanto presents agricultural biotechnology as a means of responding to the global demands for feed, food and fuel: 'Improving yields in corn and other crops on a global basis will allow farmers to meet global demand for feed, fuel, and food while minimizing the need to bring large amounts of new land into crop production.'[104] This announcement has alarmed some environmental groups, such as the African Centre for Biosafety.[105]

Monsanto has a research agreement with the plant genomics company, Ceres Inc., which has also been a significant actor in the marketplace. The company comments upon its portfolio of patents:

> Our intellectual property position has been enabled by our capability to sequence full-length genes and attribute specific utilities to them. We test multiple versions of the same gene, often from more than one plant species, to maximize the depth and breadth of our intellectual property position. Our intellectual property approach and portfolio have been validated by industry-leading licensing agreements.[106]

The company holds exclusive rights to 30 issued patents both for food crops and energy crops for biofuels.[107] The applications relate to 'methods of

---

101 Monsanto and BASF (2009), 'Gene Provides Yield Stability During Periods of Inadequate Water Supply', Press Release, 9 June, http://www.basf.com/group/pressrelease/P-09–274.

102 Ibid.

103 Edgerton, M. (2009), 'Increasing Crop Productivity to Meet Global Needs for Feed, Food, and Fuel', *Plant Physiology*, **149**, 7–13.

104 Ibid., 12.

105 African Centre for Biosafety (2009), *Patents, Climate Change and African Agriculture: Dire Predictions*, September, http://www.biosafety-info.net/file_dir/17225619254ac3067689f7a.PDF.

106 Ceres Inc., Intellectual Property, http://www.ceres.net/Technology/Tech-IP.html.

107 Ceres Inc., http://www.ceres.net.

making improved plants, and encompassed 70,000 full-length functionally annotated cDNA sequences from several species, as well as hundreds of gene-trait associations and promoters with specialized expression patterns'.[108] Ceres Inc. has a research agreement with Monsanto.

D Mendel Biotechnology

Mendel Biotechnology Inc. promotes itself as 'seeding a sustainable future'.[109] The company is squarely focused upon servicing the bioenergy sector:

> During the next 20 years, with an expanding global population, improved diets and rapidly growing energy demands, society will need to produce plants enhanced for food, feed, fiber, energy and aesthetic benefit without large increases in production acreage [...] Mendel Biotechnology's vision is that our knowledge about regulation of plant gene and pathway function will enable accelerated improvement in plant varieties, and the delivery of associated services, to meet global agricultural and energy production needs.[110]

The company holds a number of patents in the area.[111] In February 2010, the company was awarded a patent relating to 'Plant Transcriptional Regulators' – which confer improved tolerance in genetically engineered plants for drought, shade and low nitrogen conditions.[112] Mendel Biotechnology Inc.'s strategic partners include Monsanto, a long-term technology collaborator; BP, a collaborator on the development of a BioEnergy Seeds and Feedstocks business; and Bayer CropScience, a partner jointly developing chemical products to regulate plant stress tolerance.[113]

---

108 Ceres Inc., Intellectual Property, http://www.ceres.net/Technology/Tech-IP.html.

109 Mendel Biotechnology Inc., http://www.mendelbio.com/index.php.

110 Mendel Biotechnology Inc., http://www.mendelbio.com/aboutus/index.php.

111 Mendel Biotechnology Inc. (2003), 'Plant Transcriptional Regulators of Drought Stress', United States Patent Application No. US20070240243A9; Mendel Biotechnology, Inc. (2005), 'Plant Stress Tolerance from Modified AP2 Transcription Factors', World Intellectual Property Organization Patent Cooperation Treaty Application, WO06069201A3; and Mendel Biotechnology, Inc. (2006), 'Stress Tolerance in Plants may Include Salt, Hyperosmotic Stress, Heat, Cold, Drought, or Low Nitrogen Conditions', World Intellectual Property Organization Patent Cooperation Treaty Application, WO07028165C1.

112 Heard, J. and others (2006), 'Plant Transcriptional Regulators', United States Patent No. 7,663,025.

113 Mendel Biotechnology Inc., http://www.mendelbio.com/strategicpartners/index.php.

E EvoGene and Its Partners

The Israeli company Evogene has also focused upon developing plants, with tolerance to abiotic stress:

> Abiotic stress conditions such as drought, salty water, soil salinity and cold stress have a huge impact on crop productivity and agricultural supply, especially due to global warming, erratic climate changes and loss of agriculture farming land. The interest in developing plants displaying improved tolerance to abiotic stress is driven not only by the enormous economic potential, but also by the humanitarian impact it may have on agricultural supply in third world countries. Evogene has discovered key genes for improving plant yield and tolerance to abiotic stress, currently being introduced into field crops.[114]

The company has emphasised that its patented genes are being incorporated into commercial crops through collaboration with Bayer CropScience in rice, Mertec LLC in soybean, Biogemma in corn, Viterra in canola and CIRAD in cotton.

Evogene has filed a blockbuster patent claiming a large number of gene sequences relating to stress tolerance, biomass accumulation and yield in plants.[115] The invention defines 'plants' in a broad, open-ended fashion: 'The term "plant" as used herein encompasses whole plants, ancestors and progeny of the plants and plant parts, including seeds, shoots, stems, roots (including tubers), and plant cells, tissues and organs.'[116] It notes: 'According to some embodiments of the invention, the plant used by the method of the invention is a crop plant such as rice, maize, wheat, barley, peanut, potato, sesame, olive tree, palm oil, banana, soybean, sunflower, canola, sugarcane, alfalfa, millet, leguminosae (bean, pea), flax, lupinus, rapeseed, tobacco, poplar and cotton.'[117] In its description, the patent application notes that the invention relates to methods for 'increasing abiotic stress tolerance, water use efficiency (WUE), fertilizer use efficiency (FUE), biomass, vigor and/or yield of a plant'.[118] The application notes: 'Abiotic stress conditions such as salinity, drought, flood, suboptimal temperature and toxic chemical pollution, cause substantial damage to agricultural

---

114 Evogene, http://www.evogene.com/project.asp?project_id=4.

115 Evogene (2009), 'Polynucleotides, Plolypeptides Encoded Thereby, and Methods of Using Same for Increasing Abiotic Stress Tolerance and/or Biomass and/or Yield in Plants Expressing Same', World Intellectual Property Organization Patent Cooperation Treaty Application, WO/2009/013750.

116 Ibid.

117 Ibid.

118 Ibid.

plants.'[119] The international searching authority suggests, however, that a number of the claims (1–4, 6–11 and 16–20) lack novelty and inventive step because of a patent granted to Monsanto in 2004 – US 2004/0123343 A1.[120]

DuPont have also entered into a collobarative agreement with Evogene[121] to improve drought stress tolerance in corn and soybeans.[122] Under the agreement Pioneer Hi-Bred acquired exclusive rights to several genes discovered by Evogene for improving these traits. William S. Niebur, Vice President of DuPont Crop Genetics Research and Development, commented: 'Drought has an enormous impact on crop productivity and quality as well as cost of production. It's the one agronomic pressure that negatively affects growers in every country around the world, every year.'[123]

F Bayer CropScience

The Chairman of Bayer CropScience has emphasised that its crop protection program is focused on larger challenges such as climate change and global warming:

> Success tomorrow demands investment today. The time to compete for tomorrow's best products and technologies is now. But we must not lose sight of the long-term challenges either. By these I mean the major issues such as demographic change, global food supplies, energy conservation and climate protection. Bayer's climate program is already making very encouraging headway. We have good answers that address long-term challenges and at the same time take advantage of related business opportunities for our enterprise.[124]

The 2009 annual report promotes the company's commitment to sustainability as a model: 'Our objective is to develop innovative solutions that will help address global challenges such as providing sustainable health care, feeding a steadily growing world population, combating climate change and overcoming the scarcity of natural resources.'[125]

---

119 Ibid.

120 Monsanto (2004), 'Rice Nucleic Acid Molecules and Other Molecules Associated with Plants and Uses Thereof for Plant Improvement', United States Patent Application No. US2004/0123343 A1.

121 EvoGene, http://www.evogene.com/.

122 DuPont and EvoGene (2007), 'DuPont and Evogene Collaborate to Increase Drought Tolerance in Corn and Soybeans', Press Release, 30 October.

123 Ibid.

124 Bayer AG (2008), Chairman's Letter – *Annual Report*, http://www.annualreport2008.bayer.com/en/chairmans-letter.aspx.

125 Bayer AG (2009), *Annual Report*, Leverkusen: Bayer AG, 109, http://www.bayer.com/en/Bayer-Annual-Report-2009.pdfx.

Bayer CropScience has boasted that it is 'an international leader in its sector with respect to its achievements in climate protection'.[126] The company has emphasised that it has invested in 'stress-tolerant plants':

> Climate change is leading to worsening growing conditions in agriculture. Heat, drought, cold and soil salinity place food crops under extreme stress. Bayer is working to boost the stress tolerance of rice, corn and other cereals and thereby safeguard crop yields even under extreme conditions.[127]

The company hoped to increase the efficiency of agricultural production to meet the demands for food by a growing global population: 'Thanks to its state-of-the-art crop protection products and new solutions in the field of plant breeding and biotechnology, Bayer is helping to achieve sustainable increases in yields all over the world.'[128]

The company stressed: 'Bayer CropScience is developing plant biotechnology techniques to immunize crops against stress due to climatic and environmental factors.'[129] It noted: 'The company's scientists are working to develop varieties of cereals, maize and rice that can be grown in extreme regions that were not previously suitable for farming.'[130] The company observed: 'Examples include plants with a high resistance to salt and drought, that tolerate very hot or cold conditions, that can survive on very little nutrient or water input, or that can grow in heavily polluted soil.'[131] Bayer CropScience researcher Wolfgang Thielert received the Otto Bayer Medal for his work in respect of plants and stress-tolerance. The scientist observed: 'In view of worldwide climate change, it is increasingly important to enhance the ability of plants to withstand heat, lack of water and other abiotic stress factors and thus help to safeguard harvests.'[132] He commented: 'At the time, there were early signs that our tried-and-tested insecticides Confidor, Admire and Gaucho help crop plants to cope better with moderate stress than untreated plants.'[133] The researcher and his

---

126 Ibid., 28.

127 Ibid., 28.

128 Ibid., 28.

129 Bayer AG (2008), 'The Second Green Revolution: Climate Change', Leverkusen: Bayer AG, http://www.bayercropscience.com/bcsweb/cropprotection.nsf/id/EN_Climate_Change.

130 Ibid.

131 Ibid.

132 Bayer Crop Science (2008), 'Defying the Consequences of Climate Change: Tried-and-tested Insecticide Makes Plants More Resistant to Stress', 23 October, http://www.bayercropscience.com/BCSWeb/CropProtection.nsf/id/EN_2008-NST-042_Wolfgang_Thielert.

133 Ibid.

colleagues hoped that the research would be translatable to other plants: 'It is now our goal to continue building on the new Stress Shield-potential of other Bayer CropScience products.'[134]

BayerCropScience has acquired a family of patents in respect of stress tolerant plants.[135]

G Syngenta

The agricultural company, Syngenta, also recognises the challenges presented by climate change:

> Reducing greenhouse gas emissions is the key challenge, but innovation is also crucial to enable agriculture to adapt to an already changing climate. We are investing in developing new crop varieties, crop protection products and agricultural processes that will help farmers grow more resilient crops that can withstand more extreme weather and temperatures. We are also developing new crop varieties and techniques that will make producing fuel from plants more efficient. This will help make biofuels a viable alternative to fossil fuels.[136]

The company is committed to 'developing new crop varieties and techniques that will make producing fuel from plants more efficient and biofuels a viable alternative to fossil fuels' and 'enabling agriculture to adapt to a changing climate'.[137]

H Dow Agrosciences

Dow Agrosciences, a subsidiary of the Dow Chemical Company, is also focused upon intellectual property and climate change. The company insists that to 'provide humanity with a sustainable energy supply while addressing climate change is the most urgent environmental issue our society faces'.[138] The company comments:

---

134 Ibid.

135 Bayer Bioscience N.V. (2005), 'Stress Tolerant Cotton Plants', World Intellectual Property Organization Patent Cooperation Treaty Application No. WO06045633A1; and Bayer CropScience GmbH (2006), 'Active Substances for Increasing the Stress Defense in Plants to Abiotic Stress, and Methods of Finding them', United States Patent Application No. US20070124839A1.

136 Syngenta, http://www.syngenta.com/en/corporate_responsibility/climate change.html.

137 Ibid.

138 Dow (2008), 'Introduction to Dow's Position on Energy and Climate Change', http://www.dow.com/commitments/pdf/OurPosition_EngClimate_FINAL.pdf.

> As a world leader in chemistry, Dow is uniquely positioned to continue to provide innovations that lead to energy alternatives, less carbon intensive raw material sources, and other solutions not yet imagined. In fact, our science and technology has been contributing solutions to the global climate change and energy challenges since 1990. Our science has led to the development of alternative energy sources such as biofuels, photovoltaics and wind. Our products contribute to reduced energy consumption.[139]

The Dow Conglomerate promises to 'leverage the strength of the human element in our laboratories around the world and make unprecedented financial investments in R&D to achieve breakthrough solutions that will slow, stop and reverse global warming'.[140] The company holds patents on drought- and heat-resistant corn lines.[141]

Summary

Such climate-ready crops will raise a host of legal questions including issues relating to patentable subject matter, such as in the case of *JEM Ag Supply* v. *Pioneer Hi-Bred International*, *Association for Molecular Pathology* v. *the United States Patent and Trademark Office*, and *Bilski* v. *Kappos*,[142] the application of patent criteria such as inventive step, like *KSR International Co.* v. *Teleflex Inc.* and *In re Kubin*;[143] and the demands of utility, as explored in the case of *in re Fisher*.[144] Such patents will also raise questions of patent infringement, farmers' rights, and innocent bystanders like in the case of *Monsanto* v. *Schmeiser*,[145] and the appropriate remedies to be awarded in the case of infringement.[146] Exports of patented climate-ready

---

139 Ibid.

140 Ibid.

141 Agrigenetics, Inc. Mycogen Seeds (Dow Agrosciences) (2003), 'Inbred Corn Line BE1146BMR', United States Patent No. US7273970; and Agrigenetics, Inc., Mycogen Seeds (Dow Agrosciences) (2003), 'Inbred Corn Line 4VP500 Drought Heat', United States Patent Application No. US20050076401A1.

142 *JEM Ag Supply* v. *Pioneer Hi-Bred International Inc* 534 US 124 (2001); *Association for Molecular Pathology et al.* v. *United States Patent and Trademark Office, Myriad Genetics, et al.* 702 F. Supp. 2d 181, (S.D.N.Y. 2010), *appeal docketed,* No. 2010–1406 (Fed. Cir. June 22, 2010); *Association for Molecular Pathology et al.* v. *United States Patent and Trademark Office, Myriad Genetics, et al.* 653 F.3d 1329, 2011 WL 3211513 (C.A. Fed. (N.Y.)), 99 USPQ 2d 1398; and *Bilski* v. *Kappos* 130 S. Ct. 3218 (2010).

143 *KSR International Co.* v. *Teleflex Inc.* 550 US 398 (2007); and *In re Kubin* 561 F. 3d 1351 (Fed. Cir. 2009).

144 *In re Fisher* 421 f.3d 1365 (C.A.Fed., 2005).

145 *Monsanto Canada Inc.* v. *Schmeiser* (2004) SCC 34; 2004 SCC 34.

146 *Rivett* v. *Monsanto Canada Inc.*, 2010 FCA 207 (2010).

crops and related food products may raise complex legal issues.[147] There could also be disputes over the ownership of patents relating to climate-ready crops – especially if there is a disintegration of trust between collaborators from the private sector and public sector.

III FARMERS' RIGHTS

There has been much debate about the validity of the claims of agricultural biotechnology companies to develop 'climate-ready' crops. In the Report of the Special Rapporteur on the Right to Food, Olivier De Schutter expressed concerns that the use of plant breeders' rights, together with patents, contracts and genetic use restriction technologies, could have an adverse impact upon farmers.[148] He complained that there were also issues about the strengthening of plant breeders' rights:

> The strengthening of breeders' rights in the *UPOV Convention* 1991 is also a concern in this regard. This convention prohibits the commercialization of varieties which are essentially derived from a PVP-protected variety (article 14 (5)), and farmers are now prohibited from exchanging or selling seeds saved from the harvest of protected varieties (article 15).[149]

Olivier De Schutter emphasises the 'urgent need to support traditional knowledge, innovations and practices, as a source of adaptation of local communities to new threats such as climate change, soil erosion or outbreaks of new crop diseases'.[150] He fears that 'Intellectual property rights reward and encourage standardization and homogeneity, when what should be rewarded is agrobiodiversity, particularly in the face of the emerging threat of climate change.'[151] (The Secretariat to the *UPOV Convention* has taken umbrage at such claims, and issued a rejoinder.[152])

---

147 *Monsanto Technology LLC* v. *Cefetra BV and others*, No. C-428/08, European Court of Justice (6 July 2010).

148 De Schutter, Olivier (2009), *Seed Policies and the Right to Food: Enhancing Agrobiodiversity and Encouraging Innovation*, United Nations General Assembly, A/64/170, http://www.ip-watch.org/weblog/wp-content/uploads/2009/11/n0942473.pdf.

149 Ibid., 15.

150 Ibid., 5.

151 Ibid., 14.

152 Secretariat of the *UPOV Convention* (2009), 'Note on the Report of the Special Rapporteur on the Right to Food', http://upov.int/export/sites/upov/en/news/2009/un_ga_note.pdf.

During its deliberations in Copenhagen, the Ad Hoc Working Group on Long-term Co-operative Action under the *United Nations Framework Convention on Climate Change* 1992 identified five key options to address issues pertaining to intellectual property and climate change – including technology transfer; specific measures such as a patent pool and public sector licensing; patent subject matter exclusions; compulsory licensing; and a technology mechanism.[153] While the *Copenhagen Accord* 2009 has some language on a technology mechanism, the agreement is silent on the question of intellectual property.[154] In preparation for COP 16, members of the Ad Hoc Working Group on Long-term Cooperative Action under the *United Nations Framework Convention on Climate Change* 1992 continued their discussions over intellectual property and climate change.[155] The 10 February 2010 annex on enhanced action on technology development and transfer canvassed two options in bracketed text – 'Option 1' that there be 'no reference to Intellectual Property Rights in the text' and 'Option 2' that there should be reference to intellectual property and climate change in the text. A number of clauses were contemplated under 'Option 2', including measures relating to technology transfer, patent pools, public sector licensing, patent subject matter exclusions, compulsory licensing, and Climate Innovation Centres.[156] However, as there was a lack of consensus on such options, there is no specific, explicit mention of intellectual property in the *Cancún Agreements* 2010.[157]

In the more circumscribed context of agricultural intellectual property, there have been a range of similar options presented, such as excluding certain plants from intellectual property protection; promoting exceptions for farmers' rights; flexible licensing; patent pools; compulsory licensing; and competition safeguards.

---

153 Ad Hoc Working Group on Long-term Co-operative Action under the *United Nations Framework Convention on Climate Change* (2009), 'Report', Seventh Session held in Bangkok from 28 September to October 2009, and Barcelona from 2 to 6 November, FCCC/AWGLCA/2009/14, 20 November.

154 The *Copenhagen Accord* 2009, UN Doc. FCCC/KP/CMP/2009/L.9 (December 18, 2009).

155 Ad Hoc Working Group on Long-term Co-operative Action under the *United Nations Framework Convention on Climate Change* (2010), 'Work Undertaken by the Conference of the Parties at its Fifteenth Session on the Basis of the Report of the Ad Hoc Working Group on long-Term Cooperative Action Under the Convention', FCCC/CP/2010/2, 11 February, http://unfccc.int/resource/docs/2010/cop16/eng/02.pdf.

156 Ibid., 24.

157 The *Cancún Agreements* 2010, *Outcome of the Work of the Ad Hoc Working Group on Long-term Co-operative Action under the Convention*, -/CP.16, http://unfccc.int/files/meetings/cop_16/application/pdf/cop16_lca.pdf.

A Patentable Subject Matter

There has been fierce debate over the subject matter protected by intellectual property rights.

Under Article 27 (3)(b) of the *TRIPS Agreement* 1994, members of the World Trade Organization (WTO) must provide for the protection of plant varieties either by patents, by an effective *sui generis* system or by any combination thereof.[158] There remains scope, though, under Article 27 (2) to exclude from patentability inventions to 'avoid serious prejudice to the environment'.[159]

A number of landmark decisions in superior courts have recognised that patent protection extends to plants. In the matter of *Diamond* v. *Chakrabarty*, the majority of the Supreme Court of the United States recognised that the *Plant Patent Act* 1930 (US), the *Plant Variety Protection Act* 1970 (US), and the general regime of patents could co-exist alongside one another.[160] In the case of *JEM Ag Supply Inc* v. *Pioneer Hi-Bred International Inc.*, the majority of the Supreme Court of the United States held that utility patents could be granted in respect of plants in addition to plant patents and plant variety rights.[161] In the case of *Monsanto* v *Schmeiser*, McLachlin J and Fish J of the Supreme Court of Canada emphasised: 'Under the present Act, an invention in the domain of agriculture is as deserving of protection as an invention in the domain of mechanical science.'[162] In *Grain Pool Of Western Australia* v. *Commonwealth*, the High Court of Australia held that the intellectual property power of the *Australian Constitution* supported both the patent regime and the plant breeders' rights system.[163] The European Patent Office has been considering the merits of patents being granted in respect of foods such as broccoli and tomatoes.[164]

---

158 *Marrakesh Agreement Establishing the World Trade Organization*, opened for signature 15 April 1994, 1867 UNTS 3, annex 1C (Agreement on Trade-Related Aspects of Intellectual Property Rights) (entered into force 1 January 1995) ('*TRIPS Agreement*').

159 Ibid.

160 *Diamond* v. *Chakrabarty* 447 US 303 (1980).

161 *JEM Ag Supply Inc* v. *Pioneer Hi-Bred International Inc* 534 US 124 (2001).

162 *Monsanto Canada Inc.* v. *Schmeiser* (2004) SCC 34; 2004 SCC 34.

163 *Grain Pool Of Western Australia* v. *Commonwealth* (2000) 46 IPR 515.

164 *State of Israel – Ministry of Agriculture* v. *Unilever* T1242/06 – 3.3.04 (4 April 2008), EPO Boards of Appeal Decision, http://legal.european-patent-office.org/dg3/biblio/t061242ex1.htm; currently under review by the Enlarged Board of Appeals. The patent in question is Schaffer, A. (1998), 'Method for

There has been the odd decision that runs against this trend. For instance, in a heterodox decision involving Myriad Genetics' patents over BRCA1 and BRCA2, Sweet J expressed reservations in the United States District Court about such a broad approach to patentable subject matter, suggesting that genes are mere products of nature.[165] The judge attempted to distinguish past rulings: 'While the Supreme Court in *J.E.M. Ag Supply* noted the USPTO's practice of issuing patents on sexually reproducing plants in concluding that such plants represented patentable subject matter under § 101, that passing observation was neither dispositive nor central to the Court's holding and does not establish a rule of judicial deference to the USPTO's practices.'[166] However, the majority of the Court of Appeals for the Federal Circuit rejected this approach, and supported a broad, expansionist reading of patentable subject matter.[167]

Nonetheless, a number of opponents have persisted with environmental objections to patents being extended to plants, agriculture, food and the environment. Some objections are based upon concerns about the commercialisation and monopolisation of food and genetic material.[168] The anti-biotechnology organisation, the ETC Group, has protested about the agricultural biotechnology firms building up large patent portfolios in respect of plants[169] – particularly in respect of climate-ready crops. With inflammatory rhetoric, Silvia Ribeiro of the ETC Group protested:

> The Gene Giants are stockpiling patents that threaten to put a choke-hold on the world's biomass and our future food supply. The breadth of many patent claims on climate ready crop genes is staggering. In many cases, a single patent or patent application claims ownership of engineered gene sequences that could be

---

Breeding Tomatoes having Reduced Water Content and Product of the Method', European Patent Application No. 00940724.

165 *Association for Molecular Pathology et al.* v. *United States Patent and Trademark Office, Myriad Genetics, et al.* 702 F. Supp. 2d 181, (S.D.N.Y. 2010), appeal docketed, No. 2010–1406 (Fed. Cir. June 22, 2010).

166 *Association for Molecular Pathology et al.* v. *United States Patent and Trademark Office, Myriad Genetics, et al.* 702 F. Supp. 2d 181, (S.D.N.Y. 2010), appeal docketed, No. 2010–1406 (Fed. Cir. June 22, 2010).

167 *Association for Molecular Pathology et al.* v. *United States Patent and Trademark Office, Myriad Genetics, et al.* 653 F.3d 1329, 2011 WL 3211513 (C.A. Fed (N.Y.)), 99 USPQ 2d 1398.

168 Anna Shih notes that 'patenting food products and food-processing techniques is not new, or even uncommon'. Shih, A. (2002), 'The Patented Peanut Butter and Jelly Sandwich: Food as Intellectual Property', *Gastronomica*, 23–7.

169 ETC Group, 'Intervenor in *Monsanto Canada Inc.* v. *Schmeiser* (2004) SCC 34', http://www.canadians.org/food/documents/ETC_Affidavit.pdf.

> deployed in virtually all major crops – as well as the processed food and feed products derived from them.[170]

In 2008, the ETC Group contended that 'the world's largest seed and agrochemical corporations are stockpiling hundreds of monopoly patents on genes in plants that the companies will market as crops genetically engineered to withstand environmental stresses such as drought, heat, cold, floods, saline soils and more'.[171] The ETC Group contends: 'Monsanto, Bayer, Syngenta, Dupont and biotech partners have filed 532 patent documents (a total of 55 patent families) on so-called "climate ready" genes at patent offices around the world.'[172] In 2010, the ETC Group updated this study, and observed that '262 patent families (subsuming 1663 patent documents worldwide) published between June 2008 – June 2010 make specific claims to abiotic stress tolerance (such as drought, heat, flood, cold and salt-tolerance) in plants'.[173]

Vandana Shiva of the Indian-based Research Foundation for Science, Technology and Ecology, has also shown a strong interest in the interlinkage of agricultural intellectual property, environmental justice and climate change.[174] She observed of the patents in respect of climate-ready crops identified by the ETC Group:

> These patents are the latest form of biopiracy. Farmers have bred seeds for drought, flood and salt tolerance over millennia. Climate resilience ultimately depends on farmers' innovation, biodiversity and agro-ecological processes staying in the hands of farming communities.[175]

---

170 ETC Group (2010), 'Gene Giants Stockpile Patents on "Climate-Ready" Crops in Bid to Become Biomassters', Press Release, 25 October, http://www.etcgroup.org/en/node/5220.

171 ETC Group (2008), 'Patenting the "Climate Genes"... And Capturing the Climate Agenda', *ETC Group Communique*, May/ June, Issue 99, http://www.etcgroup.org/en/materials/publications.html?pub_id=687.

172 Ibid.

173 ETC Group (2010), 'Gene Giants Stockpile Patents on "Climate-Ready" Crops in Bid to Become "Biomassters": Patent Grab Threatens Biodiversity, Food Sovereignty', *ETC Group Communique*, November, Issue 102, http://www.etcgroup.org/en/node/5220.

174 Shiva, Vandana (2008), *Soil Not Oil: Environmental Justice in an Age of Climate Crisis*, South End Press.

175 ETC Group (2010), 'Gene Giants Stockpile Patents on "Climate-Ready" Crops in Bid to Become Biomassters', Press Release, 25 October, http://www.etcgroup.org/en/node/5220.

Shiva observed in her affidavit in the *Monsanto* v. *Schmeiser* case: 'Free exchange of seed among farmers has been the basis of maintaining biodiversity as well as food security.'[176] She suggested that the case went to the 'heart of the debate about the effect of intellectual property rights on biodiversity and food security'.[177] Vandana Shiva has written elsewhere: 'Patents on seed are a direct undermining of the human rights of farmers to save, exchange, and multiply seed.'[178] She contends: 'When those patents are broad species patents, it is not just the human rights of farmers but everyone's right to food security that is threatened.'[179]

Some farmers' groups[180] and environmental groups such as Greenpeace[181] and Maude Barlow's Council of Canadians[182] are also hostile to the concentration of agricultural intellectual property in respect of climate-ready crops. Gwynne Dyer sardonically observes in *Climate Wars*: 'The recently revived claims that genetically modified crops will help to "fight world hunger" are as false and cynical as ever: GM food crops that have been engineered for drought resistance, salt tolerance, increased yield, etc. simply do not exist.'[183]

---

176 Research Foundation for Science, Technology, and Ecology, 'Intervenor in *Monsanto Canada Inc.* v. *Schmeiser* (2004) SCC 34', http://www.canadians.org/food/documents/RSTE_Affidavit.pdf.

177 Ibid., 5.

178 Shiva, Vandana (2003), 'Food Rights, Free Trade, and Fascism', in Matthew Gibney (ed.) *Globalizing Rights: The Oxford Amnesty Lectures 1999*, Oxford: Oxford University Press, 87–108 at 97.

179 Ibid.

180 Bell, E. and S. Smith (2010), 'Future of Food Now A Global Battle', *ABC News*, April, http://www.abc.net.au/news/events/gene-wars/future-of-food.htm.

181 Greenpeace, 'Patents on Life', http://www.greenpeace.org/international/en/campaigns/agriculture/problem/genetic-engineering/ge-agriculture-and-genetic-pol/patents-on-life/.

182 Council of Canadians, 'Intervenor in *Monsanto Canada Inc.* v. *Schmeiser* (2004) SCC 34', http://www.canadians.org/food/documents/COC_Affidavit.pdf.

183 Dyer, Gwynne (2010), *Climate Wars: The Fight for Survival as the World Overheats*, Melbourne: Scribe, 52.

B Farmers' Rights to Save Seed

There has been much debate about whether farmers' rights to save seed will be adversely affected and threatened by plant breeders' rights, agricultural patents, technology user agreements, and genetic use restriction technologies.[184]

Frederick Abbott reflects upon the system of exceptions operating in respect of plant breeders' rights: 'Pursuant to the UPOV system of protection, national governments are permitted to authorize "farmer's rights" under which seeds may be replanted, so long as this does not unreasonably interfere with the legitimate commercial opportunities of the certificate holder.'[185] Such concerns about farmers' rights have been re-awakened in the new context of plant breeders' rights and patents relating to climate adaptation and climate mitigation.

In 2009, researchers from the International Institute for Environment and Development (IIED) and partner organisations in China, India, Kenya, Panama and Peru complained that farmers in developing countries were struggling to mitigate the impact of climate change because of the corporate control of plants and seeds. Project leader Krystyna Swiderska of IIED contended:

> Where farming communities have been able to maintain their traditional varieties, they are already using them to cope with the impacts of climate change. But more commonly, these varieties are being replaced by a smaller range of 'modern' seeds that are heavily promoted by corporations and subsidized by governments. These seeds have less genetic diversity yet need more inputs such as pesticides and fertilizers and more natural resources such as land and water.[186]

Similarly, Michael Pimbert of the IIED argues that there should be an exclusion of genetic resources from patent protection and plant breeders' rights protection: 'The patenting and commercialisation of genetic

---

184 Rimmer, Matthew (2008), *Intellectual Property and Biotechnology: Biological Inventions*, Cheltenham and Northampton, MA: Edward Elgar, 71–3.

185 Abbott, F. (2009), 'Innovation and Technology Transfer to Address Climate Change: Lessons from the Global Debate on Intellectual Property and Public Health', International Centre for Trade and Sustainable Development Platform on Climate Change, Trade Policies and Sustainable Energy, Issue Paper No. 24, June, 6.

186 Staff Writer (2009), 'Researchers: Farmers' Rights to Adapt to Climate Change Ignored', *China Daily*, *Xinhua*, 7 September.

resources by private companies must be prohibited.'[187] He maintains: 'Farming communities have the right to freely use and protect the diverse genetic resources, including seeds and livestock breeds, which have been developed by them throughout history.'[188]

Jingsong Li of the Center for Chinese Agricultural Policy in China promoted Participatory Plant Breeding research partnerships between farmers and breeders in which there was equitable benefit-sharing:

> Traditional seed varieties are critical to help Chinese farmers adapt to climate change. At the same time, this biological diversity is under threat from problems such as drought, floods, pests and diseases, which climate change may promote. For these reasons, farmers are keen to improve their varieties through Participatory Plant Breeding.[189]

The World Seed Conference is intended to raise awareness of the importance of new plant varieties and high quality seed in this context and considers how governments can develop an enabling environment to encourage plant breeding and the production and distribution of high quality seed.

C Patent Pools

Olivier De Schutter has suggested that flexible licensing mechanisms – such as patent pools, clearinghouses and open source mechanisms – could be useful in respect of plants. He called upon governments to 'Seek to overcome the problems of delayed or blocked access to needed research tools and plant material, and encourage innovative mechanisms such as patent pools, clearing houses and open source experiments in order to overcome barriers to research on patented material, particularly where multiple patents exist on a single plant variety.'[190]

Anatole Krattiger, a Swiss policy-maker and former farmer, masterminded the development of the *Intellectual Property Management in Health*

---

187 Pimbert, Michael (2009), *Towards Food Sovereignty: Reclaiming Autonomous Food Systems*, London: The International Institute for Environment and Development, 45.

188 Ibid., 45.

189 Staff Writer (2009), 'Researchers: farmers' rights to adapt to climate change ignored', *China Daily*, *Xinhua*, 7 September.

190 De Schutter, Olivier (2009), *Seed Policies and the Right to Food: Enhancing Agrobiodiversity and Encouraging Innovation*, United Nations General Assembly, A/64/170, 21, http://www.ip-watch.org/weblog/wp-content/uploads/2009/11/n0942473.pdf.

and Agricultural Innovation: A Handbook of Best Practices.[191] Famously, the handbook highlights the example of Golden Rice (or beta-carotene-containing rice). Much work was done to identify any patent barriers in respect of the technology:

> The development of Golden Rice led to a significant change in the relationship between the public sector and intellectual property. A freedom to operate review of pro-Vitamin A-containing Golden Rice was commissioned by the International Rice Research Institute, a center of the Consultative Group on International Agricultural Research (CGIAR), with funding from the Rockefeller Foundation. The review showed that about 70 patents and patent applications were applicable to the improved rice when *all* patents issued in or applied for in *all* countries, including patents on commercially accessed research tools, were considered.[192]

Krattiger comments that the patent owners were willing to enable humanitarian licensing: 'All of these companies, including Zeneca (which, coincidentally, almost immediately merged with Novartis Agribusiness to form Syngenta), provided access to their technologies, free of charge, for defined humanitarian research and use of Golden Rice in developing countries.'[193] He reflects that, within a short time, licences were granted to public sector licensees, and national programs were established in Bangladesh, China, India, Indonesia, the Philippines, South Africa and Vietnam.

If there was a global patent pool created in respect of clean technologies, it could certainly include agricultural inventions within its remit.

D Public Sector Licensing

There has been much debate as to whether public sector research entities should engage in flexible licensing practices in respect of plant breeders'

---

[191] Krattiger, Anatole et al. (eds) (2007), *Intellectual Property Management in Health and Agricultural Innovation: A Handbook of Best Practices,* Oxford: Centre for the Management of Intellectual Property in Health Research and Development.

[192] Krattiger, Anatole and Ingo Potrykus (2007). 'Golden Rice: A Product-Development Partnership in Agricultural Biotechnology and Humanitarian Licensing', in Anatole Krattiger et al. (eds), *Executive Guide to Intellectual Property Management in Health and Agricultural Innovation: A Handbook of Best Practices*, Oxford: Centre for the Management of Intellectual Property in Health Research and Development, http://www.iphandbook.org/handbook/case_studies/cs03/. See also Potrykus, I. (2001), 'Golden Rice and Beyond', *Plant Physiology*, **125**, 1157–61.

[193] Ibid.

rights and patents – particularly where there are larger humanitarian concerns at stake.

The United States Department of Agriculture supports a wide range of research relating to climate change.[194] The Department notes that 'Areas of emphasis include evaluating risks to natural resources, estimating the role of forestry and agricultural activities in greenhouse gas emissions and carbon sequestration, and developing practical management strategies and approaches to manage emissions and adapt to changes.'[195]

In the European Union, there has been work done on the role of agriculture in climate change mitigation.[196] The Farm Commissioner Mariann Fischer Boel noted that 'climate change is now a "given" in agricultural policy-making'.[197] In the United Kingdom, the John Innes Centre in Norwich located a key gene – dubbed the 'thermometer gene' – that enabled plants to sense and react to temperatures around them.[198] Dr Philip Wigge, one of the researchers, observed of the discovery: 'Climate change will have a huge effect on crop productivity and that's something we feel gives added impetus to our research.'[199]

In Australia, the public sector research organisation CSIRO has been focused upon identifying the likely impacts of climate change on plant growth and production and its effects on farming.[200] The research body has been 'modelling the effect of climate change on wheat to see how it responds under different climatic conditions'.[201] It hopes that such work 'could lead to the development of wheat which is adapted to cope with, or even benefit from, climate change and increased atmospheric carbon dioxide concentration'.[202] CSIRO scientist Dr T.J. Higgins is optimistic about the future of crop biodiversity.

---

194 United States Department of Agriculture (2009), *Our Changing Planet: The U.S. Climate Change Science Program for Fiscal Year 2009*, http://www.usgcrp.gov/usgcrp/Library/ocp2009/ocp2009-usda.htm.

195 Ibid.

196 Commission of the European Communities (2009), 'The Role of European Agriculture in Climate Change Mitigation', 23 July, http://ec.europa.eu/agriculture/climate_change/sec2009_1093_en.pdf.

197 Boel, M.F. (2007), 'Farming's Role in Mitigating Climate Change', Member of the European Commission responsible for Agriculture and Rural Development, Press Release, 3 July.

198 Staff Writer (2010), 'Climate Change Resistant Crops Move Nearer after Gene Breakthrough', *The Telegraph*, 8 January.

199 Ibid.

200 CSIRO, Plants and Climate Change, http://www.csiro.au/science/pshy.html.

201 Ibid.

202 Ibid.

> We have access to 300 different crops at least for food. Intellectual property rights are probably concentrated on five or six out of that group so I don't see it as a major threat [...] but from the point of food security for the future I don't see it as playing a major role. It's much more important that we get stuck into research and development to feed the nine billion people.[203]

CSIRO has become a fierce defender of its intellectual property rights, having recently been involved in litigation over its patents relating to its Wi-Fi technology.[204]

The Australian Research Council Centre of Excellence in Plant Energy Biology contends that 'knowing the factors that control biogenesis of plant energy organelles will better enable the breeding of plants with specified metabolic traits'.[205] The Centre predicts that 'benefits include enhanced control of biomass and yield, water and nutrient efficiency, and plants better able to tolerate the environmental stresses of Australia'.[206] The Director, Professor Ian Small, observes that the Centre has engaged in 'several technology transfer projects [...] covering areas of major agricultural interest such as drought tolerance and nutrient use efficiency'.[207] The Centre has an ongoing relationship with Agilent Technologies Australia to develop proteomic technologies in plants. It has established a relationship with the plant growth imaging company based in Germany, Lemna Tec GmbH. As a result of work carried out by the Centre on drought tolerance, a provisional patent in 2005 for a SAL1 Arabidopsis mutant was filed.[208] In 2009, the Centre commenced a collaborative project with the Grains Research and Development Corporation and CSIRO to develop non-transgenic drought-tolerant wheat plants.[209]

The Public Intellectual Property Resource for Agriculture was established to help facilitate public sector licensing.[210] It serves to provide

---

203 Bell, E. and S. Smith (2010), 'Future of Food Now A Global Battle', *ABC News*, April, http://www.abc.net.au/news/events/gene-wars/future-of-food.htm.

204 *Intel* v. *CSIRO* 455 F.3d 1364, 79 U.S.P.Q.2d 1508, C.A.Fed. (Cal.) (July 14, 2006); *Microsoft* v. *CSIRO* 2007 WL 4376104, E.D.Tex.; and *CSIRO* v. *Buffalo Technology Inc*. 542 F.3d 1363 (2008).

205 The Australian Research Centre of Excellence for Plant Energy Biology (2009), *Genes to Energy: Annual Report*, Crawley: The University of Western Australia, iv, http://www.plantenergy.uwa.edu.au/aboutus/reports/ARCCPEB-2009.pdf.

206 Ibid.

207 Ibid., 7.

208 Ibid., 24.

209 Ibid.

210 Public Intellectual Property Resource for Agriculture, http://www.pipra.org/.

assistance in facilitating access to intellectual property: 'From design to delivery, complex decisions affect whether the technology will be useful and adopted by target households, whether sustainability can be achieved, and ultimately whether the technology can have the greatest possible impact on poverty.'[211] Such an entity could play an important practical role in facilitating technology transfer from universities and public research entities to developing countries, least developed countries, and small island states.

E Compulsory Licensing and Competition Law

There is scope for compulsory licensing under both patent law and plant breeders' rights – subject, of course, to international rules under Article 31 of the *TRIPS Agreement* 1994. Such measures could be particularly pertinent when dealing with food shortages, famines and other threats to food security.[212]

The Special Rapporteur on the Right to Food, Olivier De Schutter, has suggested that compulsory licensing could be an appropriate option in certain circumstances to address concerns about food security and climate change: 'Where patents restrict research in ways which may have an impact on food security and are an obstacle to face situations of "national emergency" or other "extreme urgency", for instance in the face of declining crop productivity, article 31 of the *TRIPS Agreement* allows compulsory licensing.'[213] He referred to the *Patents and Plant Variety Rights (Compulsory Licensing) Regulations* 2002 (UK). In the United Kingdom matter of *Sacker Potatoes Ltd* v *C Meijer BV*, an application under the *Plant Varieties Act* 1997 (UK) for a compulsory licence in respect of a variety of potato 'Lady Rosetta' used in the manufacture of crisp chips was refused.[214] Margaret Llewelyn and Mike Adcock caution that 'Whilst there is the notion of a compulsory licence present, its use is so rare, and politically sensitive, that its role as an actual instrument to ensure protection of the public interest looks marginal.'[215] In the United States, Ohio Democrat Marcy Kaptur introduced the *Seed Availability and Competition Act* of 2009 (US) HR 3299, in order to create a system of compulsory

---

[211] Ibid.

[212] Keneally, Thomas (2010), *Three Famines*, North Sydney, NSW: Knopf.

[213] De Schutter, Olivier (2009), *Seed Policies and the Right to Food: Enhancing Agrobiodiversity and Encouraging Innovation*, United Nations General Assembly, A/64/170, 12, http://www.ip-watch.org/weblog/wp-content/uploads/2009/11/n0942473.pdf.

[214] *Sacker Potatoes Ltd* v *C Meijer BV* (Unreported, October 31, 2001).

[215] Llewelyn, Margaret and Michael Adcock (2006), *European Plant Intellectual Property*, Oxford and Portland: Hart Publishing, 524.

licensing of patents on seeds. This bill, however, has not been passed by the United States Congress.

Moreover, the Special Rapporteur has maintained that there is a need to 'consider using antitrust legislation in order to combat excessive concentration in the input providers' market, which entails the risk of abuse of dominant position by the seed companies concerned and the setting of prices at levels which may be unjustifiably high and unaffordable for poor farmers'.[216]

In respect of genetically modified crops, there have been a number of unsuccessful class actions by organic farmers against agricultural biotechnology companies.[217] Given such precedents, it may well be difficult to rely upon competition law to regulate the use of intellectual property rights by agricultural biotechnology companies.

CONCLUSION

The emergence of climate-ready crops poses a number of challenges for an array of intellectual property regimes. The long-established *sui generis* system of plant breeders' rights needs to be modernised to better deal with larger public concerns relating to food security, climate change and nutrition. Moreover, the institution of the UPOV should be modernised, and the *UPOV Convention* 1991 revised. The appearance of climate-ready crops also raises questions for the patent system. A range of agricultural biotechnology companies have been investing heavily in research and development in respect of climate-ready crops. There will be patent races for the control of both food and fuel crops. Exclusive rights being granted in respect of climate-ready crops will raise issues about access to inventions by farmers, researchers and consumers. In international climate negotiations, there has been a fierce debate about intellectual property and climate change. A number of options have been mooted – including exclusions for patentable subject matter; exceptions to patents; public sector licensing; patent pools; compulsory licensing; and competition safeguards. As the report by the Special Rapporteur for Food, Olivier De Schutter, makes clear, there is a

---

216 De Schutter, Olivier (2009), *Seed Policies and the Right to Food: Enhancing Agrobiodiversity and Encouraging Innovation*, United Nations General Assembly, A/64/170, http://www.ip-watch.org/weblog/wp-content/uploads/2009/11/n0942473.pdf.

217 *Sample* v. *Monsanto* (United States District Court for the Eastern District of Missouri, 2003), https://ecf.moed.uscourts.gov/documents/opinions/BLADES_V_MONSANTO_CO-RWS-397.PDF; and *Hoffman* v. *Monsanto Canada Inc.*, 2007 SKCA 47 (CanLII).

need to contemplate such exceptions in the context of agricultural intellectual property. There is a need for future international climate negotiations to focus, more specifically, upon the question of the relationship between intellectual property, agriculture and climate change.

13. The Doomsday Vault: seed banks, food security and climate change

Matthew Rimmer

Hunger allows no choice

W.H. Auden[1]

> It could easily provide the back-drop for a James Bond movie. Deep inside a mountain near the North Pole, down a fortified tunnel, and behind airlocked doors in a vault frozen to -18 degrees Celsius, scientists are squirreling away millions of seed samples. The samples constitute the very foundation of agriculture, the biological diversity needed so the world's major food crops can adapt to the next pest or disease, or to climate change. It's little wonder that the Svalbard Global Seed Vault has captured the public's imagination more than almost any agricultural topic in recent years. Popular press reports about the 'Doomsday Vault,' however, typically mask the complexity of the endeavor and, if anything, underestimate its practical utility.

Cary Fowler[2]

There has been concern about the threat posed to plant genetic resources by global warming and climate change.

The United Nations Food and Agriculture Organization has long been concerned with seed banks, food security and climate change. There have been perennial debates in this forum about the underlying questions of intellectual property rights. In a 1981 conference, the Mexican delegation proposed an international seed bank. As John Seabrook noted, 'it would contain seeds from national and international seed banks, and patented seeds created by private seed companies. The North could have free access to the seeds from the centers of diversity only if the South could have free

1 Auden, W.H. (1940), '1 September 1939', *Another Time*, London and New York: Random House.

2 Fowler, C. (2008), 'The Svalbard Seed Vault and Crop Security', *BioScience*, **58** (3), 190–91.

access to the patented seeds.'[3] The American Seed Trade Association said that the Undertaking 'strikes at the heart of free enterprise and intellectual property rights'.[4]

The *International Treaty on Plant Genetic Resources for Food and Agriculture* 2001 – which entered into force in 2004 – aims to promote 'the conservation and sustainable use of plant genetic resources for food and agriculture and the fair and equitable sharing of the benefits arising out of their use, in harmony with the *Convention on Biological Diversity* 1992, for sustainable agriculture and food security'.[5]

The preamble recognises 'that plant genetic resources for food and agriculture are a common concern of all countries, in that all countries depend very largely on plant genetic resources for food and agriculture that originated elsewhere'. The preamble emphasises 'that plant genetic resources for food and agriculture are the raw material indispensable for crop genetic improvement, whether by means of farmers' selection, classical plant breeding or modern biotechnologies, and are *essential in adapting to unpredictable environmental changes and future human needs*' (my emphasis). The parties affirm that 'the past, present and future contributions of farmers in all regions of the world, particularly those in centres of origin and diversity, in conserving, improving and making available these resources, is the basis of Farmers' Rights'. It also recognises that 'the rights recognized in this Treaty to save, use, exchange and sell farm-saved seed and other propagating material, and to participate in decision-making regarding, and in the fair and equitable sharing of the benefits arising from, the use of plant genetic resources for food and agriculture, are fundamental to the realization of Farmers' Rights, as well as the promotion of Farmers' Rights at national and international levels'.

Providing an overview of the treaty,[6] Part II provides general obligations in relation to conservation of plant genetic resources for food and

---

3 Seabrook, J. (2007), 'Sowing for Apocalypse: The Quest For a Global Seed Bank', *The New Yorker*, 27 August, http://www.newyorker.com/reporting/2007/08/27/070827fa_fact_seabrook.

4 Ibid.

5 Article 1 (1) of the *International Treaty on Plant Genetic Resources for Food and Agriculture*, adopted Nov. 3, 2001, S. Treaty Doc. No. 110–19, (entered into force 29 June 2004), ftp://ftp.fao.org/ag/cgrfa/it/ITPGRe.pdf.

6 For an overview, see Tansey, Geoff and Tamsin Rajotte (eds) (2008), *The Future Control of Food: A Guide to International Negotiations and Rules on Intellectual Property, Biodiversity, and Food Security*, London: Earthscan.

agriculture,[7] sustainable use of plant genetic resources,[8] and national commitments and international cooperation.[9] Part III recognises the concept of farmers' rights – 'The Contracting Parties recognize the enormous contribution that the local and indigenous communities and farmers of all regions of the world, particularly those in the centres of origin and crop diversity, have made and will continue to make for the conservation and development of plant genetic resources which constitute the basis of food and agriculture production throughout the world.'[10] Part IV establishes a multilateral system for access to plant genetic resources, and benefit sharing. Part V provides for supporting components to provide a coherent framework for 'capacity building, technology transfer and exchange of information'.[11] Article 15 of the Treaty recognises the 'importance to this Treaty of the ex situ collections of plant genetic resources for food and agriculture held in trust by the International Agricultural Research Centres (IARCS) of the Consultative Group on International Agricultural Research (CGIAR)'. Article 16 encourages the development of international plant genetic resources. Article 17 promotes the development of a 'global information system to facilitate the exchange of information, based on existing information systems, on scientific, technical and environmental matters related to plant genetic resources for food and agriculture, with the expectation that such exchange of information will contribute to the sharing of benefits by making information on plant genetic resources for food and agriculture available to all Contracting Parties'. Annex 1 to the Treaty provides a list of crops covered under the multilateral system.

In his book, *Seed Wars*, Keith Aoki documents some of the controversies over intellectual property and plant genetic resources.[12] He suggests that there are a number of outstanding issues in respect of the Treaty – including the Terms of the Material Transfer Agreements; the list of crops and forages listed in Annex 1; the role of private entities in the multilateral process; the scope of intellectual property rights in material derived from

---

7 Article 5 of the *International Treaty on Plant Genetic Resources for Food and Agriculture* 2001.

8 Article 6 of the *International Treaty on Plant Genetic Resources for Food and Agriculture* 2001.

9 Article 7 of the *International Treaty on Plant Genetic Resources for Food and Agriculture* 2001.

10 Article 9 of the *International Treaty on Plant Genetic Resources for Food and Agriculture* 2001.

11 Article 14 of the *International Treaty on Plant Genetic Resources for Food and Agriculture* 2001.

12 Aoki, Keith (2008), *Seed Wars: Controversies and Cases on Plant Genetic Resources and Intellectual Property*, Durham, NC: Duke University Press, 60.

plant genetic resources; the vindication of plant genetic resources; and compliance with obligations. Aoki suggests that 'the plethora of international actors, fora, principles, interests, and commitments have produced dynamic, unfolding, and contradictory consequences as the "Seed Wars" continue in the 21st century'.[13]

Somewhat bleakly, Charles Lawson argues that 'patents and [plant breeders' rights] are also likely to undermine the objectives of the [*International Treaty on Plant Genetic Resources for Food and Agriculture* 2001], which seeks to provide "facilitated access" to its Multilateral System, and the incentive for those holding valuable plant genetic resources for food and agriculture to contribute them to the Multilateral System'.[14] Somewhat more hopefully, Charles McManis and Eul Soo Seo comment that the Food and Agriculture Organization Multilateral System 'is particularly noteworthy because it represents the first international attempt to combine an open source system of facilitated access to [plant genetic resources] with a mandatory system of benefit-sharing, including mandatory sharing of monetary and other benefits arising out of commercialization of certain patent-protected plant innovation, thus implicitly recognizing a role for intellectual property protection and proprietary plant innovation in generating monetary benefits to maintain the open source system'.[15]

Marte Qvenild comments that the international legal framework established by the *International Treaty on Plant Genetic Resources for Food and Agriculture* 2001 enabled the development of global facilities, such as the Svalbard Global Seed Vault:

> The realisation of the Svalbard Global Seed Vault has been a complicated process of putting together the right political and legal components. Within an international context of conflicting values and interests, a non-profit conservation effort such as the Svalbard initiative was impossible to take forward in 1989/90, owing to international suspicion in the absence of legal institutions to regulate access and rights to plant genetic resources. The International Treaty has to some extent reframed the international debate on plant genetic resources

---

13 Ibid., 90.

14 Lawson, C. (2004), 'Patents and Plant Breeder's Rights Over Plant Genetic Resources for Food and Agriculture', *Federal Law Review*, **32** (1), 107–39; see also Lawson, C. (2009), 'Intellectual Property and the Material Transfer Agreement under the International Treaty on Plant Genetic Resources for Food and Agriculture', *European Intellectual Property Review*, **31** (5), 244–54.

15 McManis, C. and E. Soo Seo (2009), 'The Interface of Open Source and Proprietary Agricultural Innovation: Facilitated Access and Benefit-Sharing Under the New FAO Treaty', *Washington University Journal of Law and Policy*, **30**, 405–64.

> for food and agriculture, by establishing a legal institution for the most important plant genetic material. It has also made possible a greater emphasis on conservation, access, and availability, rather than ownership and financial profit. Unfortunately, the Treaty is not an ideal tool for ensuring that the plant genetic resources remain in the public domain, since it lacks the necessary mechanisms to sanction infringements of the Standard Material Transfer Agreement. Despite this weakness, the Treaty has reduced the uncertainties and risks for depositors wishing to store their seeds on foreign territory, and has shifted the focus of the international community towards the benefits of access to and conservation of these resources. In this context, a global seed vault at Svalbard is welcomed as a much-needed 'safety net' which will secure the valuable international collections of plant genetic resources from the potential risk of extinction.[16]

Cary Fowler echoes these sentiments, observing that the Treaty provided an important legal foundation for the development of global seed banks such as Svalbard Global Seed Vault.

> Despite the gathering together of so many seeds in Svalbard, no legal basis existed for the sharing of crop diversity until as recently as 2001, when the *International Treaty on Plant Genetic Resources for Food and Agriculture* was adopted. International agricultural research centers, as well as some countries, have traditionally been willing to give samples from their collections to others. Yet over the past 20 years, many countries have closed their borders to outgoing samples for fear of 'biopiracy,' the idea that a recipient might acquire intellectual property rights through such a sample and reap undeserved benefits. Consequently, some of the world's most important gene banks, holding tens of thousands of unique varieties, have not provided a single sample to a foreigner in years, even though gene banks are highly dependent on crops and crop diversity that originate elsewhere. The jealous guarding of such collections has rarely been matched by funding adequate for their conservation. National gene banks have languished, and the biological diversity in their care has deteriorated or even been lost.[17]

Fowler notes: 'Self-funded national projects typically serve national self-interest, and international support is hard to generate when countries act solely in self-interest.'[18] He observes: 'The idea of having a secure global backup system for hundreds of individual gene banks makes sense, but it is fundable only if the gene banks agree to cooperate and to address more than their immediate national concerns: that is, they must agree to a

---

16 Qvenild, M. (2008), 'Svalbard Global Seed Vault: A "Noah's Ark" for the World's Seeds', *Development in Practice*, **18** (1), 110–16.

17 Fowler, C. (2008), 'The Svalbard Seed Vault and Crop Security', *BioScience*, **58** (3), 190–91.

18 Ibid.

common standard of access and benefit sharing.'[19] In his view, 'The international treaty provided this standard, making it possible to think of constructing a communal safety net for all crop diversity held in gene banks.'[20]

This chapter considers the use of seed banks to address concerns about intellectual property, climate change and food security. It has a number of themes. First of all, it is interested in the use of 'Big Science' projects to address pressing global scientific concerns and Millennium Development Goals.[21] Second, it highlights the increasing use of banks as a means of managing both property and intellectual property across a wide range of fields of agriculture and biotechnology.[22] Third, it considers the linkage of intellectual property, access to genetic resources and benefit sharing.[23] There are a variety of positions in this debate. Some see requirements in respect of access to genetic resources and benefit sharing as an inconvenient burden for science and commerce. Others defend access to genetic resources and benefit sharing as meaningful and productive. Those inclined to

---

19 Ibid.

20 Ibid.

21 'Big Science' is a term used by historians of science to refer to large-scale scientific projects that require massive government and industrial investment. World War II provided an important impetus for such endeavours as the Manhattan Project and the Space Race. Galison, Peter and Bruce Hevly (1992), *Big Science: The Growth of Large-Scale Research*, Stanford, CA: Stanford University Press. The field of biology has featured 'Big Science' projects – such as the Human Genome Project, the Genographic Project, and the Sorcerer II Expedition: Rimmer, M. (2007), 'The Genographic Project: Traditional Knowledge and Population Genetics', *Australian Indigenous Law Review*, **11** (2) 33–55; and Rimmer, M. (2009), 'The Sorcerer II Expedition: Intellectual Property and Biodiscovery', *Macquarie Journal of International and Comparative Environmental Law*, **6**, 147–89.

22 See, for instance, Knoppers, B. (2005), 'Biobanking: International Norms', *Journal of Law, Medicine, and Ethics*, **33**, 7–14; Andrews, L. (2005), 'Harnessing the Benefits of Biobanks', *Journal of Law, Medicine, and Ethics*, **33**, 22–30; McLennan, A. (2007), 'Which Bank? A Guardian Model for Regulation of Embryonic Stem Cell Research in Australia', *Journal of Law and Medicine*, **15** (1), 45–76; and Simon, B. (2009), 'How to Get a Fair Share: IP Policies for Publicly Supported Biobanks', *Stanford Journal of Law, Science, and Policy*, **1**, 65–79.

23 Ten Kate, Kerry and Sarah Laird (1999), *The Commercial Use of Biodiversity: Access to Genetic Resources and Benefit-Sharing*, London: Earthscan; Phillips, Peter and Chika Onwuekwe (eds) (2007), *Accessing and Sharing the Benefits of the Genomics Revolution*, AA Dordrecht: Springer; and Kamau, Evanson C. and Gerd Winter (2009), *Genetic Resources, Traditional Knowledge and the Law – Solutions for Access and Benefit Sharing*, London: Earthscan.

somewhat more conspiratorial views suggest that access to genetic resources and benefit sharing are a ruse to facilitate biopiracy.

This chapter has a number of components. Section I focuses upon the Consultative Group on International Agricultural Research (CGIAR) network – often raised as a model for Climate Innovation Centres. Section II considers the Svalbard Global Seed Vault – the so-called Doomsday Vault. After a consideration of the World Summit on Food Security in 2009, it is concluded in this chapter that any future international agreement on climate change needs to address intellectual property, plant genetic resources and food security.

I THE GREEN REVOLUTION

A The Consultative Group on International Agricultural Research

The Consultative Group on International Agricultural Research (CGIAR) is a network of agricultural research centres.[24] Aoki considers the evolution of the system:

> CGIAR is the parent institution of an international network of agricultural research centers [...] The CGIAR is governed by a board representing donors of funds such as the Rockefeller and Ford Foundations. These foundations, in part, created and funded the CGIAR to try and unite privately or nationally funded gene banks into an international network.[25]

Aoki suggests that the organisation was subject to philosophical disagreements and conflicts – with a shift from treating plant genetic resources as the common heritage of mankind to allowing them to be protected under plant breeders' rights and patent law.

Lord Nicholas Stern is laudatory of the role played by the network in the 'Green Revolution' – the notable research and development initiatives in the 1960s and 1970s, which increased agricultural production around the world. He suggests that the organisation can perform an equally instrumental role in respect of agriculture and climate change:

> For agriculture, a particularly important challenge is to develop climate-resilient crop varieties and techniques. As well as significant investment, progress will

---

24 The Consultative Group on International Agricultural Research, http://www.cgiar.org/.

25 Aoki, Keith (2008), *Seed Wars: Controversies and Cases on Plant Genetic Resources and Intellectual Property*, Durham, NC: Duke University Press, 66–7.

> depend on international agricultural research systems and stations making it a priority. The group of bodies associated with the Consultative Group on International Agricultural Research (CGIAR) has played an outstanding role in crop development and the improvement of practices – for example, in their development and dissemination through the 1970s of the agricultural techniques associated with the Green Revolution – and can and should play a leading role in developing new, climate-resilient crop varieties and agricultural techniques.[26]

Stern is impressed by the Group's 'long experience from the Green Revolution of the 1960s onwards in developing agricultural crops and techniques suitable for environmental, social and economic conditions in developing countries'.[27]

In the conference report, *Who Owns the Clean Revolution?*, Robert Collier and Josephine Mutugu suggest that CGIAR may be a useful model for the creation of an international institution managing research and development in respect of clean technology.[28] They observe the strengths of the organisation:

> With a 2008 budget of $542 million, CGIAR represents the world's single largest investment to mobilize science for the benefit of the rural poor. When founded in 1971 by the Rockefeller Foundation, the World Bank and several U.N. agencies, CGIAR was conceived as an outgrowth of the Green Revolution, and its mission was tightly scripted as a top-down process in which experts from developed nations spread their knowledge to the poor nations. Gradually, however, its mission was democratized, and it has been widely praised for bringing together thousands of scientific researchers from all over the world to conduct anti-poverty research. It thus constitutes the world's most ambitious international scientific collaboration project.[29]

Collier and Mutugu, though, recognise that the CGIAR system has been increasingly weakened in recent years: 'CGIAR's budgets have declined in real terms since 1990, endangering its research and even its abilities to maintain its valuable gene banks.'[30] Moreover, the authors note some

---

26 Stern, Nicholas (2009), *A Blueprint for a Safer Planet: How to Manage Climate Change and Create a New Era of Progress and Prosperity*, London: The Bodley Head, 65.

27 Ibid., 172.

28 Collier, Robert and Josephine Mutugu (2009), *Who Owns the Clean Tech Revolution? Intellectual Property Rights and International Cooperation in the U.N. Climate Negotiations*, Center for Environmental Public Policy, Goldman School of Public Policy, University of California, Berkeley, 26–27 October, http://energyandclimate.berkeley.edu/sites/default/files/collier-Who%20Owns%20the%20Clean%20Tech%20Revolution.pdf.

29 Ibid., 14.

30 Ibid.

disenchantment with the regime: 'Some funders have complained that CGIAR has a low return on investment – an outcome that, at least in part, results from its mission to provide open source R&D for the poor rather than to earn profits.'[31]

Anne Moorehead was commissioned by the Alliance of the Consultative Group on International Agricultural Research centres to consider 'Climate, Agriculture and Food Security'.[32] She comments:

> Climate change promises serious negative impacts on agricultural systems. These same systems and the natural resources that support them are already under severe strain from over-exploitation, the current climate, and multiple other stresses. Many of the world's most vulnerable people depend directly on these systems for their food and livelihoods; and many countries' economies are also highly dependent on them. Agriculture is also adding to the climate change problem.[33]

The Challenges program, in particular, is designed to consider the question of climate change: 'As well as developing new approaches and new responses to the climate change–food security problem, it will also provide an overarching programme to support the CGIAR centres in their climate change-related endeavours, and help consolidate their efforts.'[34]

Moorehead concluded:

> Building on several decades of research by the CGIAR centres and their partners, we know how to make agricultural and other natural resource-based systems more productive and more sustainable. Even without climate change, we have a moral imperative to turn this knowledge into action. Climate change adds urgency to the situation, but it also provides an opportunity. The products of agricultural research are ready to be implemented in adaptation and mitigation strategies that will help people build successful livelihoods despite changing conditions. As the future climate unfolds, more will be needed. Agriculture – and agricultural research – face a race against time.[35]

---

31 Ibid.

32 Moorhead, Anne (2009), *Climate, Agriculture and Food Security: A Strategy for Change*, The Alliance of the Consultative Group on International Agricultural Research centres, http://www.cgiar.org/pdf/CCAFS_Strategy_december2009.pdf.

33 Ibid., 45.

34 Ibid., 6.

35 Ibid., 45.

CGIAR and the Earth System Science Partnership (ESSP) launched a ten-year Challenge program on climate change, agriculture and food security.[36] It 'seeks to overcome the threats to agriculture and food security in a changing climate, exploring new ways of helping vulnerable rural communities adjust to global changes in climate'.[37] The program observes:

> The research programme explores new ways of helping farmers to adjust to global changes in climate and helping decision makers to weigh up the pros and cons of different policies. Inevitably, both at grass-roots levels and in the corridors of power, adjusting to climate change will mean making complex tradeoffs between food security, livelihoods and the environment. This important research will help policymakers, farmers and others affected by climate change understand the implications of their decisions when making difficult compromises.[38]

The program has three main objectives – 'to close critical gaps in knowledge of how to enhance food security, livelihood and environmental goals in the face of a changing climate'; 'to develop and evaluate options for adapting to a changing climate to inform agricultural development, food security policy and donor investment strategies'; and 'to enable and assist farmers, policymakers, researchers and donors to track, assess and adjust the actions they take to deal with changes in climate'.[39]

In an effort to modernise the system of CGIAR, a Change Steering Team put forward a reform proposal in 2008.[40] This report stresses that the network of organisations has to be reshaped to deal with a new era in agricultural research:

> Rising food prices, increasing concern over global climate change, and the energy crisis have ushered in a new era of challenge and opportunity for agriculture and agricultural research. According to the World Bank, food price inflation threatens to push at least 100 million people back into poverty. Scientists estimate that rising temperatures and changing rainfall patterns could cause

---

36 The Program on Climate Change, Agriculture and Food Security (CCAFS) is a new ten-year research initiative launched by the Consultative Group on International Agricultural Research (CGIAR) and the Earth System Science Partnership: http://www.ccafs.cgiar.org/index.php.

37 Climate Change, Agriculture and Food Security, Research, http://www.ccafs.cgiar.org/research.

38 Ibid.

39 Ibid.

40 Consultative Group on International Agricultural Research Change Steering Team (2008), 'A Revitalized CGIAR – A New Way Forward: The Integrated Reform Proposal', http://www.cgiar.org/pdf/agm08/agm08_reform_proposal.pdf.

> agriculture production to drop by as much as 50 percent in many African countries and by 30 percent in Central and South Asia. Meanwhile, growth in middle income countries is increasing food demand, while natural resources are over stressed and readily available arable land, which could support increased food production, is increasingly hard to find. After nearly two decades of neglect, the role of agriculture and agricultural research in poverty reduction is once again receiving high-level political recognition.[41]

The report emphasised that 'even more productive and resilient crop varieties, sustainable technologies and farming systems must be developed to counter balance food price inflation and increasingly extreme and erratic weather'.[42] The Change Steering Team emphasised that 'investment in such research will result in technologies that serve the dual purpose of both abating hunger and helping farmers cope with the impacts of global climate change'.[43] The reform proposal noted: 'Many of the traits required to increase crop productivity in poor rural communities, such as drought and flood tolerance, are also those needed for adaptation to climate change.'[44] The report stressed that there was an opportunity for CGIAR, with its network of scientists and extensive genebanks, to play an important role in this context.[45]

The report observed, notwithstanding its past successes, CGIAR had fallen short of its 'potential in addressing the new and emerging global challenges, and risks rapid loss of relevance'.[46] Several problems were identified with CGIAR. First, there had been 'mission creep', with a significantly expanded mandate. Second, there were increased problems with overlapping mandates and increased complexity, owing to complicated governance structures.[47] Third, there had been a stagnation of resources and a lack of donor coordination. Fourth, CGIAR was struggling to keep up with the globalisation of agricultural research.'[48]

In its strategic plan, the Change Steering Team proposed that a 'revitalized CGIAR will be driven by a unifying vision' – namely 'to reduce poverty and hunger, improve human health and nutrition, and enhance ecosystem resilience through high-quality international agricultural research, partnership and leadership'.[49] The Change Steering Team recommended that

---

41 Ibid., 3.
42 Ibid.
43 Ibid.
44 Ibid.
45 Ibid.
46 Ibid.
47 Ibid.
48 Ibid.
49 Ibid., 4.

CGIAR focus on three high level Strategic Objectives – including 'food for people', 'environment for people', and 'policies for people'.[50] CGIAR would seek to 'create and accelerate sustainable increases in the productivity and production of healthy food by and for the poor'; 'conserve, enhance and sustainably use natural resources and biodiversity to improve the livelihoods of the poor in response to climate change and other factors'; and 'promote policy and institutional change that will stimulate agricultural growth and equity to benefit the poor, especially rural women and other disadvantaged groups'.[51]

In March 2010, there was a significant public debate over the future direction and funding of CGIAR at the Global Conference on Agricultural Research for Development.[52] In the journal *Nature*, Natasha Gilbert commented:

> The Montpellier conference marks their first chance to influence the CGIAR's research plan, a key draft of which was released on 20 March. Under the proposed reforms, donor contributions would go into a common pot, which would then be distributed among eight broad research areas, or 'mega programmes'. These include 'climate change and agriculture' and 'mobilizing agricultural biodiversity for food security and resilience'. (By contrast, donors currently fund individual centres directly, either through specific projects or as a lump sum.) The idea is to cut out research overlap between centres, create a clear mission and refocus research on the questions and problems donors want tackled.[53]

Carlos Pérez del Castillo, chair of CGIAR's consortium board, informed a press briefing that changes would be made to the mega program to incorporate concerns expressed at the conference. In particular, CGIAR would focus research on specific regions, and target the more vulnerable groups of people for its programs.[54]

The ETC Group has expressed reservations that CGIAR has been co-opted by biotechnology companies and large philanthropic organisations, like the Gates Foundation: 'Publicly funded developers of freely

---

50 Ibid.

51 Ibid.

52 Gilbert, N. (2010), 'Future Funding for Agricultural Research Uncertain: Financial Donors Wrangle Over Global Research Group's Strategy', *Nature*, 31 March.

53 Ibid.

54 The Consultative Group on International Agricultural Research, http://www.cgiar.org/changemanagement/index.html.

accessible plant varieties could succumb to biotech's market dominance.'[55] The ETC Group fears: 'Facing big budget cuts from its traditional funders, CGIAR is now a central player in the Gates-funded collaboration with Monsanto and BASF – a project that a CGIAR spokesman defended as a "global public good".'[56]

B The Global Crop Diversity Trust

The Global Crop Diversity Trust is an independent international organisation established in October 2004 by the United Nations Food and Agriculture Organization and Bioversity International.[57] The policy framework for the Trust is provided by the *International Treaty on Plant Genetic Resources for Food and Agriculture* 2001. The Trust has a particular interest in addressing Millennium Development Partnership Goals 1, 7 and 8: to eradicate extreme poverty and hunger, to ensure environmental sustainability and to develop a global partnership for development respectively.

The Trust has commented upon the impact of climate change in the field of agriculture:

> Forecasts for declines in the yields of staple crops show that climate change will place unprecedented pressures on our ability to grow the food we require, and these impacts will be particularly severe in developing countries. All IPCC scenarios show warming over the next several decades will take place irrespective of any action taken today. The same models show conditions for agriculture will be dramatically different from those which dominate today. Adapting agriculture to these future conditions is therefore essential. The need for new crop varieties that can withstand these challenges is now widely recognized and is frequently cited in climate change discussions. These are essential not only to reduce hunger but also to strengthen global food security in the medium and long term.[58]

---

55 Weiss, R. (2008), 'Firms Seek Patents on "Climate Ready" Altered Crops', *The Washington Post*, 13 May, A04.

56 Ibid.

57 The Global Crop Diversity Trust, http://www.croptrust.org/main/ and on the role of the trust see http://www.croptrust.org/documents/web/RoleofTrust Sept08.pdf.

58 The Global Crop Diversity Trust, 'Preparing for Climate Change', http://www.croptrust.org/main/climatechange.php.

The Global Crop Diversity Trust concludes that 'the development of crop varieties that can cope with heat, drought, flood and other extremes may well be the single most important step we can take to adapt to climate change'.[59]

The Executive Director of the Trust is Cary Fowler. He is an intriguing figure – after a long history as a civil society activist with Patrick Mooney (the founder of the ETC Group), he has become a member of the international institutional framework.[60] Nonetheless, he still has a touch of radicalism, being described as a 'biodiversity warrior'. The executive board is the principal decision-making body of the trust. The members are elected by the governing body of the *International Treaty on Plant Genetic Resources for Food and Agriculture* 2001.[61]

The Trust seeks to direct donations from private companies, nation governments and philanthropic foundations towards addressing food security. The Trust has been supported by several governments including the United States Government, the Australian Government, who through AusAID have pledged US$16 million to the trust, Spain, the Slovak Republic, Norway and Sweden. Seeking to establish an endowment fund, the Trust has received contributions from Australia, DuPont/Pioneer Hi-Bred, Ethiopia, the Gates Foundation, Germany, India, the International Seed Federation, Ireland, New Zealand, Norway, the Slovak Republic, Spain, Sweden, Switzerland, Syngenta, the United Kingdom and the United States.

The Global Crop Diversity Trust played a key role in establishing the Svalbard Global Seed Vault. Moreover, 'the Trust contributes to the Seed Vault through technical advice and assistance, and by providing annual operating support, and by facilitating and sponsoring the shipment of the majority of seeds to the Vault'.[62] Moreover, 'safety duplicates of the seeds regenerated with funding from the Trust are shipped to the Vault, as are unique accessions from the international collections' held by CGIAR.[63]

---

59 Ibid.

60 Lappe, Frances Moore, Joseph Collins and Cary Fowler (1977), *Food First: Beyond the Myth of Scarcity*, Boston: Houghton-Mifflin; Fowler, Cary and Patrick Mooney (1990), *Shattering: Food, Politics, and The Loss of Genetic Diversity*, Tucson: University of Arizona Press; Fowler, Cary (1994), *Unnatural Selection: Technology, Politics, and Plant Evolution*, Yverdon: Gordon and Breach; and Fowler, C. (2000), 'The Plant Patent Act of 1930: A Sociological History of Its Creation', *Journal of the Patent and Trademark Office Society*, **82** (9) 62.

61 Global Crop Diversity Trust (2009), *Annual Report*, Rome: Global Crop Diversity Trust, 15.

62 Ibid., 6.

63 Ibid.

C The Gates Foundation

The Gates Foundation is informed by a particular credos, 'Guided by the belief that every life has equal value, the Bill & Melinda Gates Foundation works to reduce inequities and improve lives around the world.'[64] The philanthropy is two tiered: the Gates Foundation distributes money to grantees[65] and the Gates Foundation Trust manages the endowment and makes contributions to fund the Gates Foundation's grant-making operations. The combined entities have US$38.7 billion as endowment assets available for charitable activities. The Gates Foundation has benefited particularly from the gift of shares worth approximately US$30.7 billion from the American businessman and philanthropist, Warren Buffett.[66] There is a US$4.4 billion liability for future year payments on already approved grants. On a cash basis, the combined entities paid approximately US$2.0 billion in grants.

The Gates Foundation plays a significant role in respect of agriculture and development.[67] The Foundation contributed US$29 million to 'protect the genetic diversity of 21 critical crops for food security and poverty alleviation, by supporting national genebanks, the Svalbard Global Seed Vault, and the Global Crop Diversity Trust'.[68] In alliance with the Rockefeller Foundation, Bill Gates has promoted the need for a new 'green revolution'.[69] The Gates Foundation is optimistic about such possibilities:

> During the Green Revolution, which took place from the 1960s to the 1980s, improvements in staple crops including maize, wheat, and rice helped double the amount of food produced, save hundreds of millions of lives, and drive broader development throughout much of Asia and Latin America. There were also some serious unintended consequences – particularly regarding the environment – that left us with important lessons for today. But the Green Revolution demonstrated that large-scale progress against hunger and poverty is possible.

---

64 The Gates Foundation, http://www.gatesfoundation.org/AboutUs/.

65 Ibid.

66 Lowenstein, Roger (2008), *Buffett: The Making of an American Capitalist*, New York: Broadway Books.

67 The Gates Foundation, 'Agricultural Development', http://www.gatesfoundation.org/agriculturaldevelopment/Pages/default.aspx.

68 Gates Foundation (2007), 'Grant to the United Nations Foundation to Protect the Genetic Diversity of 21 Critical Crops for Food Security and Poverty Alleviation, by Supporting National Genebanks, the Svalbard Global Seed Vault, and the Global Crop Diversity Trust', http://www.gatesfoundation.org/Grants-2007/Pages/United-Nations-Foundation-OPP45782.aspx.

69 The Rockefeller Foundation, 'Strengthening Food Security: Alliance for a Green Revolution in Africa', http://www.rockefellerfoundation.org/what-we-do/current-work/strengthening-food-security-alliance/.

> Today, many people in countries such as India and China are reaping the benefits of agricultural investments and innovations made a generation ago. Unfortunately, the benefits of the Green Revolution did not extend to Sub-Saharan Africa. Over the past 25 years, as the rest of the world has doubled, tripled, and even quadrupled productivity, agricultural productivity in Sub-Saharan Africa has stagnated. This stagnation has come at a tremendous cost. In the 1960s, Africa was a net exporter of food. Today, it's a net importer, and one in three people in Sub-Saharan Africa lives in chronic hunger.[70]

The Gates Foundation insists: 'We believe the best way for the foundation to address climate change is to help poor farmers adapt.'[71]

Outlining a 'new approach to capitalism', Gates discusses related concerns about access to food, water and energy: 'There are roughly a billion people in the world who don't get enough food, who don't have clean drinking water, who don't have electricity – the things that we take for granted.'[72] He observed: 'Climate change will have the biggest effect on people who have done the least to cause it.'[73]

The Gates Foundation has funded projects in Generation Challenge Programme,[74] such as Water Efficient Maize for Africa.[75] Richard Trethowan says of the Generation Challenge Programme:

> Everything we find, and everything we do, is publicly available, and that will be available to researchers and ultimately to farmers in developing countries. And that, after all, is the focus of the Generation Challenge Programme. It's to provide those enabling technologies to researchers on the ground in developing countries so that the farmers ultimately benefit.[76]

Advocating a credos of 'creative capitalism', Gates has argued that there is a significant role to be played by corporations in addressing questions of development – including food security:

---

70 Gates Foundation (2009), Agricultural Development: Strategy Overview, http://www.gatesfoundation.org/agriculturaldevelopment/Documents/agricultural-development-strategy-overview.pdf.

71 Ibid.

72 Gates, B. (2008), 'Speech to the World Economic Forum', The Gates Foundation, http://www.gatesfoundation.org/speeches-commentary/Pages/bill-gates-2008-world-economic-forum-creative-capitalism.aspx.

73 Ibid.

74 Generation Challenge Programme, http://www.generationcp.org/.

75 Water Efficient Maize for Africa, http://www.aatf-africa.org/wema.

76 Bell, E. and S. Smith (2010), 'Future of Food Now A Global Battle', *ABC News*, April, http://www.abc.net.au/news/events/gene-wars/future-of-food. htm.

> Capitalism has improved the lives of billions of people – something that's easy to forget at a time of great economic uncertainty. But it has left out billions more. They have great needs, but they can't express those needs in ways that matter to markets [...] Governments and nonprofit groups have an irreplaceable role in helping them, but it will take too long if they try to do it alone. It is mainly corporations that have the skills to make technological innovations work for the poor. To make the most of those skills, we need a more creative capitalism: an attempt to stretch the reach of market forces so that more companies can benefit from doing work that makes more people better off. We need new ways to bring far more people into the system – capitalism – that has done so much good in the world.[77]

Gates suggests that companies should be rewarded by some kind of return for their participation in development projects: 'It's not just about doing more corporate philanthropy or asking companies to be more virtuous.'[78] In his view, 'it's about giving them a real incentive to apply their expertise in new ways, making it possible to earn a return while serving the people who have been left out'.[79]

Commenting on Gates's proposal, Elizabeth Stuart suggests that 'creative capitalism' is essential in addressing climate change: 'The true creative capitalists, having recognized that global warming threatens their business model just as it threatens progress in poor countries, would look for much more fundamental ways to engage with the issue.'[80] Indeed, they would 'join with climate change activities in lobbying Congress'; 'invest more in their supply chains'; and 'working with communities in poor countries, they would develop technologies such as cheap, solar-powered stoves – technologies that might also have applications in the rich world'.[81]

II THE SVALBARD GLOBAL SEED VAULT

The nicknames of the Svalbard Global Seed Vault are intriguing, the semiotics suggestive of the larger objectives and appeal of the facility. The 'Doomsday Vault' is an eerie eschatological nickname, suggesting the Last Judgment. The title also evokes the Domesday Book – the 1086 record made of the extent, value, ownership and liabilities of land made in

---

77 Gates, B. (2008), 'Making Capitalism More Creative', *Time Magazine*, 31 July.
78 Ibid.
79 Ibid.
80 Kinsley, Michael (ed.) (2009), *Creative Capitalism: A Conversation with Bill Gates, Warren Buffett and Other Economic Leaders*, New York: Simon & Schuster, 106.
81 Ibid., 108.

England by the order of William I.[82] The nickname of 'Noah's Ark' also has rather religious overtones. The Svalbard Global Seed Vault is also often characterised with the language and rhetoric of emergencies – disasters, catastrophes, and so forth. The rhetoric of 'insurance' is also prominent.

A The Origins of the Vault

In 2004, the Norwegian Ministry of Foreign Affairs and the Ministry of Agriculture commissioned a 'Study to Assess the Feasibility of Establishing a Svalbard Arctic Seed Depository for the International Community'.[83] The report observed that there was a need to provide for an international seed bank to duplicate the holdings of gene banks around the world:

> International standards promulgated in 1994 for the conservation of plant genetic resources for food and agriculture call for each 'unique' accession to be duplicated and stored in at least one additional location, ideally in a different country, because, as FAO's State of the World's Plant Genetic Resources (1998) documented, individual genebanks are vulnerable to a host of problems that can endanger their collections, including poor management, lack of adequate funding, equipment failures, and natural catastrophes. In addition to these threats, we now recognize the need to provide protection against potentially more cataclysmic dangers: civil strife, war, and acts of terrorism.[84]

The report suggested that there was a need to establish a facility 'to provide an ultimate safety net to the international community for the world's unique plant genetic resources for food and agriculture'.[85]

The Committee sought to 'assess whether a facility located in Svalbard might provide ultimate "fail-safe" protection for the world's most valuable

---

82 The historian Simon Schama depicts the Domesday Book as a proto database: 'The Domesday Book was much more than a glorified audit. It was a complete inventory of the kingdom up to the Tyne, shire by shire, hundred by hundred: who owned what on the eve of the Conquest and who owned what now; how much it had been worth then and how much it was worth now. Beyond the immediate pragmatic need for money, William's instinct [...] was that knowledge was also power. William the Conqueror was also the first data-base king.' Schama, Simon (2000), *A History of Britain I: At the Edge of the World? 3000BC–AD1603*, London: BBC, 109–10.

83 Center for International Environment and Development Studies (Noragric), Agricultural University of Norway and Nordic Gene Bank (2004), 'Study to Assess the Feasibility of Establishing a Svalbard Arctic Seed Depository for the International Community Prepared for the Ministry of Foreign Affairs and the Ministry of Agriculture', 14 September, http://www.regjeringen.no/Upload/LMD/kampanjeSvalbard/Vedlegg/Frohvelv_Study_to_assess.pdf.

84 Ibid., 2.

85 Ibid., 6.

natural resources, and whether it might be able to do so in a manner that is efficient, sustainable, inexpensive, and politically and legally acceptable'.[86]

The report concluded 'that a Svalbard facility can provide all of these things, and can thus make a major contribution to food and environmental security and to the safety and well-being of human beings for as far into the future as we can see'.[87]

In 2004, the Norwegian Government presented the initiative to a meeting of the Commission for Genetic Resources of the Food and Agricultural Organization of the United Nations (FAO) in October 2004. The Commission supported the initiative and Norway was encouraged to proceed with the project. The building of the Svalbard facility took place between 2007 and 2008 at a cost of NOK 50 million.

The Svalbard Global Seed Vault was opened in February 2008.[88] At the opening ceremony, the Prime Minister of Norway, Jens Stoltenberg, and the African Nobel Peace Prize-winning environmentalist Wangari Maathai placed the first seeds in the vault. Stoltenberg observed of the initiative: 'With climate change and other forces threatening the diversity of life that sustains our planet, Norway is proud to be playing a central role in creating a facility capable of protecting what are not just seeds, but the fundamental building blocks of human civilization.'[89] Maathai reflected: 'We now understand that along with international movements to save endangered species and the rainforests of the world, it is just as important for us to conserve the diversity of the world's crops for future generations.'[90]

The Svalbard Global Seed Vault is managed by the Norwegian Government, the Global Crop Diversity Trust, and the Nordic Genetic Resource Center.

The building for the Svalbard Global Seed Vault featured in the book *Extreme Architecture*.[91] Depicting the building, Ruth Slavid observes:

> This planetary insurance policy is basically a set of giant vaults, bored into the rock of a hillside behind Longyearbyen airport and accessed by a tunnel driven largely through permafrost. There is a small administration office, just off the

---

86 Ibid., 2.

87 Ibid., 2.

88 Svalbard Global Seed Vault (2008), 'Svalbard Global Seed Vault: Arctic Seed Vault Opens Doors for 100 Million Seeds', Ministry of Agriculture and Food, 26 February, http://www.regjeringen.no/en/dep/lmd/campain/svalbard-global-seed-vault/news/arctic-seed-vault-opens-doors-for-100-mi.html?id=501721.

89 Ibid.

90 Ibid.

91 Slavid, Ruth (2009), *Extreme Architecture: Building for Challenging Environments*, Laurence King Publishers.

> tunnel, that is heated to a higher temperature [...] The portal is a simple, stark wedge protruding from the mountainside, its sharp form differentiating itself from the softer contours of nature around it.[92]

The Norwegian Prime Minister, Jens Stoltenberg, said: 'This is a Noah's Ark for today, securing global diversity for future generations and, if crop species are lost in a global disaster, we can start all over again, here in Svalbard.'[93]

B The Mechanics of the Vault

In his role as the executive director of the Global Crop Diversity Trust, Fowler has this take on the role of intellectual property in the Svalbard Global Seed Vault:

> I understand why intellectual property rights exist, to protect inventors and support innovation. But, the important thing in our field is to make sure the base stock remains available to use. The support that we give for collecting and conservation is all directed towards supporting diversity in the public domain. We think of it as the common heritage of mankind. This heritage dates back a long, long time, before the nation-state even existed. Rice dates back 12,000 years. This is not a crop that any individual or corporation can claim as their own. But, agriculture is a big field and there are many fights. In most of those, we don't have a dog in the fight, so to speak. We have a long, long time horizon. We're planning hundreds of thousands of years in advance, so a lot of today's fights are kind of peripheral to our longer-term goal.[94]

This interesting statement is in some respects nostalgic for past approaches to genetic resources; yet it is also quietist in avoiding some of the intense conflicts over intellectual property.

Fowler discusses how the Svalbard Global Seed Vault sought to overcome legal and political issues in respect of the ownership of the property and intellectual property rights in respect of deposited seeds:

> The Svalbard initiative thus took on the challenge of attracting biological diversity and managing it properly and cost-efficiently, while sidestepping political and property-rights issues. The latter could be accomplished only if deposits of seeds in Svalbard did not involve the transfer of legal ownership. Organizers opted for a 'black box' arrangement wherein a 'deposit agreement' would be

---

92 Ibid., 73–5.

93 Jacob, C. (2008), 'Doomsday Vault is Open for Business', *Sky News Channel*, 27 February.

94 TED (2009), 'Q&A with Cary Fowler: Saving Seeds to Protect Our Food Supply', September, http://blog.ted.com/2009/09/02/qa_with_cary_fo/.

> concluded between the depositing gene bank and the Norwegian government. The agreement would recognize that legal ownership remained vested in the depositor, in much the same way that a bank recognizes that a customer owns the contents of a safety deposit box held by the bank.[95]

Fowler says of the management plan: 'The seed vault would store deposited materials, but it would not be responsible for monitoring their viability – opening boxes to do so would undermine depositors' confidence that their material was legally inviolate.'[96] He noted: 'Moreover, the vault would not be responsible for regenerating materials when germination rates dipped, a procedure best carried out by the depositor under conditions similar to those where the sample was originally collected.'[97] He observed: 'Rather, when gene banks regenerate their own stocks, they will simply resupply the vault with fresh seed at the same time.'[98]

The Svalbard Global Seed Vault has emphasised that depositors will need to comply with appropriate international laws and national laws: 'The Seed Vault will only agree to receive seeds that are shared under the Multilateral System or under Article 15 of the *International Treaty on Plant Genetic Resources for Food and Agriculture* 2001 or seeds that have originated in the country of the depositor.'[99] The 'Black Box' arrangements mean 'that the deposit of the seeds will not affect any property or other rights pertaining to the material'; 'that the deposited seeds will remain in sealed envelopes, unless otherwise agreed with the Depositor'; and 'that the Svalbard Global Seed Vault will take no action to further transfer the material except back to the original Depositor or the Depositor's successor in title, or in accordance with the Depositor's instructions'.[100]

In terms of management and operations, the Nordic Genebank has developed a standard depositor agreement.

Clause 1 (1) provides: 'The Government of the Kingdom of Norway has established the Svalbard Global Seed Vault to provide a safety net for the international conservation system of plant genetic resources, and to contribute to securing the maximum amount of plant genetic diversity of

---

95 Fowler, C. (2008), 'The Svalbard Seed Vault and Crop Security', *BioScience*, **58** (3), 190–91.

96 Ibid.

97 Ibid.

98 Ibid.

99 Svalbard Global Seed Vault: Management and Operations, http://www.regjeringen.no/en/dep/lmd/campain/svalbard-global-seed-vault/management-and-operations.html?id=462223.

100 Ibid.

importance to humanity for the long-term in accordance with the latest scientific knowledge and most appropriate techniques.'

Under a Standard Depositor's Agreement, Clause 2 (1) declares: 'The act of depositing the Deposited Materials in the Svalbard Global Seed Vault shall have no effect whatsoever on the nature and extent of any property rights pertaining to the Deposited Materials.' Article 2 (2) notes: 'In particular and without prejudice to the generality of the above, the act of deposit shall not act in any way to convey any property rights over the Deposited Materials to the Nordic Gene Bank or the Royal Norwegian Ministry of Agriculture and Food.' Article 3 stresses that 'the Depositor shall deposit only samples of plant genetic resources that: a. are, to the best of the Depositor's knowledge, i. of importance to food security and sustainable agriculture; ii. samples of plant genetic resources that have not yet been deposited in the Svalbard Global Seed Vault; b. have been safety duplicated in a suitable genebank'.

Fowler was hopeful that the site would be able to withstand the worst impacts of climate change:

> Global warming will certainly diminish the utility of the permafrost cooling at the new seed-vault site, though the seeds will be stored so deep in the mountain that permafrost should persist there for some 200 years even in a worst-case climate change scenario. The site is also situated so high above sea level that even if all the glaciers and ice caps melted, the water would not rise to its doors.[101]

Fowler observes: 'The Svalbard Global Seed Vault, conceived as protection against the loss of diversity that takes place routinely across the world, is an insurance policy for a global system that doesn't quite exist yet.'[102] He recognises: 'No single facility can promise a complete, fail-safe solution for all threats to crop diversity – such guarantees are unavailable in this world.'[103] Nonetheless, he is of the view that 'the Svalbard Global Seed Vault is a key component of the global system' and 'it is the best solution we can devise at this time'.[104] Fowler concludes: 'Fulfilling [our responsibility to conserve and share crop diversity] will not necessarily ensure that our agricultural systems will be able to provide food security for a growing world population or to cope with climate change, but without crop diversity, meeting either of those goals is inconceivable.'[105]

---

101 Fowler, C. (2008), 'The Svalbard Seed Vault and Crop Security', *BioScience*, **58** (3), 190–91.
102 Ibid., 191.
103 Ibid.
104 Ibid.
105 Ibid.

The storage of genetically modified seeds remains a touchy issue. The operators of the Svalbard Global Seed Vault explain:

> Import and storage of GMO seeds according to Norwegian legislation will require advance approval. Certain other criteria will apply to 'sealed internal use' for research purposes and indoor storage of GMO, for example with regard to the risk of spreading GMO. Norwegian gene technology legislation was formulated before the Svalbard Global Seed Vault (SGSV) was set up, and therefore fails to take into account the vault's special status, or the low risk related to handling seeds in sealed packaging. Until changes can be made to the rules or exemptions can be provided from them, long-term storage of GMO seeds in the SGSV will not be approved.[106]

The operators comment: 'If it becomes apparent at a later date that modification of the facilities to include GMO seeds can be necessary in order to comply with the purpose of the vault, Norway will revise its policy and rules, and consider ways to upgrade the facility.'[107] Moreover, 'As part of this process, the Norwegian authorities will attach great importance to the recommendations of the international council being established to advise on policy for the Svalbard Global Seed Vault.'[108]

John Seabrook comments that the Svalbard Global Seed Vault is governed by quite conservative rules:

> The vault is not the seed bank that Fowler envisaged back in the eighties – a common bowl from which the world could feed as one. Far from it: each nation will have access to only its own seeds. Nor does the vault much resemble the world bank that Vavilov envisioned – a kind of breeders' utopia, in which the seeds of every kind of food plant in the world are available. In 2006, thirty million new acres of G.M.O.s were planted, bringing the total worldwide acreage to two hundred and fifty-two million acres – about seven per cent of the world's cropland. Altogether, more than ten million farmers from twenty-two countries planted biotech crops in 2006. But few, if any, of these seeds are going into the Svalbard vault. Seed companies have shown little interest in putting them there, and, in any case, Norway severely restricts bringing G.M.O.s into the country.[109]

---

106 Svalbard Global Seed Vault, Frequently Asked Questions, http://www.regjeringen.no/en/dep/lmd/campain/svalbard-global-seed-vault/frequently-asked-questions.html?id=462221.

107 Ibid.

108 Ibid.

109 Seabrook, J. (2007), 'Sowing for Apocalypse: The Quest for a Global Seed Bank', *The New Yorker*, 27 August, http://www.newyorker.com/reporting/2007/08/27/070827fa_fact_seabrook.

Nordgen maintains a dynamic website, documenting the samples and accessions to the Svalbard Global Seed Vault.[110] As at February 2011, the Svalbard Global Seed Vault contained 643 132 seed samples; covering 7033 taxon names; 3987 species; and 749 genus. The seeds originated from 227 countries and 8 continents. There were 34 depositor institutes making contributions to the Svalbard Global Seed Vault.

C Criticism of the Svalbard Global Seed Vault

In her book, *Uncertain Peril*, Claire Hope Cummings discusses the development of the Doomsday Vault.[111] She reflects: 'The Doomsday Vault is the world's newest and most innovative seed bank, but it's based on the very old idea that we should save seeds in times of calamity.'[112] Interviewing Cary Fowler, she questions the governance of the facility:

> The [Global Crop Diversity] trust is governed by a board that is appointed by international groups, mainly those involved in the *International Treaty on Plant Genetic Resources for Food and Agriculture* 2001, as well as the United Nations and the Consultative Group on International Agricultural Research (CGIAR). So far it appears to be operating in a fairly transparent manner, posting information about its donors and governing mechanisms on its website. Most of the funding for the trust comes from institutions like the World Bank, CGIAR, participating countries, a few very large foundations, and some corporations. DuPont/Pioneer Hi-Bred and Syngenta each gave $1 million, and $5 million came from Australian grain growers. The Bill and Melinda Gates Foundation gave the trust $30 million, and Norway contributed another $7.5 million to fund a separate crop-diversity preservation initiative [...] When the suggestion that whoever pays the piper calls the tune comes up, Fowler bristles. He says that anyone who suggests this collection could in any way fall prey to private interests is misinformed, and he challenges them to demonstrate how that might happen. For now, it appears to be the situation in which the corporate donors are afforded an opportunity to do a bit of green washing.[113]

Cummings also notes that 'Critics of the Doomsday Vault say that all the millions being spent on it and all the money directed at ex-situ seed banking should be directed toward preserving in-situ collections and more funding should go toward protecting seed savings by traditional farmers.'[114] She

---

110 Nordgen, http://www.nordgen.org/sgsv/index.php?page=welcome&PHPSESSID=m6jdn1j59nar67imgvrom7ck36.

111 Cummings, Claire Hope (2008), *Uncertain Peril: Genetic Engineering and the Future of Seeds*, Beacon Press.

112 Ibid., 107.

113 Ibid., 108–9.

114 Ibid., 114.

contends: 'In the long run, in-situ strategies are the most cost-effective way to save diversity.'[115]

Elizabeth Finkel has reported that there have been problems in encouraging certain signatories to the *International Treaty on Plant Genetic Resources for Food and Agriculture* 2001 to contribute seeds to the Svalbard Global Seed Bank. She reflects in 2009 that some member states showed signs of recalcitrance and uncooperative behaviour:

> Free and timely access to seeds is enshrined in the *International Treaty on Plant Genetic Resources for Food and Agriculture*, which came into force in 2004. But 'free and timely' is not how plant breeders describe their experience with five of the treaty's 120 signatories: China, Ethiopia, India, Iran, and Turkey. 'Basically, they're closed gene banks,' says Kenneth Street, a legume curator at ICARDA. That's alarming because many seed collections are vulnerable. In the past decade, Afghanistan's collection was lost when the Taliban dumped seeds to scavenge their airtight containers; Iraq's was destroyed in the most recent war; and the Philippine seed bank sustained heavy damage from a typhoon in 2006. The treaty's governing body met in Tunis to review implementation and coax reluctant nations to open their seed banks and contribute to a 'doomsday vault' on Norway's Svalbard island. At the forum, Ethiopia and other developing countries asserted that they should be compensated for custodianship and ongoing cultivation of landraces: traditional varieties adapted to local conditions.[116]

Melaku Worede, founder of the Ethiopian Seed Bank, argued: 'Seed banks like Svalbard are invaluable for preserving landraces, but so are the skills and experience that traditional farmers bring to bear in selecting seeds to breed landraces season after season.'[117]

Thomas van Dooren, an environmental philosopher, suggests that there is a need to think about the relationship between seed banks and living communities: 'While these banked resources are invariably important for present and future crop development, the loss of other diversities – and with them often subsistence autonomy and food security for local communities – needs to become a higher international conservation priority.'[118]

Van Dooren contends that '*ex situ* banking projects genes, seeds and reproductive material more generally, are being used as a kind of "proxy"

---

115 Ibid.

116 Finkel, E. (2009), 'Plant Breeding: Scientists Seek Easier Access to Seed Banks', *Science*, **324** (5933), 1376.

117 Ibid.

118 van Dooren, T. (2009), 'Banking Seed: Use and Information in the Conservation of Agricultural Diversity', *Science as Culture*, **18** (4), 373–95 at 391.

for the plant varieties or species that are being lost'.[119] He contends that there is a need to rethink the role of seed banks in protecting agricultural biodiversity:

> In the context of the current incredible loss of the diversity of life, we cannot really afford to do away with any of the 'conservation' options open to us, however limited they may be. We need these banking projects – perhaps now more than ever. We need them both as a risk mitigation strategy in a world that is starting to look more and more like a badly conducted science project, and as a resource for adaptation in the face of the massive environmental change that humankind is now causing. What this paper points to, however, is the fact that no matter how dire the situation, the need to conserve diversity cannot be allowed to cover over more critical discussions about *how* and *for whom* this 'conservation' work is to be done.[120]

In his view, there is a need to rethink the role of these resources, so that 'the seed banks are no longer primarily stockpiles of genetic resources, but rather become central hubs in a project of conservation as diversity sharing'.[121]

The ETC Group commented upon the initiative:

> While the Global Seed Vault is a step in the right direction, many vital *ex situ* gene banks are in desperate straits. As much as half of the world's crop diversity may still be in farmers' fields protected only by the family and the community – who are up against industrial agriculture, plant patenting, global trade rules, corporate concentration and climate change. Global food security depends upon a coherent *in situ* (on-farm) and *ex situ* (gene bank) strategy.[122]

Anna Lappe suggests that the Svalbard Global Seed Vault could only contain 'a fraction of the planet's biodiversity that is quickly disappearing – thanks in part of the consolidation of the seed industry, driven by Monsanto'.[123] The environmentalist notes: 'Ironically, Monsanto is one of the major funders of Norway's Svalbard Global Seed Vault.'[124]

---

119 van Dooren, T. (2009), 'Genetic Conservation in a Climate of Loss: Thinking with Val Plumwood', *Australian Humanities Review*, **46**, 101–12 at 103.

120 Ibid., 107.

121 Ibid., 108.

122 ETC Group (2008), 'Svalbard's Doomsday Vault: The Global Seed Vault Raises Political/Conservation Debate', 28 February, http://www.etcgroup.org/en/materials/publications.html?pub_id=674.

123 Lappe, Anna (2010), *Diet for a Hot Planet: The Climate Crisis at the end of Your Fork and What You Can Do About It*, New York: Bloomsbury USA, 190.

124 Ibid., 190.

Michael Hopkin has written about the limits of what such seed banks can achieve.[125] Reflecting upon the Svalbard vault, he observes:

> But no matter how well intentioned and executed, Svalbard cannot single-handedly save the world's threatened biological resources, edible or otherwise. A seed vault won't help to conserve the threatened livestock breeds that billions of people depend on for both dietary and financial sustenance, or indeed the complex ecosystems that exist alongside both crops and livestock and in many cases allow them to thrive. Conserving these, and hence securing humanity's food supply, will require schemes even more impressive than an Arctic dugout. Nevertheless, the global approach behind the Doomsday vault may offer inspiration to those aiming to conserve these other resources.[126]

Hopkin comments that 'the banks will still preserve only a fraction of the world's plant species'.[127] In his view, 'farms thus represent the real battleground where the fight to save food diversity will be won: it is there that the many varieties of food species grow today – and it is there, ideally, that they will be preserved intact'.[128]

III INDIA

Inspired by the example of Norway, India has commissioned its own Doomsday Vault in the Himalayas.[129] India has established a seed vault at Changa-Law on the Siachen Glacier to preserve the country's biological diversity and wealth for future generations. W. Selvamurthy, a scientist and Chief Controller of Research and Development at the Ministry of Defence, explained the need for such a facility:

> India is a herbal garden [...]. It is now getting threatened or endangered or extinct [...]. There is a dire necessity to preserve this rich wealth of biodiversity, which exists in India to give it to the next generation.[130]

A journalist for *The Hindu* reflected: 'If the land lies cracked and barren waiting for the rain, if the sea swallows coastal paddy fields, if plagues of

---

125 Hopkin, M. (2008), 'Biodiversity: Frozen Futures', *Nature*, **452**, 404–5.

126 Ibid.

127 Ibid.

128 Ibid.

129 Jayaraman, K.S. (2010), 'India's Doomsday Vault in Frozen Himalayas', *Nature India*, 10 March, http://www.nature.com/nindia/2010/100310/full/nindia.2010.29.html.

130 Hiddleton, S. (2010), 'A Repository of Seeds on Cliff Top of Himalayas', *The Hindu*, 17 February.

pests wipe crops out of existence, India has an insurance policy – a set of seeds carefully preserved in permafrost, ready to be restored, that can be cultivated and sown to feed its people.'[131]

CONCLUSION

At an international level, there have been ongoing discussions about the relationship between intellectual property, food security and climate change. In February 2009, Cary Fowler of the Global Crop Diversity Trust held a meeting to help further refine the priorities for selecting crop varieties held in the so-called 'Doomsday Vault' and in seed collections around the world.[132] He commented: 'Human beings are not going to adapt to climate change unless agriculture adapts.'[133]

On the first anniversary of the building of the Svalbard Global Seed Vault, the Global Crop Diversity Trust organised a conference on the topic of *Frozen Seeds in a Frozen Mountain: How to Feed a Warming World?* The summary statement for the event observed:

> Food security is threatened. Forecasts for declines in the yields of staple crops show that climate change will place unprecedented pressures on our ability to grow the food we require, particularly in developing countries. All Intergovernmental Panel on Climate Change scenarios show warming over the next several decades will take place irrespective of any action taken today. The same models show conditions for agriculture will be dramatically different from those which dominate today. Adapting agriculture to these future conditions is therefore essential. The need for new crop varieties that can withstand these challenges is now widely recognized and is frequently cited in climate change discussions.[134]

The event emphasised that 'the development of crops that can cope with heat, drought, flood and other extremes will likely be the single most important action we can take to adapt to climate change'.[135] This was an 'urgent need requiring action now, given both the serious threat to food

131 Ibid.

132 Kostel, K. (2009), 'The Language of Change', *Nature Reports Climate Change*, 29 January, http://www.nature.com/climate/2009/0902/full/climate.2009.10.html.

133 Ibid.

134 Svalbard Global Seed Vault (2009), *Frozen Seeds in a Frozen Mountain: How to Feed a Warming World? – Anniversary Seminar,* February. http://www.regjeringen.no/upload/LMD/kampanjeSvalbard/Vedlegg/Svalbard_Statement_270208.pdf.

135 Ibid.

security and the time required to breed new crop varieties'.[136] Moreover, it noted: 'Our ability to breed these new varieties cannot be taken for granted, as it is undermined by the loss of the biological basis of our food supply – the genetic diversity of crops.'[137]

In November 2009, there was a discussion about climate change and food security at the World Summit on Food Security in Rome.[138] In his address, United Nations Secretary-General Ban Ki Moon noted:

> The food crisis of today is a wake-up call for tomorrow. By 2050, our planet may be home to 9.1 billion people, over two billion more than now. At a time when the global population is growing, our global climate is changing. By 2050 we know we will need to grow 70 per cent more food. Yet, weather is becoming more extreme and unpredictable. In many parts of the world water supplies are declining. Agricultural land is drying out. Food security and climate change are deeply interconnected.[139]

Illustrating his point, Moon observed: 'If the glaciers of the Himalaya melt, it will affect the livelihoods and survival of three hundred million people in China and up to a billion people throughout Asia.'[140] Moreover, he feared: 'Africa's small farmers, who produce most of the continent's food and depend mostly on rain, could see harvests drop by 50 per cent by 2020.'[141]

José Manuel Barroso, President of the European Commission, spoke of the importance of crop diversity in tackling climate change and called for more high-level attention to this subject, stating that 'bio-diversity is also an important part of the solution, and I think we have undervalued its contribution both in tackling climate change and food insecurity'.[142]

---

136 Ibid.

137 Ibid.

138 The United Nations Food and Agriculture Organization (2009), *The World Summit on Food Security*, http://www.fao.org/wsfs/world-summit/en/.

139 Moon, Ban Ki (2009), 'Opening Remarks at World Summit on Food Security', November, http://www.fao.org/fileadmin/templates/wsfs/Summit/Statements_PDF/Monday_16_AM/01b_BAN_KI_MOON_SG_Speech_161109AM_01.pdf.

140 Ibid.

141 Ibid.

142 Barroso, José Manuel (2009), 'Remarks at World Summit on Food Security', November, http://www.fao.org/fileadmin/templates/wsfs/Summit/Statements_PDF/Monday_16_AM/12_EU_Barroso_speech_161109AM.pdf.

The Global Crop Diversity Trust coordinated the drafting and release of a Statement on Food Security and Climate Change.[143] Prominent agricultural scientists along with distinguished climate change experts joined together to underscore how the almost total absence of agriculture in the agreement could lead to widespread famine and food shortages in the years ahead. The Statement noted that 'farmers will encounter problems they have never before experienced: much greater weather variability, higher average temperatures, increased numbers of extremely hot days, shorter growing seasons, much greater moisture stress, added salinity from salt water incursion and irrigation systems, and new combinations of pests and diseases'.[144]

The Statement stressed that climate change adaptation was a tough task:

> Getting agriculture ready for such dramatically new growing environments is not a trivial matter. No one should assume that success is guaranteed. For agriculture to adapt, crops must adapt, but there is no 'climate change gene,' no single characteristic, that can ensure that they will retain, much less increase, their productivity in new climates. Concerted adaptation efforts will be required crop-by-crop, country-by-country and internationally. Crop diversity is the raw material for crop adaptation. This diversity, primarily found today in seedbanks, contains the traits that plant breeders and farmers will need to incorporate into tomorrow's resilient, climate-ready crop varieties. Agriculture – and people – cannot do without it.[145]

The statement lamented: 'Current institutional and financial arrangements, however, are inadequate to guarantee conservation of this priceless resource.'[146] It noted that 'diversity is being lost – diversity that almost certainly holds the key to future crop adaptation'.[147] The statement feared that 'the time required to integrate new traits into crop varieties can be a decade or more' and 'we cannot wait for disaster before initiating action'.[148] The statement called for greater investment in crop diversity: 'Small investments made now will soon generate enormous and recurring benefits even if climate change is far less pronounced than is now widely anticipated.'[149]

---

143 World Summit on Food Security (2009), *Food Security and Climate Change: A Call for Commitment and Preparation*, November.

144 Ibid.

145 Ibid.

146 Ibid.

147 Ibid.

148 Ibid.

149 Ibid.

The statement concluded: 'No credible or effective agreement to address the challenges of climate change can ignore agriculture and the need for crop adaptation to ensure the world's future food supplies.'[150]

The discord of the Copenhagen negotiations highlights the difficulties in resolving some of the key issues – including food security and agriculture. In the 2009 annual report for the Global Crop Diversity Trust, Chair Margaret Catley-Carlson and Executive Director Cary Fowler emphasised the need for any future international agreement on climate change to address the question of agriculture and crop diversity:

> Yet while agriculture remained unacknowledged in the run up to Copenhagen, in an apparently parallel universe, countries committed to achieve food security at the World Summit on Food Security. At the Trust, we tried to bring these two worlds together, issuing a statement to the world's media signed by a remarkable group of over 70 leading thinkers in development: 'No credible or effective agreement to address the challenges of climate change can ignore agriculture and the need for crop adaptation to ensure the world's future food supplies'. Crops, of course, cannot be adapted to meet climate change or any other challenge without crop diversity, and we continue to lay the groundwork for future food security, whatever the climate may be.[151]

The executives concluded: 'So as international political agreements very slowly begin to join the dots between agriculture and climate change, our work continues to gather pace.'[152] The topic of food security and agriculture also received short shrift during the Cancún negotiations.

Future international negotiations need to consider the nexus between intellectual property, agriculture, food security, and climate change.

---

150 Ibid.

151 Global Crop Diversity Trust (2009), *Annual Report*, Rome: Global Crop Diversity Trust, 2.

152 Ibid.

Bibliography

BOOKS, MONOGRAPHS, JOURNAL ARTICLES AND INTERNET MATERIALS

23andMe Inc. https://www.23andme.com/.

The 1000 Genomes Project Consortium (2010), 'A Map of Human Genome Variation from Population-Scale Sequencing', *Nature*, 467 (7319), 1061–73.

The 1000 Genomes Project, 'About the 1000 Genomes Project: Project Overview', http://www.1000genomes.org/about.

The 1000 Genomes Project, 'Data Release Policy', http://www. 1000 genomes.org/data#DataReleasePolicy.

The 1000 Genomes Project, 'How to Access 1000 Genomes Data', http:// www.1000genomes.org/data#DataAccess.

The 1000 Genomes Project, 'Use of the Project Data, Presentations and Publications, and Authorship', http://www.1000genomes.org/data# DataUse.

Abbott, F. (2009), 'Innovation and Technology Transfer to Address Climate Change: Lessons from the Global Debate on Intellectual Property and Public Health', International Centre for Trade and Sustainable Development Platform on Climate Change, Trade Policies and Sustainable Energy, Issue Paper No. 24, June.

Abbott, K., G. Marchant and D. Sylvester (2008), 'A Framework Convention for Nanotechnology', *Environmental Law Reporter*, **38**, 10507–17.

Aboriginal Bush Traders, http://www.aboriginalbushtraders.com/.

Advisory Council on Intellectual Property (2010), *A Review of Enforcement of Plant Breeder's Rights*, Canberra: Commonwealth of Australia, http:// www.acip.gov.au/library/ACIP_PBR_Enforcement_Final_Report.pdf.

African Centre for Biosafety (2009), *Patents, Climate Change and African Agriculture: Dire Predictions*, September, http://www.biosafety-info.net/ file_dir/17225619254ac3067689f7a.PDF.

Agovic, Amina (2010), 'Patent Morality – the Little Battler' in Niklas Bruun (ed.), *In Search of New IP Regimes*, Oy Nord Print Ab: IPR University Center.

Ahn A.C., M. Tewari, C.-S. Poon and R.S. Phillips (2006), 'The Limits of Reductionism in Medicine: Could Systems Biology Offer an Alternative?' *Public Library of Science Medicine,* **3** (6), 709–13.

Akcomak, I.S. and B. ter Weel (2008), 'Social Capital, Innovation and Growth: Evidence from Europe', *European Economic Review*, **53**, 544–67.

Alberts, Bruce (2008), *Molecular Biology of the Cell*, 5th edn, San Francisco: Garland Science.

Altman, J. and P. Whitehead (2003), 'Caring for Country and Sustainable Indigenous Economic Development: Opportunities, Constraints and Innovation', *CAEPR Working Paper* 20/2003, http://caepr.anu.edu.au/Publications/WP/2003WP20.php.

Alves, R. and I. Rosa (2007), 'Biodiversity, Traditional Medicine and Public Health: Where Do They Meet?' *Journal of Ethnobiology and Ethnomedicine*, **3**, 14, 1–19.

Amaral P.P., M.E. Dinger, T.R. Mercer and J.S. Mattick (2008), 'The Eukaryotic Genome as an RNA Machine', *Science*, **319**, 1787–9.

American Civil Liberties Union (2009), 'BRCA: Genes and Patents', 27 May, http://www.aclu.org/free-speech/brca-genes-and-patents#12.

American Medical Association, Code of Medical Ethics, http://www.ama-assn.org/ama1/pub/upload/mm/369/ceja_3i07.pdf.

Anderlik, M.R. (2003), 'Commercial Biobanks and Genetic Research: Ethical and Legal Issues', *American Journal of Pharmacogenomics*, **3**, 203–15.

Andrews, L. (2005), 'Harnessing the Benefits of Biobanks', *Journal of Law, Medicine, and Ethics*, **33**, 22–30.

Andrews, L. and J. Paradise (2005), 'Gene Patents: The Need for Bioethics Scrutiny and Legal Change', *Yale Journal of Health Policy Law and Ethics*, **5**, 403–12.

Angrist, M. and R.M. Cook-Deegan (2006), 'Who Owns the Genome?' *The New Atlantis*, **11** (Winter), 87–96.

Aoki, Keith (2008), *Seed Wars: Controversies and Cases on Plant Genetic Resources and Intellectual Property*, Durham, NC: Duke University Press.

Apple Retail Store, http://www.apple.com/au/ipodnano/.

Arnold, David (1996), *The Problem of Nature: Environment, Culture and European Expansion*, London: Blackwell.

Association of University Technology Managers (2007), *In the Public Interest: Nine Points to Consider in Licensing University Technology*, http://www.autm.net/source/NinePoints/ninepoints_endorsement.cfm.

Auden, W.H. (1940), '1 September 1939', *Another Time*, London and New York: Random House.

Australian Department of Sustainability, Environment, Water, Population and Communities, *Environment Protection and Biodiversity Conservation Act 1999: Frequently Asked Questions*, http://www.environment.gov.au/epbc/publications/epbc-act-fact-sheet.html.

Australian Government (2007), *Australian Nanotechnology: Capability and Commercial Potential*, 3rd edn, Invest Australia, http://www.investaustralia.gov.au/media/IR_Nano_NanotechReport.pdf.

Australian Human Rights Commission (2010), *The Community Guide to the UN Declaration on the Rights of Indigenous Peoples*, Sydney: The Australian Human Rights Commission.

Australian Law Reform Commission (2003), *Gene Patenting and Human Health, Issues Paper 27*, Sydney: Australian Commonwealth, http://www.austlii.edu.au/au/other/alrc/publications/issues/27/.

Australian Law Reform Commission (2004), *Gene Patenting and Human Health, Discussion Paper 68*, Sydney: Australian Commonwealth, http://www.austlii.edu.au/au/other/alrc/publications/dp/68/.

Australian Law Reform Commission (2004), *Genes and Ingenuity, Report 99*, Sydney: Australian Government.

Australian Law Reform Commission and Australian Health Ethics Committee (2003), *Essentially Yours, the Protection of Human Genetic Information in Australia, Report No 96*, Sydney: Australian Government.

The Australian Research Centre of Excellence for Plant Energy Biology (2009), *Genes to Energy: Annual Report*, Crawley: The University of Western Australia, iv, http://www.plantenergy.uwa.edu.au/aboutus/reports/ARCCPEB-2009.pdf.

Bagley, M. (2004), 'Patent First, Ask Questions Later: Morality and Biotechnology in Patent Law', *William & Mary Law Review*, **45**, 469–547.

Baker, D., G. Church, J. Collins, D. Endy, J. Jacobson, J. Keasling, P. Modrich, C. Smolke and R. Weiss (The Bio Fab Group) (2006), 'Engineering Life: Building a FAB for Biology', *Scientific American*, June, 34–9.

Baldwin, C.Y. and E. von Hippel (2009), 'Modeling a Paradigm Shift: From Producer Innovation to User and Open Collaborative Innovation', *Harvard Business School Working Paper 10–048*, http://ssrn.com/abstract=1502864.

Ball, Philip (2004), *Critical Mass: How One Thing Leads to Another*, London: William Heinemann.

Ballardini, R.M. (2008), 'Software Patents in Europe: The Technical Requirement Dilemma', *Journal of Intellectual Property Law and Practice*, **3** (9), 563–75.

Barnes, S.B. and J.A. Dupré (2008), *Genomes and What to Make of Them*, Chicago: University of Chicago Press.

BASF Corporation (2009), *Annual Report: Economic, Environmental and Social Performance*, http://www.report.basf.com/2009/en/servicepages/downloads/files/BASF_Report_2009.pdf?cat=b.

Baslar, Kemal (1998), *The Concept of the Common Heritage of Mankind in International Law*, The Hague: Martinus Nijhoff Publishers.

Bavikatte, K. and H. Jonas (2010), 'Bio-cultural Community Protocols Enforce Biodiversity Benefits: A Selection of Cases and Experiences', *Endogenous Development Magazine*, **6**, 4–6, http://www.compasnet.org/afbeeldingen/Magazines/ED%20Magazine%206/Magazine_ED6.html.

Bawa, R. (2007), 'Nanotechnology Patent Proliferation and the Crisis at the U.S. Patent Office', *Albany Law Journal of Science & Technology*, **17** (3), 699–736.

Bayer AG (2008), Chairman's Letter, *Annual Report*, http://www.annualreport2008.bayer.com/en/chairmans-letter.aspx.

Bayer AG (2009), *Annual Report*, Leverkusen: Bayer AG, 109, http://www.bayer.com/en/Bayer-Annual-Report-2009.pdfx.

Bayer AG, (2008) 'The Second Green Revolution: Climate Change', Leverkusen: Bayer AG, http://www.bayercropscience.com/bcsweb/crop protection.nsf/id/EN_Climate_Change.

Bayer CropScience (2008), 'Defying the Consequences of Climate Change: Tried-and-tested insecticide makes plants more resistant to stress', 23 October, http://www.bayercropscience.com/BCSWeb/CropProtection.nsf/id/EN_2008-NST-042_Wolfgang_Thielert.

Beal, J. (2005), 'Silence is Golden: Can RNA Interference Therapeutics Deliver? *Drug Discovery Today*, **10** (3), 169–72.

Becerra, X. (2010), Representative Becerra Applauds Decision on Gene Patenting, http://becerra.house.gov/index.php?option=com_content&view=article&id=412%3Arep-becerra-applauds-court-decision-on-gene-patenting&catid=3%3Apress-releases&Itemid=26.

Bechtel, William (2006), *Discovering Cell Mechanisms: The Creation of Modern Cell Biology*, New York: Cambridge University Press.

Bell, E. and S. Smith (2010), 'Future of Food Now a Global Battle', ABC News, April, http://www.abc.net.au/news/events/gene-wars/future-of-food.htm.

Bentley, D. (1996), 'Genomic Sequence Information Should Be Released Immediately and Freely in the Public Domain', *Science*, **274** (5287), 533–4.

Bently, Lionel and Brad Sherman (2009), *Intellectual Property*, 3rd edn, Oxford: Oxford University Press.

Bentwich, I. et al. (2005), 'Identification of Hundreds of Conserved and Nonconserved Human MicroRNAs', *Nature Genetics*, **37**, 766–70.
Benyus, Janine (1997), *Biomimicry: Innovation Inspired by Nature*, New York: Harper Perennial.
Berrier Jr, E.F. (1995), 'Global Patent Must Be Reduced', *IDEA*, **36**, 473–99.
Bertone, P. et al. (2004), 'Global Identification of Human Transcribed Sequences with Genome Tiling Arrays', *Science*, **306**, 2242–6.
Beyleveld, D. and R. Brownsword (2002), 'Is Patent Law part of the EC Legal Order? A Critical Commentary on the Interpretation of Article 6(1) of Directive 98/44/EC in Case C-377/98', *Intellectual Property Quarterly*, **1**, 97–110.
Biobanking and Biomolecular Resources Research Infrastructure, http://www.bbmri.eu/.
The BioBricks Foundation, 'FAQ', http://www.biobricks.org/FAQ.php.
The BioBricks Foundation, 'Our Goals', http://www.biobricks.org/Our_Goals.php.
The BioBricks Foundation, 'The BioBricks Foundation', http://www.biobricks.org/index.php.
Biodesic (2010), 'Company', http://www.biodesic.com/Company.html.
Birney, E. and T. Hudson et al. (2009), 'Prepublication Data Sharing', *Nature,* **461** (7261), 168–70.
Boel, M.F. (2007), 'Farming's Role in Mitigating Climate Change', Member of the European Commission responsible for Agriculture and Rural Development, Press Release, 3 July.
Boldt, Joachim and Oliver Müller (2008), 'Newtons of the Leaves of Grass: Certain Ethical Implications of Synthetic Biology Research Go Beyond Those of Genetic Engineering', *Nature Biotechnology*, **26** (4), 387–9.
Bostanci, A. and J. Calvert (2008), 'Invisible Genomes: The Genomics Revolution and Patenting Practice', *Studies in the History and Philosophy of the Biological and Biomedical Sciences*, **39**, 109–19.
Bourdieu, Pierre (2004), *Science of Science and Reflexivity*, Cambridge: Polity Press.
Boyle, James (2008), *The Public Domain: Enclosing the Commons of the Mind*, New Haven, CT and London: Yale University Press.
Bran, M. (2010), 'Gas Hub Threatens Gubinge: Traditional Owner', ABC Rural, 24 December, http://www.abc.net.au/rural/nt/content/201012/s3101215.htm?site=kimberley.
Brennan, P. (2007), 'Plant Variety Intellectual Property Rights in a Changing and Challenging Environment', 20 Years of Plant Breeder's Rights Symposium, Canberra, 8 June, http://www.cropgeninternational.com/media/PBR_Future_070506_v2.pdf.

Brewer, John and Susan Staves (eds) (1996), *Early Modern Conceptions of Property*, London: Routledge.

Brody, B. (2006), 'Intellectual Property and Biotechnology: The U.S. Internal Experience – Part I', *Kennedy Institute of Ethics Journal*, **16**, 1–37.

Brody, B. (2006), 'Intellectual Property and Biotechnology: The U.S. Internal Experience – Part II', *Kennedy Institute of Ethics Journal*, **16** (2), 105–28.

Brooke, John Hedley (1991), *Science and Religion: Some Historical Perspectives*, Cambridge: Cambridge University Press.

Brown, Michael F. (2003), *Who Owns Native Culture?* Cambridge, MA: Harvard University Press.

Brown, P. and K. Kleiner (1994), 'Patent Row Splits Breast Cancer Researchers', *New Scientist*, 44, 24 September.

Brush, S.B. (1993), 'Indigenous Knowledge of Biological Resources and Intellectual Property Rights: The Role of Anthropology', *American Anthropologist*, **95** (3), 653–71.

Bruun, Niklas (ed.) (2010), *In Search of New IP Regimes*, Oy Nord Print Ab: IPR University Center.

Bühler, M. (2009), 'RNA turnover and chromatin-dependent gene silencing', *Chromosoma,* **118**, 141–51.

Bullis, K. (2007), 'Key Nanotech Patents Licensed', *Technology Review*, 6 June, http://www.technologyreview.com/biomedicine/18829/.

Burd, A. (2007), 'Stem Cell Strategies in Burns Care', *Burns*, **33**(3), 282–91.

Burian, R.M. (2008), 'Is Molecular Genetics Becoming Less Reductionistic? Notes from recent case studies on mapping *C. elegans* and the discovery of microRNA', Invited Lecture, Institut d'histoire et de Philosophie des Sciences et des Techniques, Paris, 4 December.

Burk, D. (2000), 'Patenting Speech', *Texas Law Review*, **79**, 99–162.

Burk, Dan and Mark Lemley (2009), *The Patent Crisis and How the Courts Can Solve It*, Chicago and London: The University of Chicago Press.

Burns, L. (2009), 'You Are Our Only Hope: Trading Metaphorical "Magic Bullets" for Stem Cell Superheroes', *Theoretical Medicine and Bioethics*, **30**, 427–42.

California Institute of Technology 2010 iGEM Team (2010), 'Team:Caltech/Acknowledgments', http://2010.igem.org/Team:Caltech/Acknowledgments.

Calvert, J. (2007), 'Patenting Genomic Objects: Genes, Genomes, Function and Information', *Science as Culture*, **16** (2), 207–23.

Calvert, J. (2010), 'Synthetic Biology: Constructing Nature?' *The Sociological Review*, **58** (1), 95–112.

Calvert, J. and P. Joly (2011), 'How did the Gene Become a Chemical Compound? Shifting Ontologies of the Gene and the Patenting of DNA', *Social Science Information*, **50** (2), 157–77.

Cambon-Thomsen, A. (2003), 'Assessing the Impact of Biobanks', *Nature Genetics*, **34**, 25–6.

Campbell, A. (2007), 'The Ethical Challenges of Genetic Databases: Safeguarding Altruism and Trust', *King's Law Journal*, **18**, 227–46.

Canadian Biotechnology Advisory Committee (2002), *Patenting Of Higher Life Forms*, Ottawa: Canadian Biotechnology Advisory Committee, June.

Canadian Biotechnology Advisory Committee (2003), *Advisory Memorandum: Higher Life Forms and the Patent Act*, Ottawa: Canadian Biotechnology Advisory Committee, 24 February.

Canadian Biotechnology Advisory Committee (2006), *Human Genetic Materials, Intellectual Property, and the Health Sector*, Ottawa: Canadian Biotechnology Advisory Committee, http://cbac-cccb.ca/epic/internet/incbac-cccb.nsf/en/ah00578e.html.

Canadian Intellectual Property Office, *Manual of Patent Office Practice*, http://www.ic.gc.ca/eic/site/cipointernet-internetopic.nsf/eng/h_wr00720.html.

Cao, Q., R. Benton and S. Whittemore (2002), 'Stem Cell Repair of Central Nervous System Injury', *Journal of Neuroscience Research*, **68** (5), 501–10.

Carbone, J., E.R. Gold, B. Sampat, S. Chandrasekharan, L. Knowles, M. Angrist and R. Cook-Deegan (2010), 'DNA Patents and Diagnostics: Not a Pretty Picture', *Nature Biotechnology*, **28**, 784–91.

Carlson, Robert (2003), 'The Pace and Proliferation of Biological Technologies', *Biosecurity and Bioterrorism: Biodefense Strategy, Practice, and Science*, **1**, 203–14.

Carlson, Robert (2010), *Biology is Technology: The Promise, Peril, and New Business of Engineering Life*, Cambridge, MA and London: Harvard University Press.

Carney, J. (2009), 'Retreat From The Brink of Clarity: Why the Federal Circuit Got *In Re Bilski* Wrong, and What Can Be Done About It', *University of Illinois Journal of Law Technology and Policy*, 473–91.

CARTaGENE, http://www.cartagene.qc.ca/.

Carter, T. (2010), 'A Patent on Problems', *ABA Journal*, **96** (3), 7.

Casid, Jill H. (2005), *Sowing Empire: Landscape and Colonization*, Minneapolis: University of Minnesota Press.

Castanotto, D. and J.J. Rossi (2009), 'The Promises and Pitfalls of RNA-interference-Based Therapeutics', *Nature*, **457**, 426–33.

Castle, David (ed.) (2009), *The Role of Intellectual Property Rights in Biotechnology Innovation*, Cheltenham and Northampton, MA: Edward Elgar.

Caulfield, T., R. Cook-Deegan, F.S. Kieff and J. Walsh (2006), 'Evidence and Anecdotes: An Analysis of Human Gene Patenting Controversies', *Nature Biotechnology*, **24** (9), 1091–4.

ccMixter, http://ccmixter.org/about.

Center for International Environment and Development Studies (Noragric), Agricultural University of Norway and Nordic Gene Bank (2004), 'Study to Assess the Feasibility of Establishing a Svalbard Arctic Seed Depository for the International Community Prepared for the Ministry of Foreign Affairs and the Ministry of Agriculture', 14 September, http://www.regjeringen.no/Upload/LMD/kampanjeSvalbard/Vedlegg/Frohvelv_Study_to_assess.pdf.

Central Land Council, 'Project Principles and Special Purpose Permits', http://www.clc.org.au/onlinepermits/specialpurposeprinciples.aspx.

Central Land Council, 'Special Purpose Permit Application Form', http://www.clc.org.au/OnlinePermits/ApplicantDetails.aspx?0=Special.

Ceres Inc., http://www.ceres.net.

Ceres Inc., Intellectual Property, http://www.ceres.net/Technology/Tech-IP.html.

Chambers, J. (2002–3), 'Patent Eligibility of Biotechnological Inventions in the United States, Europe, and Japan: How Much Patent Policy Is Public Policy?' *George Washington International Law Review*, **34**, 223–46.

Chapman, A. (2009), 'The Ethics of Patenting Human Embryonic Stem Cells', *Kennedy Institute of Ethics Journal*, **19** (3), 261–88.

Chatterjee, A. (2004), 'Europe Struggles Over Software Patents', *IEEE Spectrum* (September), http://spectrum.ieee.org/at-work/innovation/europe-struggles-over-software-patents.

Chatzimarkakis, J. (2009), 'Getting an Appetite for Biotechnology', Viewpoint, *BBC News*, 24 February, http://news.bbc.co.uk/2/hi/sci/tech/7905567.stm.

Chen, H., M. Roco, X. Lin and Y. Lin (2008), 'Trends in Nanotechnology Patents', *Nature Nanotechnology*, **3**, 123–5.

Clapp, Jennifer and Doris Fuchs (eds) (2009), *Corporate Power in Global Agrifood Governance*, Cambridge, MA: MIT Press.

Clarkson, G. and D. Dekorte, (2006) 'The Problem of Patent Thickets in Convergent Technologies', *Annals of the New York Academy of Sciences*, **1093**, 180–200.

Climate Change, Agriculture and Food Security, Research, http://www.ccafs.cgiar.org/research.

Clinton, W. and T. Blair (2000), 'Joint Statement by President William Clinton and Prime Minister Tony Blair of the United Kingdom', 14 March, http://ipmall.info/hosted_resources/ippresdocs/ippd_44.htm.

Collier, Robert and Josephine Mutugu (2009), *Who Owns the Clean Tech Revolution? Intellectual Property Rights and International Cooperation in the U.N. Climate Negotiations*, Center for Environmental Public Policy, Goldman School of Public Policy, University of California, Berkeley, 26–27 October, http://energyandclimate.berkeley.edu/sites/default/files/collier-Who%20Owns%20the%20Clean%20Tech%20Revolution.pdf.

Collins, Francis (2010), 'Has the Revolution Arrived?' *Nature*, **464,** 674–5.

Collins, F., E. Green, A. Guttmacher and M. Guyer (2003), 'A Vision for the Future of Genomics Research', *Nature*, **422**, 835–47.

Collins, F., N. Morgan and A. Patrinos (2003), 'The Human Genome Project: Lessons from Large-Scale Biology', *Science*, **300** (5617), 11 April, 286–90.

Collins, K. (2010), 'Semiotics 101: Taking the Printed Matter Doctrine Seriously', *Indiana Law Journal*, **85**, 1379–443.

Commission of the European Communities (2009), 'The Role of European Agriculture in Climate Change Mitigation', 23 July, http://ec.europa.eu/agriculture/climate_change/sec2009_1093_en.pdf.

Conley, J.M. and R. Makowski (2003), 'Back to the Future: Rethinking the Product of Nature Doctrine as a Barrier to Biotechnology Patents', *Journal of the Patent and Trademark Office Society*, **85**, 301–34 (Part I), 371–98 (Part II).

The Consultative Group on International Agricultural Research, http://www.cgiar.org/.

The Consultative Group on International Agricultural Research Change Steering Team (2008), 'A Revitalized CGIAR – A New Way Forward: The Integrated Reform Proposal', http://www.cgiar.org/pdf/agm08/agm08_reform_proposal.pdf.

Cook-Deegan, R. (2001), 'Hype And Hope', *American Scientist*, **89**, 62.

Costa, F. (2008), 'Noncoding RNAs, Epigenetics and Complexity', *Gene*, **410** (1), 9–17.

Coulter, J. (1990), 'Parliamentary Debates', Australian Parliament, Senate, 22 August.

Coulter, Maureen (1991), *Property in Ideas: The Patent Question in Mid-Victorian Britain*, Kirksville: The Thomas Jefferson University Press.

The Council for Biotechnology (2008), 'Agricultural Biotechnology: Benefits for Biofuels and U.S. Energy Security', http://www.whybiotech.com/resources/briefs_benefitsforbiofuels.asp.

The Council for Biotechnology (2008), 'Agricultural Biotechnology: Benefits for Water Scarcity in the United States', http://www.whybiotech.com/resources/briefs_benefitsforwaterscarcity.asp.

Council of Canadian Academies (2008), *Small is Different: A Science Perspective on the Regulatory Challenges of the Nanoscale*, the Expert Panel on Nanotechnology, Ottawa: Council of Canadian Academies, http://www.scienceadvice.ca/en/assessments/completed/nanotechnology.aspx.

Cozzens, S. et al. (2010), 'Emerging Technologies: Quantitative Identification and Measurement', *Technology Analysis and Strategic Management*, **22** (3), 361–76.

Crane, Andrew and Dirk Matten (2004), *Business Ethics*, Oxford: Oxford University Press.

Creative Commons, http://creativecommons.org/about/.

Crichton, M. (2007), 'Patenting Life', *The New York Times*, 13 February.

Crosby, Alfred (1986), *Ecological Imperialism: The Biological Expansion of Europe, 900–1900*, Cambridge: Cambridge University Press.

CSIRO, 'Plants and Climate Change', http://www.csiro.au/science/pshy.html.

Cummings, Claire Hope (2008), *Uncertain Peril: Genetic Engineering and the Future of Seeds*, Beacon Press.

Curtin, C. (2010), 'When to Share', Genomeweb, May, http://webcache.googleusercontent.com/search?q=cache:y3_E1riwjCEJ:www.genomeweb.com/when-share+what+is+a+marker+paper+toronto+statement&cd=8&hl=en&ct=clnk&gl=us

Dakis, S. (2008), 'PBR Helping with Climate Change', *ABC News*, 27 August, http://www.abc.net.au/rural/sa/content/2006/s2348159.htm.

The Danish Council of Ethics (2004), 'The Ethics of Patenting Human Genes and Stem Cells', Copenhagen, 28 September.

Davis, Michael (2009), *Maintain and Strengthen your Culture: Handbook for Working with Indigenous Ecological Knowledge and Intellectual Property*, commissioned by the Natural Resource Management Board NT, Darwin, http://www.nrmbnt.org.au/files/iek/Handbook%20for%20Working%20with%20IEK%20&%20IP.pdf.

De Beer, J. (2007), 'Biotrespass', *Bulletin of Science Technology & Society*, **27** (4), 287–99.

Dean, Warren (1987), *Brazil and the Struggle for Rubber: A Study in Environmental History*, Cambridge: Cambridge University Press.

DeGiulio, J. (2010), '*The Genomic Research and Accessibility Act*: More Science Fiction than Fact', *Northwestern Journal of Technology and Intellectual Property,* **8**, 292–306.

Demaine, L.J. and A.X. Fellmeth (2002), 'Reinventing the Double Helix: A Novel and Nonobvious Reconceptualization of the Biotechnology Patent', *Stanford Law Review*, **55** (2), 303–462.

Department of Agriculture and Food (2009), 'Picking Gubinge – the Right Way', http://wn.com/Kakadu_plum.

Department of Agriculture, Fisheries and Forestry (2002–3), 'Ecologically Sustainable Development and Environmental Performance', in Annual Report, Canberra: Department of Agriculture, Fisheries and Forestry, 297, http://www.daff.gov.au/__data/assets/pdf_file/0012/5700/annual_report0203.pdf.

Department of the Environment, Water, Heritage, and the Arts (2008), 'Indigenous Knowledge Management Systems (Databases): Guide for Indigenous Communities', http://www.environment.gov.au/indigenous/publications/pubs/knowledge-management-guide.pdf.

Di Giorgio, F. et al. (2007), 'Non-cell Autonomous Effect of Glia on Motor Neurons in an Embryonic Stem Cell-Based ALS model', *Nature Neuroscience*, **10** (5), 608–14.

Dinwoodie, Graeme and Rochelle Cooper Dreyfuss (2006), 'Patenting Science: Protecting the Domain of Accessible Knowledge,' in Bernt Hugenholtz and Lucie Guibault (eds), *The Future of the Public Domain*, Amsterdam: Kluwer Law International, 191–221.

Doench, J.G. and P.A. Sharp (2004), 'Specificity of MicroRNA Target Selection in Translational Repression', *Genes and Development*, **18**, 504–11.

Dorries, H. Ulrich (2001), *Patentanspruche auf DNA-Sequenzen: ein Hindernis fur die Forschung? Anmerkungen zum Regierungsentwurf fur ein Gesetzzur Umsetzung der Richtlinie* 98/44/EG, MITT.

Dow (2008), 'Introduction to Dow's Position on Energy and Climate Change', http://www.dow.com/commitments/pdf/OurPosition_Eng Climate_FINAL.pdf.

Drahos, P. (1999), 'Biotechnology Patents, Markets and Morality', *European Intellectual Property Review*, **21** (9), 441–9.

Drahos, Peter and John Braithwaite (2002), *Information Feudalism: Who Owns the Knowledge Economy?*, London: Earthscan.

Drahos, Peter and Ruth Mayne (eds) (2002), *Global Intellectual Property Rights: Knowledge, Access, and Development*, New York: Palgrave Macmillan.

Drayton, Richard (2000), *Nature's Government: Science, Imperial Britain, and the 'Improvement' of the World*, Cambridge, MA: Harvard University Press.

Drexler, K. Eric (1986), *Engines of Creation: The Coming Era of Nanotechnology*, Garden City: Anchor Press/DoubleDay.

Drexler, K.E. (2004). 'Nanotechnology: From Feynman to Funding', *Bulletin of Science, Technology & Society*, **24** (1), 21–7.

Dreyfuss, R.C. (2004), 'Protecting the Public Domain of Science: Has the Time for an Experimental Use Defense Arrived?' *Arizona Law Review*, **46**, 457–72.

Du Mont, J. (2008), 'Trademarking Nanotechnology: Nano-Lies and Federal Trademark Registration', *AIPLA Quarterly Journal*, **36** (2), 148–78.

Duffy, J. (2009), 'Rules and Standards on the Forefront of Patentability', *William & Mary Law Review*, **51**, 609–53.

Dukes, Graham and Frederick Abbott (2009), *Global Pharmaceutical Policy: Ensuring Medicines for Tomorrow's World*, Cheltenham and Northampton, MA: Edward Elgar.

DuPont Australia (2006), 'Drought Tolerance a Target of New International Crop Research Agreement', Press Release, 6 April, http://www2.dupont.com/Media_Center/en_AU/australia_news/2006/article20060406.html.

DuPont, 'Statement on Climate Change', http://www2.dupont.com/Media_Center/en_US/position_statements/global_climate.html.

DuPont (2009), 'Biotechnology Can Address Global Mega Trends, DuPont Leader Says: Biotechnology Stewardship is Critical to Meeting Global Needs for Increased Food and Energy', BioJapan 2009, 7 October.

DuPont and EvoGene (2007), 'DuPont and Evogene Collaborate to Increase Drought Tolerance in Corn and Soybeans', Press Release, 30 October.

Dutfield, G. (2003), 'Should We Terminate Terminator Technology?' *European Intellectual Property Review*, **25** (11), 491–5.

Dutfield, Graham (2008), 'Turning Plant Varieties into Intellectual Property: The UPOV Convention', in Geoff Tansey and Tamsin Rajotte (eds), *The Future Control of Food: A Guide to International Negotiations and Rules on Intellectual Property, Biodiversity, and Food Security*, London: Earthscan, 27–47.

Dutfield, Graham (2009), *Intellectual Property Rights and the Life Science Industries: Past, Present and Future*, Singapore: World Scientific.

Dutra, T. (2011), 'Myriad, ACLU Battle Renewed in Fed. Cir. Oral Arguments with DOJ Joining the Fray', *Patent, Trademark & Copyright Journal*, **81**, 1–2.

Dyer, Gwynne (2010), *Climate Wars: The Fight for Survival as the World Overheats*, Melbourne: Scribe.

The Economist (1999), 'Drowning in Data', *The Economist*, 26 June, 97–8.

Edgerton, M. (2009), 'Increasing Crop Productivity to Meet Global Needs for Feed, Food, and Fuel', *Plant Physiology*, **149**, 7–13.

Editor (2010), 'Lagging Laws', *Nature Neuroscience*, **13** (1), http://www.nature.com/neuro/journal/v13/n1/full/nn0110–1.html.

Edwards, A.M., C. Bountra, D.J. Kerr and T.M. Willson (2009), 'Open Access Chemical and Clinical Probes to Support Drug Discovery', *Nature Chemical Biology*, **5**, 436–40.

Eisenberg, R. (1996), 'Public Research and Private Development: Patents and Technology Transfer in Government-Sponsored Research', *Virginia Law Review*, **82**, 1663–727.

Eisenberg, R. (1997), 'Structure and Function in Gene Patenting', *Nature Genetics,* **15**, 125–30.

Eisenberg, R. (2008), 'Noncompliance, Nonenforcement, Nonproblem? Rethinking the Anticommons in Biomedical Research', *Houston Law Review*, **45**, 1059–99.

Eisenberg, R. and R. Nelson (2002), 'Public vs. Proprietary Science: A Fruitful Tension', *Daedalus*, **131**, 89–101.

Elbashir, S.M., J. Harborth, W. Lendeckel, A. Yalcin, K. Weber and T. Tuschl (2001), 'Duplexes of 21±Nucleotide RNAs mediate RNA Interference in Cultured Mammalian Cells', *Nature,* **411**, 494–8.

Endy, D. (2005), 'Foundations for Engineering Biology', *Nature*, **438** (7067), 449–53.

Endy, D. (2009), 'Open Biotechnology and the BioBrick Public Agreement', http://openwetware.org/images/f/fd/Why_the_BPAv1.pdf.

Endy, D. (2010), 'Evidence to the Presidential Commission for the Study of Bioethical Issues, Synthetic Biology', Washington, DC, 8 July.

Endy, D. et al. (2010), 'The BioBrick™ Public Agreement: DRAFT Version 1a', http://dspace.mit.edu/bitstream/handle/1721.1/50999/BPA_draft_v1a.pdf?sequence=1.

ETC Group, 'BANG: Converging Technology', http://www.etcgroup.org/en/issues/bangconvergence.

ETC Group (2002), 'Patenting Elements of Nature', 25 March, http://www.etcgroup.org/upload/publication/220/01/nanopatentsgeno.rtf.pdf.

ETC Group (2003), *The Big Down: From Genomes to Atoms; Atomtech: Technologies Converging at the Nano-Scale,* Ottawa: ETC Group, 8, http://www.etcgroup.org/upload/publication/171/01/thebigdown.pdf.

ETC Group (2004), 'Down on the Farm: The Impact of Nano-scale Technologies on Food and Agriculture', Ottawa: ETC Group, http://www.etcgroup.org/upload/publication/80/02/etc_dotfarm2004.pdf.

ETC Group (2007), 'Press Release: Broad International Coalition Issues Urgent Call for Strong Oversight of Nanotechnology', 31 July, http://www.etcgroup.org/en/node/651.

ETC Group (2008), 'Svalbard's Doomsday Vault: The Global Seed Vault Raises Political/Conservation Debate', 28 February, http://www.etcgroup.org/en/materials/publications.html?pub_id=674.

ETC Group (2009), 'Who Will Feed Us? Questions for the Food and Climate Crises', *The ETC Group Communique*, November, Issue 102, 1, http://www.etcgroup.org/upload/publication/pdf_file/ETC_Who_Will_Feed_Us.pdf.

ETC Group (2010), *The Big Downturn: Nanogeopolitics*, Ottawa: ETC Group, http://www.etcgroup.org/upload/publication/pdf_file/nano_big4web.pdf.

ETC Group (2010), 'Gene Giants Stockpile Patents on "Climate-Ready" Crops in Bid to Become Biomassters', Press Release, 25 October, http://www.etcgroup.org/en/node/5220.

ETC Group (2010), 'Gene Giants Stockpile Patents on "Climate-Ready" Crops in Bid to Become "Biomassters": Patent Grab Threatens Biodiversity, Food Sovereignty', *The ETC Group Communique*, November, Issue 102, http://www.etcgroup.org/en/node/5220.

Europa (2005), 'A Major Trading Power', Europa, http://europa.eu/abc/keyfigures/tradeandeconomy/tradingpower/index_en.htm.

European Commission (2007), 'Nanotechnology', http://cordis.europa.eu/nanotechnology/.

European Group on Ethics in Science and New Technologies (2010), *Ethical Aspects of Patenting Inventions Involving Human Stem Cells*, Opinion No. 16 EGE, http://ec.europa.eu/europan_group_ethics/docs/avis16_en.pdf.

European Life Science Infrastructure for Biological Information, *ELIXIR: Data for Life*, http://www.elixir-europe.org/bcms/elixir/Documents/Elixir_brochure.pdf.

European Patent Office, 'Nanotechnology in European Patents: Opportunity and Challenges', http://www.epo.org/news-issues/issues/nanotechnology.html.

European Patent Office (2007), 'Interview with Patent Expert Greenpeace International Dr Christoph Then', Open Ended Interviews, http://documents.epo.org/projects/babylon/eponet.nsf/0/106FF770E1FA4E02C12572D8003A49FD/$File/Interview_Then.pdf.

European Patent Office (2007), *Scenarios for the Future: How might IP Regimes evolve by 2025? What Global Legitimacy might such Regimes Have?* Munich: European Patent Office, http://documents.epo.org/projects/babylon/eponet.nsf/0/63A726D28B589B5BC12572DB00597683/$File/EPO_scenarios_bookmarked.pdf.

European Patent Office (2011), 'EPO classification work leads to new international standard', Press Release, 28 January, http://www.epo.org/news-issues/news/2011/20110128.html.

EvoGene, http://www.evogene.com/.

ExxonMobil, 'Algae Biofuels', http://www.exxonmobil.com/Corporate/energy_climate_con_vehicle_algae.aspx.

Farnley, S., P. Morey-Nase and D. Sternfeld (2004), 'Biotechnology – A Challenge to the Patent System', *Current Opinion in Biotechnology*, **15**, 254.

Faunce, Thomas (2010), 'Exploring the Safety of Nanoparticles in Australian Sunscreens', *International Journal of Biomedical Nanoscience and Nanotechnology*, **1** (10), 87–94.

Fecteau, L. (2001), 'The Ayahuasca Patent Revocation: Raising Questions About Current U.S. Patent Policy', *Boston College Third World Law Journal*, **21** (1), 69–104.

Federal Trade Commission (2011), *The Evolving IP Marketplace: Aligning Patent Notice and Remedies with Competition*, Washington, DC: Federal Trade Commission.

Federico, P. (1937), 'Louis Pasteur's Patents', *Science*, **86**, 327 (8 October).

Federico, P. ([1952] 1993), 'Commentary on the New Patent Act', *Journal of the Patent and Trademark Office Society*, **75**, 161–231.

Feldman, R. (2004), 'The Open Source Biotechnology Movement: Is It Patent Misuse?', *Minnesota Journal of Law, Science and Technology*, **6**, 117–67.

Finkel, E. (2009), 'Plant Breeding: Scientists Seek Easier Access to Seed Banks', *Science*, **324** (5933), 1376.

Fire A., S. Xu, M.K. Montgomery, S.A. Kostas, S.E. Driver and C.C. Mello (1998), 'Potent and Specific Genetic Interference by Double-stranded RNA in Caenorhabditis elegans', *Nature*, **19,** 391 (6669), 806–11.

Fitt, R. and E. Nodder (2009), 'The Industrial Applicability of Biotechnology Patents – A New Test for Europe', *Biotechnology Law Report*, **28** (2), 151–8.

Flemings Nurseries, http://www.flemings.com.au/topten.asp?type=7.

Flicek, P. (2010), 'Challenges for the Data Management and Analysis of Large-Scale Human Genome Sequencing', Advances in Genomics conference, 29 January, http://www.advances-in-genomics.org/presentations/Flicek.pdf.

Foladori, G. and N. Invernizzi, (2005), 'Nanotechnology for the Poor?' *Public Library of Science Medicine*, **2** (8), e280. doi:10.1371/journal.pmed.0020280.

Fortin, S., S. Pathmasiri, R. Grintuch and M. Deschenes (2010), '"Access Arrangements" for Biobanks: A Fine Line between Facilitating and

Hindering Collaboration', *Public Health Genomics*, 1–11, DOI. 10.1159/000309852.

Fourmile, Henrietta (2000), 'Indigenous Interests in Biological Resources in Commonwealth Areas', in John Voumard, *Commonwealth Public Inquiry Into Access To Biological Resources In Commonwealth Areas*, Canberra: Environment Australia http://www.ea.gov.au/biodiversity/science/access/inquiry/index.html, 24.

Fowler, C. (2000), 'The Plant Patent Act of 1930: A Sociological History of Its Creation', *Journal of the Patent and Trademark Office Society*, **82** (9), 621–44.

Fowler, C. (2008), 'The Svalbard Seed Vault and Crop Security', *BioScience*, **58** (3), 190–91.

Fowler, Cary (1994), *Unnatural Selection: Technology, Politics, and Plant Evolution*, Yverdon: Gordon and Breach.

Fowler, Cary and Patrick Mooney (1990), *Shattering: Food, Politics, and the Loss of Genetic Diversity*, Tucson: University of Arizona Press.

Free Software Foundation Inc (2010), 'The Free Software Definition', http://www.gnu.org/philosophy/free-sw.html.

Free Software Foundation Inc, 'The GNU General Public License (GPL) Version 2', http://www.opensource.org/licenses/gpl-2.0.php.

Friedel, Robert (2007), *A Culture of Improvement: Technology and the Western Millennium*, Cambridge and London: MIT Press.

Fusco, S. (2010), '*In re Bilski*: A Conversation with Judge Randall Rader and a First Look at the BPAI's Patent Cases', *Albany Law Journal of Science and Technology*, **20**, 123–58.

Gajarsa, A. (2002), 'Quo Vadis? The Fifth Annual Honorable Helen Wilson Nies Memorial Lecture in Intellectual Property Law', *Marquette Intellectual Property Law Review*, **6**, 1–9.

Galison, Peter and Bruce Hevly (1992), *Big Science: The Growth of Large-Scale Research*, Stanford, CA: Stanford University Press.

Gamow, G. (1955), 'Information Transfer in the Living Cell', *Scientific American*, **193**, 70–78.

Gardiner, H. (2010), 'US Judge Rules Against Obama's Stem Cell Policy', *New York Times*, 23 August.

Garnaut, Ross (2008), *The Garnaut Climate Change Review: Final Report*, Cambridge: Cambridge University Press, http://www.garnautreview.org.au.

Garner, Helen (1992), *Cosmo, Cosmolino*, Melbourne: McPhee Gribble and Penguin Books.

Gates, B. (2008), 'Speech to the World Economic Forum', The Gates Foundation, http://www.gatesfoundation.org/speeches-commentary/Pages/bill-gates-2008-world-economic-forum-creative-capitalism.aspx.

Gates, B. (2008), 'Making Capitalism More Creative', *Time Magazine*, 31 July.

Gates Foundation (2007), 'Grant to the United Nations Foundation to Protect the Genetic Diversity of 21 Critical Crops for Food Security and Poverty Alleviation, by Supporting National Genebanks, the Svalbard Global Seed Vault, and the Global Crop Diversity Trust', http://www.gatesfoundation.org/Grants-2007/Pages/United-Nations-Foundation-OPP45782.aspx.

Gates Foundation (2009), Agricultural Development: Strategy Overview, http://www.gatesfoundation.org/agriculturaldevelopment/Documents/agricultural-development-strategy-overview.pdf.

Gates Foundation, http://www.gatesfoundation.org/AboutUs/.

Gendreau, Ysolde (ed.) (2008), *An Emerging Intellectual Property Paradigm: Perspectives from Canada*, Cheltenham and Northampton, MA: Edward Elgar.

Generation Challenge Programme, http://www.generationcp.org/.

Generation Scotland, http://www.generationscotland.co.uk/.

Genetic Technologies Limited (2008), 'Genetic Technologies to Enforce Rights to BRCA Patents', Press Release, 11 July, http://www.gtglabs.com.au/announcements/genetic-technologies-to-enforce-rights-to-brca-patents.

Genetic Technologies Limited (2008), 'New Position re BRCA Testing', Press Release, 2 December, http://www.gtglabs.com.au/announcements/new-position-re-brca-testing.

Genetic Technologies Limited (2010), 'Genetic Technologies Moves to Assert Its Intellectual Property Rights and Files Patent Infringement Suit in USA', Press Release, 16 February, http://www.gtglabs.com.au/images/stories/pdf/ASX%20Announcements/Assertion_of_property_rights_February_16_2010.pdf.

Genetic Technologies Limited (2011), 'Genetic Technologies files Second Patent Infringement Suit in the USA', Press Release, 20 January.

Genome Web (2010), 'US DoJ Argues Against Patenting Isolated Genes, But USPTO Will Maintain Status Quo', *Pharmacogenetics Reporter*, 3 November, http://www.genomeweb.com/print/953728?page=show.

Gepts, P. (2004), 'Who Owns Biodiversity, and How Should the Owners Be Compensated?' *Plant Physiology*, **134**, 1295–307.

Ghildiyal M. and P. Zamore (2009), 'Small Silencing RNAs: An Expanding Universe', *Nature Reviews Genetics*, **10** (2), 94–108.

Gibbs, W.W. (2004), 'Synthetic Life', *Scientific American*, May, 75–81.

Gibney, Matthew (ed.) (2003), *Globalizing Rights: The Oxford Amnesty Lectures 1999*, Oxford: Oxford University Press.

Gibson, Daniel G. et al. (2010), 'Creation of a Bacterial Cell Controlled by a Chemically Synthesized Genome', *Science*, **329** (5987), 52–6.

Gilbert, N. (2010), 'Future Funding for Agricultural Research Uncertain: Financial Donors Wrangle Over Global Research Group's Strategy', *Nature*, 31 March.

Gilbert, N. et al. (2005), 'Multiple Fates of L1 Retrotransposition Intermediates in Cultured Human Cells', *Molecular and Cellular Biology*, **25**, 7780–95.

Gitter, D. (2001), 'International Conflicts Over Patenting Human DNA Sequences in the United States and the European Union: An Argument for Compulsory Licensing and a Fair-Use Exemption', *New York University Law Review*, **76** (6), 1623–91.

Gitter, D. (2007), 'Resolving the Open Source Paradox in Biotechnology: A Proposal for a Revised Open Source Policy for Publicly Funded Genomic Databases', *Houston Law Review*, **43** (4), 1476–521.

Gitter, D. (2010), 'The Challenges of Achieving Open Source Sharing of Biobank Data', *Biotechnology Law Report*, **29** (6), 623–35.

Global Crop Diversity Trust (2009), *Annual Report*, Rome: Global Crop Diversity Trust.

Global Crop Diversity Trust, 'Preparing for Climate Change', http://www.croptrust.org/main/climatechange.php.

Global Crop Diversity Trust, http://www.croptrust.org/main/.

Gold, Richard (2009), 'Avoiding the Mistakes of Biotech: How Intellectual Property Can Be Better Managed to Advance Nanotechnology Research', *Studies in Ethics, Law, and Technology*, **3** (3), Article 4.

Gold R., Y. Joly and T. Caulfield (2005), 'Genetic Research Tools, the Research Exception and Open Science', *GenEdit*, **3** (2) 1–8.

Gold, Richard and Bartha Knoppers (eds) (2009), *Biotechnology, IP and Ethics*, Markham, ON: LexisNexis Canada.

Gordon, W. (2002), 'Authors, Publishers, and Public Goods: Trading Gold for Dross', *Loyola of Los Angeles Law Review*, **36**, 159–98.

Gorman, Julian and Peter Whitehead (2006), *Small-scale Commercial Plant Harvests by Indigenous Communities*, Canberra: Rural Industries Research and Development Corporation.

Government of Western Australia Department of Health (2010), *Guidelines for Human Biobanks, Genetic Research Databases and Associated Data*, Perth: Government of Western Australia, http://www.genomics.health.wa.gov.au/publications/index.cfm.

GRAIN (1998), 'Ten Reasons Not to Join UPOV: Global Trade and Biodiversity in Conflict', *Seedling*, http://www.grain.org/seedling/?id=10.

Grandin, Greg (2010), *Fordlandia: The Rise and Fall of Henry Ford's Forgotten Jungle City*, London: Icon Books.

Grant Allen, J. (2010), 'Group Consent and the Nature of Group Belonging: Genomics, Race and Indigenous Rights', *Journal of Information, Law, and Science*, **20** (2), 28–59.

Greenpeace, 'Patents on Life', http://www.greenpeace.org/international/en/campaigns/agriculture/problem/genetic-engineering/ge-agriculture-and-genetic-pol/patents-on-life/.

Griffith University, Intellectual Property Policy, http://www62.gu.edu.au/policylibrary.nsf/binders/8a581b3be85b69e34a2570530063fdcf?opendocument.

Griffiths, P.E. (2001), 'Genetic Information: A Metaphor in Search of a Theory', *Philosophy of Science*, **68**, 394–412.

Grimm, D. et al. (2006), 'Fatality in Mice due to Oversaturation of Cellular MicroRNAs/Short Hairpin RNA Pathways', *Nature*, **441**, 537–41.

Grishok, A., A.E. Pasquinelli, D. Conte, N. Li, S. Parrish, I. Ha, D.L. Baillie, A. Fire, G. Ruvkun and C.C. Mello (2001), 'Genes and Mechanisms Related to RNA Interference Regulate Expression of the Small Temporal RNAs that Control *C. elegans* Developmental Timing', *Cell*, **106** (1), 23–34.

Haddow, G., S. Cunningham-Burley, A. Bruce and S. Parry (2008), 'Generation Scotland: Consulting Publics and Specialists at an Early Stage in a Genetic Database's Development', *Critical Public Health*, **18**, 139–49.

Hahlbrock, Klaus (2009), *Feeding the Planet: Environmental Protection Through Sustainable Agriculture*, London: Haus Publishing Ltd.

Halbert, Debora J. (2005), *Resisting Intellectual Property*, London and New York: Routledge.

Hamilton, A.J. and D.C. Baulcombe (1999), 'A Species of Small Antisense RNA in Posttranscriptional Gene Silencing in Plants', *Science*, **286**, 950–52.

Hammond, S.M. (2005), 'MicroRNAs as Tumor Suppressors', *Nature Genetics*, **39**, 582–3.

Hamson, C.J. (1930), *Patent Rights for Scientific Discoveries*, Indianapolis: Bobbs-Merill Co.

Haraway, Donna J. (1997), *Modest_Witness@Second_Millennium. Femaleman_Meets_Oncomouse*, New York: Routledge.

Haussecker, D. (2008), 'The Business of RNAi Therapeutics', *Human Gene Therapy*, **19** (5), 451–62.

Haussecker, D. (2010), 'Tuschl Litigation Decidedly Shifting into Max Planck-Alnylam's Favor', http://rnaitherapeutics.blogspot.com/2010/10/tuschl-litigation-decidedly-shifting.html.

Haverkort, Bertus and Stephen Rist (eds) (2007), *Endogenous Development and Bio-Cultural Diversity: the Interplay of Worldviews, Globalisation and Locality*, Leusden, Netherlands and Bern, Switzerland: Compas and Centre for Development and Environment.

Hawke, Allan et al. (2009), *The Australian Environment Act – A Report of the Independent Review of the Environment Protection and Biodiversity Conservation Act 1999*, Canberra: Commonwealth of Australia, http://www.environment.gov.au/epbc/review/publications/final-report.html.

Hawkins, N. (2010), 'Human Gene Patents and Genetic Testing in Europe: A Reappraisal', *Script-ed*, **7** (3), 453–73, http://www.law.ed.ac.uk/ahrc/script-ed/vol7–3/hawkins.asp.

Hayden, C. (2007), 'Taking as Giving: Bioscience, Exchange, and the Politics of Benefit-Sharing', *Social Studies of Science*, **37**, 729–58.

Hayden, Cori (2003), *When Nature Goes Public: The Making and Unmaking of Bioprospecting in Mexico*, Princeton: Princeton University Press.

Hazuka, C. (2002), 'Supporting the Work of Lesser Geniuses: An Argument for Removing Obstructions to HESC Research', *University of Miami Law Review*, **57**, 157–220.

Headrick, Daniel (1981), *The Tools of Empire: Technology and European Imperialism in the Nineteenth Century*, Oxford: Oxford University Press.

Heller, M. (1998), 'The Tragedy of the Anticommons: Property in the Transition from Marx to Markets', *Harvard Law Review*, **111** (3), 621–88.

Heller, Michael (2008), *The Gridlock Economy: How Too Much Ownership Wrecks Markets, Stops Innovation and Costs Lives*, New York: Basic Books.

Heller, Michael (ed.) (2010), *Commons and Anticommons*, vols I and II, Cheltenham and Northampton, MA: Edward Elgar.

Heller, M. and R. Eisenberg (1998), 'Can Patents Deter Innovation? The Anticommons in Biomedical Research', *Science*, **280**, 698–701.

Helwegen, W. (2010), 'The Research Exemption from a Nanotechnology Perspective', *European Intellectual Property Review*, **32** (7), 341–51.

Henaghan, Mark (ed.) (2006), *Choosing Genes for Future Children: Regulating Preimplantation Genetic Diagnosis*, Dunedin: Human Genome Research Project and New Zealand Law Foundation.

Henaghan, Mark (ed.) (2007), *Genes, Society, and the Future*, Dunedin: Human Genome Research Project and New Zealand Law Foundation.

Henaghan, Mark (ed.) (2009), *Findings from the Law Foundation Sponsored Human Genome Research Project*, Dunedin: Human Genome Research Project and New Zealand Law Foundation.

Henkel, J. and S. Maurer (2007), 'The Economics of Synthetic Biology', *Molecular Systems Biology*, **3** (117), 1–4.

Henkel, J. and S. Maurer (2009), 'Parts, Property and Sharing', *Nature Biotechnology*, **27** (12), 1095–8.

The Henry J Kaiser Family Foundation (2003), 'National Politics and Policy, Lawmakers Reach Agreement on Appropriations Bill Language Barring Patents on Human Organisms', Kaisernetwork.org, 25 November, http://www.kaisernetwork.org/daily_reports/rep_index.cfm?%20hint=2&DR_ID=21029.

Hermeren, Göran (2000), 'Patents and Licensing, Ethics, International Controversies', in Thomas Murray and Maxwell Mehlman (eds), *Encyclopedia of Ethical, Legal and Policy Issues in Biotechnology*, New York: Wiley, 817–25.

Herrmann, J.R. and M. Rowlandson (2008), 'The Role of Ethics and Morality in EU Law', *Journal of International Biotechnology Law*, **5** (6), 241–51.

Hiddleton, S. (2010), 'A Repository of Seeds on Cliff Top of Himalayas', *The Hindu*, 17 February.

Hilgartner, S. (2009), 'Intellectual Property and the Politics of Emerging Technology: Investors, Citizens, and Powers to Shape the Future', *Chicago-Kent Law Review*, **84**, 197−244.

Hill, D. (2004), 'Latin America Shows Rapid Rise in S&E Articles', *Info-Brief*, National Science Foundation, August, 1.

Hippel, Eric von (2005), *Democratizing Innovation*, Cambridge, MA and London: The MIT Press.

Hobhouse, Henry (2002), *Seeds of Change: Six Plants That Transformed Mankind*, London: Pan.

Hodge, Graeme, Diana Bowman and Andrew Maynard (eds) (2011), *International Handbook on Regulating Nanotechnologies*, Cheltenham and Northampton, MA: Edward Elgar.

Holcombe, Sarah (2009) *Indigenous Ecological Knowledge and Natural Resources in the Northern Territory: Guidelines for Indigenous Ecological Knowledge Management (including Archiving and Repatriation)*, Darwin: Natural Resource Management Board NT, 12, http://www.nrmbnt.org.au/files/iek/IEK%20&%20NRM%20NT%20Guidelines%20final.pdf.

Holcombe, S. and N. Gould (2010), 'A Preliminary Review of Ethics Resources, with a Particular Focus of those Available On-line from Indigenous Organisations in WA, NT and Qld', *Australian Aboriginal Studies Journal*, No. 2., 107–25.

Holcombe, S., P. Yates and F. Walsh (2011), 'Reinforcing Alternative Economies: Self-motivated Work by Central Anmatyerr People to Sell Akatyerr (Desert Raisin) in Central Australia', *Rangelands Journal* (forthcoming).

Holman, C. (2010), 'On Remand, Federal Circuit (Once Again) Decides *Prometheus* v. *Mayo* in Favor of Patent Eligibility for Methods of Treatment and Diagnostic Tests', Holman's Biotech IP Blog, 17 December, http://holmansbiotechipblog.blogspot.com/2010/12/on-remand-federal-circuit-once-again.html.

Holmes, H.M. (1932), 'Patent Rights for Scientific Discoveries by C.J. Hamson', *Harvard Law Review*, **45**, 1431–3.

Holzapfel, H. and J. Sarnoff (2008), 'A Cross-Atlantic Dialog on Experimental Use and Research Tools', *IDEA: Intellectual Property Law Review*, **48**, 123–4.

Honingsbaum, Mark (2001), *The Fever Trail: The Hunt for the Cure for Malaria*, London: Macmillan.

Hope, Janet (2008), *Biobazaar: The Open Source Revolution and Biotechnology*, Cambridge, MA: Harvard University Press.

Hopkin, M. (2008), 'Biodiversity: Frozen Futures', *Nature*, **452**, 404–5.

Hornyak, Gabor Louis, H.F. Tibbals and Joydeep Dutta (2008), *Introduction to Nanoscience and Nanotechnology*, London: Taylor and Francis.

Hubicki, S. and B. Sherman (2005), 'The Killing Fields: Intellectual Property and Genetic Use Restriction Technologies', *The University of New South Wales Law Journal*, **28** (3), 740–58.

Hugenholtz, Bernt and Lucie Guibault (eds) (2006), *The Future of the Public Domain*, Amsterdam: Kluwer Law International.

Human Genome Project and United States Department of Energy Office of Science (1996), *Summary of Principles Agreed at the First International Strategy Meeting on Human Genome Sequencing*, http://www.ornl.gov/sci/techresources/Human_Genome/research/bermuda.shtml #1.

Hunt, Geoffrey (2006), 'The Global Ethics of Nanotechnology' in Geoffrey Hunt and Michael Mehta (eds), *Nanotechnology: Risks, Ethics and Law*, London: Earthscan.

Hunt, Geoffrey and Michael Mehta (eds) (2006), *Nanotechnology: Risks, Ethics and Law*, London: Earthscan.

Hunter, Graeme (2000), *Vital Forces: The Discovery of the Molecular Basis of Life*, London and San Diego: Academic Press.

Hunter, Kathryn and Graeme Laurie (2009), 'Involving Publics in Biobank Governance: Moving Beyond Existing Approaches', in Heather Widdows and Caroline Mullen (eds), *The Governance of Genetic Information*, Cambridge: Cambridge University Press.

Ilag, L.L., L.M. Ilag and L.L. Ilag (2002), 'From Patenting Genes to Proteins: The Search for Utility via Function', *TRENDS in Biotechnology*, **20** (5), 197–9.

IMS Health (2010), 'Top-Line Industry Data', http://www.imshealth.com/portal/site/imshealth/menuitem.a46c6d4df3db4b3d88f611019418c22a/

?vgnextoid=e599410b6c718210VgnVCM100000ed152ca2RCRD&cpsextcurrchannel=1.

The International Cancer Genome Consortium, http://www.icgc.org/.

International Centre for Trade and Sustainable Development and the Quaker United Nations Office (2010), 'The Future of UPOV In A Changing World: Issues and Challenges', 25 March, http://ictsd.org/i/events/dialogues/72761/.

International Expert Group on Biotechnology, Innovation and Intellectual Property (2008), *Toward a New Era of Intellectual Property: From Confrontation to Negotiation*, Montreal: International Expert Group on Biotechnology, Innovation and Intellectual Property.

International HapMap Consortium (2004), 'Integrating Ethics and Science in the International HapMap Project', *Nature Reviews Genetics*, **5**, 467–75.

International HapMap Consortium (2005), 'A Haplotype Map of the Human Genome', *Nature*, **437** (7063), 1299–320.

International Human Genome Sequencing Consortium (2001), 'Initial Sequencing and Analysis of the Human Genome', *Nature*, **409**, 860–921.

IP Australia (2010), *Dream Shield: A Guide to Protecting Designs, Brands and Inventions for Aboriginal and Torres Strait Islanders*, http://www.ipaustralia.gov.au/resources/dream_shield.shtml.

IP Australia (2010), *Nanga Mai Arung – Dream Shield: A Guide to Protecting Designs, Brands and Inventions for Aboriginal and Torres Strait Islanders*, Canberra: Commonwealth of Australia, http://www.ipaustralia.gov.au/pdfs/news/dream_shield/print/dreamshield.pdf.

Jacob, C. (2008), 'Doomsday Vault is Open for Business', *Sky News Channel*, 27 February.

Janke, Terri (2009), *Indigenous Ecological Knowledge and Natural Resources in the Northern Territory: Report on the Current Status of Indigenous Intellectual Property*, Darwin: Natural Resource Management Board NT.

Japan Patent Office, http://www.jpo.go.jp.

Japan Patent Office (2007), *Annual Report*, http://www.jpo.go.jp/shiryou_e/toushin_e/kenkyukai_e/pdf/annual_report2007/part1.pdf.

Jarcho, Saul (1993), *Quinine's Predecessor: Francesco Torti and the Early History of Cinchona*, Baltimore and London: The Johns Hopkins University Press.

Jarlmadangah Community Inc, http://www.jarlmadangah.com/1.htm.

Jasanoff, Sheila (2007), *Designs on Nature: Science and Democracy in Europe and the United States*, Princeton: Princeton University Press.

Jayaraman, K.S. (2010), 'India's Doomsday Vault in Frozen Himalayas', *Nature India*, 10 March, http://www.nature.com/nindia/2010/100310/full/nindia.2010.29.html.

Jeffrey, M. (2002), 'Bioprospecting: Access to Genetic Resources and Benefit-Sharing under the Convention on Biodiversity and the Bonn Guidelines', *Singapore Journal of International and Comparative Law*, **6**, 747–808.

Johnson, R. (2009), 'Evidence to The US National Academies, Opportunities and Challenges in the Emerging Field of Synthetic Biology', Washington, DC, 10 July.

Joly, Yann (2009), 'Patents on Genetic Research Tools: Recent Trends and Future Outlook', in Richard Gold and Bartha Knoppers (eds), *Biotechnology, Intellectual Property and Ethics*, Markham: LexisNexis Canada, 39–57.

Joly, Yann (2010), 'Open Biotechnology: Licenses Needed', *Nature Biotechnology*, **28** (5), 417–19.

Jones, R. (2007), 'Can Nanotechnology Ever Prove That It Is Green?' *Nature Nanotechnology*, **2**, 71–2.

Jopling, C., M. Yi, A. Lancaster, S. Lemon and P. Sarnow (2005), 'Modulation of Hepatitis C Virus RNA Abundance by a Liver-Specific Micro-RNA', *Science*, **309** (5740), 1577–81.

Kadidal, S. (1993), 'Plants, Poverty, and Pharmaceutical Patents', *The Yale Law Journal*, **103** (1), 223–58.

Kaiser, J. (2008), 'A Plan to Capture Human Diversity in 1000 Genomes', *Science*, **319** (5862), 395.

Kaiser, Mario, Monika Kurath, Sabine Maasen and Christoph Rehmann-Sutter (eds) (2010), *Governing Future Technologies Nanotechnology and the Rise of an Assessment Regime*, Dordrecht: Springer.

Kallinger, C. et al. (2008), 'Patenting Nanotechnology: A European Patent Office Perspective', *Nanotechnology Law and Business*, **5**, 95–105.

Kamau, Evanson C. and Gerd Winter (2009), *Genetic Resources, Traditional Knowledge and the Law – Solutions for Access and Benefit Sharing*, London: Earthscan.

Kappos, D. (2011), 'Statement on the *America Invents Act*', Subcommittee on Intellectual Property, Competition and the Internet, Committee on the Judiciary, United States House of Representatives, 30 March, http://www.uspto.gov/news/speeches/2011/kappos-_house_testimony.jsp.

Kapranov, P. et al. (2007), 'RNA Maps Reveal New RNA Classes and a Possible Function for Pervasive Transcription', *Science*, **316**, 1484–8.

Karjala, D. (2011), 'Protecting Innovation in Computer Software, Biotechnology, and Nanotechnology', *Virginia Journal of Law and Technology*, **16** (1), 42–65.

Karow, J. (2009), '1000 Genomes Project to Sequence Nearly 1,000 More Samples by Early 2010; New Samples Collected', Genomeweb, http://webcache.googleusercontent.com/search?q=cache:pU05R7uEqyMJ:www.genomeweb.com/sequencing/1000-genomes-project-sequence-nearly-1000-more-samples-early-2010-new-samples-co+1000+Genomes+Project+african-american+hispanic&cd=1&hl=en&ct=clnk&gl=us.

Kass, L. (1981), 'Patenting Life', *Journal of the Patent Office Society*, **63**, 571–600.

Kay, Lily E. (2000), *Who Wrote the Book of Life? A History of the Genetic Code*, Stanford: Stanford University Press.

Kaye, J. (2005), 'Do We Need a Uniform Regulatory System for Biobanks across Europe?' *European Journal of Human Genetics*, **14**, 245–8.

Kaye, Jane (2009), 'Biobank Networks – What Are the Governance Challenges?', in Jane Kaye and Mark Stranger (eds), *Principles and Practice of Biobank Governance*, Cheltenham: Ashgate, 201–13.

Kaye, Jane and Mark Stranger (eds) (2009), *Principles and Practice of Biobank Governance*, Cheltenham: Ashgate.

Keasling, J. (2008), 'Synthetic Biology for Synthetic Chemistry', *ACS Chemical Biology*, **3** (1), 64–76.

Keasling, J. (2010), 'Evidence to Committee on Energy and Commerce, United States House of Representatives, Effects of Developments in Synthetic Genomics and Implications for Energy and Health', Washington, DC, 27 May.

Keim, B. (2007), 'Patent Eligible Subject Matter in the Biotechnological Arts', Berkeley Electronic Press, http://works.bepress.com/benjamin_keim/1.

Keller, Evelyn Fox (2009), 'What Does Synthetic Biology have to do with Biology', *BioSocieties*, **4** (2–3), 291–302.

Kellert, Stephen, Helen Longino and C. Kenneth Waters (eds) (2006), *Scientific Pluralism: Minnesota Studies in the Philosophy of Science*, vol. 19, Minneapolis: University of Minnesota Press, 190–214.

Keneally, Thomas (2010), *Three Famines*, North Sydney, NSW: Knopf.

Kennerdell, J.R. and R.W. Carthew (1998), 'Use of Double-stranded RNA-Mediated Genetic Interference to Demonstrate that Frizzled and Frizzled 2 Act in the Wingless Pathway', *Cell*, **95**, 1017–26.

Kerr, I. and G. Bassie (2004), 'Not *That* Much Room? Nanotechnology, Networks and the Politics of Dancing', *Health Law Journal*, **12**, 103–23.

Kevles, D. (2002), 'Of Mice and Money: The Story of the World's First Animal Patent', *Daedalus*, **131** (2), 78–88.

Kevles, Daniel (2002), *A History of Patenting Life in the United States with Comparative Attention to Europe and Canada*, Brussels: European Group on Ethics in Science and New Technology, European Commission.

Khan, B. Zorina (2005), *The Democratisation of Inventions: Patents and Copyrights in American Economic Development, 1790–1920*, Cambridge: Cambridge University Press.

Kieff, F.S. (2001), 'Facilitating Scientific Research: Intellectual Property Rights and the Norms of Science – A Response to Rai and Eisenberg', *Northwestern University Law Review*, **95**, 691–705.

Kinsley, Michael (ed.) (2009), *Creative Capitalism: A Conversation with Bill Gates, Warren Buffett and Other Economic Leaders*, New York: Simon & Schuster.

Kitano, H. (2008), 'Synthetic Biology: A Brief Overview', *Science*, **295**, 1662–4.

Klein, R. and M. Mahoney (2008), 'LabCorp v. Metabolite Laboratories: The Supreme Court Listens but Declines to Speak', *Journal of Law, Medicine & Ethics*, **36** (1), 141–9.

Kleinman, M.E. et al. (2008), 'Sequence- and Target-independent Angiogenesis Suppression by siRNA via TLR3', *Nature*, **452**, 591–7.

Kloppenburg Jr., Jack (1988), *First the Seed: The Political Economy of Plant Biotechnology 1492–2000*, Cambridge: Cambridge University Press.

Knoppers, B. (2005), 'Biobanking: International Norms', *Journal of Law, Medicine, and Ethics*, **33**, 7–14.

Knoppers, Bartha (2009), 'Biotechnology and Ethics', in Richard Gold and Bartha Knoppers (eds), *Biotechnology, IP and Ethics*, Markham, ON: LexisNexis Canada, 59–67.

Knoppers, B. and L. Sheremeta (2003), 'Beyond the Rhetoric: Population Genetics and Benefit-sharing', *Health Law Journal*, **11**, 89–117.

Konski, A. and D. Spielthenner (2009), 'Stem Cell Patents: A Landscape Analysis', *Nature Biotechnology*, **27** (8), 722–6.

Kopinski, N. (2004), 'Human-Nonhuman Chimeras: A Regulatory Proposal on the Blurring of Species Lines', *Boston College Law Review*, **45**, 619–66.

Koppikar, V., S. Maebius and J.S. Rutt (2004), 'Current Trends in Nanotech Patents: A View From Inside the Patent Office', *Nanotechnology Law and Business*, **1**, 24–30.

Korobkin, Russel and Stephen Munzer (2007), *Stem Cell Century: Law and Policy for a Breakthrough Technology*, New Haven, CT: Yale University Press.

Kostel, K. (2009), 'The Language of Change', *Nature Reports Climate Change*, 29 January, http://www.nature.com/climate/2009/0902/full/climate.2009.10.html.

Krattiger, Anatole and Ingo Potrykus (2007), 'Golden Rice: A Product-Development Partnership in Agricultural Biotechnology and Humanitarian Licensing', in Anatole Krattiger et al. (eds), *Executive Guide to Intellectual Property Management in Health and Agricultural Innovation: A Handbook of Best Practices*, Oxford: Centre for the Management of Intellectual Property in Health Research and Development, http://www.iphandbook.org/handbook/case_studies/cs03/.

Krattiger, Anatole et al. (eds) (2007), *Intellectual Property Management in Health and Agricultural Innovation: A Handbook of Best Practices,* Oxford: Centre for the Management of Intellectual Property in Health Research and Development.

Kulinowski, K. (2004) 'Nanotechnology: From "Wow" to "Yuck"?', *Bulletin of Science, Technology & Society,* **24** (1), 13–20.

Lagos-Quintana, M., R. Rauhut, W. Lendeckel and T. Tuschl (2001), 'Identification of Novel Genes Coding for Small Expressed RNAs', *Science*, **294**, 853–8.

Lakoff, George and Mark Johnson (1980), *Metaphors We Live By*, Chicago: Chicago University Press.

Lanford, R.E. et al. (2010), 'Therapeutic Silencing of MicroRNA-122 in Primates with Chronic Hepatitis C Virus Infection', *Science*, **327** (5962), 198–201.

Lappe, Anna (2010), *Diet for a Hot Planet: The Climate Crisis at the end of Your Fork and What You Can Do About It*, New York: Bloomsbury USA.

Lappe, Frances Moore, Joseph Collins and Cary Fowler (1977), *Food First: Beyond the Myth of Scarcity*, Boston: Houghton-Mifflin.

Latour, Bruno (1987), *Science in Action: How to Follow Scientists and Engineers Through Society*, Cambridge, MA: Harvard University Press.

Latour, Bruno and Steven Woolgar (1979), *Laboratory Life: The Social Construction of Scientific Facts*, London: Sage.

Laurie, Graeme, Ann Bruce and Catherine Lyall (2009), 'The Roles of Values and Interests in the Governance of the Life Sciences: Learning Lessons from the "Ethics+" Approach of UK Biobank', in Catherine Lyall, Theo Papaioannou and James Smith (eds), *The Limits to Governance: The Challenge of Policy-Making for the New Life Sciences*, Aldershot: Ashgate, 51–77.

Laury, E. (2010), 'Patentable Subject Matter Paradigm Evolves: Bilski's Impact on Nanotechnology', NanoLaw Blog, 1 July, http://blogs.law.widener.edu/nanolaw/2010/07/01/patentable-subject-matter-paradigm-evolves-bilskis-impact-on-nanotechnology/.

Lawson, C. (2004), 'Patents and Plant Breeder's Rights Over Plant Genetic Resources for Food and Agriculture', *Federal Law Review*, **32** (1), 107–39.

Lawson, C. (2009), 'Intellectual Property and the Material Transfer Agreement under the International Treaty on Plant Genetic Resources for Food and Agriculture', *European Intellectual Property Review*, **31** (5), 244–54.

Ledford, H. (2010), 'Drug Giants turn their Backs on RNA Interference', *Nature*, **468**, 487.

Lee, A. (2006), 'Examining the Viability of Patent Pools for the Growing Nanotechnology Patent Thicket', *Nanotechnology Law and Business*, **3**, 317–27.

Lee, P.Y. (2005), 'Inverting the Logic of Scientific Discovery: Applying Common Law Patentable Subject Matter Doctrine to Constrain Patents on Biotechnology Research Tools', *Harvard Journal of Law and Technology*, **19** (1).

Legislative Analysts' Office (2004), *Proposition 71 Stem Cell Research. Funding. Bonds. Initiative Constitutional Amendment and Statute*, http://www.lao.ca.gov/ballot/2004/71_11_2004.htm.

LeGuyader, J. (2004), 'Patenting Interfering RNA. (USPTO Presentation)', www.cabic.com/bcp/020304/JLeGuyader_PIRNA.ppt.

Lemke, A., W. Wolf, J. Herbert-Beirne and M.E. Smith (2010), 'Public and Biobank Participant Attitudes towards Genetic Research Participation and Data Sharing', *Public Health Genomics* **13**, 368–77.

Lemley, M. (2005), 'Patenting Nanotechnology', *Stanford Law Review*, **58,** 601–30.

Lemley, M. (2011), 'Point of Novelty', Stanford Public Law Working Paper, 2, http://papers.ssrn.com/sol3/papers.cfm?abstract_id=1735045.

Lemley, M., M. Risch, T. Sichelman and R.P. Wagner (2011), 'Life After *Bilski*', *Stanford Law Review*, **63**, 1315–47.

Lerner, J. (1995), 'Patenting in the Shadow of Competitors', *Journal of Law and Economics*, **38**, 463–95.

Levitt, M. and S. Weldon (2005), 'A Well Placed Trust? Public Perceptions of the Governance of DNA Databases', *Critical Public Health*, **15**, 311–21.

Lewin, T. (1996), 'Move to Patent Cancer Gene is Called Obstacle to Research', *The New York Times*, 21 May, 14.

Lewis, B.P., C.B. Burge and D.P. Bartel (2005), 'Conserved Seed Pairing, Often Flanked by Adenosines, Indicates that Thousands of Human Genes are MicroRNA Targets', *Cell*, **120**, 15–20.

Lewontin, R. (2001), 'They got the Wrong Key of Life', *The Sunday Times*, 8 July.

Li, L.-C. et al. (2006), 'Small dsRNAs induce transcriptional activation in human cells PNAS', **103**, 17337–42.

Liddell, K. and A. Hall (2005), 'Beyond *Bristol* and *Alder Hey*: The Future Regulation of Human Tissue', *Medical Law Review*, **13**, 170–223.

Liivak, O. (2010), 'Rescuing the Invention from the Cult of the Claim', working paper, http://www.law.berkeley.edu/files/bclt_IPSC2010_Liivak%282%29.pdf..

Llewelyn, Margaret (2005), 'Schrodinger's Cat: An Observation on Modern Patent Law', in Peter Drahos (ed.), *Death of Patents*, Witney: Lawtext Publications, 39–45.

Llewelyn, Margaret and Mike Adcock (2006), *European Plant Intellectual Property*, Oxford and Portland: Hart Publishing.

Lok, L. (2002), 'Software for Signaling Networks, Electronic and Cellular', *Science's STKE*, **122**, PE11.

Lowenstein, Roger (2008), *Buffett: The Making of an American Capitalist*, New York: Broadway Books.

Lowrance, W. (2006), 'Access to Collections of Data and Materials for Health Research', A Report to the Medical Research Council and Wellcome Trust, 1–36, http://www.wellcome.ac.uk/stellent/groups/corporatesite/@msh_grants/documents/web_document/wtx030842.pdf.

Luk, E. (2006), 'The United Kingdom and Germany: Differing Views on Therapeutic Cloning and How the Belgian Resolution Brings Them Together', *Michigan State University Journal of Medicine and Law*, **10**, 523–56.

Lyall, Catherine, Theo Papaioannou and James Smith (eds) (2009), *The Limits to Governance: The Challenge of Policy-Making for the New Life Sciences*, Aldershot: Ashgate.

Lyon, Z. (2008), 'Bartier Perry give community helping hand', *Lawyers Weekly*, 21 November.

Mack, G.S. (2007), 'MicroRNA gets down to Business', *Nature Biotechnology*, **25** (6), 631–8.

MacQueen, Hector, Charlotte Waelde and Graeme Laurie (2008), *Contemporary Intellectual Property: Law and Policy*, Oxford: Oxford University Press.

Magliocca, G. (2009), 'Patenting the Curve Ball: Business Methods and Industry Norms', *Brigham Young University Law Review*, 875–904.

Magnus, David, Arthur Caplan and Glenn McGee (eds) (2002), *Who Owns Life?* Amherst, NY: Prometheus Books.

Malin, B. et al. (2010), 'Technical and Policy Approaches to Balancing Patient Privacy and Data Sharing in Clinical and Translational Research', *Journal of Investigative Medicine*, **58** (1), 11–18.

Mandel, Gregory (2011), 'Regulating Nanotechnology Through Intellectual Property Rights', in Graeme Hodge, Diana Bowman and Andrew Maynard (eds), *International Handbook on Regulating Nanotechnologies*, Cheltenham and Northampton, MA: Edward Elgar, 388–407.

Mara, K. (2010), 'Change Coming to Quiet UN Plant Variety Protection Agency', *Intellectual Property Watch*, 26 October, http://www.ip-watch.org/weblog/2010/10/26/change-coming-for-quiet-un-plant-variety-protection-agency/.

Marchant, G., D. Sylvester and K. Abbott (2009), 'What Does the History of Technology Regulation Teach Us About Nano Oversight?' *Journal of Law, Medicine, and Ethics*, **37**, 724–31.

Marden, E. (1999), 'The Neem Tree Patent: International Conflict over the Commodification of Life', *Boston College International and Comparative Law Review*, **22** (2), 279-–95.

Marques, J.T.B. and R.G. Williams (2005), 'Activation of the Mammalian Immune System by siRNAs', *Nature Biotechnology*, **23,** 1399–405.

Marshall, E. (2001), 'Bermuda Rules: Community Spirit, with Teeth', *Science*, **291** (5507), 1192.

Martinezserrano, A. and A. Bjorklund (1996), 'Protection of the Neostriatum against Excitotoxic Damage by Neurotrophin-producing, Genetically Modified Neural Stem Cells', *Journal of Neuroscience*, **16** (15), 4604–16.

Maskus, Keith and Jerome H. Reichman (eds) (2005), *International Public Goods and Transfer of Technology under a Globalized Intellectual Property Regime*, Cambridge: Cambridge University Press.

Massachusetts Institute of Technology Institute for Soldier Nanotechnologies, 'Institute for Soldier Nanotechnologies', http://web.mit.edu/isn/.

Massachusetts Institute of Technology Institute for Soldier Nanotechnologies, 'Research', http://web.mit.edu/isn/research/index.html.

Matthijs, G. (2006), 'The European Opposition against the BRCA Gene Patents', *Familial Cancer*, **5**, 95.

Mattick, J.S. (2003), 'Challenging the Dogma: The Hidden Layer of Non-protein-coding RNAs in Complex Organisms', *Bioessays*, **10**, 930–39.

Mattick, J.S. (2004), 'RNA Regulation: A New Genetics?' *Nature Reviews Genetics*, **5**, 316–23.

McGuire, A.L. and L.M. Beskow (2010), 'Informed Consent in Genomics and Genetic Research', *Annual Review of Genomics and Human Genetics*, **11**, 361–81.

McLennan, A. (2007), 'Which Bank? A Guardian Model for Regulation of Embryonic Stem Cell Research in Australia', *Journal of Law and Medicine*, **15** (1), 45–76.

McManis, C. and S. Noh (2006), 'The Impact of the Bayh-Dole Act on Genetic Research and Development: Evaluating the Arguments and Empirical Evidence to Date', University of California, Berkeley, 13 August, 1, www.law.berkeley.edu/files/mcmanis(1).doc.

McManis, C. and E. Soo Seo (2009), 'The Interface of Open Source and Proprietary Agricultural Innovation: Facilitated Access and Benefit-Sharing Under the New FAO Treaty', *Washington University Journal of Law and Policy*, **30**, 405–64.

McSherry, Corynne (2001), *Who Owns Academic Work? Battling for Control of Intellectual Property*, Cambridge, MA: Harvard University Press.

Medical Research Council, 'Medical Research Council Policy on Data Sharing and Preservation Policy', http://www.mrc.ac.uk/Ourresearch/Ethicsresearchguidance/Datasharinginitiative/Policy/index.htm#P16_1349.

Medical Research Council, http://www.mrc.ac.uk/About/Structure/index.htm.

Meijers, Anthonie (ed.) (2009), *Philosophy of Technology and Engineering Sciences*, Amsterdam and London: Elsevier.

Mendel Biotechnology Inc., http://www.mendelbio.com/index.php.

Merton, Robert K. (1973), *The Sociology of Science: Theoretical and Empirical Investigations*, Chicago: University of Chicago Press.

Metlay, G. (2006), 'Reconsidering Renormalization: Stability and Change in 20th-Century Views on University Patents', *Social Studies of Science*, **36** (4), 565–97.

Mgbeoji, I. (2003), 'The Juridical Origins of the International Patent System: Towards a Historiography of the Role of Patents in Industrialization', *Journal of the History of International Law*, **5**, 403–22.

Mgbeoji, Ikechi (2006), *Global Biopiracy: Patents, Plants, and Indigenous Knowledge*, Vancouver: University of British Columbia Press.

Milanovic F., D. Pontille and A. Cambon-Thomsen (2007), 'Biobanking and Data Sharing: A Plurality of Exchange Regimes', *Genomics, Society and Policy*, **3**, 17–30.

Miles, J. (2008), 'Aborigine, Scientist find Pain Relief in Marjarla tree', *Courier Mail*, 21 October, http://www.couriermail.com.au/news/queensland/native-tree-bark-eases-pain/story-e6freoof-1111117814741.

Milford, P. (2008), 'Elan Wins $55.2 Million Jury Award From Abraxis', Bloomberg, 13 June.

Miller, Philip David and Peter Hanns Reill (eds) (1996), *Visions of Empire: Voyages, Botany, and Representations of Empire*, Cambridge: Cambridge University Press.

Mills, Oliver (2005), *Biotechnological Inventions: Moral Restraints and Patent Law*, Aldershot: Ashgate Publishing.

Minssen, T. and R. Schwartz (2011), 'US Patent Eligibility in the Wake of *Bilski* v. *Kappos:* "Business as Usual" in an Age of New Technologies?' *Biotechnology Law Report*, **30** (1), 3–56.

Moazed, D. (2009), 'Small RNAs in Transcriptional Gene Silencing and Genome Defence', *Nature*, **457**, 413–20.

Mohammed, E.A.C. (2008), 'What Is an Invention? A Review of the Literature on Patentable Subject Matter', *Richmond Journal of Law and Technology*, **15**, 1–36.

Mohan-Ram, V., R. Peet and P. Vlaemminck, 'Biotech Patent Infringement in Europe: The "Functionality" Gatekeeper', *John Marshall Review of Intellectual Property Law*, **10**, 540–52.

Mohan-Ram, V. and J. Waxman (2008), 'Synthetic Biology Patent Applications Expected to Present New Challenges', *Life Sciences Law & Industry Report*, **2** (12), 1–3.

Mohanty, C. et al. (2009), 'Nanobiotechnology: Application of Nanotechnology in Therapeutics and Diagnosis', *International Journal of Green Nanotechnology: Biomedicine*, **1**(1), B24–B38.

Monotti, Anne and Sam Ricketson (2003), *Universities and Intellectual Property: Ownership and Exploitation*, Oxford: Oxford University Press.

Monsanto (2008), 'Agriculture Can Help Keep Carbon in Balance', http://www.monsanto.com/responsibility/our_pledge/healthier_environment/climate_change.asp (available on the Internet Archive).

Monsanto and BASF (2009), 'Gene Provides Yield Stability During Periods of Inadequate Water Supply', 9 June, http://www.basf.com/group/pressrelease/P-09–274.

Monsanto and BASF (2009), 'Monsanto, BASF scientists disclose discovery of gene conferring drought tolerance in corn plants', Press Release, 9 June, http://www.basf.com/group/pressrelease/P-09–274.

Mooallem, J. (2010), 'Do-It-Yourself Genetic Engineering', *The New York Times Online*, 10 February.

Moore, S. (2002), 'Challenge To the Biotechnology Directive', *European Intellectual Property Review*, **24** (3), 149–54.

Moorhead, Anne (2009), *Climate, Agriculture and Food Security: A Strategy for Change*, The Alliance of the Consultative Group on International Agricultural Research Centres, http://www.cgiar.org/pdf/CCAFS_Strategy_december2009.pdf.

Morange, Michel (1998), *A History of Molecular Biology*, Cambridge: Harvard University Press.

Morse, J. (2011), 'Nurturing Nature, Nurturing Knowledge: The Nagoya Protocols on Access and Benefit Sharing', Terri Janke and Company, January (in press).

Morse, Jeremy (2011), *Know your Rights to your Indigenous Plant Knowledge – A Guide for Indigenous Knowledge Holders on Recording and Commercializing Indigenous Plant Knowledge*, Darwin: Aboriginal Bush Traders, http://www.aboriginalbushtraders.com/.

Murray, Thomas and Maxwell Mehlman (eds) (2000), *Encyclopedia of Ethical, Legal and Policy Issues in Biotechnology*, New York: Wiley.

Myers, G. (1995), 'From Discovery to Invention: The Writing and Rewriting of Two Patents', *Social Studies of Science*, **25** (1), 57–105.

Nano-Terra Inc., http://www.nanoterra.com/.

NanoBusiness Alliance (2002), 'US Patent Examiners May Not Know Enough About Nanotech', *Small Times*, http://www.smalltimes.com/articles/stm_print_screen.cfm?ARTICLE_ID=267691.

Nanobusiness Alliance (2009), 'Federal Trade Commission Hearing on Evolving IP Marketplace – Evolution of Remedies: Comment Number 540872–00017', http://www.ftc.gov/os/comments/iphearings/540872–00017.htm.

Nanosys (2009), 'Nanosys Reaches Settlement of Patent Infringement Lawsuit Against Nano Technologies for Quantum Dot Technology', Press Release, 23 July.

Nanotechnology Development Blog (2007), 'Harvard Nanotechnology Patents to Nano-Terra', 6 June, http://www.nanotechnologydevelopment.com/news/harvard-nanotechnology-patents-to-nano-terra.html.

Nasu, H. and T. Faunce (2009/2010), 'Nanotechnology and the International Law of Weaponry: Towards International Regulation of Nano-Weapons', *Journal of Law, Information and Science*, **20**, 21–54.

National Health and Medical Research Council (2010), *Biobanks Information Paper 2010*, Canberra: Commonwealth of Australia, http://www.nhmrc.gov.au/your_health/egenetics/practitioners/practitioners.htm ('NHMRC Biobanks Information Paper').

National Heritage Trust (2003), *Indigenous Knowledge Forum, Alice Springs 28–29 May 2003: Workshop Outcomes*, Canberra: Commonwealth of Australia, www.environment.gov.au/indigenous/publications/pubs/workshop.pdf.

National Human Genome Research Institute and National Institutes of Health, 'International HapMap Project Overview', http://www.genome.gov/10001688.

National Institutes of Health (1999), 'Principles and Guidelines for Recipients of NIH Research Grants and Contracts on Obtaining and Disseminating Biomedical Research Resources: Final Notice', *Federal Register*, **64**, 72090.

National Institutes of Health (NIH) (2003), *Statement on Sharing Research Data*, http://grants.nih.gov/grants/policy/data_sharing/.

National Institutes of Health (2003), *Final NIH Statement on Sharing Research Data*, http://grants.nih.gov/grants/guide/notice-files/NOT-OD-03–032.html.

National Institutes of Health (2007), 'Policy for Sharing of Data Obtained in NIH Supported or Conducted Genome-wide Association Studies (GWAS)', 28 August, http://grants.nih.gov/grants/guide/notice-files/NOT-OD-07–088.html.

National Institutes of Health (2008), 'Background Fact Sheet on GWAS Policy Update', 28 August, http://grants.nih.gov/grants/gwas/.

National Institutes of Health (2008), 'International Consortium Announces the 1000 Genomes Project', Press Release, 22 January, http://www.nih.gov/news/health/jan2008/nhgri-22.htm.

National Institutes of Health (2008), 'Three Sequencing Companies Join 1000 Genomes Project', Press Release, 11 June, http://www.genome.gov/27526680.

National Institutes of Health (2008), *Modifications to Genome-Wide Association Studies (GWAS) Data Access*, 28 August, http://grants.nih.gov/grants/gwas/.

National Institutes of Health (2011), 'Legislative Updates: Cloning', Office of Legislative and Policy Analysis Legislative Updates, 3 February, http://olpa.od.nih.gov/legislation/107/pendinglegislation/7cloning.asp; 148 *Congressional Record* S5579 (daily ed. June 14, 2002).

National Nanotechnology Initiative, 'Funding', http://www.nano.gov/html/about/funding.html.

National Research Council (1997), *Intellectual Property Rights and the Dissemination of Research Tools in Molecular Biology*, Washington, DC: National Academy Press.

National Research Council of the National Academies (2009), *A New Biology for the 21st Century*, Washington, DC: The National Academies Press.

National Science and Technology Council (2011), *The National Nanotechnology Initiative: Strategic Plan*, Washington, DC: Executive Office of the President, The White House, 4 February.

Natural Heritage Trust (2004), *Working with Indigenous Knowledge in Natural Resource Management: Guidelines for Regional Bodies*, Canberra: Commonwealth of Australia, http://www.environment.gov.au/indigenous/publications/pubs/guidelines.pdf.

Natural Justice (2009), 'Submission Information and Views in Preparation for the Meeting of the Expert Group on Traditional Knowledge Associated with Genetic Resources. Answers to Questions Posed to the Expert

Group on Traditional Knowledge Associated with GR as Specified in COP Decision IX/12', 9.

Natural Justice (2009), *Bio-cultural Community Protocols: A Community Approach to Ensuring the Integrity of Environmental Law and Policy*, Nairobi: United Nations Environment Programme (UNEP) and Natural Justice.

Natural Justice (2009), *Imagining a Traditional Knowledge Commons: A Community Approach to Sharing Knowledge for Non-commercial Research*, Rome: International Development Law Organisation.

Natural Resource Management Ministerial Council (2002), *Nationally Consistent Approach for the Access to and Utilisation of Australia's Native Genetic and Biochemical Resources*, Canberra: Department of the Environment and Heritage, http://www.environment.gov.au/biodiversity/publications/access/nca/pubs/nca.pdf.

Nelson, B. (2009), 'Data Sharing: Empty Archives', *Nature*, **461** (7261), 160–63.

New South Wales Parliament Legislative Council Standing Committee on State Development (2008), *Nanotechnology in New South Wales*, Sydney: New South Wales Parliament, 29 October.

Ng, P., S. Murray, S. Levy and J.C. Venter (2009), 'An Agenda for Personalized Medicine', *Nature*, **461**, 724–6.

Nicol, D. (2004), 'Property in Human Tissue and the Right of Commercialisation: The Interface between Tangible and Intellectual Property', *Monash University Law Review*, **30**, 139–64.

Nicol, D. (2006), 'Public Trust, Intellectual Property and Human Genetic Databanks: The Need to Take Benefit Sharing Seriously', *Journal of International Biotechnology Law*, **3**, 89–103.

Nicol, Dianne and Christine Critchley (2009), 'What Benefit Sharing Arrangements Do People Want from Biobanks? A Survey of Public Opinion in Australia', in Jane Kaye and Mark Stranger (eds), *Principles and Practice of Biobank Governance*, Cheltenham: Ashgate, 18–31 ('Nicol and Critchley Benefit Sharing Arrangements').

Nicol, D. and C. Critchley (2011), 'Benefit Sharing and Biobanking in Australia', *Public Understanding of Science*, 21 April, 1–22.

Nicol, Dianne and Jane Nielsen (2003), *Patents and Medical Biotechnology: An Empirical Analysis of Issues Facing the Australian Industry,* Hobart: the Centre for Law and Genetics Occasional Paper No. 6.

Nicol, D., M. Otlowski and D. Chalmers (2001), 'Consent Commercialisation and Benefit Sharing', *Journal of Law and Medicine*, **9**, 80–94.

Nilsen, T.W. (2008), 'Endo-siRNAs: Yet Another Layer of Complexity in RNA Silencing', *Nature Structural and Molecular Biology*, **15**, 546–8.

Nordgen, http://www.nordgen.org/sgsv/index.php?page=welcome&PHPSESSID=m6jdn1j59nar67imgvrom7ck3.

Nuffield Council on Bioethics (2004), 'The Ethics of Patenting DNA', Discussion Paper, Nuffiled Council on Bioethics, 12 August.

Nuffield Council on Bioethics (2011), *Emerging Biotechnologies*, London: Nuffield Council on Bioethics, April, http://www.nuffieldbioethics.org/sites/default/files/files/Nuffield_Council_Bioethics_Emerging_Biotechnologies_consultation_April-June_2011.pdf.

Nurton, J. (2008), 'EPO Rulings Clarify Biotech Protection', *Managing Intellectual Property*, **185**, 14.

Obama, B. (2011), 'President Obama Signs *America Invents Act*, Overhauling the Patent System to Stimulate Economic Growth, and Announces New Steps to Help Entrepreneurs Create Jobs', The White House Press Office, http://www.whitehouse.gov/the-press-office/2011/09/16/president-obama-signs-america-invents-act-overhauling-patent-system-stim.

Oguamanam, Chidi (2006), *International Law and Indigenous Knowledge: Intellectual Property, Plant Biodiversity, and Traditional Knowledge*, Toronto: University of Toronto Press.

Okamura, K. and E.C. Lai (2008), 'Endogenous Small Interfering RNAs in Animals', *Nature Reviews Molecular Cell Biology*, **9**, 673–8.

O'Neill, S., K. Hermann, M. Klein, J. Landes and R. Bawa (2007), 'Broad Claiming in Nanotechnology Patents: Is Litigation Inevitable?' *Nanotechnology Law and Business*, **4**, 595–606.

Orlic, D. et al. (2001), 'Bone Marrow Cells Regenerate Infarcted Myocardium', *Nature*, **410** (6829) 701–5.

Ormond, K.E., M.E. Smith and W.A. Wolf (2010), 'The Views of Participants in DNA Biobanks', *Stanford Journal of Law, Science and Policy*, **1**, 80–87.

Osborne, R. (2007), 'Companies Jostle for Lead in RNAi, Despite Uncertainties', *Nature Biotechnology*, **25**, 1191–2.

Ostrom, Elinor (1990), *Governing the Commons: The Evolution of Institutions for Collective Action*, Cambridge: Cambridge University Press.

Otlowski, Margaret (2009), 'Developing an Appropriate Consent Model for Biobanks: In Defence of "Broad" Consent', in Jane Kaye and Mark Stranger (eds), *Principles and Practice of Biobank Governance*, Cheltenham: Ashgate, 79–92.

P3G, http://www.p3g.org/secretariat/index.shtml.

Palsson, B. (2000), 'The Challenges of in Silico Biology', *Nature Biotechnology*, **18**, 1147–50.

Paradise, J. (2004), 'European Opposition to Exclusive Control Over Predictive Breast Cancer Testing and the Inherent Implications for U.S.

Patent Law and Public Policy: A Case Study of the Myriad Genetics' BRCA Patent Controversy', *Food and Drug Law Journal*, **59**, 133–54.

Parry, Bronwyn (2004), *Trading the Genome: Investigating the Commodification of Bio-Information*, New York: Columbia University Press.

Parthasarathy, S. (2010), 'Breaking the Expertise Barrier: Understanding Activist Strategies in Science and Technology Policy Domains', *Science and Public Policy*, **37**, 355–67.

Patrick, A. (2009), 'Patent Eligibility and Computer-Related Processes: A Critique of *In re Bilski* and the Machine-or-Transformation Test', *Virginia Journal of Law and Technology*, **14**, 181–211.

Payne, M. (2007), 'Matters of Urgency: United Nations Declaration on the Rights of Indigenous Peoples', Senate Hansard, Australian Parliament, 10 September.

Pearson, H. (2006), 'What is a Gene?' *Nature*, **441**, 399–401.

Pecorino, Lauren (2008), *Molecular Biology of Cancer: Mechanisms, Targets, and Therapeutics*, Oxford: Oxford University Press.

Pernick, Ron and Clint Wilder (2008), *The Clean Tech Revolution: Discover the Top Trends, Technologies and Companies to Watch*, New York: Harper Collins Business.

Perry, Mark (2008), 'From Pasteur to Monsanto: Approaches to Patenting Life in Canada' in Ysolde Gendreau (ed.), *An Emerging Intellectual Property Paradigm: Perspectives from Canada*, Cheltenham and Northampton, MA: Edward Elgar, 67–80.

Peterson, C. (2004), 'Nanotechnology: From Feynman to the Grand Challenge of Molecular Manufacturing', *IEEE Technology and Society Magazine*, **23** (4), 9–15.

Pettitt, Clare (2004), *Patent Inventions: Intellectual Property and the Victorian Novel*, Oxford: Oxford University Press.

Pharmakina, www.pharmakina.com.

Phillips, Peter and Chika Onwuekwe (eds) (2007), *Accessing and Sharing the Benefits of the Genomics Revolution*, AA Dordrecht: Springer.

Pickler, N. (2010), 'Stem Cell Research Funding Appeal: Court OK's Federal Support While Appeal Continues', *The Huffington Post*, September.

Pimbert, Michael (2009), *Towards Food Sovereignty: Reclaiming Autonomous Food Systems*, London: The International Institute for Environment and Development.

Pioneer HiBred International (2009), 'Public-Private Partnerships Key to Achieving Global Food Security, Says DuPont Leader', 2009 Crawford Annual Fund Conference, Canberra, Australia, 27 October.

Plant Breeders Rights Office, 'Drysdale', http://www.ipaustralia.gov.au/pbr/drysdale.shtml.

Plas, J.V. (2008), 'Patent Office Upholds Remaining WARF Stem Cell Patents', *Wisconsin Technology Network*, 11 March, http://wistechnology.com/articles/4601/.

Plasterk, R.H.A. (2002), 'RNA Silencing: The Genome's Immune System', *Science,* **296,** 1263–5.

Pleiss, J. (2006), 'The Promise of Synthetic Biology', *Applied Microbiological Biotechnology*, **73**, 735–9.

Plomer, Aurora (2006), *Stem Cell Patents: European Patent Law and Ethics Report* (FP6), 'Life Sciences, Genomics and Biotechnology for Health', SSA LSSB-CT-2004- 005251, http://www.nottingham.ac.uk/~llzwww/StemCellProject/project.report.pdf.

Plomer, Aurora (2009), 'Human Dignity, Human Rights, and Article 6 (1) of the EU Directive on Biotechnological Inventions', in Aurora Plomer and Paul Torremans (eds), *Embryonic Stem Cell Patents: European Patent Law and Ethics*, Oxford: Oxford University Press, 203–26.

Plomer, Aurora and Paul Torremans (eds) (2009), *Embryonic Stem Cell Patents: European Patent Law and Ethics*, Oxford: Oxford University Press.

Pogge, Thomas, Matthew Rimmer and Kim Rubenstein (eds) (2010), *Incentives for Global Public Health: Patent Law and Access to Essential Medicines*, Cambridge: Cambridge University Press.

Porter, A. and J. Youtie (2009), 'Where Does Nanotechnology Belong in the Map of Science?' *Nature Nanotechnology*, **4**, 534–6.

Posey, Darrell and Graham Dutfield (1996), *Beyond Intellectual Property: Toward Traditional Resource Rights for Indigenous Peoples and Local Communities*, Ottawa: International Development Research Centre.

Potrykus, I. (2001), 'Golden Rice and Beyond', *Plant Physiology*, **125**, 1157–61.

Pottage, Alain and Brad Sherman (2010), *Figures of Invention: A History of Modern Patent Law*, Oxford: Oxford University Press.

Potter, K.G. (2009), 'Getting Written Description Right in the Biotechnology Arts: A Realist Approach to Patent Scope', *Biotechnology Law Report*, **28** (1), 1–17.

Powell, R. and L. Murdoch (2010), 'Patent Fight Erupts over Kakadu plum', *The Sydney Morning Herald*, 4 December, 11.

Pradeep, T. (2008), *Nano: The Essentials*, New York: McGraw-Hill.

Pratt, Mary Louise (1992), *Imperial Eyes: Travel Writing and Transculturation*, London: Routledge.

The Presidential Commission for the Study of Bioethical Issues (2010), *New Directions: The Ethics of Synthetic Biology and Emerging Technologies*, Washington, DC: The Presidential Commission for the Study of

Bioethical Issues, http://www.bioethics.gov/documents/synthetic-biology/PCSBI-Synthetic-Biology-Report-12.16.10.pdf.

The Presidential Commission for the Study of Bioethical Issues (2010), 'Synthetic Biology Meeting Transcripts', http://www.bioethics.gov/transcripts/synthetic-biology/.

The President's Council on Bioethics (2004), *Monitoring Stem Cell Research*, Washington, DC: The President's Council on Bioethics, http://bioethics.georgetown.edu/pcbe/reports/stemcell/.

Pressman, L., R. Burgess and R. Cook-Deegan (2006), 'The Licensing of DNA Patents by US Academic Institutions: An Empirical Survey', *Nature Biotechnology*, **24**, 31–9.

The Program on Climate Change, Agriculture and Food Security (CCAFS), http://www.ccafs.cgiar.org/index.php.

Public Intellectual Property Resource for Agriculture, http://www.pipra.org/.

Pure Wellbeing, http://www.purewellbeing.com/contents/en-uk/d180.html.

Qvenild, M. (2008), 'Svalbard Global Seed Vault: a "Noah's Ark" for the World's Seeds', *Development in Practice*, **18** (1), 110–16.

Radin, Margaret (1996), *Contested Commodities: The Trouble with Trade in Sex, Children, Body Parts and Other Things*, Cambridge, MA: Harvard University Press.

Rai, A. (1999), 'Regulating Scientific Research: Intellectual Property Rights and the Norms of Science', *Northwestern University Law Review*, **94**, 77–152.

Rai, A. (2001), 'Evolving Scientific Norms and Intellectual Property Rights: A Reply to Kieff', *Northwestern University Law Review*, **95**, 707–14.

Rai, Arti (2005), 'Proprietary Rights and Collective Action: The Case of Biotechnology Research with Low Commercial Value', in Keith Maskus and Jerome H. Reichman (eds), *International Public Goods and Transfer of Technology under a Globalized Intellectual Property Regime*, Cambridge: Cambridge University Press, 288–306.

Rai, A. (2010), 'Evidence to The Presidential Commission for the Study of Bioethical Issues, Synthetic Biology', Philadelphia, 13 September.

Rai, Arti (ed.) (2011), *Intellectual Property Law and Biotechnology: Critical Concepts*, Cheltenham and Northampton, MA: Edward Elgar.

Rai, A. and R. Eisenberg (2003), 'Bayh-Dole Reform and the Progress of Biomedicine', *Law and Contemporary Problems*, **66** (1&2), 289–314.

Rajan, Kaushik Sunder (2006), *Biocapital: The Constitution of Post-genomic Life*, Durham: Duke University Press.

Registry of Standard Biological Parts, 'Help: About the Registry', http://partsregistry.org/Help:About_the_Registry.

Registry of Standard Biological Parts, 'Help: Requesting Parts', http://partsregistry.org/Help:Requesting_Parts.

Registry of Standard Biological Parts, 'iGEM 2011: Regions/iGEM Growth', http://2011.igem.org/Regions/iGEM_Growth.

Registry of Standard Biological Parts, 'iGEM 2011: Regions/Overview', http://2011.igem.org/Regions/Overview.

Registry of Standard Biological Parts, 'iGEM: Synthetic Biology Based on Standard Parts', http://ung.igem.org/Main_Page.

Registry of Standard Biological Parts, 'Part:BBa_I742152: Get Part', http://partsregistry.org/partsdb/get_part.cgi?part=BBa_I742152.

Registry of Standard Biological Parts, 'Welcome to the Registry of Standard Biological Parts', http://partsregistry.org/Main_Page.

Reichman, J.H. and P. Uhlir (2003), 'The Public Domain: A Contractually Reconstructed Research Commons for Scientific Data in a Highly Protectionist Intellectual Property Environment', *Law and Contemporary Problems*, **66** (1&2), 315–462.

Reisenfeld, S. (1954), 'The New United States Patent Act in the Light of Comparative Law I', *University of Pennsylvania Law Review*, **102**, 291–322.

Rejeski, D. (2008), 'Comment on a Framework Convention for Nanotechnology', *Environmental Law Reporter*, **38**, 10518–19.

Rettberg, R. (2009), 'Evidence to the US National Academies Opportunities and Challenges in the Emerging Field of Synthetic Biology', Washington, DC, 10 July.

Rewerski, P. (2007), 'The Need for a New US Stem Cell Research Policy: A Comparative Look at International Stem Cell Research Laws', *University of Illinois Journal of Law, Technology and Policy Review*, **1**, 415–31.

Rimmer, M. (2004), 'The Race To Patent The SARS Virus: The TRIPS Agreement And Access To Essential Medicines', *Melbourne Journal of International Law*, **5** (2), 335–74.

Rimmer, M. (2005), 'Japonica Rice: Intellectual Property, Scientific Publishing, and Data-Sharing', *Prometheus*, **23** (3), 325–47.

Rimmer, M. (2007), 'The Genographic Project: Traditional Knowledge and Population Genetics', *Australian Indigenous Law Review*, **11** (2), 33–55.

Rimmer, Matthew (2008), *Intellectual Property and Biotechnology: Biological Inventions*, Cheltenham and Northampton, MA: Edward Elgar.

Rimmer, M. (2009), 'The Sorcerer II Expedition: Intellectual Property and Biodiscovery', *Macquarie Journal of International and Comparative Environmental Law*, **6**, 147–89.

Rimmer, Matthew (2011), *Intellectual Property and Climate Change: Inventing Clean Technologies*, Cheltenham and Northampton, MA: Edward Elgar.

Risch, M. (2009–10), 'Forward to the Past', *Cato Supreme Court Review*, 333–68.

Ritter, M. (2008), 'Scientists Make Human Embryo Clones', *Huffington Post*, 17 January, http://www.huffingtonpost.com/huff-wires/20080117/cloned-embryos.

Ritvo, Harriet (1996), 'Possessing Mother Nature: Genetic Capital in Eighteenth Century Britain', in John Brewer and Susan Staves (eds), *Early Modern Conceptions of Property*, London: Routledge, 413–26.

Robert, J.S. and F. Baylis (2003), 'Crossing Species Boundaries', *American Journal of Bioethics*, **3**, 1–13.

Robertson, J. (1999), 'Two Models of Human Cloning', *Hofstra Law Review*, **27**, 609–38.

Robinson, D.F. (2010), 'Traditional Knowledge and Biological Product Derivative Patents: Benefit Sharing and Patent Issues relating to Camu Camu, Kakadu Plum and Acai Plant Extracts', Traditional Knowledge Bulletin – Topical Issues Series, United Nations University, Institute of Advanced Studies, http://www.unutki.org/news.php?doc_id=174.

Rocco, Fiammetta (2003), *The Miraculous Fever-Tree: Malaria and the Quest for a Cure That Changed the World*, New York: HarperCollins.

The Rockefeller Foundation, 'Strengthening Food Security: Alliance for a Green Revolution in Africa', http://www.rockefellerfoundation.org/what-we-do/current-work/strengthening-food-security-alliance/.

Roco, Mihail and William Sims Bainbridge (2003), *Converging Technologies for Improving Human Performance: Nanotechnology, Biotechnology, Information Technology and Cognitive Science*, AA Dordrecht: Kluwer Academic.

Roco, Mihail and William Sims Bainbridge (eds) (2010), *Societal Implications of Nanoscience and Nanotechnology*, London: Springer.

Rogan, J. (2003), 'Letter to the United States Senate Committee and House Appropriations committees', United States Patent and Trademark Office, 20 November.

Rogers, Ben, Sumita Pannathur and Jesse Adams (2011), *Nanotechnology: Understanding Small Systems*, Taylor and Francis.

Rose, Steven (2005), *Lifelines: Life Beyond the Gene*, London: Vintage.

Rowen., L., G. Wong, R. Lane and L. Hood (2000), 'Publication Rights in the Era of Open Data Release Policies', *Science*, **289** (5486), 1881.

The Royal Society of the United Kingdom (2003), *Keeping Science Open: The Effects Of Intellectual Property Policy on the Conduct of Science*, London: The Royal Society, April, http://www.royalsoc.ac.uk/files/statfiles/document-221.pdf.

The Royal Society of the United Kingdom and the Royal Academy of Engineering (2004), *Nanoscience and Nanotechnologies: Opportunities and Uncertainties*, London: The Royal Society, http://www.nanotec.org.uk/finalReport.htm.

Ruse, Michael (2010), *Science and Spirituality: Making Room for Faith in the Age of Science*, New York: Cambridge University Press.

Rutt, J.S. (2010), 'Recent Nanotech Patent Trends: Top Ten Observations', Nanocleantech Blog, http://www.nanocleantechblog.com/tags/cleantech-patent-landscape-rep/.

Safrin, S. (2004), 'Hyperownership in a Time of Biotechnological Promise: The International Conflict to Control the Building Blocks of Life', *The American Journal of International Law*, **98** (4), 641–85.

Sage Bionetworks, 'The Sage Commons', http://sagebase.org/commons/index.php.

Sahai, Suman (2002), 'India's Plant Variety Protection and Farmers' Rights Legislation', in Peter Drahos and Ruth Mayne (eds), *Global Intellectual Property Rights: Knowledge, Access, and Development*, New York: Palgrave Macmillan, 214–23.

Salamanca-Buentello, F. et al. (2005) 'Nanotechnology and the Developing World', *Public Library of Science Medicine*, **2** (5), e97, 302.

Salter, B. (2007), 'Patenting, Morality and Human Embryonic Stem Cell Science: Bioethics and Cultural Politics in Europe', *Regenerative Medicine*, **2** (3), 301–11.

Samuelson, P. and J. Schultz (2011), '"Clues" for Determining Whether Business and Service Innovations Are Unpatentable Abstract Ideas', *Lewis and Clark Law Review*, **15**, 110–31.

Sanderson, J. (2006), 'Essential Derivation, Law and the Limits of Science', *Law in Context*, **24** (1), 34–53.

Sanderson, J. and K. Adams (2008), 'Are Plant Breeder's Rights Outdated? A Descriptive and Empirical Assessment of Plant Breeder's Rights in Australia, 1987–2007', *Melbourne University Law Review*, **32** (3), 980–1006.

Sanderson, Katharine (2011), 'Curation Generation', *Nature*, **470**, 295–6.

Sanhai, W., J. Sakamoto, R. Canady and M. Ferrari (2008), 'Seven Challenges for Nanomedicine', *Nature Nanotechnology*, **3**, 242–4.

Sarkar, Sahotra (1996), 'Biological Information: A Sceptical Look at some Central Dogmas of Molecular Biology', in Sahotra Sarkar (ed.), *The Philosophy and History of Molecular Biology: New Perspectives*, Dordrecht: Kluwer Academic, 187–232.

Sarkar, Sahotra (ed.) (1996), *The Philosophy and History of Molecular Biology: New Perspectives*, Dordrecht: Kluwer Academic.

Sarnoff, J. (2011), 'Patent Eligible Inventions after *Bilski:* History and Theory', *Hastings Law Journal*, **63** (forthcoming).

Sarnoff, J. (2011), 'Patent Eligible Inventions after *Bilski, Prometheus*, and *Myriad*', *Texas Intellectual Property Law Journal*, **19** (3), 393–418.

Sarnoff, J. (2011), 'Constitutional Hard Limits on the Patent Power', unpublished draft.

Schadt, E. (2011), 'First Steps on a Long Road', *Science*, **331** (6018), 691.

Schama, Simon (2000), *A History of Britain I: At the Edge of the World? 3000BC–AD1603*, London: BBC.

Schechter, J. (2010), 'Promoting Human Embryonic Stem Cell Research: A Comparison of Policies in the United States and the United Kingdom and Factors Encouraging Advancement', *Texas International Law Journal*, **45** (3), 603–29.

Schiebinger, Londa (2004), *Plants and Empire: Colonial Bioprospecting in the Atlantic World*, Cambridge, MA: Harvard University Press.

Schmidt, C. (2007), 'Negotiating the RNAi Patent Thicket', *Nature Biotechnology*, **25** (3), 273–5.

Schofield, P. et al. (2009), 'Post-Publication Sharing of Data and Tools', *Nature*, **461** (7261), 171–3.

Scott, A. (1999), 'The Dutch Challenge to the Bio-Patenting Directive', *European Intellectual Property Review*, **21** (4), 212–15.

Scrinis, G. and K. Lyons (2007), 'The Emerging Nano-Corporate Paradigm: Nanotechnology and the Transformation of Nature, Food and Agri-Food Systems', *International Journal of Sociology of Food and Agriculture*, **15** (2), 22–44.

Seabrook, J. (2007), 'Sowing for Apocalypse: The Quest for a Global Seed Bank', *The New Yorker*, 27 August, http://www.newyorker.com/reporting/2007/08/27/070827fa_fact_seabrook.

Secretary's Advisory Committee on Genetics, Health and Society (2010), *Gene Patents and Licensing Practices and their Impact on Patient Access to Genetic Tests*, Bethesda, MD: Secretary's Advisory Committee on Genetics, Health, and Society, http://oba.od.nih.gov/oba/sacghs/reports/SACGHS_patents_report_2010.pdf.

Sell, Susan (2009), 'Corporations, Seeds, and Intellectual Property Rights Governance', in Jennifer Clapp and Doris Fuchs (eds), *Corporate Power in Global Agrifood Governance*, Cambridge, MA: MIT Press, 187–224.

Sellenthin, Mark O. (2006), *Beyond the Ivory Tower: A Comparison of Patent Rights Regimes in Sweden and Germany*, Linköping: Linköping Studies in Arts and Science, No. 355.

Senate Standing Committees on Community Affairs Reference Committee (2010), *Gene Patents*, Canberra: Australian Parliament, http://www.aph.gov.au/senate/committee/clac_ctte/gene_patents_43/report/report.pdf.

Senate Standing Committees on Legal and Constitutional Affairs (2011), *Patent Amendment (Human Genes and Biological Materials) Bill* 2010, Canberra: Australian Parliament, http://www.aph.gov.au/senate/committee/legcon_ctte/patent_amendment/index.htm

Shih, A. (2002), 'The Patented Peanut Butter and Jelly Sandwich: Food as Intellectual Property', *Gastronomica*, 23–7.

Shiva, Vandana (2001), *Protect or Plunder? Understanding Intellectual Property Rights*, London: Zed Books.

Shiva, Vandana (2003), 'Food Rights, Free Trade, And Fascism', in Matthew Gibney (ed.), *Globalizing Rights: The Oxford Amnesty Lectures 1999*, Oxford: Oxford University Press, 87–108.

Shiva, Vandana (2008), *Soil Not Oil: Environmental Justice in an Age of Climate Crisis*, South End Press.

Shulman, S. (2001), 'The Morphing Patent Problem', Owning the Future, *Technology Review*, November.

Silver, L. (2007), 'Life 2.0', *Newsweek*, 4 June, 41–5.

Simon, B. (2009), 'How to Get a Fair Share: IP Policies for Publicly Supported Biobanks', *Stanford Journal of Law, Science, and Policy*, **1**, 65–79.

Simondon, G. (1989), *L'individuation Psychique et Collective*, Paris: Aubier.

Skene, L. (2002), 'Ownership of Human Tissue and the Law', *Nature Reviews Genetics*, **3**, 145–8.

Skoblov, M.Y. (2009), 'Prospects of Antisense Therapy Technologies', *Molecular Biology,* **43** (6), 917–29.

Slavid, Ruth (2009), *Extreme Architecture: Building for Challenging Environments*, London: Laurence King Publishers.

Smith, D. (2000), 'Adventures in Lilliput', *The Sydney Morning Herald*, 11 December.

Smith, D. (2008), 'Scientists Attempt Stem Cell Breakthrough', *Sydney Morning Herald*, 17 September.

Smolke, C. (2009), 'Evidence to the US National Academies, Opportunities and Challenges in the Emerging Field of Synthetic Biology', Washington, DC, 9 July.

Sörlin, Sverker and Otto Fagerstedt (2004), *Linné och hans apostlar*, Stockholm: Natur och Kultur.

Spary, E.C. (2003), 'Peaches Which the Patriarchs Lacked: Natural History, Natural Resources, and the Natural Economy in France', *History of Political Economy*, **35**, 15–16.

Staff Writer (2009), 'Researchers: Farmers' Rights to Adapt to Climate Change Ignored', *China Daily*, *Xinhua*, 7 September.

Staff Writer (2010), 'Climate Change Resistant Crops Move Nearer after Gene Breakthrough', *The Telegraph*, 8 January.

Stedman's Medical Dictionary (2005) (28th edn).
Stern, Nicholas (2009), *A Blueprint for a Safer Planet: How to Manage Climate Change and Create a New Era of Progress and Prosperity*, London: The Bodley Head.
Stevens, A. (2010), 'Evidence to The Presidential Commission for the Study of Bioethical Issues, Synthetic Biology, Philadelphia', 13 September.
Stott, M. and J. Valentine (2004), 'Gene Patenting and Medical Research: A View from a Pharmaceutical Company', *Nature Reviews Drug Discovery*, **3**, 364–8.
Stratton, M.R., P. Campbell and A. Futreal (2009), 'The Cancer Genome', *Nature*, **458** (8), 719–24.
Svalbard Global Seed Vault (2008), 'Svalbard Global Seed Vault: Arctic Seed Vault Opens Doors for 100 Million Seeds', Ministry of Agriculture and Food, 26 February, http://www.regjeringen.no/en/dep/lmd/campain/svalbard-global-seed-vault/news/arctic-seed-vault-opens-doors-for-100-mi.html?id=501721.
Svalbard Global Seed Vault (2009), *Frozen Seeds in a Frozen Mountain: How to feed a Warming World? – Anniversary Seminar,* February, http://www.regjeringen.no/upload/LMD/kampanjeSvalbard/Vedlegg/Svalbard_Statement_270208.pdf.
Svalbard Global Seed Vault, 'Frequently Asked Questions', http://www.regjeringen.no/en/dep/lmd/campain/svalbard-global-seed-vault/frequently-asked-questions.html?id=462221.
Svalbard Global Seed Vault, 'Management and Operations', http://www.regjeringen.no/en/dep/lmd/campain/svalbard-global-seed-vault/management-and-operations.html?id=462223.
Swanson, K. (2007), 'Biotech in Court: A Legal Lesson on the Unity of Science', *Social Studies of Science*, **37** (3), 357–84.
Swiderska, Krystyna (2007), 'Protecting Traditional Knowledge: A Framework Based on Customary Laws and Bio-cultural Heritage', in Bertus Haverkort and Stephen Rist (eds), *Endogenous Development and Bio-Cultural Diversity: The Interplay of Worldviews, Globalisation and Locality*, Leusden, Netherlands and Bern, Switzerland: Compas and Centre for Development and Environment, 358–65.
Sylvester, D. and D.M. Bowman (2010), 'English Garden or Tangled Grounds? Navigating the Nanotechnology Patent Landscape', 12 January, http://ssrn.com/abstract=1535650.
Syngenta, http://www.syngenta.com/en/corporate_responsibility/climate change.html.
Synthetic Biology Research Community, 'FAQ – What is Synthetic Biology?' http://syntheticbiology.org/FAQ.html.

Taft, K. and S. Webb (2005), 'United States: Stem Cells: Their Promise, Their Problems', *Mondaq*, 9 March.

Taft, R.J. et al. (2010), 'Non-coding RNAs: Regulators of Disease', *Journal of Pathology*, **220**, 126–39.

Tansey, Geoff and Tamsin Rajotte (eds) (2008), *The Future Control of Food: A Guide to International Negotiations and Rules on Intellectual Property, Biodiversity, and Food Security*, London: Earthscan.

TED (2009), 'Q&A with Cary Fowler: Saving Seeds to Protect Our Food Supply', September, http://blog.ted.com/2009/09/02/qa_with_cary_fo/.

Ten Kate, Kerry and Sarah Laird (1999), *The Commercial Use Of Biodiversity: Access To Genetic Resources And Benefit-Sharing*, London: Earthscan.

The Third World Network (2010), 'Call to Support Farmer and Civil Society Organisations' Application to UPOV For Observer Status', 11 March, http://www.twnside.org.sg/title2/intellectual_property/info.service/2010/ipr.info.100305.htm.

Thomson, J. (2007), 'The Example of Stem Cells', *Hastings Centre Report*, **37** (1), 12–13.

Thomson, J. et al. (1998), 'Embryonic Stem Cell Lines Derived from Human Blastocysts', *Science*, **282** (5391), 1145–7.

Thomson, P. (2007), 'Ethical Aspects and Patents in Lifescience', Paper for the WIPO symposium on IP and Bioethics, Geneva, 4 September, http://www.wipo.int/meetings/en/doc_details.jsp?doc_id=85513.

Thorisson, G. and A.J. Brookes (2009), *IRBW2009 Workshop*, 13–14 May, Meeting Minutes, http://www.gen2phen.org/system/files/private/IRBW2009%20meeting%20minutes%20v2.pdf.

Toronto International Data Release Workshop Authors (2009), 'Prepublication Data Sharing', *Nature*, **461**, 168–70.

Torrance, A. (2010), 'Synthesizing Law for Synthetic Biology', *Minnesota Journal of Law, Science & Technology,* **11** (2), 629–65.

Truss, W. (2002), 'The Plant Breeder's Rights Amendment Bill 2002 (Cth)', House of Representatives, Hansard, 12 December, 10589.

Tullis, T. (2005–2006), 'Application of the Government License Defense to Federally Funded Nanotechnology Research: The Case for a Limited Patent Compulsory Licensing Regime', *University of California, Los Angeles Law Review*, **53,** 279–314.

Twyman, R.M. (2003), 'Silencing is Golden: RNA Interference in Research and Medicine', *Wellcome News,* **36** (Q3) 10–11, http://www.wellcome.ac.uk/News/2003/Features/WTD004556.htm.

United Kingdom Biobank, http://www.ukbiobank.ac.uk/assessment/takepart.php.

United Kingdom Biobank (2011), *Access Procedures: Application and Review Procedures for Access to the UK Biobank Resource,* 31 May, http://www.ukbiobank.ac.uk/procedures/.

United Nations University and the Institute of Advanced Studies (2004), *The Role of Registers and Databases in the Protection of Traditional Knowledge: A Comparative Analysis*, Japan: United Nations University and the Institute of Advanced Studies, http://www.ias.unu.edu/binaries/UNUIAS_TKRegistersReport.pdf.

United States Department of Agriculture (2009), Our Changing Planet: *The U.S. Climate Change Science Program for Fiscal Year 2009*, http://www.usgcrp.gov/usgcrp/Library/ocp2009/ocp2009-usda.htm.

United States Department of Health and Human Services (2010), '1000 Genomes Project Publishes Analysis of Completed Pilot Phase', National Institutes of Health News, Press Release, 27 October, http://www.genome.gov/27541917.

United States Patent and Trademark Office, 'New Cross-Reference Digest for Nanotechnology', http://www.uspto.gov/web/patents/biochempharm/crossref.htm.

United States Patent and Trademark Office (2001), 'Utility Examination Guidelines', *Federal Register*, **66** (4), 1092–9, http://www.uspto.gov/web/offices/com/sol/notices/utilexmguide.pdf.

United States Patent and Trademark Office (2009), 'Pilot Program for Green Technologies Including Greenhouse Gas Reduction', *Federal Register*, **74** (234), 8 December, 64666–9.

University of California Berkeley 2007 iGEM Team (2007), 'Bactoblood', http://parts.mit.edu/igem07/index.php/Berkeley_UC.

Vaidhyanathan, Siva (2006), 'Nanotechnology and the Law of Patents: A Collision Course', in Geoffrey Hunt and Michael Mehta (eds), *Nanotechnology: Risk, Ethics, Law*, London: Earthscan, 225–36.

Van den Belt, Henk (2009), 'Philosophy of Biotechnology', in Anthonie Meijers (ed.) (2009), *Philosophy of Technology and Engineering Sciences*, Amsterdam and London: Elsevier, 1301–40.

Van Dooren, T. (2009), 'Banking Seed: Use and Information in the Conservation of Agricultural Diversity', *Science as Culture*, **18** (4), 373–95.

Van Dooren, T. (2009), 'Genetic Conservation in a Climate of Loss: Thinking with Val Plumwood', *Australian Humanities Review*, **46**, 101–12.

Van Overwalle, Geertrui (2005), 'Legal and Ethical Aspects of Bio-Patenting: the EU Biotechnology Directive', in Peter Drahos (ed.), *Death of Patents*, London: Lawtext Publications, 212–27.

Van Overwalle, G. (ed.) (2009), *Gene Patents and Collaborative Licensing Models: Patent pools, Clearinghouses, Open Source Models and Liability Regimes*, Cambridge: Cambridge University Press.

Van Overwalle, G. (2010), 'Turning Patent Swords into Shares', *Science*, **330**, 1630–31.

Van Overwalle, Geertrui and Esther van Zimmeren (2009), 'Chapter 22 – Functions and Limits of Patent Law', in Erik Claes, Wouter Devroe and Bert Keirsbilck (eds), *Facing the Limits of the Law*, Berlin: Springer.

Van Overwalle, G., E. Van Zimmeren, B. Verbeure and G. Matthijs (2006), 'Models for Facilitating Access to Patents on Genetic Inventions', *Nature Review Genetics*, **7**, 143.

Vaver, D. (2003), 'Invention in Patent Law: A Review and Proposal', *International Journal of Law and Information Technology*, **11**, 286–307.

Veash, N. (2000), 'Biopiracy – A New Threat to Amazon Rainforest's Treasures', *The Independent*, London, 16 October.

Venter, J.C. et al. (2001), 'The Sequence of the Human Genome', *Science*, **291**, 16 February, 1301–4.

Venter, J.C. (2007), *A Life Decoded*, London: Allen Lane and Penguin Books.

Venter, J.C. (2011), 'The Human Genome at 10: Successes and Challenges', *Science*, **331** (6017), 546–7.

Voumard, John (2000), *Commonwealth Public Inquiry into Access to Biological Resources in Commonwealth Areas*, Canberra: Environment Australia, http://www.ea.gov.au/biodiversity/science/access/inquiry/index. html.

Waldby, C. (1999), 'Iatrogenesis: The Visible Human Project and the Reproduction of Life', *Australian Feminist Studies*, **14**, 77–90.

Walsh, F. and J. Douglas (2009), 'Aboriginal Harvesters who sell Bush Foods and Seeds from Central Australia for Food and Mining Revegetation Markets: Their Practices, Priorities and Opinions', Alice Springs: Desert Knowledge Co-operative Research Centre Report.

Walterscheid, Edward (2002), *The Nature of the Intellectual Property Clause: A Study in Historical Perspective*, Buffalo: William S. Hein & Co.

Walterscheid, E.C. (2002), 'To Promote the Progress of Science and Useful Arts: The Anatomy of a Congressional Power', *IDEA*, **43**, 1–82.

Washburn, Jennifer (2005), *University Inc.: The Corporate Corruption of Higher Education*, New York: Basic Books.

Water Efficient Maize for Africa, http://www.aatf-africa.org/wema.

Waters, C. Kenneth (2006), 'A Pluralist Interpretation of Gene-centered Biology', in Stephen Kellert, Helen Longino and C. Kenneth Waters (eds), *Scientific Pluralism: Minnesota Studies in the Philosophy of Science*, vol. 19, Minneapolis: University of Minnesota Press, 190–214.

Webb, James L.A. (2002), *Tropical Pioneers: Human Agency and Ecological Change in the Highlands of Sri Lanka, 1800–1900*, Athens: Ohio University Press.

Webster, Andrew, Nik Brown, Conor Douglas, Graham Lewis, Jane Kaye, Richard Tutton and Nick Williams (2008), *Public Attitudes to Third Party Access and Benefit Sharing: Their Application to UK Biobank,* Final Report to the UK Biobank Ethics and Governance Council, University of York: Science and Technology Studies, University of York, http://www.egcukbiobank.org.uk/assets/wtx052208.pdf.

Weiss, R. (2008), 'Firms Seek Patents on "Climate Ready" Altered Crops', *The Washington Post*, 13 May, A04.

Wellcome Trust (2003), 'Sharing Data from Large-Scale Biological Research Projects: A System of Tripartite Responsibility', Fort Lauderdale, January, http://www.genome.gov/Pages/Research/WellcomeReport0303.pdf.

Wellcome Trust (2010), 'Policy on Data Management and Sharing', August, http://www.wellcome.ac.uk/About-us/Policy/Policy-and-position-statements/wtx035043.htm.

Wells, Spencer (2006), *Deep Ancestry: Inside the Genographic Project*, New York: Penguin Books and National Geographic.

White, A. (2000–1), 'Gene and Compound Per Se Claims: An Appropriate Reward?' *Bio-Science Law Review*, **6**, 239–48.

White, A. (2004), 'The Ethics – Gene Patenting and Human Health', *Australian Intellectual Property Journal*, **15** (1), 6–17.

Wianny, F. and M. Zernicka-Goetz (2000), 'Specific Interference with Gene Function by Double-Stranded RNA in early mouse development', *Nature Cell Biology*, **2**, 70–75.

Widdows, Heather and Caroline Mullen (eds) (2009), *The Governance of Genetic Information*, Cambridge: Cambridge University Press.

Williams-Jones, B. (2002), 'History of a Gene Patent: Tracing the Development and Application of Commercial BRCA Testing', *Health Law Journal*, **10**, 123–46.

Willinsky, John (2006), *The Access Principle: The Case for Open Access to Research and Scholarship*, Cambridge, MA: The MIT Press.

Winickoff, David (2009), 'From Benefit Sharing to Power Sharing: Partnership Governance in Population Genomics Research', in Jane Kaye and Mark Stranger (eds), *Principles and Practice of Biobank Governance*, Cheltenham: Ashgate, 53–66.

Winickoff, D. et al. (2009), 'Opening Stem Cell Research and Development: A Policy Proposal for the Management of Data, Intellectual Property, and Ethics', *Yale Journal Health Policy, Law & Ethics*, **9**, 52–127.

Winston, E. (2011), 'The Technological Edge', *Akron Intellectual Property Law Journal*, 1–25, http://papers.ssrn.com/sol3/papers.cfm?abstract_id=1692836.

Wolfson, M. et al. (2010), 'DataSHIELD: Resolving a Conflict in Contemporary Bioscience – Performing a Pooled Analysis of Individual-Level Data Without Sharing the Data', *International Journal of Epidemiology*, **39**, 1372–82.

Würtenberger, G. (2009), 'An Analysis of the European and International Plant Variety Protection Systems', *Journal of Intellectual Property Law and Practice*, **4**, 914–15.

Wynberg, R. and S. Laird (2007), '"Less is often more" Governance of a Non-timber Forest Product, Marula (Sclerocarya birrea subsp. Caffra) in Southern Africa', *International Forestry Review*, **9** (1), 475–89.

Wynberg, Rachel, Doris Schroeder, Samantha Williams and Saskia Vermeylen (2009), 'Sharing Benefits Fairly: Decision Making and Governance', in Rachel Wynberg, Doris Schroder and Roger Chennells (eds), *Indigenous Peoples, Consent and Benefit Sharing: Lessons from the San Hoodia Case*, Dordrecht: Springer, 231–60.

Yates, P. (2009), 'The Bush Foods Industry and Poverty Alleviation in Central Australia', *Dialogue*, **28** (2), 47–56.

You, E. (2010), 'Evidence to the Presidential Commission for the Study of Bioethical Issues, Synthetic Biology', Washington, DC, 9 July.

Yuille, M. et al. (2007), 'Biobanking for Europe', *Briefings in Bioinformatics*, **9**, 14–24.

Yuille, M. et al. (2010), 'The UK DNA Banking Network: A "Fair Access" Biobank', *Cell Tissue Bank*, **11**(3), 241–51.

Zamore, P.D. and B. Haley (2005), 'Ribo-gnome: The Big World of Small RNAs', *Science*, **309**, 1519–24.

Zamore, P.D., T. Tuschl, P.A. Sharp and D.P. Bartel (2000), 'RNAi: Double-Stranded RNA Directs the ATP-Dependent Cleavage of mRNA at 21 to 23 Nucleotide Intervals', *Cell*, **101**, 25–33.

Zovko, N. (2006), 'Nanotechnology and the Experimental Use Defense to Patent Infringement', *McGeorge Law Review*, **37** (1), 129–56.

PATENT AND TRADEMARK APPLICATIONS

Agrigenetics, Inc. and Mycogen Seeds (Dow Agrosciences)(2003), 'Inbred corn line BE1146BMR', United States Patent No. 7,273,970.

Agrigenetics, Inc., and Mycogen Seeds (Dow Agrosciences) (2003), 'Inbred corn line 4VP500 Drought Heat', United States Patent Application No. US20050076401A1.

Anderson D. and D. Stemple (1995), 'Mammalian Multipotent Neural Stem Cells', European Patent No. EP0658194.
Anderson, D. and D. Stemple (1995), 'Mammalian Multipotent Neural Stem Cells', European Patent No. EP 93921175 (21 June 1995).
Bayer Bioscience N.V. (2005), 'Stress tolerant cotton plants', World Intellectual Property Organisation Patent Cooperation Treaty Application No. WO06045633A1.
Bayer CropScience GmbH (2006), 'Active substances for increasing the stress defense in plants to abiotic stress, and methods of finding them', United States Patent Application No. US20070124839A1.
Bodnar, A. (2004), 'Methods and materials for the growth of primate-derived primordial stem cells in feeder-free culture', United States Patent No. 6,800,480, 5 October.
Brüstle, O. (2010), 'Neural Precursor Cells, Method for the Production and Use Thereof in Neural Defect Therapy', European Patent No. EP1040185.
Da Costa, S. (2007), 'Transcription Factor Stress-Related Proteins and Methods of Use in Plants', United States Patent No. 7,619,137.
Evogene (2009), 'Polynucleotides, Plolypeptides Encoded Thereby, and Methods of Using Same for Increasing Abiotic Stress Tolerance and/or Biomass and/or Yield in Plants Expressing Same', World Intellectual Property Organization Patent Cooperation Treaty Application No. WO/2009/013750.
Heard, J. and others (2006), 'Plant Transcriptional Regulators', United States Patent No. 7,663,025.
Jarlmadangah Burru Aboriginal Corporation and Griffith University (2008), 'Novel Compounds', Australian Patent Application No. 2008905126 (patent has since lapsed).
Jarlmadangah Burru Aboriginal Corporation and Griffith University (2008), 'Novel Compounds', Australian Patent Application No. 2004293125.
Jarlmadangah Burru Aboriginal Corporation and Griffith University (2008), 'Novel Compounds', Australian Patent Application No. 2003906558 (lapsed).
Jarlmadangah Burru Aboriginal Corporation and Griffith University (2009), Australian Patent Application No. 2009904829 (filed and under examination).
Jones, B., D. Gan, J. Aravena and M. Hines (2007), 'Compositions Comprising Kakadu Plum Extract or Acai Berry Extract', Australian Patent Application No. 2007205838.

Kaufman, D. and J. Thomson (2001), 'Hematopoietic Differentiation of Human Pluripotent Embryonic Stem Cells', United States Patent No. 6,280,718.

Leder, P. and T. Stewart (1988), 'Transgenic Non-Human Mammals', United States Patent No. 4,736,866.

Leder, P. and T. Stewart, Harvard College (1986), 'Method for Producing Transgenic Animals', European Patent No. EP 0169672.

Mandalam, R. and C. Xu, Geron Corporation, 'Medium for Growing Human Embryonic Stem Cells', United States Patent No. 7,297,539 ('Geron patent').

Mary Kay Inc. et al. (2006), 'Compositions Comprising Kakadu Plum Extract or Acai Berry Extract', World Intellectual Property Organization Patent Cooperation Treaty Application, No. WO/2007/084998 filed on 19 January 2007, priority date 19 January 2006 possible owing to a United States patent.

Mendel Biotechnology Inc. (2003), 'Plant Transcriptional Regulators of Drought Stress', United States Patent Application No. US20070240243A9.

Mendel Biotechnology, Inc. (2005), 'Plant Stress Tolerance from Modified AP2 Transcription Factors', World Intellectual Property Organization Patent Cooperation Treaty Application No. WO06069201A3.

Mendel Biotechnology, Inc. (2006), 'Stress Tolerance in Plants may Include Salt, Hyperosmotic Stress, Heat, Cold, Drought, or Low Nitrogen Conditions', World Intellectual Property Organization Patent Cooperation Treaty Application No. WO07028165C1.

Monsanto (2004), 'Rice Nucleic Acid Molecules and Other Molecules Associated with Plants and Uses Thereof for Plant Improvement', United States Patent Application No. US 2004/0123343 A1.

Monsanto (2009), 'Drought Tolerant Corn with Reduced Mycotoxin', World Intellectual Property Organization Patent Cooperation Treaty Application No. WO/2009/049110.

Petitte, J. and Z. Yang (2004), 'Method of Producing an Avian Embryonic Stem Cell Culture and the Avian Embryonic Stem Cell Culture Produced by the Process', United States Patent No. 5,340,740.

Schaffer, A. (1998), 'Method for Breeding Tomatoes having Reduced Water Content and Product of the Method', European Patent Application No. 00940724.

Seaborg, G. (1964), 'Element 95 [Americium] and Method of Producing Said Element', United States Patent No. 3,156,523.

Seaborg, G. (1964), 'Element 96 [Curium] and Compositions Thereof', United States Patent No. 3,161,462.

Seidman, E. and Y. Theoret (1999), 'Method of Treating IBD/Crohn's Disease and Related Conditions wherein Drug Metabolite Levels in Host Blood Cells Determine Subsequent Dosage', United States Patent No. 6,355,623.

Seidman, E. and Y. Theoret (2001), 'Methods of Optimizing Drug Therapeutic Efficacy for Treatment of Immune-mediated Gastrointestinal Disorders', United States Patent No. 6,680,302.

Simons, M. (1992), 'Intron Sequence Analysis Method for Detection of Adjacent and Remote locus alleles as Haplotypes', United States Patent No. 5,612,179.

Simons, M. (1994), 'Genomic Mapping Method by Direct Haplotyping using Intron Sequence Analysis', United States Patent No. 5,851,762.

Sivasankar, S., T. Helentjaris and D. Xu (2004), 'Transcriptional Activators involved in Abiotic Stress Tolerance', United States Patent No. 7,253,000.

Sivasankar, S., T. Helentjaris, D. Xu and A. Lyznik (2006), 'Transcriptional Activators involved in Abiotic Stress Tolerance', United States Patent No. 7,317,141.

Sivasankar, S., D. Selinger and N. Brugiere (2009), 'Transcriptional Activators Involved in Abiotic Stress Tolerance', WIPO Publication No: WO/2009/094527.

Smith, A. (1996), 'Isolation, Selection And Propagation Of Animal Transgenic Stem Cells', European Patent No. EP695351 ('Edinburgh patent').

Thomson, J. (1997), 'Primate Embryonic Stem Cells', European Patent No. EP 0770125.

Thomson, J. (1998), 'Primate Embryonic Stem Cells', United States Patent No. 5,843,780.

Thomson, J. (2001), 'Primate Embryonic Stem Cells', United States Patent No. 6,200,806.

Thomson, J. (2006), 'Primate Embryonic Stem Cells', United States Patent No. 7,029,913.

CASE LAW

Altana Pharma AG v. *Teva Pharmaceuticals USA, Inc.*, 566 F.3d 999 (2009).

Amazon.com v. *Canada (Attorney General)* (2010), F.C.J. 1209.

American Fruit Growers, Inc. v. *Brogdex Co.*, 283 U.S. 1 (1931).

American Wood-Paper Co. v. *Fibre Disintegrating Co.*, 90 U.S. (23 Wall.) 566, 593–94 (1874).

Anderson's-Black Rock, Inc. v. *Pavement Salvage Co.*, 396 U.S. 57 (1969).

Ansonia Brass & Copper Co. v. *Elec. Supply Co.*, 144 U.S. 11, 18 (1892).

Ariad Pharmaceuticals., Inc. v. *Eli Lilly & Co.*, 598 F.3d 1336 (Fed. Cir. 2010) (en banc).
Association for Molecular Pathology et al. v. *United States Patent and Trademark Office, Myriad Genetics, et al.* 702 F. Supp. 2d 181, (S.D.N.Y. 2010), *appeal docketed,* No. 2010–1406 (Fed. Cir. June 22, 2010).
Association for Molecular Pathology et al. v. *United States Patent and Trademark Office, Myriad Genetics, et al.* 653 F.3d 1329, 2011 WL 3211513 (C.A. Fed. (N.Y.)), 99 USPQ 2d 1398.
Bilski v. *Kappos*, 130 S. Ct. 3218 (2010).
Bonito Boats v. *Thunder Craft Boats*, 489 U.S. 141 (1989).
Boulton v. *Bull* (1795) 2 H. Bl. 463.
Brenner v. *Manson*, 383 U.S. 519 (1966).
California/Stem Cells (T 522/04) 2009 EPOR 45.
Cancer Voices Australia et al. v. *Myriad Genetics Inc. et al.* (2010), Federal Court of Australia, Statement of Claim, 8 June.
Carnegie Mellon University v. *Hoffman-La Roche* 541 F 3d. 1115 (2008).
Classen v. *Biogen*, WDQ-04–2607 (2006).
Classen v. *Biogen*, 304 Fed. Appx. 866 (2008).
Classen v. *Biogen*, 2010 U.S. LEXIS 5533 (2010).
Cochrane v. *Badische Anilin & Soda Fabrik*, 111 U.S. 293 (1884).
Cochrane v. *Deener*, 94 U.S. 780 (1887).
CSIRO v. *Buffalo Technology Inc*. 542 F.3d 1363 (2008).
Cultivaust v. *Grain Pool Pty Ltd* [2004] FCA 638.
Cultivaust v. *Grain Pool Pty Ltd* [2005] FCAFC 223.
Daiichi Sankyo Co., Ltd. v. *Matrix Labs., Ltd.*, 619 F.3d 1346 (Fed. Cir. 2010).
Dash v. *Van Kleeck*, 7 Jonhs 477 (Sup Ct of Judicature, 1811).
Decision Revoking the European Patent (art. 102(1), (3) EPC) (Eur. Pat. Office May 17, 2004) (revoking European Patent No. 0699754).
Decision of the Technical Board of Appeal T1213/05 -3.3.04 (arts. 52(2), 53(a), 54, 56, 57, 83, 84, 87 to 89, 111(1),112(1)(a), 123(2)(3) EPC) (European Patent Office Sept. 27, 2007) (reviewing the Interlocutory decision of the Opposition Division of the European Patent Office of 19 September 2005 concerning European patent No. 0705902, maintained in amended form).
Diamond v. *Chakrabarty*, 447 U.S. 303 (1980).
Diamond v. *Diehr*, 450 U.S. 175 (1981).
Dolbear v. *American Bell Telephone Co.*, 126 U.S. 1 (1888).
Edinburgh case T1079/03.
Eisai Co. Ltd. v. *Dr. Reddy's Labs., Ltd.*, 533 F.3d 1353, 1359 (Fed. Cir. 2008).
Eldred v. *Ashcroft*, 537 U.S. 186 (2002).

Eli Lilly and Co. v. *Human Genome Sciences, Inc.*, [2008] EWHC 1903 (High Court of Justice – Patents Court).

European Court of Justice (2011), 'Opinion of the Advocate General in Case C-34/10', Press Release, No 18/11, 10 March, 1.

Euthanasia Compositions/ Michigan State University (2005), T 0866/01; http://www.epo.org/law-practice/case-law-appeals/recent/t010866eu1.html.

Evans v. *Eaton*, 8 F. Cas. 846 (C.C.D. Pa. 1816) (No. 4,559).

Foundation for Taxpayer and Consumer Rights and the Public Patent Foundation, *Request for re-examination in respect of US Patent No 5,843,780* (2006) PUBPAT http://www.pubpat.org/assets/files/warfstem cell/780Request.pdf.

Foundation for Taxpayer and Consumer Rights v. *Patent of Wisconsin Alumni Research Foundation* (Appeal 2010–001854), the Board of Patent Appeals and Interference http://www.consumerwatchdog.org/resources/WARFDecision042910.pdf.

Foundation for Taxpayer and Consumer Rights v. *Wisconsin Alumni Research Foundation*, BPAI No 2010–001854 (28 April 2010).

Funk Brothers Seed Co. v. *Kalo Inoculant Co.*, 333 U.S. 127 (1948).

Genetic Technologies Limited v. *Berkman Coulter Inc. et al.* (2010), United States District Court for the Western District of Wisconsin, Civil Action No. 10 CV 0069.

Genetic Technologies Ltd. v. American Esoteric Laboratories et al. (2011) United States District Court for the Western District of Texas, 1:11-cv-00057.

Golan v. *Holder*, 609 F.3d 1076 (10th Cir. 2010), 501 F.3d 1179 (10th Cir. 2009), *cert granted*, 131 S.Ct. 1600 (2011).

Gottschalk v. *Benson*, 409 U.S. 63 (1972).

Graham v. *John Deere*, 383 U.S. 1 (1966).

Grain Pool Of Western Australia v. *Commonwealth* (2000) 46 IPR 515.

Greenpeace v. *Oliver Brüstle* Decision of the German Bundespatentgericht (BPatG) of 5 Dec 2006, 3 Ni 42/04.

Hartranft v. *Wiegmann,* 121 U.S. 609 (1887).

Harvard/Onco-mouse [1989] OJ EPO 451.

Harvard/Onco-mouse [1990] OJ EPO 490 (TBA).

Harvard/Transgenic Animals, T315/03 [2006] OJ EPO 15.

Harvard College v. *Canada (Commissioner of Patents)* 2002 SCC 76.

Hoffman v. *Monsanto Canada Inc.*, 2007 SKCA 47 (CanLII).

Howard Florey Institute/Relaxin. Decision of Technical Board of Appeal 3.3.1 dated 30 August 1994 (1995) *Office Journal of the EPO* 6: 373–455, 400.

In re Alappat, 33 F.3d 1526 (Fed. Cir. 1994).

In re Bilski, 545 F.3d 943 (Fed. Cir., 2008).
In re Fisher, 421 F.3d 1365 (Fed. Cir. 2005).
In re Grams, 888 F.2d 835 (1989).
In re Kubin 561 F. 3d 1351 (Fed. Cir. 2009).
In re Meyer, 688 F.2d 789, (CCPA 1982).
In re Thuau, 135 F.2d 344 (C.C.P.A. 1943).
Industries Ltd. v. *Commissioner of Patents*, [1986] 3 F.C. 40.
Integra Life Sciences I Ltd. v. *Merck KgA* 331 F. 3d 860 (2003).
Intel v. *CSIRO* 455 F.3d 1364, 79 U.S.P.Q.2d 1508, C.A.Fed. (Cal.), July 14, 2006).
JEM Ag Supply Inc v. *Pioneer Hi-Bred International Inc* 534 US 124 (2001).
Jungerson v. *Ostby and Barton Co*, 335 US 560, 80 USPQ 32 (1948).
KSR International Co. v. *Teleflex Inc.* 550 US 398 (2007).
Laboratory Corporation of American Holdings v. *Metabolite Laboratories, Inc.* 548 U.S. 124 (2006).
Le Roy v. *Tatham*, 55 U.S. (14 How.) 156 (1853).
Loctite Corp. v. *Ultraseal Ltd.* 781 F.2d 861 (Fed. Cir. 1985).
Lowell v. *Lewis*, 15 Fed Cas 1018 (1817).
Mackay Radio & Tel. Co. v. *Radio Corp. of America*, 306 U.S. 86 (1939).
Mayo Collaborative Services v. *Prometheus Laboratories, Inc.*, *cert. granted*, __ U.S. __, 2011 WL 973139 (U.S. June 20, 2011) (No. 10–1150).
Metabolite Laboratories, Inc. v. *Laboratory Corporation of America Holdings*, 370 F.3d 1354 (Fed. Cir. 2004).
Microsoft v. *CSIRO* 2007 WL 4376104, E.D.Tex.
Monsanto Canada Inc. v. *Schmeiser* (2004) SCC 34; 2004 SCC 34.
Monsanto Technology LLC v. *Cefetra BV and others*, No. C-428/08, ¶¶ 33–50, European Court of Justice (6 July 2010).
Morton v. *New York Eye Infirmary,* 17 F. Cas. 879, 881–82 (C.C.S.D.N.Y. 1862) (No. 9,865).
National Research Development Corporation v. *Commissioner of Patents* (1959) 102 CLR 279.
Netherlands v. *European Parliament* (2001) 3 CMLR 49.
Nicotinic Acid Compositions for Treating Hyperlipidemia, Case G 0002/08, EPO Enlarged Board of Appeal ¶ 5.3 (2010).
O'Reilly v. *Morse*, 56 U.S. (15 How.) 62 (1854).
Oliver Brüstle v. *Greenpeace eV* (Case C-34/10) (Opinion of Advocate General Bot) (10 March 2011).
Oliver Brüstle v. *Greenpeace eV* (Case C-34/10), Judgment of the European Court of Justice (Grand Chamber) (18 October 2011).
Opinion of the Enlarged Board of Appeal of 12 May 2010 in relation to a Point of Law referred by the President of the European Patent Office Pursuant to Article 112(1)(b) of the EPC, Case G 003/08.

Parke-Davis & Co. v. *H. K. Mulford Co.,* 189 F. 95 (S.D.N.Y. 1911), affirmed 196 F. 496 (2d Cir. 1912).
Parker v. *Flook*, 437 U.S. 584 (1978).
Plant Genetic Systems/Glutamine Synthetase Inhibitors [1995] EPOR 357.
Prometheus Laboratories, Inc. v. *Mayo Collaborative Services*, No. 04-CV-1200, 2008 WL 878910 (S.D. Cal. Mar. 28, 2008).
Prometheus Laboratories, Inc. v. *Mayo Collaborative Services*, 581 F.3d 1336 (2009).
Prometheus Laboratories Inc. v. *Mayo Collaborative Services* No. 2008–1403, 2010 WL 5175124 (Dec. 17, 2010).
Research Corp. Techs., Inc. v. *Microsoft Corp.*, 627 F.3d 859, 868 (Fed. Cir. 2010).
Rivett v. *Monsanto Canada Inc.*, 2010 FCA 207 (2010).
Rubber-Tip Pencil Co. v. *Howard,* 87 U.S. (20 Wall.) 498 (1874).
Sacker Potatoes Ltd v *C Meijer BV* (Unreported, October 31, 2001).
Sakraida v. *Ag-Pro, Inc*. 425 U.S. 273 (1976).
Sample v. *Monsanto* (United States District Court for the Eastern District of Missouri, 2003), https://ecf.moed.uscourts.gov/documents/opinions/BLADES_V_MONSANTO_CO-RWS-397.PDF.
State of Israel – Ministry of Agriculture v. *Unilever* T1242/06 – 3.3.04 (4 April 2008), EPO Boards of Appeal Decision, http://legal.european-patent-office.org/dg3/biblio/t061242ex1.htm; currently under review by the Enlarged Board of Appeals.
Stem Cells/WARF (T 1374/04) EPO OJ 2007, 313 ('*Stem Cells/WARF*').
Tennessee Eastman Co. v. *Commissioner of Patents*, [1974] S.C.R. 111.
Terrett v. *Taylor*, 9 Cr 43 (1815).
Tol-O-Matic Inc v. *Promo Produkt-und Marketing Gesellschaft MbH*, 945 F 2d 1546 (1991).
Transgenic animals/Harvard (T 315/03) [2006] 1 OJ EPO 15, [2005] EPOR 31.
Wisconsin Alumni Research Foundation (G 0002/06) EPO OJ 2009, 306.

APPELLATE AND AMICUS CURIAE BRIEFS

American Association for Retired Persons (2005), 'Brief Amicus Curiae of AARP in Support of Petitioner in *Laboratory Corp. of America Holdings v. Metabolite Laboratories, Inc.*', 2005 WL 3597809, 23 December.
American College of Medical Genetics et al. (2009), 'Corrected Brief for the American College of Medical Genetics in *Prometheus Laboratories, Inc.* v. *Mayo Collaborative Services.*, No. 2008–1403, 2009 WL 1307167, 6 April.

American Medical Association *et al.* (2010), 'Brief for Amici Curiae American Medical Association. *et al.* In Support of Plaintiffs' Opposition to Defendants' Motion to Dismiss and in Support of Plaintiffs' Motion for Summary Judgment', in *Association for Molecular Pathology et al.* v. *United States Patent and Trademark Office et al.* (S.D.N.Y. Apr. 5), 8–9, ('AMA *Myriad* D.Ct. Brief').

Appellate Brief (2009), 'The Brief of the Appellant in *Prometheus Labs., Inc.* v. *Mayo Collaborative Services.*, No. 2008–1403, 2009 WL 327984, 9 January' ('Prometheus Appellants Brief').

Appellate Brief (2010), 'Brief for the Appellants in the *Association for Molecular Pathology et al.* v. *United States Patent and Trademark Office et al.*, No. 2010–1406 (Fed. Cir), 2010 WL 4600106, 22 October' ('*Myriad* Appellants Brief').

Appellate Brief (2011), 'Brief of Nine Law Professors as *Amici Curiae* in Support of Petitioners', *Prometheus Laboratories, Inc.* v. *Mayo Collaborative Services*, No. 10–1150, 9 September.

Appellate Brief (2011), 'Plaintiffs-Appellees' Petition for Panel Rehearing', *Association for Molecular Pathology et al.* v. *United States Patent and Trademark Office et al.* (Fed. Cir.) (No. 2010–1406), 25 August.

BASF Corporation, 'Brief in Support of the Respondent in *J.E.M. Ag Supply, Inc.* v. *Pioneer Hi-Bred International, Inc.*', 2001 WL 689271 (U.S.) (Appellate Brief).

Biotechnology Industry Organization, 'Brief in Support of the Respondent in *J.E.M. Ag Supply, Inc.* v. *Pioneer Hi-Bred International Inc.*', 2001 WL 689273 (U.S.) (Appellate Brief).

Council of Canadians, 'Intervenor in *Monsanto Canada Inc.* v. *Schmeiser* (2004) SCC 34', http://www.canadians.org/food/documents/COC_Affidavit.pdf.

Eleven Law Professors and the American Association for Retired Persons (2009), 'Brief of Eleven Law Professors and AARP as *Amici Curiae* in Support of Respondent in *Bilski* v. *Kappos*', 2009 WL 3167954, 2 October, ('AARP *Bilski* Brief').

ETC Group, 'Intervenor in *Monsanto Canada Inc.* v. *Schmeiser* (2004) SCC 34', http://www.canadians.org/food/documents/ETC_Affidavit.pdf.

The Foundation for Taxpayer and Consumer Rights and the Public Patent Foundation (2006), 'Request for *Ex Parte* Re-examination' in respect of US Patent No. 5,843,780, http://www.pubpat.org/assets/files/warfstemcell/780Request.pdf.

Gilead Sciences, Inc., Biogenerator, and Elan Pharmaceuticals, Inc. (2010), 'Brief of Amici Curiae Gilead Sciences, Inc., Biogenerator, and Elan Pharmaceuticals, Inc. in Support of Defendants-Appellants and Urging Reversal' in *Association for Molecular Pathology et al.* v. *United States*

Patent and Trademark Office et al.', No. 2010–1406, 2010 WL 5306804 (Fed. Cir.), 29 October.

Interested Patent Law Professors (2003), 'Brief of *Amici Curiae* Interested Patent Law Professors in Support of Neither Part in *Prometheus Laboratories, Inc.* v. *Mayo Collaborative Services*', No. 2008–1403, 2009 WL 462602, 21 January.

Research Foundation for Science, Technology, and Ecology, 'Intervenor in *Monsanto Canada Inc.* v. *Schmeiser* (2004) SCC 34', http://www.canadians.org/food/documents/RSTE_Affidavit.pdf.

Ten Law Professors (2008), 'Brief of *Amici Curiae* Ten Law Professors in Support of Appellee Director of the United States Patent and Trademark Office in the matter of *In re Bilski*', 2008 WL 1842283 (Fed. Cir. 2008) (en banc).

United States (2010), 'Brief for the United States as Amicus Curiae in Support of Neither Party at 18, *Association for Molecular Pathology et al.* v. *United States Patent and Trademark Office et al.*', No. 2010–1406 2010 WL 4853320 (Fed. Cir.), 29 October, 18, ('USG *Myriad* Brief').

LEGISLATION

21st Century Nanotechnology Research and Development Act 2003 (US)

Aboriginal Land Rights Act 1974 (NT).

Aboriginal Land Rights (Northern Territory) Act 1976 (Cth).

African Model Legislation for the Protection of the Rights of Local Communities, Farmers and Breeders, and for the Regulation of Access to Genetic Resources 2000 http://www.cbd.int/doc/measures/abs/msr-abs-oau-en.pdf.

An Act Concerning Monopolies and Dispensations with Penal Laws and Forfeitures Thereof 2 1 Jac. I c. 3 9 (the *Statute of Monopolies* 1624 (UK)).

Australian Constitution.

Biodiscovery Act 2004 (Qld).

Biological Resources Act 2006 (NT).

Constitution of the United States of America.

Environment Protection and Biodiversity Conservation Act 1999 (Cth).

Genomic Research and Accessibility Act 2007 (US).

Human Tissue Act 2004 (UK).

Intellectual Property Code of the Philippines (Republic Act 8293) (Philippines).

Land Rights Act 1976 (NT).

Leahy-Smith America Invents Act, Pub. L. 112–29, §§ 14(a), 125 Stat. 284, 327 (2011).
Native Title Act 1993 (Cth).
Patent Act 1952 (US).
Patent Act 1985 (RSC).
Patent Amendment (Human Genes and Biological Materials) Bill 2010 (Cth).
Patent Law of the People's Republic of China 2000 (China).
Patent Reform Act of 2007 (US).
The Patent Reform Act of 2011 (US).
Patents Act 1970 (India).
Patents Act 1990 (Cth).
Patents and Plant Variety Rights (Compulsory Licensing) Regulations 2002 (UK).
Patents Bill 2008 (NZ).
Plant Breeder's Rights Act 1994 (Cth).
Plant Breeder's Rights Amendment Act 2002 (Cth).
Plant Patent Act 1930 (US).
Plant Varieties Act 1997 (UK).
Plant Variety Protection Act 1970 (US).
Plant Variety Protection and Farmers' Rights Act 2001 (India).
Plant Variety Rights Act 1987 (Cth).
Privacy Act 1988 (Cth).
Stem Cell Research and Cures Act 2004 (California).
Territory Parks and Wildlife Act 2010 (NT).
United States Code (U.S.C.) Title 35 – Patents.
University and Small Business Patent Procedures Act 1980 (United States) (the '*Bayh-Dole Act*') 35 U.S.C. § 200–212 (1980).

INTERNATIONAL MATERIALS

Ad Hoc Working Group on Long-term Co-operative Action under the *United Nations Framework Convention on Climate Change* (2009), 'Report', Seventh Session held in Bangkok from 28 September to October 2009, and Barcelona from 2 to 6 November, FCCC/AWGLCA/2009/14, 20 November.
Ad Hoc Working Group on Long-term Co-operative Action under the *United Nations Framework Convention on Climate Change* (2010), 'Work Undertaken by the Conference of the Parties at its Fifteenth Session on the Basis of the Report of the Ad Hoc Working Group on Long-term

Cooperative Action Under the Convention', FCCC/CP/2010/2, 11 February, http://unfccc.int/resource/docs/2010/cop16/eng/02.pdf.

The Australian Government (2005), 'Submission to the *Ad Hoc Open-ended Working Group on Access and Benefit Sharing* under the CBD', UNEP/CBD/WG-ABS/3/INF/1/Add.1, 25 January.

The Australian Government (2009), 'Submission on Development of Elements of sui generis Systems of the Protection of Traditional Knowledge, Innovations and Practices', Ad Hoc Open Ended Working Group Inter-sessional working group on Article 8(j) and related provisions of the Convention on Biological Diversity, Sixth meeting, Montreal 2–6 November.

Barroso, José Manuel (2009), 'Remarks at World Summit on Food Security', November, http://www.fao.org/fileadmin/templates/wsfs/Summit/Statements_PDF/Monday_16_AM/12_EU_Barroso_speech_161109 AM.pdf.

Council of Europe Recommendation 1213 on developments in biotechnology and the consequences for agriculture [1993]. Assembly debate on 12 May 1993 (34th sitting) (see Doc. 6780, Report of the Committee on Agriculture, Rapporteur: Mr Gonzalez Laxe). Text adopted by the Assembly on 13 May 1993 (36th Sitting).

De Schutter, Olivier (2009), *Seed Policies and the Right to Food: Enhancing Agrobiodiversity and Encouraging Innovation*, United Nations General Assembly, A/64/170, http://www.ip-watch.org/weblog/wp-content/uploads/2009/11/n0942473.pdf.

European Parliament Resolution on the Patenting of BRCA1 and BRCA2 ('Breast Cancer') Genes, Texts Adopted by Parliament, Provisional Edition: 04/10/2001, B5–0633, 0641, 0651 and 0663/2001.

European Parliament Resolution on Patents for Biotechnological Inventions, P6_TA (2005) 0407, 26 October 2005, http://eur-lex.europa.eu/LexUriServ/LexUriServ.do?uri=OJ:C:2006:272E:0440:0442:EN:PDF.

European Parliament Resolution on Trade and Climate Change, 29 November 2007, (2007/2003 (INI)), http://www.europarl.europa.eu/sides/getDoc.do?Type=TA&Reference=P6-TA-2007–0576&language=EN.

European Parliament Resolution on Regulatory Aspects of Nanomaterials, 24 April 2009, http://www.europarl.europa.eu/sides/getDoc.do?type=TA&reference=P6-TA-2009–0328&language=EN.

Gurry, F. (2008), 'Statement to the Forty-Second Ordinary Session of the International Union for the Protection of New Varieties of Plants', 30 October, http://www.upov.int/en/documents/c/42/c_42_21.pdf.

Human Genome Organization (1995), 'HUGO Statement on Patenting of DNA Sequences', http://www.hugo-international.org/PDFs/Statement%20on%20Patenting%20of%20DNA%20Sequences%202000.pdf.

Human Genome Organization (1996), 'Summary of Principles Agreed at the First International Strategy Meeting on Human Genome Sequencing', http://www.ornl.gov/sci/techresources/Human_Genome/research/bermuda.shtml#.

International Union for the Protection of New Varieties of Plants (2005) *UPOV Report on the Impact of Plant Variety Protection*, Geneva: International Union for the Protection of New Varieties of Plants, http://www.upov.int/en/publications/pdf/353_upov_report.pdf.

Moon, Ban Ki (2009), 'Opening Remarks at World Summit on Food Security', November, http://www.fao.org/fileadmin/templates/wsfs/Summit/Statements_PDF/Monday_16_AM/01b_BAN_KI_MOON_SG_Speech_161109AM_01.pdf.

Organisation for Economic Co-operation and Development (OECD) (2004), *Patents and Innovation: Trends and Policy Changes*, Paris: OECD.

Organisation for Economic Co-operation and Development (OECD) (2006), *Creation and Governance of Human Genetic Research Databases*, Paris: OECD Publishing, 35, http://www.oecd.org/document/50/0,3343,en_2649_34537_37646258_1_1_1_1,00.html ('OECD Creation and Governance Report').

Organisation for Economic Co-operation and Development (OECD), (2006), *Guidelines for the Licensing of Genetic Inventions,* Paris: OECD.

Organisation for Economic Co-operation and Development (OECD) (2008), 'Compendium of Patent Statistics 2008,' http://www.oecd.org/dataoecd/5/19/37569377.pdf.

Organisation for Economic Co-operation and Development (OECD) (2009), *Guidelines on Human Biobanks and Genetic Research Databases*, Paris: OECD Publishing, http://www.oecd.org/sti/biotechnology/hbgrd ('OECD Biobank Guidelines').

Organisation for Economic Co-operation and Development (OECD) (2010), *The Patent System and Research Freedom: a Comparative Study*, Working Paper, Doc. No. SCP/15/3, Annex VI.

Oxford Statement on International Data Sharing 2010 (draft), http://helexoxford.com/content/oxford-statement-data-sharing.

Secretariat of the Convention on Biological Diversity (2008), *Access and Benefit-sharing in Practice: Trends in Partnerships Across Sectors,* Montreal: CBD Technical Series No. 38.

Secretariat of the UPOV Convention (2009), 'Note on the Report of the Special Rapporteur on the Right to Food', http://upov.int/export/sites/upov/en/news/2009/un_ga_note.pdf.

United Nations (2011), 'UN Protocol on Sharing the World's Genetic Wealth Opens for Signing', Press Release, UN News Centre, 2 February http://www.un.org/apps/news/story.asp?NewsID=37453&Cr=&Cr1.

United Nations Educational, Scientific and Cultural Organization (UNESCO) (2006), *The Ethics and Politics of Nanotechnology*, Paris: UNESCO, http://unesdoc.unesco.org/images/0014/–1459/145951e.pdf.

The United Nations Food and Agriculture Organization (2009), *The World Summit on Food Security*, http://www.fao.org/wsfs/world-summit/en/.

World Health Organization (2005), *World Malaria Report 2005*, Geneva: World Health Organization, http://rbm.who.int/wmr2005/html/exsummary_en.htm.

World Summit on Food Security (2009), *Food Security and Climate Change: A Call for Commitment and Preparation*, November.

INTERNATIONAL INSTRUMENTS

The *Cancún Agreements* 2010, *Outcome of the Work of the Ad Hoc Working Group on Long-Term Cooperative Action Under the Convention*, -/CP.16, http://unfccc.int/files/meetings/cop_16/application/pdf/cop16_lca.pdf.

Convention on Biological Diversity 1992, opened for signature 5 June 1992, 1760 UNTS 143 (entered into force 29 December 1993).

Convention on the Grant of European Patents ('European Patent Convention') of 5 Oct 1973 Official Journal Special Edition 1/2007 (entered into force 7 October 2000).

Copenhagen Accord 2009, UN Doc. FCCC/KP/CMP/2009/L.9 (December 18, 2009).

European Convention for the Protection of Human Rights and Fundamental Freedoms of 1950.

European Parliament Directive 98/44/EC of the European Parliament and of the Council on the Legal Protection of Biotechnological Inventions [1998] OJ (L 213) 13, art 6(1) ('*European Biotechnology Directive*').

International Convention for the Protection of New Varieties of Plants, opened for signature 2 December 1961, 815 UNTS 89 (entered into force 24 April 1968), as revised on 10 November 1972, 23 October 1978, and 19 March 1991 ('*UPOV Convention*').

International Treaty on Plant Genetic Resources for Food and Agriculture, adopted 3 November 2001, S. Treaty Doc. No. 110–19, ftp://ftp.fao.org/ag/cgrfa/it/ITPGRe.pdf (entered into force 29 June 2004).

Marrakesh Agreement Establishing the World Trade Organization, opened for signature 15 April 1994, 1867 UNTS 3 (entered into force 1 January 1995) annex 1C ('Agreement on Trade-related Aspects of Intellectual Property Rights' – *TRIPS Agreement* 1994).

Nagoya Protocol on Access to Genetic Resources and the Fair and Equitable Sharing of Benefits Arising From Their Utilization to the Convention on Biological Diversity, COP 10 Decision, ('*Nagoya Protocol* 2010').

Patent Cooperation Treaty 1970, opened for signature 19 June 1970, 1160 *UNTS* 231 (signed and entered into force, 24 January 1978).

Treaty establishing the European Community (*EC Treaty*).

United Nations Covenant on Economic, Social and Cultural Rights 1966.

United Nations Declaration on the Rights of Indigenous Peoples 2007, Adopted by General Assembly Resolution 61/295 on 13 September 2007.

Index